D1283878

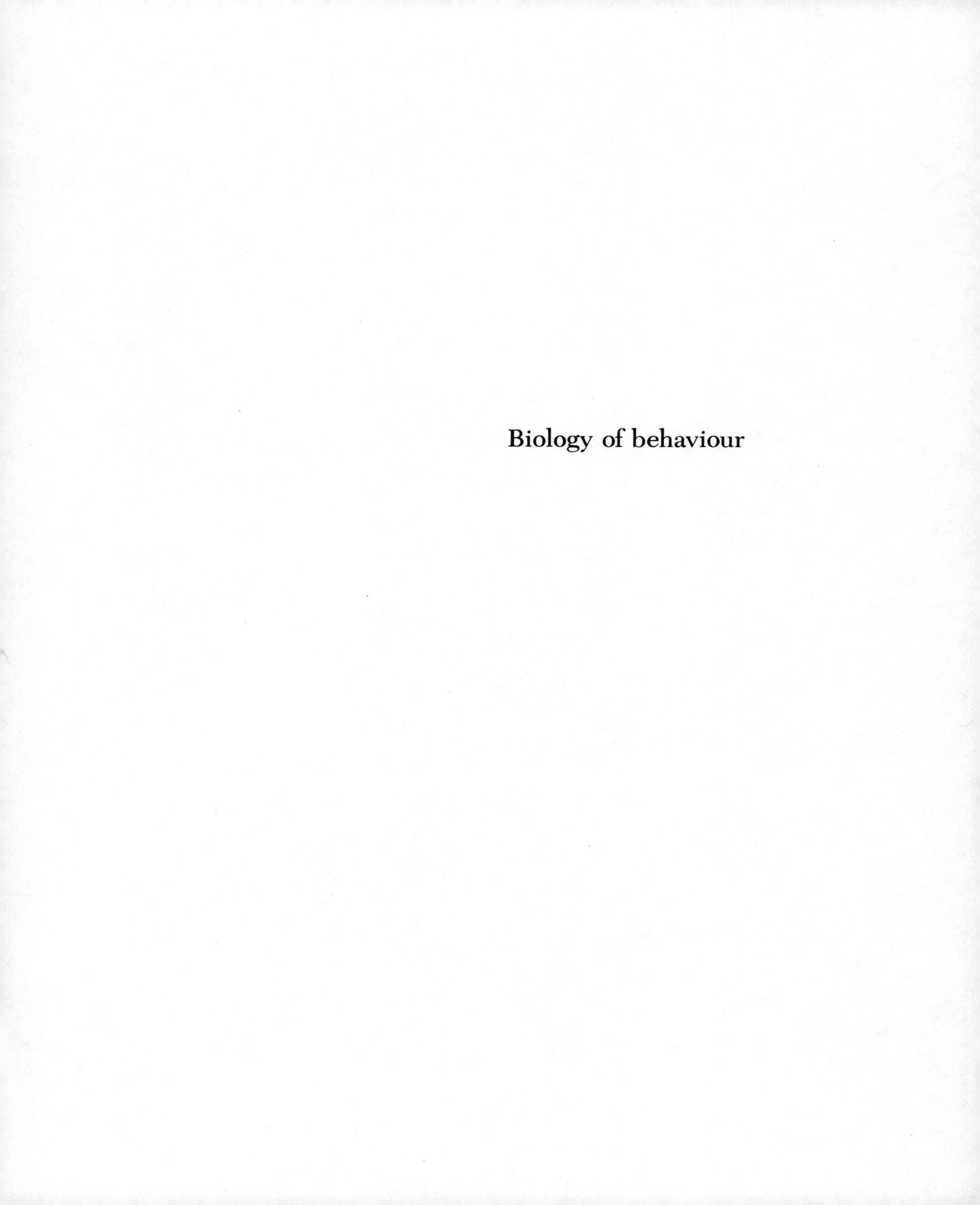

Biology of behaviour

Biology of behaviour

Mechanisms, functions and applications

DONALD M. BROOM

Senior Lecturer in Zoology, University of Reading

WITH ANIMAL DRAWINGS BY ROBERT GILLMOR

CAMBRIDGE UNIVERSITY PRESS

Cambridge

London New York New Rochelle

Melbourne Sydney

Published by the Press Syndicate of the University of Cambridge
The Pitt Building, Trumpington Street, Cambridge CB2 1RP
32 East 57th Street, New York, NY 10022, USA
296 Beaconsfield Parade, Middle Park, Melbourne 3206, Australia

First published 1981

Printed in the United States of America by
Vail-Ballou Press, Inc.,
Binghamton, N.Y.

British Library Cataloguing in Publication Data
Broom, Donald M.
Biology of behaviour

1. Social behaviour in animals
I. Title
591.51 QL775 80-42096

ISBN 0 521 23316 X hard covers
ISBN 0 521 29906 3 paperback

Contents

The behaviour of farm animals and pests

Page numbers refer to the beginning of relevant sections

FARM ANIMALS

PESTS

Preface

It is now apparent that detailed studies of behaviour are important in many areas of physiology, psychology, zoology and agriculture. In this book I have attempted to draw together the results of observational studies of behaviour and evidence from the work of those who study ecology, evolution, sociobiology, sensory and motor physiology, physiological psychology, motivation and learning. Wherever possible I have referred to examples of behaviour studies which are relevant to agricultural and medical problems, especially farm animal husbandry and pest control, and I have listed on the previous pages the sections where examples are to be found.

Some of the exciting, current ideas about behaviour are introduced in Chapter 1. Major developments include the links between physiological and behavioural studies of both sensory functioning and motor control mechanisms. The central importance, to all those who study behaviour, of work on motivation is emphasised in the chapter on the allocation of resources. Other chapters which refer to many recent ideas are those concerned with the functions of social behaviour and with the organisation of social groups. Themes which recur in each chapter are the evolution of behaviour, the role of learning and other effects of experience, the changes in behaviour which occur during development, the relevant brain mechanisms and economic applications of such studies.

I am most grateful to the following, each of whom was kind enough to read and criticise a chapter of the book: Jack Albright, Stuart Altmann, John Archer, George Barlow, Patrick Bateson, Hilary Box, Robert Drewett, Robert Elwood, Jörg-Peter Ewert, John Fentress, Robert Forrester, Jack Hailman, Toshitaka Hidaka, Robert Hinde, Jerry Hogan, Alick Jones, John Lazarus, David McFarland, Martin Potter, Francis Pring-Mill, Gillian Thompson, Larry Wolf, and Amotz Zahavi. In addition, I thank my wife Sally who read and commented on the whole book, the staff at Cambridge University Press for many helpful suggestions, people who gave me photographs, and Robert Gillmor for his excellent drawings.

Donald Broom

1

Introduction

1 This bitch is standing still but her behaviour can be described in terms of leg positions, the way that the head is held, gaze direction, face muscle and ear positions, tail posture, degree of hair erection and so on.

2 The bitch shown in Figure 1 soon started grooming her fore leg.

3 When a male dog arrived, the bitch shown in Figures 1 and 2 responded to him and started investigating him. Compare the head, tail and stance with Figure 1. The dog which arrived was unfamiliar to the bitch and the interaction lasted for several minutes. Two subsequent interactions with familiar dogs lasted for a few seconds only.

Questions about behaviour

Suppose that you start to watch a dog which, like that in Figure 1, is standing still. What will it do next? What questions can you ask about the dog's behaviour? The behaviour of the dog at any instant depends upon a great variety of factors which interact with one another in a complex way and yet you may be able to predict the behaviour which is most likely. Your ability to predict what the dog will do depends, partly, upon how often you have watched dogs. You will, however, be able to make a reasoned guess based on your experience of other species including our own. We are all expert observers of human behaviour. We can deduce a great deal about the behavioural events which are occurring in any situation and the most likely preceding and succeeding events. Such ability is very important to us in our complex society and we all possess mechanisms which enable us to make predictions about human behaviour and about the behaviour of other species. A first step in trying to understand the mechanisms which control behaviour is to consider what evidence we use when we analyse a behavioural situation. The factors listed below might affect the behaviour of the dog. We consider most of them when trying to predict behaviour. Some of the events which occurred soon after the photograph in Figure 1 was taken, are shown in Figures 2 and 3.

Motor ability

The dog's next behaviour is inevitably limited by its physical ability. It can perform only those movements which its muscles and the mechanics of its structure allow. Its movements might be described in a general way as being characteristic of dogs but they will be different according to both genetic variation, such as breed, and to the previous experience of the dog. Muscular function is modified by practice so a dog which has exercised regularly will be different from a dog which has exercised infrequently. It is already apparent that environmental factors, such as the degree and type of exercise, and genetic factors have interacted in the development of the dog to its present state. In describing motor ability it is necessary to consider some sensory and general physiological factors as well, for co-ordinated muscular movements depend upon input from sensory receptors in the muscles and upon the presence of adequate energy sources in the muscle cells.

Sensory input

The world as detected by sensory receptors is continually changing. The dog may stop standing still and initiate a new activity, the nature and timing of which is influenced by the input from a receptor. Some environmental changes might pass undetected but the odour of meat, the sound of footsteps, the sight of another dog, a tactile input due to a flea bite, a feeling of strain in a tendon, or the cooling of the ear tips might elicit different behaviours. The functioning of these receptors, like that of the motor system, will vary according to the previous experience of the dog and its interaction with genetic factors during development up to this time. Input along sensory pathways from receptor cells is not inevitable when a potentially detectable physical change occurs, for sensitivity and analysis of input can be affected at all stages by modifying commands from the central decision-taking areas of the brain. Physiological variables which alter energy availability may also affect input to central analytical areas of the brain.

Attentiveness

When a flea bites a dog the likelihood that the dog will scratch itself and the pattern of movement which it shows will differ according to whether the dog is asleep, drowsy or alert. The scratching probability and pattern will also be different in a dog which is watching an approaching cat, a dog which has heard a sound associated with the advent of food and a dog which is resting quietly. The response to input along one particular sensory channel depends upon the input which is arriving along other channels and the receptivity of central processing mechanisms. The systems concerned here are loosely referred to as attention and arousal. The differences in general responsiveness between sleepy and alert individuals are obvious to us. Attentional mechanisms may function at various levels in the analysis and decision-making systems of the brain. As mentioned above, the functioning of the receptor mechanism itself can be centrally modified. Since there must be a limit to the amount of incoming information which can be processed at any one time, it is essential for the individual to have mechanisms of selective attention. The priorities ascribed to different sorts of input will vary according to the motivational state and will probably differ according to experience during development.

Nutrient levels in the body

Irrespective of any sensory input, the dog might walk towards a place which could be a food source. The likelihood of this behaviour will depend partly on the levels of nutrients within the dog's body, which in turn depend upon the interval since the last meal and the rate of metabolism during that interval. Previous experience of feeding will alter the direction of movement and, perhaps, the posture and gait. Responsiveness to food odours and other sensory input must differ according to nutrient levels in the animal.

Water levels in the body

If internal receptors provide information to the effect that the body is short of water, the dog may go to drink. This factor will interact with others in a way similar to that of nutrient levels. If, as a result of previous drinking and kidney function, its bladder is full, the dog may walk and then urinate.

Body temperature

Like bladder distension, body temperature is monitored by receptors but a change in the input from these receptors may result from physiological changes within the animal and need not be temporally related to any external change. Metabolic activity, or external temperature, may change so that the dog becomes too cold or too warm. It will then modify its behaviour so as to correct that deviation from the optimum temperature range. As with other behaviours mentioned earlier, the precise nature of the behaviour will depend upon the precise interaction of genetic information with environmental factors during the development up to the point of observation.

Hormone levels

Hormonal influences on behaviour become apparent when, for example, responsiveness to sensory input is seen to be different according to hormone levels in the body. A bitch in oestrus would behave towards an approaching dog in a manner which is different from that of an anoestrus bitch. She may vocalise more and stand for longer when sniffed. Sometimes, however, behavioural changes might occur consequent upon hormone levels exceeding a threshold and without any sensory input occurring. In a male dog, when testosterone levels rise during courtship they may affect directly the likelihood that he will attempt to mount the bitch. Other examples include the effects on behaviour of increased adrenaline levels in a dog which had been dreaming and the moulting behaviour which is initiated in some insects by a sudden release of eclosion hormone.

Rhythms

Nutrient levels, water levels, body temperature and hormone levels all fluctuate within limits and the possible effects of such fluctuations on the dog's behaviour have been mentioned. The value to animals of predicting the necessity for action is such that timing mechanisms exist in the body which can initiate behaviour; for example the dog may start to look for food before the nutrient levels in the body initiate such behaviour. If there exists a rhythm of alternate food finding and resting which has a constant wavelength, i.e. which is periodic, then an individual may suddenly walk towards a food source because of the action of this periodic internal process. Most of us are aware of the internal clock which can tell us that it is lunch time, whether or not we have consumed food during the breakfast–lunch interval. A similar periodic process might result in our dog going to sleep at the moment of observation.

Other experiential factors

A dog which is in an unfamiliar place is likely to spend much more time looking around and preparing for rapid movement than would that same dog if the place was familiar. If the dog had been attacked recently in that place it would be much warier than if no attack had occurred. Such effects of previous experience modify behaviour via the sensory, attentional, hormonal and general decision-making systems. The various systems also interact in determining responsiveness to people or to other dogs. The previous social interactions of the dog will alter its responses towards a strange dog and previous encounters with a known dog will affect responses to it. The advent of a person who is recognised as a potential dog-kicker may result in the rapid withdrawal of the dog whereas the sight of a known provider of food may provoke the dog to approach, to bark and so on.

General themes

This book is about observable behaviour, the factors which affect it and its role in the life of the animal. How do the mechanisms which control behaviour work, and what selective pressures have resulted in the existence of all the components of these mechanisms? Many of the answers to these questions can be discovered by ethological investigations. Ethology involves observing behaviour and describing it in detail in order to find out more about how biological mechanisms function. Most people who would call themselves ethologists look for a wide range of actions when observing behaviour. They try to conduct their investigations in situations where the individual or group under observation is not so restricted that much of its behavioural repertoire is impossible. This approach can be applied in laboratory experiments or in farm situations as well as in wild, undisturbed conditions, for many important questions about behaviour cannot be answered by field work alone.

In order to use the information obtained from ethological studies it is often necessary to relate it to physiological or other biological and psychological research. Observational work must also be combined with the construction of theoretical models which attempt to explain the functional system which is being investigated. Thoughts about how a system works are often aided by considering how natural selection may have acted during the evolution of the system to its present state. It is largely as a result of theorising about the evolution of behaviour patterns in relation to the general ecology of species that the important cost–benefit approach to behaviour and other problems has developed. This approach is of use when considering motivational mechanisms as well as, for example, details of feeding behaviour. Another area in which theories about ecology and evolution have lead to exciting advances in our understanding of behaviour is the study of social relationships in groups, as will be discussed later.

One of the themes of this book is that answers to the two questions 'How does it work?' and 'Why does it exist?' are usefully considered together when dealing with any biological mechanism. A second theme is that information which results from different approaches to related problems should be pooled when trying to understand general mechanisms. Physiological studies of sensory or motor mechanisms are relevant to the work of a biologist or psychologist studying what individuals detect and how they respond. Ecological and evolutionary principles, as well as evidence from physiological and psychological experiments and knowledge of biological necessities during development, help to explain how individuals learn what and when they do. A third theme is the importance of applying the results of research on behaviour and related topics to medical, economic and social problems. A better understanding of behaviour means a better chance of preventing or curing behavioural disorders in man and in domestic animals, a better chance of maximising economically important behaviour by farm animals, and a better chance of impairing behavioural function and consequently of reducing the numbers of pest species.

Functional systems and homeostats

For convenience, all activities of animals have been classified in *functional systems* such as feeding, the regulation of body temperature and others listed in this section. Behaviour forms part of several functional systems and behavioural components of these systems are just as important as anatomical or physiological components. The behaviours which are observed during a short period may be part of several systems but, whatever the study, it is useful to consider all possible systems when trying to understand a behaviour sequence. Each of the systems listed below can be subdivided. Their relative importance in the life of an individual varies with time and from species to species.

The term *homeostat*, which is often used in physiological and behavioural studies, requires some comment here. Cannon (1929) coined the word *homeostasis* to refer to the maintenance of constant conditions or steady states in the body by co-ordinated physiological processes. A mechanism which does this is called a homeostat. It may include physiological and behavioural components and 'steady' state really means 'restricted within defined limits'. It is often assumed that homeostats operate by negative feedback, i.e. a corrective mechanism is initiated when a displacement from the steady state is detected. As Hogan (1981) points out, this is only one of several types of homeostat. Others use positive feedback or no feedback at all. In positive feedback the correction is made before the displacement occurs and is of a sufficient magnitude to ensure that an approximately steady state is maintained. An example of homeostasis which involves no feedback is the operation of storage mechanisms which are brought into operation when the concentration of a material rises above a critical level.

Body regulation and maintenance

Obtaining oxygen. All but the smallest animals on land and in water must obtain oxygen from their surroundings by breathing air or ensuring that water passes over their gills. If the oxygen level in the medium drops to a critical level, most animals use behavioural methods which increase the chance of reaching a place where there is sufficient oxygen.

Osmotic. Animals on land have to avoid dehydration whilst those in freshwater have to avoid loss of salts and other dissolved matter into the surrounding medium. Water conservation methods often involve behavioural components, for example going to a place where less water loss occurs. Acquiring water by drinking, on land inevitably involves complex behaviour. There must be receptors which detect changes towards dangerous levels of osmotic state, a decision-making mechanism which receives input from these receptors, a method of acting so as to return the state towards the optimum and receptors which provide an input so that the correcting behaviour can be terminated (see Chapter 5).

Temperature. Behavioural mechanisms and physiological mechanisms combine in regulating temperature. Although especially important in warm-blooded animals, the dangers of being unable to show activity at low temperatures have resulted in the existence of many body-temperature modifying behaviours in cold-blooded animals (Chapter 5). Temperature extremes can often be avoided by locomotion, but methods of temperature regulation will be selected according to the circumstances. For example, in the presence of a predator which can see well, physiological methods such as sweating would be less hazardous but if the predator hunts by olfaction, the prey may be detected if it sweats, so running away may be safer.

Cleaning. Birds must maintain their wing feathers in order that flight will remain possible. Grooming, preening and cleansing behaviour are also important to increase the efficiency of other forms of locomotion, sensory function and display as well as to minimise the effects of disease and parasites.

Feeding

Before they can ingest food, most animals need to search actively for, to recognize, and to acquire it (see Chapter 6). Elaborate behavioural strategies are used when hunting for animal prey or for widely distributed plant sources. Even when the animal is surrounded by food, as is a caterpillar on a host food plant, decisions about which part to eat and when must still be taken.

Hazard avoidance

Chemical. The ingestion of harmful substances and the accumulation of indigestible material or harmful metabolic products within the body must be minimised. The major components of this system are digestion, detoxification and immunological processes within the body but elimination is a frequent behaviour. The time and place of defaecation or urination are often precisely controlled, although the presence of very harmful substances in the gut may over-ride normal constraints on the occurrence of the behaviour. The avoidance of harmful substances when feeding depends, to a large extent, on experience and is discussed in Chapter 6.

Physical. Terrestrial animals encounter hazards such as falling, drowning and being squashed or buried. Aquatic animals may be stranded out of water, they may suffer lack of oxygen, or they may experience large pressure changes. Most of the methods of avoiding or minimising the effects of these hazards are behavioural. Such methods and those for avoiding noxious chemicals, other than those in food, are discussed in Chapter 5.

Predator avoidance

Predators constitute a major hazard for most animal species and their existence has had a large effect on the evolution of behaviour in all species including man (see Chapter 7). Animals avoid large predators by concealment, mimicry, keeping away from danger areas, flight, combat, or combinations of these methods. The depredations of parasites are also countered by using some or all of these methods.

Reproduction

All of the functional systems mentioned so far operate in such a way that individual survival is maximised. The functional system in which reproduction and the survival of offspring or other relatives is promoted operates at the same time as systems promoting individual survival and may conflict with them (see later in this chapter and Chapter 9). The system includes behavioural mechanisms for mate finding, mate recognition, display and other methods of persuasion, mating, parental care, and other means of increasing the chances that offspring and other close relatives will survive and breed. This last function incorporates elements of many of the functional systems which promote individual survival.

Themes of other chapters

In order that any of these functional systems can operate, animals need to detect changes in their surroundings, to make decisions and, sometimes, to modify their behaviour accordingly (Figure 4). Methods of detecting, filtering and coding the effects of physical changes in the environment are discussed in Chapter 2 and the control of movement is the subject of Chapter 3. Since these two chapters, and those which discuss functional systems, include discussions of brain mechanisms, the major parts of the brain are illustrated here (Figure 5). Glands, as well as muscles, are important effectors and the action of hormones is described in Chapters 5 and 9.

4 Some of the mechanisms within an individual and the relationship of the individual with its environment.

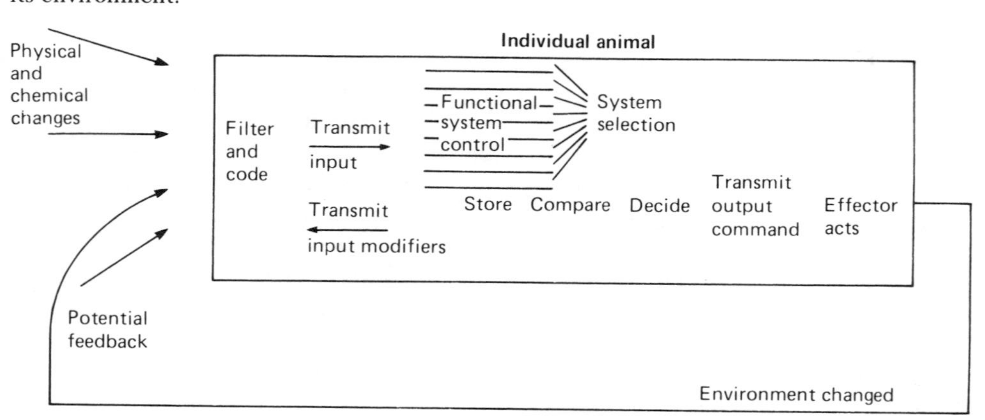

The study of motivation encompasses decisions about which activity to carry out, about the timing of activities and about the amount of energy to expend on an activity. This is a central theme in any attempt to understand behaviour, which psychology departments have long recognised. Consequent upon the work of Hinde (1970) and others, those who teach courses on behaviour for students of zoology or animal science have also come to recognise it. Recent research on motivational mechanisms has demonstrated the importance of considering their biological function and how they might have evolved (McCleery 1978; Toates and Halliday 1981). Mechanisms for deciding which functional system should be in operation and more especially, which behaviour should occur at any moment, are discussed in Chapter 4, on the allocation of resources. For details of storage, comparison and other aspects of decision mechanisms in the brain, books on physiological psychology and cognitive processes should be consulted.

As an example of the importance of biolog-

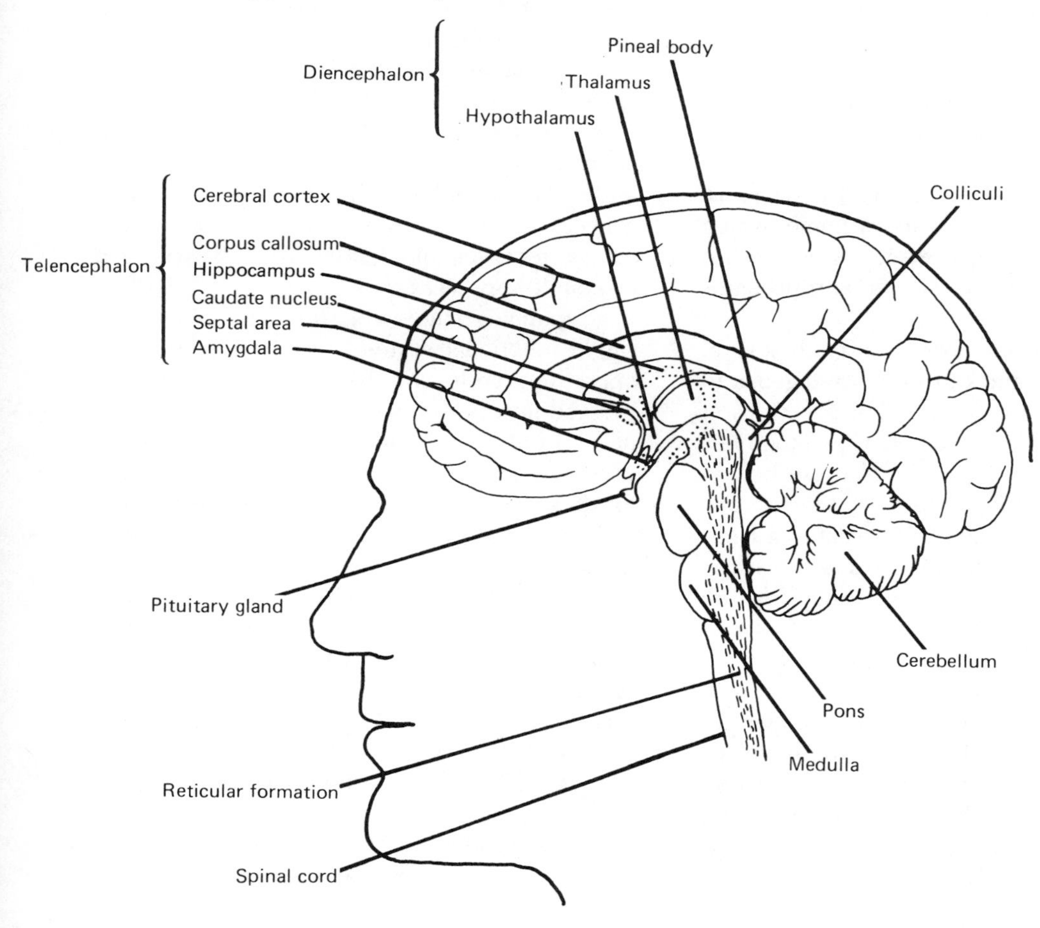

5 Sagittal section of human brain with certain areas lateral to this section also indicated (dotted). The fore brain is made up of the telencephalon (upper left labels) and the diencephalon (upper centre labels). The mid brain includes the colliculi, part of the reticular formation, the auditory tectum and the optic tectum. The lower right-hand labels refer to parts of the hind brain.

ical priorities in making decisions about behaviour, consider an animal which, while it is feeding, detects a predator. If it does not rapidly stop eating and initiate predator-avoidance behaviour then it is unlikely to survive. Suppose that there were two genotypes within a population of animals such that the possessors of one set of genes were able to switch more rapidly from feeding to predator avoidance than were the individuals with the other set of genes. Provided that a high likelihood of predation impaired individual survival more than did the loss of food, the first set of genes would be more likely to spread in the population than would the second. The decisions which are made today depend upon motivational systems which have evolved and, like any other character, have been influenced considerably in their evolution by natural selection.

Familiarity with an area often facilitates the operation of several functional systems. If some part of that familiar area is modified by secretion or artefact construction, adverse environmental effects may be reduced even more easily. The use of a hiding place or home, which is found or constructed, is discussed in several chapters, especially Chapter 7 on anti-predator behaviour.

Social behaviour forms part of several functional systems and can be advantageous to an individual in various ways. The functions of aggregation and of dispersal behaviour, together with ideas about the evolution of these behaviours, are discussed in Chapter 8. Interactions between mates and between parents and offspring are surveyed in Chapter 9 and the structure and organisation of social groups is the topic of Chapter 10.

Optimality and fitness

In 1966 MacArthur and Pianka wrote: 'There is a close parallel between the development of theories in economics and population biology'. The idea that mechanisms for assessing the costs and benefits of actions are of paramount importance in determining what an animal does at any moment and, as a consequence, how animals are distributed, has become very influential in behavioural and ecological work. This economic view of behaviour bore some resemblance to the thoughts of some psychologists about decision-making mechanisms (e.g. Edwards 1954). MacArthur and Pianka were principally concerned with the utilisation of food resources. They argued that since the present phenotype of an animal, i.e. the body form etc. which results from the expression of its genes, depends upon the results of natural selection over a long period, it is likely to include mechanisms for the 'optimal allocation of time and energy expenditures'. They presented a method of specifying the optimal diet of a predator in terms of the net amount of energy gained from prey capture and the energy expended in searching for the prey. These ideas, and similar ideas by Emlen (1966) about optimal foraging, have been developed considerably and are discussed further in Chapter 6. The general concepts are now applied to all aspects of behaviour. McFarland (1977) applies optimality concepts to mechanisms of decision making and brings together biological, psychological and economic approaches.

In many studies of situations in which a single functional system, usually feeding, was operating, costs and benefits have been expressed in terms of energy units. As an individual is hunting for and then acquiring food, it uses energy. The number of Joules required for these activities can be related to the number of Joules available from the food. When calculating the energetic costs of behaviour it is necessary to include those of basal metabolism during periods of low activity. A hunter which expends very little energy for a long period may die of starvation before catching anything. The total cost of an activity is the product of the time spent and the cost per unit time.

A further factor is the energy required for an activity in relation to the maximum possible energy expenditure. The amount of energy required may be a high proportion of that which is possible for a weak individual but may be a small proportion for a strong individual. The weak individual may not risk that activity, even though the food returns are high, if there is any possibility of failing to get the food and hence dying. The strong individual is very unlikely to die when it carries out that activity so the real cost is less. A simple assessment of the energy required for an activity is not a sufficient estimate of cost, for the risk to the individual must be considered. When predator–prey interactions are considered, the risk of dying is obvious.

The measurement of benefit in terms of the energy gain from food seemed logical in early cost-benefit analyses of feeding but food of a given energetic value might be a life-saver to one individual but not to another. A much better means of expressing costs and benefits is in terms of effect on reproductive potential. A large meal may result in an increase in the number of offspring produced. Such benefits of an action have been thought of by many authors as improvements in the fitness of an individual. This concept of Darwin's encompassed the survival of the individual and its subsequent breeding. Parental care, and other behaviour which benefits close relatives, often involves an obvious cost to the individual showing the behaviour. This behaviour is therefore an example of altruism. W. D. Hamilton's (1963, 1964*a*,*b*) ideas about the actions of genes during the evolution of altruistic behaviour lead to a further sophistication of the idea of fitness. By introducing his concept of 'inclusive fitness', Hamilton not only laid the foundations for a better understanding of the evolution of certain aspects of social behaviour but changed the ideas of many biologists about the level at which natural selection acts.

Ultimately, benefit must be assessed by considering the genes in the population. Behaviour patterns which promote individual survival usually increase the chances that the individual will produce offspring. Those offspring receive half of their genes from that parent. Hence any gene whose expression in the phenotype increases the chances of individual survival, reproduction or the survival of the offspring to breed, is more likely to be present in succeeding generations. Hamilton's example is of a gene causing its possessor to give parental care. Suppose that we are trying to calculate the costs and benefits of a particular behaviour pattern at a particular time in the life of an individual. We are, in fact, considering the many genes whose effects increase the likelihood that the behaviour pattern is shown. Our calculation must include an estimation of the effects that the action might have on the number of individuals in future generations which possess replicas of those genes. If the genes influencing the occurrence of two alternative behaviour patterns are being compared, they may affect individual survival differentially but have no other effect on future reproduction. Thus in some studies of feeding behaviour, simple estimates of energetic costs and benefits are helpful when one is trying to understand why some genes have spread in the population. Where parental care and complex social behaviour are analysed, it is more difficult to estimate costs and benefits. In altruistic behaviour there are apparent costs to one individual and benefits to others but if the net effect of a gene is to promote the spread of its replicas then, by definition, that gene will survive in the population.

One of the consequences of the arguments put forward by Hamilton has been the final demise of suggestions that any characteristic might be present in an individual solely 'for the good of the species'. Although Lack (1954) and Fisher (1958) had rejected the idea of such group selection, Wynne-Edwards (1962) ascribed to group selection a major role in population regulation. Wynne-Edwards' book brought together behaviour, ecology and population genetics and thus served all three disciplines but his theories were strongly contested by Hamilton (1963), Crook (1965), Williams (1966) and by Lack (1968). A gene will not persist in the population because its effects benefit all the members of the species, or some smaller group, unless those effects also promote the spread of that gene. The arguments about group selection and the limited situation in which it could occur are explained by Maynard Smith (1976*a*).

Maynard Smith also explains the concept of kin selection which developed from the inclusive fitness idea. Hamilton had explained that inclusive fitness would be increased if a parent behaved in a way which resulted in its death but, as a consequence, more than two offspring survived to breed. In the same way, for various degrees of relatedness, an individual should die for more than two siblings, more than four grandchildren or more than eight cousins. The degree of relatedness between siblings is 0.5 because each obtains half of its gene complement from each parent but it is not correct to say that the siblings share 50% of their genes because this depends upon the variability within the species. Many genes are the same for all individuals because they code for some essential protein which will not function if its structure is altered. Such variability, or lack of it, is discussed later in this chapter but the actual proportion of genes shared by siblings is always much higher than 50%. Inclusive fitness is a term which refers to individuals as the unit upon which selection acts but as Dawkins (1978) explains, the logical conclusion of Hamilton's arguments is to consider how selection acts on replicators. These replicators are genes or groups of genes, hence the arguments above about the ultimate criterion for assessing costs and benefits.

Evolutionarily stable strategies

It is useful to consider behaviour as being the result of the operation of a set of computer programs. This analogy is useful, for the interaction of gene complexes with the environment must be similar in many ways to a computer program which says: carry out this operation, read this environmental input, modify the next step according to which of four kinds of input are received, carry out another operation until a certain type of input is received, etc. The environmental factors which influence the result of this process may be fairly constant among individuals, or very variable, but the fact that such interaction occurs during the development of any characteristic is of fundamental importance. This point will be developed later in this chapter. Examples of such programs might be those which enable a dog which is getting hotter to sweat, or pant, or stand in the wind, or move out of the sun. The dog will utilise these abilities via a decision-making program which determines what it must do when it gets too hot. This program will result in courses of action which depend upon sensory input etc. If a certain combination of inputs exists then a certain pathway is chosen.

The environmental factors which interact with the genetic program will not be the same in each individual so there will be variation in the final form of the expression of that program. Arguments about the evolution of behavioural mechanisms refer to the average individual which has the program and should not be taken to imply that all individuals bearing a particular genetic program will be the same in respect of that behavioural mechanism. It is likely, however, that individuals with slightly different programs will, on average, differ behaviourally. Hence it is useful to consider the way in which natural selection might act in a population which includes more than one alternative program. Taking an example from the Bible, a thirsty man arriving at water might lie face down and lap the water, or scoop up the water with one hand, or scoop up the water with two hands. If the likelihood of predation by lions at water-holes is high, individuals which lie and lap may be less likely to survive until they breed because it is difficult for them to see the stalking lion when they are lying face downwards. Scooping two-handed might be the fastest way to obtain water whilst preserving some vigilance against the approach of a predator so any genetic factors which facilitated this behaviour would become commoner and those which did not would become less frequent. The use of two-handed water-scooping on all occasions is an example of a drinking strategy. A strategy may involve the use of one drinking method in one situation and a different method in another situation.

The way in which selection might act in populations where there are alternative strategies has been considered by Maynard Smith and Price (1973). The question which can be asked when a strategy for dealing with a certain type of situation is present in a population is 'Could a population of a gene-complex which, on average, results in the individuals using that strategy be invaded, during a number of generations, by a gene-complex which results in an alternative strategy?' If lying and lapping was the only strategy for drinking in a human population which was subject to lion attack, when individuals arrived who could use two-handed scooping, they would be at an advantage. Assuming no change in the behaviour of those who lie and lap or of the lions, the population would be invaded by the gene-complex which results, on average, in the new strategy. Maynard Smith and Price discussed intraspecific animal conflicts in their paper. They considered five strategies for contests between evenly matched individuals. The 'Hawk' strategy is to attack as soon as a contest is initiated, thus inevitably incurring quite high costs but sometimes benefiting by victory. 'Doves' threaten initially when a contest starts but always give in, with consequently low costs but no victories against attackers. 'Bullies' attack initially but give in if the attack is returned. 'Retaliators' threaten initially and attack only if they themselves are attacked. 'Prober-Retaliators' threaten initially, attack if threatened weakly, threaten if threatened strongly or if retaliation occurs but retreat if the contest is prolonged. When assumptions were made about the costs and benefits of serious injury, slight injury, winning, and saving or losing time, the average net result of each possible encounter could be calculated. These are summarised in Figure 6 and they indicate that whilst a population of 'Hawks' could be invaded by 'Doves' and 'Bullies', a population of 'Retaliators' would not be invaded by any one of the other strategies. Hence 'Retaliator' is an Evolutionarily Stable Strategy (ESS). The term 'invade' seems particularly appropriate in this example but in its general sense an ESS is a strategy which will persist in the population because the gene complex which usually results in it cannot be replaced due to invasion by a gene-complex for another strategy. Put in Hamilton's terms, the inclusive fitness of the ESS is higher than that of any alternative strategy.

6 Conflicts between animals of equal fighting ability. Populations in which the strategy named in the box was shown, would be invaded by gene-complexes which resulted, on average, in individuals which played the strategies at the base of the arrows. Arrow width indicates ease of invasion. 'Retaliator' is an Evolutionarily Stable Strategy but 'Dove' could co-exist with it (dashed line). Data from Maynard Smith and Price (1973).

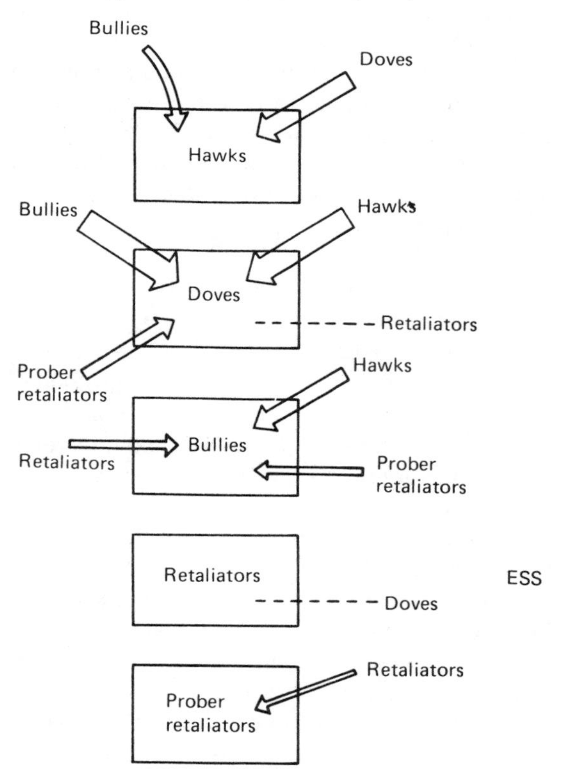

The Evolutionarily Stable Strategy in the competitive interaction example used by Maynard Smith and Price can be calculated mathematically but various assumptions must be made. The ESS idea is just as useful in the strategies for drinking mentioned above but it is more difficult to calculate the ESS. In no case is it possible to be sure that all possible alternative strategies have been considered in a real biological situation. Hence it is best to refer to a strategy as being an ESS if it cannot be replaced by any known alternative. The concept is then of value in considering any aspect of biology which has been examined sufficiently rigorously to make possible the understanding of alternative strategies. The area in which the ESS concept is particularly useful is that in which the playing of a particular strategy has consequences as to which alternative strategy by another individual is subsequently most effective. Examples of such situations, where there is interaction between individuals, include competition for food or for other resources as well as direct encounters. The optimum strategy for food acquisition from an area is likely to be different according to what strategies are being used by competitors in the same area. As Maynard Smith (1976*b*) put it 'If everyone else is eating spinach it will pay to concentrate on cabbage'.

Behavioural diversity and reproductive strategies

Man is a large, long-lived species capable of repeated reproduction but with a prolonged juvenile period under parental care. Some species of fly, on the other hand, are small, develop from egg to adult in a few days, breed once producing many eggs, and then die. Fisher (1930) wondered how 'apportionment is made between the nutriment devoted to the gonads and that devoted to the rest of the parental organism' and 'what circumstances in the life-history and environment would render profitable the diversion of a greater or lesser share of the available resources towards reproduction?' Such questions of strategies in reproduction and life history have been discussed by ecologists for some time but it was not until MacArthur and Wilson (1967) introduced their ideas of *r*-selection and *K*-selection that some order appeared in these discussions. These ideas from ecology and their developments, are also fundamental to an understanding of certain aspects of behaviour. The rate of increase, *r*, in the numbers of individuals in a population has a maximum, *r* max, in an ideal environment. If the fly mentioned above is in an ideal environment, as it might well be at the beginning of the favourable season, its population will increase at r max. There will be little competition between individuals except in the number of offspring they can produce. A gene favouring faster rate of offspring production (higher *r* max) will spread in the population but selection may not alter the proportions of alternative genes concerned with many other characteristics because all individuals survive almost equally well. Such selection is best called *r* max-selection.

If the environment is not favourable enough to make possible very rapid population growth, genetic factors which increase the chances that an individual will survive until breeding and those which promote the chances that the offspring will survive until breeding, become much more important. As soon as there is extensive competition between individuals for essential resources, a different type of selection mechanism operates. The population is maintained at or near the carrying capacity (*K*) of the environment in *K*-selection. More efficient utilisation of resources, such as greater net energy gain from food supply, may alter *K* but most of the major effects of natural selection will be in situations where one strategy might be better than another in a competitive situation. Some of the consequences of *r* max- and *K*-selection are discussed by Pianka (1970) and Brown (1975) and are summarised in Table 1. As Pianka emphasises, there is a continuum between *r* max- and *K*-selection and it is possible to have a mixed reproductive strategy. The most obvious differences in behaviour, according to the type of selection which operates, are in the timing of first reproduction, the amount of parental care and the dispersal mechanisms. If selection leads to early reproduction, no parental care and extensive dispersal, the species is very different from one like our own. Less obvious consequences for behaviour are that strategies for a variety of competitive interactions are likely to be most elaborate in *K*-selected species and, as a result, behaviour in total will be more complex.

Functional aspects of species differences in relation to life-history variables have been extensively reviewed by Wilson (1975). Differences in parental care in relation to number of offspring produced, life expectancy etc. are described further in Chapter 9. It is apparent that closely related species may differ in their strategies, for example the work on deermice (*Peromyscus*) species reviewed by Brown (1975). A dramatic contrast is that between lemmings (*Lemmus*) which breed very fast and migrate in a spectacular way, and beavers (*Castor*) which are long-lived, disperse two years or more after birth, have low birth and death rates, and extensively modify their surroundings. Any comparative psychological or biological study must take into account the fact that selection may result in changes in different directions in different species.

The universality of environmental effects on behaviour

Environmental factors and genetic programs

An important point, which is made earlier in this chapter, is that the operation of genetic programs always involves interaction with environmental variables. Research on genetic expression in cells and during morphogenesis has provided some evidence as to how genetic programs operate. The end result of the operation of a program may vary greatly from individual to individual or may be relatively constant

Table 1. *Summary of some aspects of* r *max- and* K*-selection. (Modified after Brown 1975)*

	r max-selection favouring a higher population growth rate and higher productivity	*K*-selection favouring more efficient utilisation of resources
Defining characteristics		
1	Conditions of occurrence far below *K*	At *K*
2	Little competition for resources among and within species	Much competition
3	Often transient habitats	Long-lasting habitats
4	Frequent and large population fluctuations	Infrequent and small population fluctuations
Consequences		
1	*r* max higher	*r* max often lower
2	*K* rarely approached	*K* raised
3	Higher birth and death rates	Lower birth and death rates
4	Rapid development	Slower development
5	Earlier reproduction	Later reproduction
6	Shorter life	Longer life
7	Less parental care	More parental care
8	Good dispersal	Poor dispersal

among individuals. Some of the variation will be the result of the imperfect functioning of the program, but most results from the program's interaction with environmental variables (Figure 7). The aspects of the environment with which a program interacts may vary little from individual to individual. For example, an adequate oxygen supply and a temperature somewhere within a wide range may be needed. Even these factors may sometimes be outside the tolerable range and hence affect the end result. Other programs may produce a different result according to the precise temperature and oxygen level, or according to the characteristics and concentrations of a variety of possible biochemical substrates.

Some parts of functional systems in animals can work successfully only if their components have a particular structure. The cytochromes are concerned in electron transport in cells and hence form part of the respiration functional system. Studies of the amino-acid sequence in cytochrome *c* show that it is relatively invariable within individuals, between individuals of a species and even among individuals of different, but closely-related, species such as cow, sheep and pig (Fitch & Margoliash 1967). The amino-acid sequences in some enzymes, however, are much more variable. It is presumed that cytochromes must have a certain structure in order to function but that variability is tolerable, or even desirable, in other biochemical molecules. We know much less about the control of development in more complex systems but it seems likely that natural selection has produced some programs which do not go wrong and which depend upon very constant environmental factors. It is possible that such programs are energetically costly to operate and this factor will influence the likelihood of their continued existence. Other programs may go slightly wrong and depend upon variable environmental factors. Indeed the mechanism produced may not function unless influenced by such factors. All mechanisms of behaviour control, like all other anatomical characteristics, depend on genetic programs and on the environmental influences on the expression of these programs. The developmental processes are stochastic, i.e. they always include some random factors, rather than deterministic. The interesting questions are what the programs say, how

7 Diagrammatic representation of the operation of a genetic program. No genetic program operates without interacting with the environment. Some of the environmental factors might be relatively constant from one time to another in the same or in different individuals, however, whilst others might be very variable.

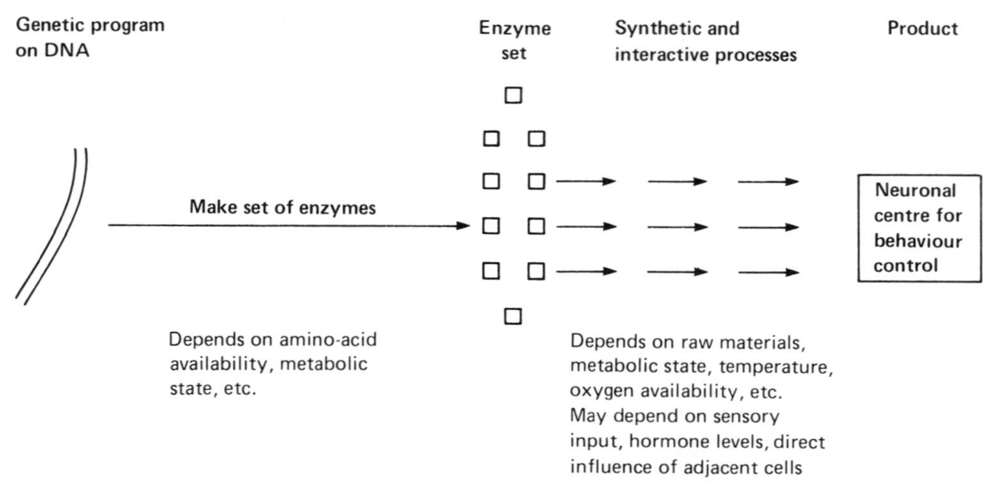

the environmental factors act and how variable are the characteristics which result from the operation of a program? Some of the answers to these questions can be obtained by behaviour-genetic analysis, as mentioned in later chapters and as reviewed by Hirsch (1967) and Manning (1976).

A consequence of the universal environmental influence is that no characteristic in an individual animal can ever be predicted with certainty just by knowing which relevant genetic program is present. Statements about the evolution of programs and strategies in behaviour, as mentioned earlier, refer to the average individual which has a particular genetic make-up. Individuals which are genetically identical, with respect to a certain type of program, may vary considerably because the environmental factors which interact with the program differ amongst individuals. This fact is often overlooked in general discussions about the ideas of Hamilton, Wilson, Dawkins (1976) and others. The fact that natural selection has acted in such a way that programs exist which are related to specific aspects of social behaviour, does not mean that the behaviour of all individuals is 'determined by their genes'. Neither does it mean that individual people should give up trying to modulate these aspects of their social behaviour.

New concepts of learning and experience

The assumptions that learning is a phenomenon which can usefully be studied without any consideration of the biology of the animal and that it is different in character from other environmental influences on behaviour, have had adverse effects on the development of our understanding of the mechanisms which control behaviour. One of the themes of this book is to consider the plasticity of behaviour, both during development and during later interactions with the environment, in relation to each major topic in each chapter. Sensory and motor systems develop as a result of interaction between genes and environment and the ways in which genetic programs interact with a variety of environmental factors are gradually being elucidated. The effects of rearing an individual in darkness are dramatic, for they involve changes in the number of cells in regions of the visual pathway (e.g. Rasch, Swift, Riesen & Chow 1961). Effects of rearing conditions on the occurrence of pattern detectors have also been found (e.g. Hirsch and Spinelli 1970, Blakemore and Cooper 1970, see Chapter 2). No doubt we shall eventually know more of the anatomical, physiological and biochemical changes which occur when an adult individual improves some aspect of perceptual ability or develops a new motor skill. The actual changes which occur in the nervous systems as a result of environmental effects must inevitably be of various types but there is no real evidence for dividing them sharply into learned and other environmental changes.

All interactions between an individual and its environment are occasions for potential modification of that individual. Most environmental changes which impinge upon the individual will not have a modifying effect but it is seldom possible to say with any certainty that a particular change has had no effect. The effect might be mediated via one or more sense organs or via other cells. It is of interest to investigate the pathway by which behaviour is affected. The use of the word *experience* is sometimes restricted to effects mediated via sense organs. This usage implies some fundamental difference between such effects and those in which sense organs are not involved. There is no justification for such a distinction so experience is used interchangeably with *environmental effects* here. It is very difficult to define the word *learning* without producing a very broad or very narrow definition. W. H. Thorpe's (1963) broad definition is 'that process which manifests itself by adaptive changes in individual behaviour as a result of experience, or, to put it another way, the organisation of behaviour as a result of individual experience'. A narrower definition is Skinner's (1953) 'the reassortment of responses in a complex situation' which referred, principally, to a certain range of experimental situations. The difficulties experienced by those who have sought to define learning must be due to the existence of a continuum of phenomena from the effects of oxygen levels on enzyme action during the development of a motor control mechanism to the effects of scarcely perceptible signals on the behaviour of an individual in a complex society. The arbitrary positioning of a dividing line between learning and non-learning does not alter the fact that the continuum exists.

It is apparent that all who study behaviour see learning, in its broad sense, occurring even if they do not always recognise it because the behaviour change is subtle or because the environmental change and its consequence are widely separated in time. As Hinde (1973) points out, in one of his contributions to a collection of papers on constraints on learning (Hinde and Stevenson-Hinde 1973), 'learning is not episodic but is occurring continuously, though not necessarily affecting behaviour immediately'. The problem facing those attempting to analyse the effects of experience on behaviour is to determine why behaviour changes so infrequently rather than just to consider what does make behaviour change. As Mackintosh (1973) asks, how do individuals learn to ignore irrelevant cues? This topic is discussed further and in different ways, in Chapters 2 and 4. The fact that species differ in their predispositions to learn particular abilities has been made by various authors, including Lorenz (1965). Lorenz also made the point that habituation to cues received from the approach of a predator is much slower than most forms of habituation. Similarly, Bolles (1975) suggests that rats learn to avoid an electric shock much more readily than they learn to press a lever because the shock elicits a variety of species-specific defence reactions. Rats which modify their behaviour rapidly when a cue indicating danger is detected will survive better and bear more offspring than rats which modify their behaviour less rapidly. All genetic programs which facilitate such ability to modify behaviour are thus more likely to spread in the population than those which do not. Another example of impressive learning ability, which is logical when the general biology of the animal is considered, is the ability of rats to avoid poison and inadequate diets. Poison-based aversion must involve relating cues from eating to later cues from the effects of eating that food. This is facilitated in rats by their habit of feeding on one type of food at a time (Rozin and Kalat 1971). Every rat is equipped, presumably, with taste preferences but it has to modify its eating behaviour according to experience. The system controlling feeding must make it possible for inadequate and poisonous dietary components to be avoided (McFarland 1973 and see Chapter 6).

The limitations as to which behaviour patterns can be modified in structure, timing or frequency by a given type of environmental change, i.e. by a *reinforcer* (see Chapter 4), are emphasised by Shettleworth (1972, 1973). Shettleworth presented food to hamsters on successive occasions when they were carrying out one of several activities and observed the subsequent frequency of the activity. The frequency of the activity was increased if the hamsters had been pressing a bar in their cage, rearing on their hind legs in the centre of the cage, scrabbling or digging at the instant of food presentation. If they had been washing their faces, scratching themselves with the hind leg or scent-marking in the cage when fed, the frequency of the activity was not changed greatly by presenting food. Clearly it is easier for hamsters to associate food with activities which might be related in some way to food acquisition than with cleaning or social activities.

The number of environmental changes which can be used as *reinforcers*, which will alter the frequency of some behaviour patterns, is very large (Hogan and Roper 1978). As well as food, water and the termination of something aversive, various correctors of homeostatic imbalance (see Chapter 4) and various opportunities for social behaviour have been used in experiments. An individual which is cold will work to get warm, a male chaffinch (*Fringilla coelebs*) will jump from perch to perch in order to hear the song of a potential rival (Stevenson-Hinde 1972) and a male fighting fish (*Betta splendens*) will swim down a maze in order to see a displaying rival (Hogan, Kleist and Hutchings 1970). The constraints which determine whether or not an environmental change will be reinforcing are summarised by Stevenson-Hinde (1973) as (1) causal factors, e.g. blood glucose level; (2) the relationship between the stimulus situation and the apparent response, e.g. chaffinches will not peck for a song as a reinforcer unless there is a food grain on the key to be pecked; (3) the relevance of the response to the reinforcer, e.g. food reinforcers are not relevant to face washing; (4) the ease of association of the stimuli with the consequence, e.g. detecting delayed poisoning; and (5) the compatibility between recent performance of the consummatory response and repetition of the operant, e.g. stickleback courtship inhibited rod-biting but not swimming through a ring, so rod-biting was less readily reinforced by courtship opportunity (Sevenster 1973). Learning is inextricably related to the biological functioning of all animals, including man, so it will be considered during discussions of all major tropics in the following chapters.

Behaviour development

In a developing individual behaviour must be adapted to deal with the problems of life at each particular stage and must also be a preparation for adulthood. Young individuals are smaller and are usually poorer at defence than adults. Hence they are subject to attack by predators, which may be different from those which attack adults. Some predators attack young individuals preferentially. The work of many naturalists on predator–prey interactions amongst large mammal species on African savannah has always emphasised the great vulnerability to predation of young animals. Recent detailed studies of hyaenas (Kruuk 1972), lions (Schaller 1972) and hunting dogs (van Lawick and van Lawick-Goodall, 1971) have verified this. With such strong predation pressure, genetic programs whose expression usually reduces the chances of death by predation for the young animals have spread in the prey population. This spread of genes must have occurred even if the characteristics concerned, behavioural or non-behavioural, could serve no useful function in adult life. Since our ancestors must have been especially vulnerable to predation during the prolonged juvenile phase the predator avoidance functional system must have an important influence on the behaviour of children (see Chapter 7).

The food of young animals is often different from that of adults and there may be differences in their thermoregulatory and osmoregulatory problems. Young and adults are considered in the discussions of the various functional systems in Chapter 4 to 7. Some of the problems of a young, developing animal can be exemplified by considering what a cod and a domestic chick must do in order to survive to adulthood and be fit to breed. The cod egg is planktonic and there is no parental care, whereas the hen behaves in a way which increases the chances that her chicks will survive. The larval cod emerges from the egg with some yolk still attached, so that it can survive for some time without feeding. Osmoregulation is principally a physiological problem but the larval fish has to avoid predation and it has to start acquiring information which it will be able to use later. The programs which are relevant to these functions will have been interacting with the environment so that adequate mechanisms are likely to exist at the time that they are needed. When the yolk is used up, mechanisms for finding and acquiring food must be added to the behavioural repertoire if the individual is to survive. Such mechanisms will continue to be adapted and improved with further experience during development. Behaviour patterns associated with competing with other fish for resources and shoaling also start to appear and, like food finding and predator avoidance, change with experience during growth.

The problems for a domestic chick in an egg are similar to those of the cod in many ways. The regulation of body temperature, the necessity to breathe air and conserve moisture and the need to interact with the hen cause the greatest differences. Some of the interrelated changes which do occur in the developing chick's body, brain anatomy, brain biochemistry, brain physiology, and behaviour are shown in Figure 8. Within the egg, environmental factors are more constant than they are after hatching but, as stressed earlier, all the genetic programs operate in cells and affect groups of cells. Their environment includes some factors which are predictable and others which are not very predictable. Whilst general anatomical changes are occurring in the body, the brain is also developing. The brain structure is established by 14 days, seven days before hatching, and there is no evidence of gross change after this. The major changes in brain biochemistry and physiology occur between five days before and two days after hatching. Sensory responsiveness, both physiological and behavioural, appears by four days before hatching. Work on the synchronisation of hatching of eggs incubated to-

8 Some changes in the domestic chick's anatomy, physiology and behaviour during development from fertilisation. Abbreviations in diagram: GABA, γ amino-butyric acid (a neurotransmitter); ChE, cholinesterase and GAD, glutamic acid decarboxylase (enzymes which break down neurotransmitters); EEG, electroencephalogram.

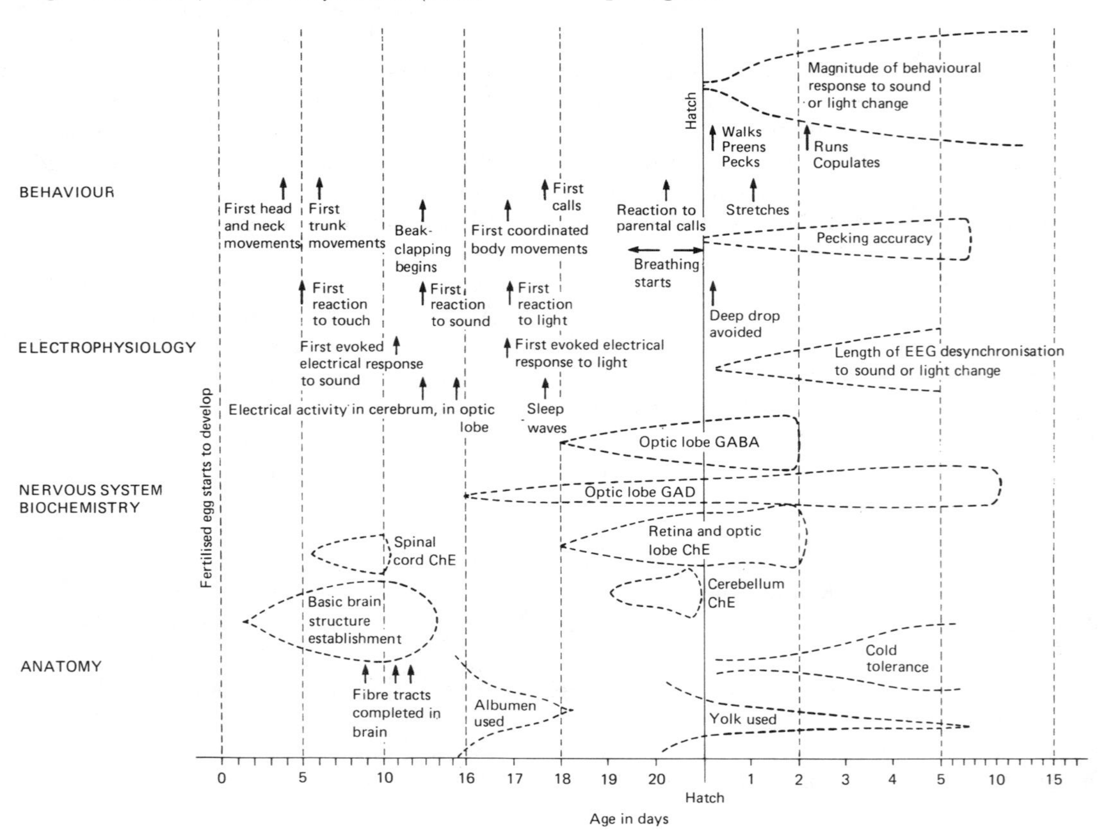

gether and on other behavioural responses to differences in sensory input before hatching (reviewed by Freeman and Vince 1974), emphasises the importance of sensory experience before hatching. Whilst considerable motor ability is present at hatching, motor ability and sensory-motor co-ordination improve during the first few days after hatching (Kruijt 1964). Pecking accuracy improves considerably during the first week after hatching (Padilla 1935) and is only slightly impaired by dark-rearing (Cruze 1935). Experiments by Hess (1956) suggested that chicks reared with prismatic spectacles which distorted their vision could not correct for this distortion, but later work by Rossi (1968) has shown that correction for such distorted vision is possible. Sensory input does affect the development of this sensory-motor ability.

The structure of the visual system with its analysers (Chapter 2) inevitably means that some aspects of the world surrounding a newly-hatched chick will be more conspicuous than others. These aspects are the most likely to be investigated by the chick so such mechanisms must form part of the neuronal basis for the preferences which are shown by newly-hatched chicks. The first problem for the chick after hatching is to find and subsequently to recognise mother and, perhaps, other chicks and the resting place. The behaviour of the mother is likely to facilitate this recognition (Chapter 9) but young chicks, ducklings, goslings and moorhens (*Gallinula chloropus*) do approach objects which are larger than the chick and which move at approximately walking speed (Lorenz 1935, Fabricius 1951, Hinde, Thorpe and Vince 1956). Subsequent studies showed that a flashing light elicited rapid approach by chicks (James 1959). Experiments to determine the optimum size, brightness, rate of flashing, or rate of movement for eliciting approach are reviewed by Smith (1962). Subsequent recognition of mother, or other aspects of the environment, after such early experience has been the subject of many experiments (Bateson 1966). The learning involved is now considered to be similar to many other forms of learning so it is dignified by the separate name 'imprinting' merely for convenience.

Chicks can utilise the yolk which remains in the body for 4–6 days after hatching so they do not need to peck for food during the first few days. Directly after hatching, however, they will approach and peck at some objects more than others. Shiny objects are readily pecked (Baeumer 1955) and in tests where there was a choice between two coloured spots, chicks, ducklings, gulls etc were found to show specific colour preferences (Hess 1956, Kear 1964, 1966). Chicks peck more frequently at three-dimensional spots than at two-dimensional spots (Fantz 1957) but Dawkins (1968) reported on some ways in which such preferences were modified by experience. Chicks reared with small hemispheres on the floor of their pen pecked at them initially, but the response waned after repeated pecks. In addition, chicks could be trained to peck at a flat disc and avoid a hemisphere if they had been presented with food, or if an overhead heater had been switched on whenever they pecked the flat disc. Other changes in behaviour during development include the appearance of social behaviours (Kruijt 1964). These become modified and their direction restricted as a result of the social experiences of that individual. Developmental aspects of social behaviour are discussed in Chapter 10.

It is valuable to study developing animals as a means of assessing how experiental factors are involved in the operation of genetic programs related to behavioural function (Gottlieb 1976 *a,b;* Broom 1981). In such studies, however, it is necessary to consider all environmental factors, not just those which have an effect after birth or hatching. As stressed above, it is also necessary to consider function in the developing animal. There are very large differences amongst species in the functional systems which operate at the time of birth or of hatching because of the differences in developmental age then. Figure 9 shows, for some mammals and birds, the differences between altricial species, which emerge into the outside world in a comparatively helpless state and precocial species, which have well-developed motor and sensory systems at emergence. The differences will be discussed further in Chapter 9 but the examples shown are of interest because man is intermediate between the rat and the goat in this series.

Rhythms of activity

The environment of every individual includes more or less regularly recurring events such as dawn, high tide, full moon or the first ground frost. If such variables, which are external to the body, result in physiological or behavioural rhythms they are said to exert exogenous control. Much research has been concerned with determining whether the control of a rhythm is exogenous or endogenous, i.e. due to some internal *zeitgeber* or time-giver (Aschoff 1960, Saunders 1976). Some of the terms used to describe biological time series will now be defined (after Broom 1980). A *rhythm* is a series of events repeated in time at intervals whose distribution is an approximation to regularity rather than being random. It is not a precise term so it is often used with qualifiers such as regular, circadian, i.e. wavelength approximately 24 h, or circannual, i.e. wavelength approximately one year. The terms which are used most by statisticians, geophysicists and econometricians are period and its derivatives. A *period* is the interval between two events repeated in time; *periodic* refers to events in a time series which are separated by equal periods; a *periodicity* is a periodic variation within a time series. The term periodicity should therefore be used instead of rhythm if regularity has been demonstrated by measurement. Another widely used word is *cycle*, a recurrent set of changes such that the state at the end point is the same as that at the starting point. This term is useful where there are several changes, as in a reproductive cycle, all of which are repeated. These terms can also be used to describe spatial series. Some methods of detecting and analysing activity rhythms are reviewed by Broom (1979).

9 Diagram of development from fertilised egg to adult showing that five species, examples of altricial animals, emerge into the outside world at an earlier stage of development than do precocial animals. The origins of arrows show the point of hatching or birth.

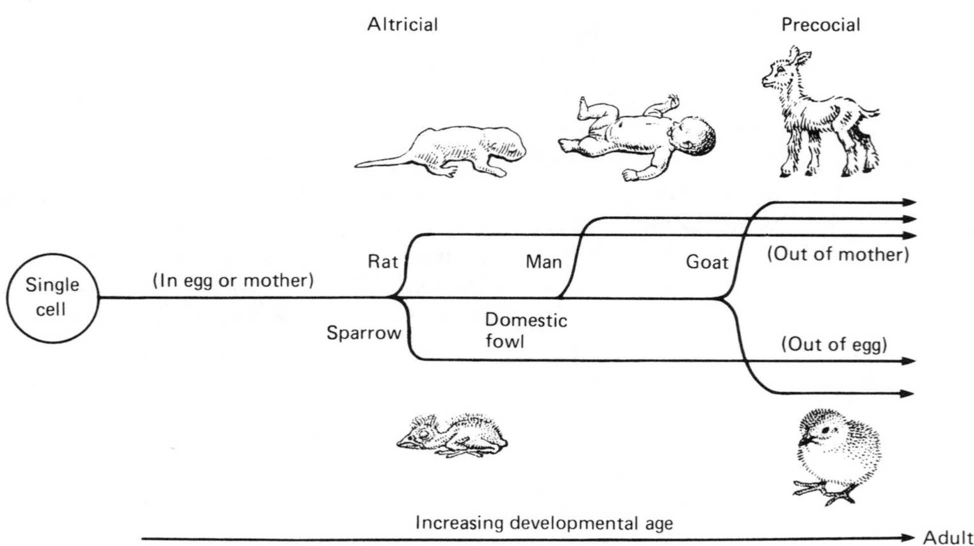

Rhythmic activities include ventilating lungs or tracheal system, movements of gills or palps creating gill-currents, swimming, walking, flying, chewing, other food acquisition movements, being active rather than resting, activities associated with tides, diurnal, nocturnal or crepuscular activities, oestrus activities, and breeding. The wavelength of the rhythms varies from 0.001 s for the flight muscles of very small insects to more than a year for some breeding cycles. Some of the activities occur with sufficient regularity to be called periodic. In motor patterns the periodicity often originates in central nervous pacemakers (Bullock 1961). Controversies over the control of locomotor patterns are discussed in Chapter 3. Many biological control systems operate by allowing a gradual change of state to occur until a critical level is reached and then initiating a behavioural or physiological correction mechanism. Examples of the latter include the utilisation of energy until immediate food reserves drop to a level at which feeding is initiated, and the gradual production of urine from the kidneys until the pressure in the bladder reaches a point at which urination becomes a likely behaviour. If negative feedback control operates at two levels the state will oscillate between these levels. As Sollberger (1965) explains at length, the consequence of such control mechanisms is a behavioural or physiological rhythm. Feedback may also act as fine tuning for a central pattern in some control systems. Rhythms are mentioned in Chapters 3, 5, 6 and 9 and possible mechanisms underlying rhythmicity are considered in Chapter 4 and by Sollberger (1965) and Rusack and Zucker (1975, 1979).

Economic and social aspects of behaviour study

Studies on the biology of behaviour are relevant to medical, social, agricultural and other economic problems. Those whose major concern is to understand the fundamental principles of behavioural biology should also be aware of the applications of such principles. As a consequence of this argument and of the fact that many useful examples in behaviour study have come from applied work, reference is made to work whose major aim is medical, agricultural etc. in each of the chapters in this book. In addition, an attempt is made to relate the general principles to the applications so that those concerned especially with applied work can understand better the factors which affect behaviour.

The diagnosis of disease by medical practitioners and by veterinarians often involves observation of the patient's behaviour and recognition of abnormalities. In order that abnormal behaviour can be assessed, it is necessary for the observer to have a precise knowledge of normal behaviour. As stressed at the beginning of this chapter, all people have some ability to observe human behaviour and many people have also observed the behaviour of other animal species. Those concerned with medical and social work and those with interests in veterinary, animal husbandry or pet-care problems often know a lot about behaviour. Much of such knowledge is, however, undocumented and very little of it is gained by systematic study in controlled conditions. Collaboration between behavioural biologists, psychologists and those listed above, does occur on a small scale but there is a need for more sharing of knowledge and more experimental investigation of methods used in behavioural diagnosis.

Ethological studies of people performing tasks and interacting with others in small or large groups provide evidence which can be used when trying to improve work efficiency, job satisfaction and social harmony. Ethology clearly overlaps here with ergonomics and social psychology. Studies of the development of behaviour in children are also important in relation to education and the establishment and maintenance of social cohesiveness. Such studies of human behaviour, and the relevance of work on the behaviour of other species to these general issues, are extensively discussed by Hinde (1974). The methods used in detailed behavioural studies of human subjects are essentially the same as those used for other species. The results of such studies must be considered together with results from work on cognitive processes, interpersonal relationships (Hinde 1979) etc in order to understand human behaviour.

Domestic animals are very convenient subjects for behaviour observation and work on them has been influential, especially on ideas about reproductive behaviour and social organisation. Many general ideas about behaviour can be obtained equally well from studies of farm animals, laboratory animals or wild animals. The spread of genes in populations of farm animals is affected by selection pressures imposed by man but evolutionary processes occur in both domestic and wild species. The proportion of the energy expenditure of an individual which is directed towards predator avoidance, food acquisition, mate finding etc is different in domestic and wild species but the strategies used in domestic species are of intrinsic as well as economic interest. Hafez (1975) provides a compendium of information on the behaviour of domestic animals. Examples of behavioural studies on animals which are of agricultural importance are used in most chapters of this book. Another sphere of work which yields examples of general interest is that concerned with pest control. Much of the best work on chemical communication has been carried out with insect pests as subjects and studies of the responses of rats to poisoned baits have provided new insight into learning mechanisms. The medical and agricultural advantages of studies on the behaviour of insect and other pests have been very great.

10 Diagram of neuron and of electrical changes detected (i) when an electrode is pushed into the cell (ii) when the cell membrane properties are altered a little and (iii) when they are altered a lot.

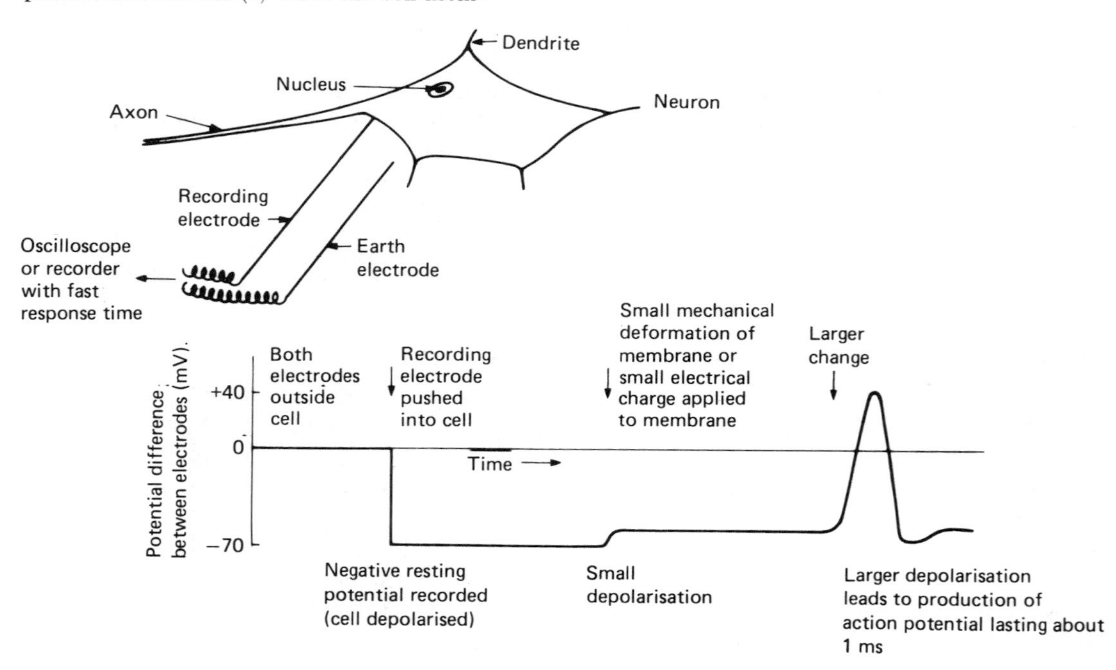

2

Sensory function: behavioural and physiological evidence

Principles of receptor function

Living cells may be altered chemically or physically when their immediate environment changes. Some environmental changes are more important to a cell than others, for example they may occur when food or danger is present. The likelihood that a unicellular animal would survive to reproduce, or that a cell in a multicellular animal would function adequately, would be increased if the cells could detect changes and modify their behaviour appropriately. Environmental changes reach cell membranes first so it is not surprising that the modification of cell membrane properties is the basis for sensory receptor mechanisms. In a multicellular animal, some changes in the membranes of the outer cells modify the internal states of those cells in such a way that the change can be communicated to other cells via their cell membranes. Thus it seems likely that, in the course of animal evolution, the first nerve cells were on the surface of the animal and functioned as receptors. Later, there were cells internal to these receptor cells which were specialised for transmission of sensory information.

Work on the nerve cells of the frog (*Rana*) and of some molluscs including the squid (*Loligo*) has provided much of our information about electrical mechanisms within cells. Studies on red blood cells have also yielded important facts. The electrical mechanisms are properties of cells in general; neurons are the cells in which they are best developed. Molluscs have been used in neurophysiological work because they have very large nerve cells, e.g. in the opisthobranchs *Aplysia* and *Tritonia* (Figure 52). The fast swimming squids have giant nerve fibres which conduct electrical change very rapidly. Vertebrate nerve cells are much smaller and thus harder to study directly, but all nerve cells share fundamental properties.

Resting potential

Nerve cells are electrically polarised for they show a *negative resting potential* (Figure 10). The reasons for this are explained by Aidley (1978) and in other physiology texts. In summary, the first reason is that there is a Donnan equilibrium between the inside and the outside of the cell. This exists because the cell membrane is impermeable to large, negatively charged ions such as proteins but permeable to small ions. The small, positively charged potassium ions, therefore, tend to diffuse out of the cell into areas of lower potassium-ion concentration. This leaves more negative than positive ions inside. Secondly, there is a sodium pump which uses metabolic energy to remove positively charged sodium ions from the cell. If the permeability of the cell membrane is altered by some external change, the most likely consequence is that positive ions will pass into the cell thus reducing the negative resting potential, i.e. *depolarising* the cell. Such a change in one region of the cell is readily transmitted to other regions so it is possible for effector systems within the cell to initiate action which is appropriate to the external change. Hence the first principle of receptor action is that animals can detect only those external changes which alter the permeability to ions of a receptor cell membrane.

Functioning of typical sense organs

In a protozoan, a localised or whole-animal response might follow any depolarisation. In a multicellular animal, however, the electrical change is transmitted, usually, from a receptor cell to a sensory neuron and thence, via other nerve cells, to an effector cell, before a response can occur. Most sense organs include accessory structures, which are affected by external energy and numerous receptor cells connected to many sensory neurons. Some, however, have little or no accessory structure or no receptor cell. The axons from sensory neurons leave some sense organs in a single bundle which is often referred to as a nerve.

The sequence of events in a typical sense organ is shown in Figure 11. A very small influx of ions may be insufficient to produce a *receptor potential* which is above the threshold necessary for transmission to the sensory neuron. It may increase, however, the likelihood that the receptor potential will reach that threshold level when a subsequent external change occurs. If the receptor potential is above the threshold, transmitter molecules will cross the synaptic cleft and cause depolarisation of the sensory neuron, thus producing a *generator potential*. If this generator potential reaches a threshold level, production of an *action potential* is initiated. The magnitude of the generator potential is important in determining the length of the delay before the first and any subsequent action potentials are produced. It may also determine the extent of any interaction, such as lateral inhibition (see later in this chapter), with nearby sensory neurons.

In most sense organs, each receptor cell is innervated by two or more sensory neurons and each sensory neuron synapses with several receptor cells. This means that none of these cells is indispensable and it makes possible some of the analytical systems described later. The accessory structures mentioned in Figure 11 include the lens and eye muscles in the vertebrate eye, the layers of fatty tissue in the mammalian Pacinian corpuscle and the cuticular hair which

11 Sequence of events in a typical sense organ. Some features shown here are absent in certain sense organs. Modified after Davis 1961.

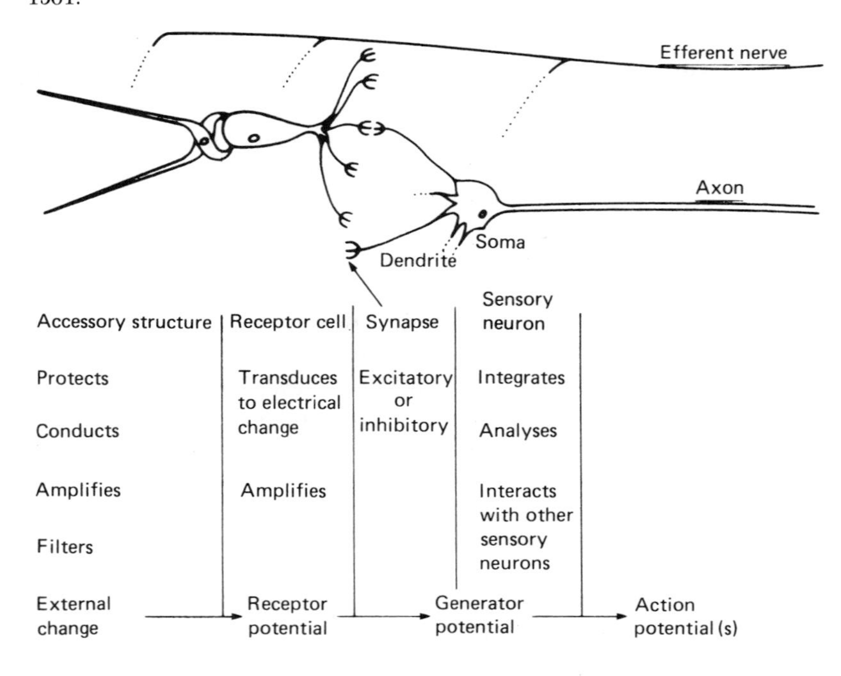

forms part of many insect mechanoreceptors. As a result of the existence of these accessory structures, the receptor cells are much more likely to be affected by some types of external change than by others and there will sometimes be differences in the ways in which different receptor cells within a sense organ respond to a change. Accessory structures are therefore important aids to sensory filtering as well as amplifying, modulating and channelling external change so that it has an effect on membrane permeability.

One study of sense organ function which has provided evidence for the general mechanisms described above is that of Loewenstein and Rathkamp (1958) on the Pacinian corpuscle in mammals. The corpuscle, which is about 10×0.6 mm, is a mechanoreceptor found in mammalian mesentery and is similar in function to the Herbst corpuscle in birds. As shown in Figure 12 it is composed of a neuron whose tip if unmyelinated and surrounded by a series of onion-skin like layers which can be dissected away from the nerve. The receptor is unusual in that it is very large and there is no separate receptor cell. The sensory neuron itself is affected by mechanical change, as can be shown in a dissected preparation. If an electrode is inserted into the nerve at *A* a generator potential is recorded when the exposed nerve is compressed. Since there is no separate receptor cell it is simplest to refer to this potential as a generator potential (g.p.). If one third of the nerve beyond *A* is removed, the g.p. is reduced by one third; if larger proportions of the nerve are removed, the g.p. is altered in direct proportion. If the recording electrode is at *B*, or further from the tip, action potentials are recorded so it is more difficult to measure the g.p. The accessory structure acts so that a small external pressure deforms a short length of neuron and produces a small generator potential. A greater external pressure deforms more of the neuron and initiates a larger g.p. and a faster rate of action potential production.

Further evidence for the relationship between the extent of any detectable external change and the magnitude of the g.p. comes from the work of Katz (1950) on the frog muscle spindle. The muscle spindle (Figure 13) is in parallel with the normal muscle fibres and is stretched when the muscle is stretched. Katz recorded from the sensory neuron which synapses with the spindle receptor cell. If the receptor potential in the receptor cell is sufficiently large, it results in the production of a g.p. and action potentials in the sensory neuron. Katz blocked the production of action potentials with the drug procaine and found that a g.p. was still produced. The extent to which the muscle was stretched could be accurately measured and the magnitude and duration of the g.p. was found to be precisely related to the rate of stretching (Figure 14). As explained later in this chapter, the action potentials produced in the

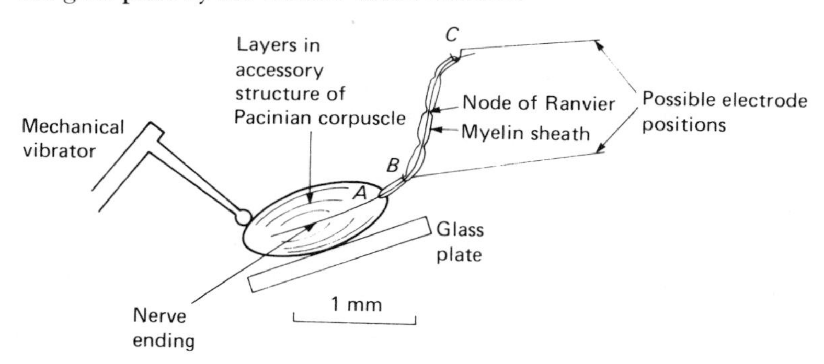

12 Pacinian corpuscle from the mesentery of a cat. The corpuscle can be compressed against the glass plate by the vibrator whilst electrical changes are recorded from *A*, *B* or *C*. Modified after Loewenstein and Rathkamp 1958.

sensory neuron are of a constant size but the rate at which they are produced can provide information about the extent of the external change.

13 Frog muscle showing experimental arrangement for recording from a muscle spindle. Modified after Katz 1950.

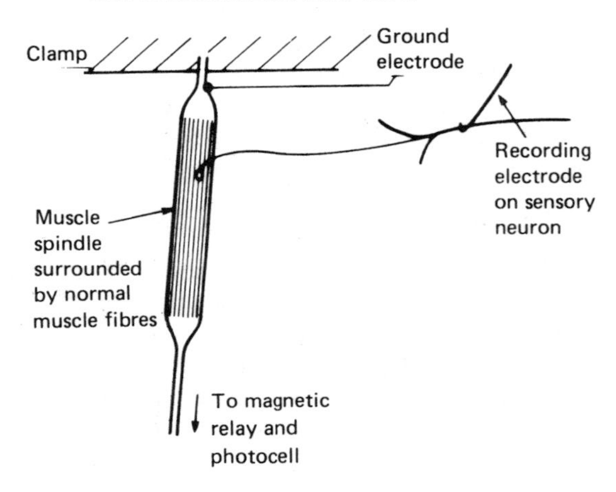

14 The generator potential in the sensory neuron of the frog muscle spindle is proportional to the rate of stretching of the muscle. Action potentials have been blocked with procaine. Modified after Katz 1950.

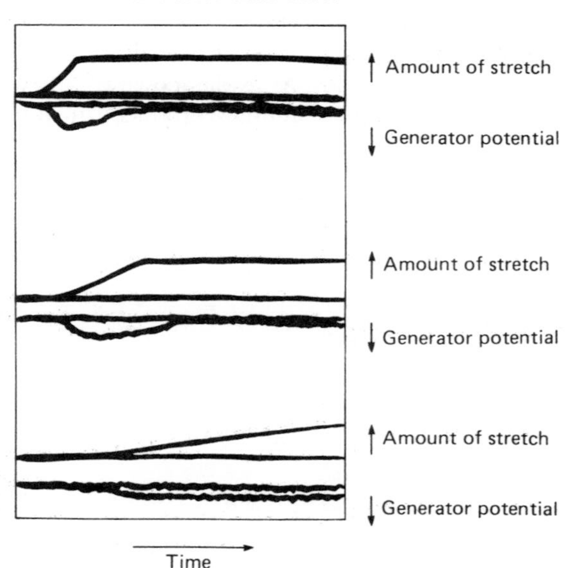

Sensory detection

A wide range of physical and chemical changes can be detected by animals. Some of these changes have a direct effect on receptor cell membranes whilst others have an indirect action via accessory structures.

1. The *direction of orientation of magnetic fields* can be detected by some animals.
2. (*a*) Most animals are sensitive to the incidence of light upon them. A change in light intensity means a change in *the rate at which photons impinge upon a receptor cell*.
   (*b*) Electrons, like photons, are very small particles and a large electric current can be detected by any animal. Many aquatic animals are capable of distinguishing *the characteristics of electrical fields*.
3. The behaviour of cells within animals and of small aquatic animals can be altered by *ion concentration and the presence of certain specific ions outside the cells*. Such effects depend upon the structures of cell membranes and the biochemical mechanisms associated with them.
4. Olfaction involves the detection of *the presence of certain molecules or sudden changes in concentration of certain molecules*.
5. A change in temperature is *a change in molecular velocity* which can alter the electrical state of all cells.
6. When *one part of an animal moves in relation to another* a mechanoreceptor may be affected by the movement. Such movements involve large aggregations of molecules and are thus on a larger scale than the changes described in 1–5. The movement may be initiated by a force from outside the animal or by the animal's own movements. It may involve a comparatively simple mechanical change, as in receptors which respond to localised pressure or stretch, or a vibration as in hearing receptors.

For any sensory modality, a high-intensity

input may cause changes which are different in character from those caused in cells by smaller changes. We call the resulting sensation *pain*. Some receptors are specific to inputs which produce pain.

What information does an animal need from its battery of receptors? Since any change in an animal's surroundings may be dangerous, the very fact that a change has occurred is significant. Once a change is detected, a general alerting mechanism, involving decreased thresholds for further changes and increased readiness for motor action, can be initiated. More specific action can be taken if the animal can determine the quality, quantity and pattern of the change. Although any change might be important to an individual, most detectable changes will not require a response, so it is just as important for an animal to avoid wasting energy by responding to every change, as it is to detect those which are relevant. Specialisation of receptors thus necessitates the exclusion of the effects of most factors which might affect a cell whilst facilitating and directing the action of others. For example, the ideal chemoreceptor cell should not be depolarised when mechanical changes occur. The next part of this chapter explains how various sorts of change can be transduced so as to cause the depolarisation of a receptor cell. Methods of coding quantitative information about changes will then be considered and finally, as an example of higher levels of processing sensory inputs, physiological and behavioural studies of visual pattern recognition will be discussed and compared.

Transduction in receptors

The principle of the mechanism of initiation of receptor potentials in mechanoreceptors is quite easy to understand. If the cell membrane of a receptor is deformed by stretching part of it, or by compressing one part and hence stretching another, positive ions, from outside the cell, flow through the membrane and the cell is depolarised. We cannot see the ions passing into the cell but it seems most likely that the difference in permeability is due to a stretched membrane having pores which are larger than, or of different shape from, those in an unstretched membrane. It is possible that permeability to ions is altered by a substance which is produced, or an active transport mechanism which is activated, when the membrane is stretched, but the pore stretching hypothesis is simpler. The Pacinian corpuscle (Figure 12), muscle stretch receptor (Figure 13) and insect mechanoreceptor hair (Figure 15, described in detail by Dethier 1963) are examples of sense organs in which the electrical output in the sensory neuron is related to the degree of mechanical deformation. The transduction mechanism in the mammalian ear is clearly more complex and is still a matter of some dispute. If the theory put forward by Davis (1965) is accepted, then the receptor potential is the large 'cochlear

15 Vertical section through tactile hair of caterpillar (*Aglais urticae*) showing the receptor cell which will be stretched when the hair moves in its socket. (Modified after Hsü 1938.)

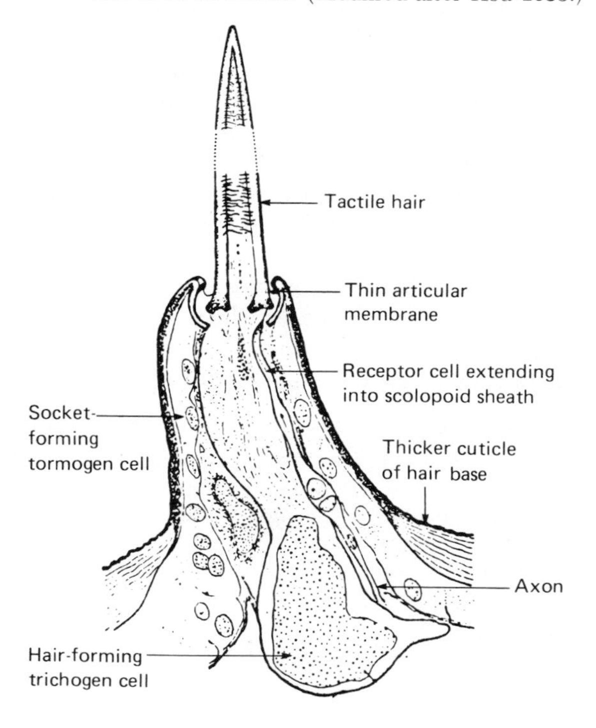

microphonic potential' which can be recorded from neurons in the inner ear and which is modulated when hair cells respond to particular sounds.

Some other receptors detect non-mechanical physical changes by means of accessory structures which convert those changes to mechanical membrane deformations. The pit organ of the rattlesnake and other members of the family Crotalidae consists of a chamber filled with gas which expands and contracts as the temperature changes. Infra-red radiant energy is absorbed by the gas molecules in the pit which convert it to heat. The heated gas expands and moves in a channel which includes mechanoreceptors. A small increase in temperature thus causes distortion of the hair cells. Acceleration due to gravity is detected in many animals by statocysts in which a high-density particle presses on hair cells when its acceleration is different from that of its surroundings. Liquid inertia effects are similarly utilised in the semicircular canals of vertebrates.

The mechanisms of transduction in chemoreceptors are still largely unknown. Whether the sense be called taste or smell, we do not know how a receptor potential is produced by the presence of some chemical substance at the surface of a receptor cell or inside it. It seems likely that specific binding sites exist on cell surfaces and that when the appropriate chemical attaches to them the structure of the membrane is altered so that ion flow and hence depolarisation can occur. Substances which smell or taste alike to us can be chemically diverse but may include in their molecular structure precise configurations of chemically active sites which act as the 'key' which fits the membrane receptor's 'lock'. The importance of the structural configuration is emphasised by the work of Russell and Hills (1971) and Friedman and Miller (1971) who showed that the two enantiomeric (mirror image) forms of carvone were interconvertible but changed in smell from spearmint (R-carvone) to caraway (S-carvone). These two very similar substances must have different effects on the receptor cells, perhaps fitting different 'locks'. It is possible that strongly ionised substances such as acids or salts may act on receptor cell membranes in a manner which is different from that of many organic molecules.

The discovery by Wald that light acts on the pigment rhodopsin in the vertebrate eye and breaks it down to a carotenoid, retinal, and a protein, opsin, showed that chemical transduction is utilised in photoreceptors (Wald 1959, 1968). The opsin produced during the photochemical breakdown has some effect on the receptor cell membrane. The part of the opsin molecule which has the effect is probably the area exposed when the retinal breaks off. A receptor potential can be produced when a very few photons hit a receptor cell (Table 2). In arthropods, the pigment molecule is rearranged rather than broken down when light acts on it but the sensitivity is still extremely high.

The two principal transduction mechanisms in sensory receptors involve either membrane distortion or the action of chemicals on the membrane. It is likely that some or all electrical receptors utilise a third transduction mechanism. Electric receptors are probably

Table 2. *Maximum sensitivity recorded for various sensory modalities*

Sensory modality	Minimum detectable change	References	Subject
Mechano/hearing	10^{-11} cm displacement	Tonndorf and Khanna 1966	Cat ear
Vision	a few photons	Pirenne 1956 Lillywhite 1977	Human eye Insect eye
Temperature	3×10^{-3} °C	Bullock and Diecke 1956	Rattlesnake pit organ
Electric	$0.01\ \mu V\ cm^{-1}$	Dijkgraaf and Kalmijn 1966	Skate
Olfactory	1 mol/10s	Schneider 1957, 1969	Silkmoth

common in aquatic animals for every muscle contraction causes an electrical disturbance and most aquatic animals would benefit from being able to detect the presence of another active animal near them. Most studies of sensitivity to electrical fields have been carried out on fish. The precise ways in which receptor potentials are produced are again not clear but, whilst very small electrical differences might act on membranes via chemical mediators, large currents must surely have a direct effect on receptor cells.

Sensitivity and specificity of receptors

The efficiency of the mechanisms for converting physical or chemical energy to electrical energy in receptors is most impressive. It is sometimes important for animals to detect very small changes and Table 2 emphasises that examples of extreme sensitivity can be found for various sensory modalities. Receptors which respond only to gross changes are also useful to animals. The specialisations of receptors for particular types of physical change are remarkable and can give the impression that receptors in complex animals do not respond to other types of change. Many studies show that receptors do respond to various sorts of physical change but have a lower threshold for one modality than for the others. The rate of firing in the sensory neuron connected to the mammalian muscle spindle is altered by fluctuations in temperature as well as in degree of stretch whilst the cutaneous receptors in mammalian skin respond to touch, temperature and pain. Early work on the ampullae of Lorenzini, which are found under the skin of the head in rays and dogfish, suggested that they are temperature receptors (Sand 1938) but later studies show them to be sensitive electro-receptors (Murray 1962). Receptors in the lateral line of the skate (*Raja*) also respond to changes in temperature and electric field. Even the receptor cells in the human eye will fire when subject to mechanical rather than to visual change, as can readily be confirmed when we 'see sparks' if banged in the eye.

The transmission of information from receptors

Once a receptor potential is produced, the normal sequence of events is the secretion of a transmitter substance and the production of an excitatory, post-synaptic generator potential in the sensory neuron. There are then various possible ways in which information about the quality and quantity of the external change can be transmitted to the brain (see Uttal 1973).

1. As has already been stressed, a simple way to ensure that the brain can identify the quality of the change which is occurring is to have *one channel for one type of information*. The information may be general, for example, input from a particular channel may indicate that a visual rather than a tactile change is occurring. It may be very specific, however, for example a variety of similar odours may each initiate the production of action potentials on a different channel.
2. The intensity or magnitude of a change could be assessed by the *number of activated receptor cells*. These receptors might all fire when the same threshold level is reached but, since some would be further from the source of physical disturbance, more would reach their threshold and initiate the production of action potentials if the change were greater. Alternatively, each receptor cell, or set of receptor cells, could have a different threshold. A simple example of this is the ear of noctuid moths (Roeder and Treat 1961) which has two receptor cells with thresholds about 20dB apart.
3. The amplitude of action potential produced in the sensory neuron could, theoretically, be related to the degree of change outside the receptor cell but, in fact, it fluctuates very little. Any fluctuations which have been described depend on metabolic factors and are not used in information coding. The *amplitude of the generator potential*, however, may be im-

portant in coding if interactions between nearby neurons occur. Such interactions are described later in this chapter.

4. In most sensory neurons the *temporal pattern of the production of action potentials* is related to the external change. The aspect of the temporal pattern which is most important in information coding is the rate of firing, but temporally localised frequency fluctuations and occasional missed pulses can also provide information. Patterns of input on different channels can be compared and this allows the detection of variations amongst sensory neurons in the delay before the first action potential is produced. The temporal pattern in the generator potential, such as abrupt or gradual change, linear or exponential change, much or little adaptation, can be important in interactions between sensory neurons.

A *tonic response*, in which there is a linear relationship between the rate of production of action potentials and the magnitude of external change was noticed first by Adrian and Zotterman (1926). A linear response change is simple to produce and simple to analyse but many physical changes in the surroundings of an animal are too large for such a system. The minimum firing rate in a nerve cell is about one action potential per second, for the noise level in the system masks any slower rate, and the maximum firing rate is about one thousand action potentials per second. Thus an animal can perceive a thousandfold change by means of a tonic response. Much greater intensity ranges are detectable, however. In man, for example, the magnitude of the detectable range of light intensity is 10^{10} whilst the sound range of the human ear for a 1 Khz tone from 2×10^{-4} dyne cm^{-2} is 140 dB, i.e., 10^{14} (Uttal 1973). Sense organs could deal with such intensity ranges. Firstly a greater increase in external change could be necessary, for a given increase in firing rate, in a receptor at high firing rates than at low firing rates. The sensory neuron in the blowfly salt receptor (Figure 16) fires at a rate which is directly proportional to the log of the molarity of the salt solution in contact with the receptor cell. Alternatively, a sense organ could include several receptors with different response ranges for the same sensory modality, as suggested as point 2

16 Diagram of blowfly sensory hair; the recording method and relationship between salt concentration and firing rate in axon from salt receptor are also shown After Gillary 1966.

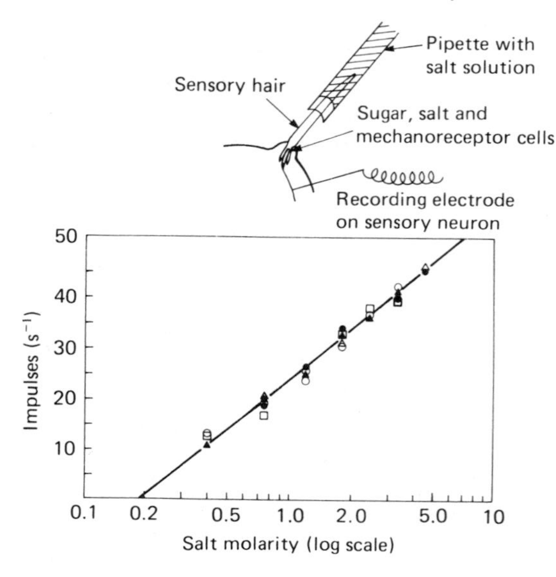

17 Rate of firing in electroreceptor of gymnotid fish when exposed to a voltage gradient. There is a resting discharge of 105 impulses s^{-1} and the response is largely tonic. Modified after Bennett 1968.

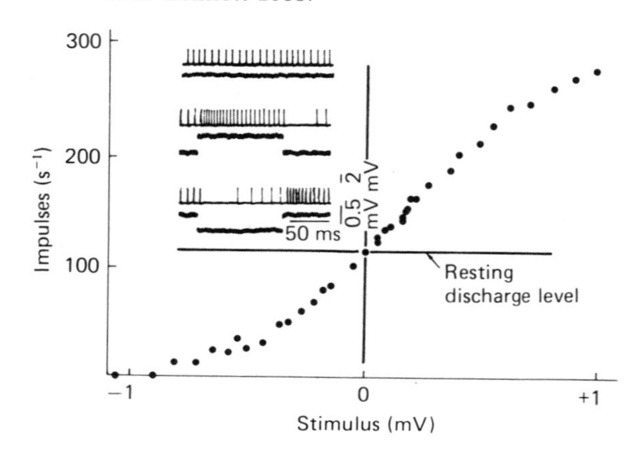

(p. 29). These receptors might have larger response ranges for the higher levels of external change. Since information about the magnitude of change could be appreciated merely by determining the channels on which firing occurred, rapid adaptation in the sensory neurons would be possible. In the extreme form of such a *phasic receptor* the sensory neurons would adapt after firing once.

Records from two electric receptors are illustrated in Figure 17, a tonic response from a gymnotid fish, and in Figure 18, a largely phasic response from a marine elasmobranch fish. These examples also emphasise the existence of a *resting discharge* which is widespread in sensory neurons. The advantages of such a resting discharge are that less energy is needed to increase the rate of firing in a neuron than to initiate firing, so the occurrence of smaller external changes can be transmitted from the receptor; and that both increases and decreases in firing rate are possible if there is a resting discharge. Figures 17 and 18 show that the electrical field near the receptor could become more positive or more negative. Both changes would be meaningful to the animal and both would be detected by the receptors.

Tonic and phasic responses are advantageous in different situations. A battery of receptors which show a phasic response is needed to provide as much information as a single receptor showing a tonic response, so, initially, a phasic response system is energetically more costly. Tonic responses can continue all the time and are therefore especially suitable for situations where continuing information is required, e.g. from a joint receptor which signals the position of a limb to the brain. Phasic receptors, however, are out of action for a few milliseconds after firing and do not provide continuous information. The disadvantages of tonic responses are as follows. (1). They must be averaged over a period in order to provide information about the magnitude of the external change, whereas a single action potential from the sensory neuron of a receptor with a threshold and a phasic response can be detected much faster. (2). Tonic receptors need more storage facility whilst simultaneous inputs from a battery of cells with phasic responses can be analysed without storage being needed. (3). A battery of cells with phasic responses can cover a wider range of variation in input than one receptor with a tonic response. As a consequence, phasic responses are commoner in situations where rapid analysis is advantageous and where widely varying external changes must be distinguished. It is seldom correct to refer to a receptor as tonic or phasic, for almost all 'tonic receptors' do adapt their rates of firing and very few 'phasic receptors' produce only one action potential in their sensory nerve whatever the input. Receptors monitoring the position of the body, or other input to which the animal must often refer in order to control its behaviour, are more tonic in their response. Most other receptors adapt much more rapidly.

18 Rate of firing in an electroreceptor (ampullae of Lorenzini) of a ray when exposed to a small voltage gradient (64 μV cm). There is a resting discharge of 25 impulses s^{-1} and the response is largely phasic. After Murray 1962.

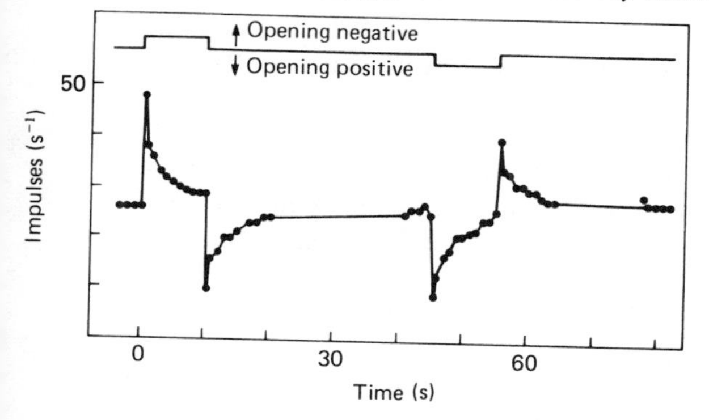

Behavioural studies and sensory function

Throughout the first part of this chapter, the fact that animals use their receptors to detect *changes* in their surroundings has been emphasised. Changes in light or sound level, odours, pressures, or body position are more important to animals than are external conditions which are not changing. The term 'stimulus' is often used to refer to a detectable characteristic of an animal's surroundings. It can be misleading, however, because it is often taken to mean a characteristic which is constant rather than changing and because it is also used to refer to an action potential or to a group of action potentials in a neuron or in a group of neurons. Behavioural evidence for the existence of mechanisms for detecting more or less complex patterns in the environment will now be discussed.

How do we know what environmental characteristics can be detected? In the first place we know from consideration of our own abilities that we can recognise certain features of our surroundings. This is one of the situations where William James' introspective psychology has provided useful information about the mechanisms which control behaviour. When subjective impressions are considered together with observations of the behaviour of individuals of our own and other species, we can theorise on the type of mechanisms which must exist in the sensory part of the brain. In the 1930s Köhler and others wrote about their ideas of the recognition of complex patterns as a whole, i.e. a *gestalt*, rather than as the sum of a large number of parts. Some aspects of their ideas persisted in the writings of Lorenz and Tinbergen during the following 25 years. Lorenz (1935) observed that there were occasions during an animal's life when exposure to a specific environmental feature was followed by a specific and often elaborate response. He used the word Auslöser, usually translated as *releaser*, for characteristics of one animal which elicited a social response in another member of its species. For example, nestling birds often have brightly coloured spots in their mouths which are exposed when they gape. Such spots were called releasers of feeding by parents. The parent carrying food will place food in the gaping mouths of their young. Other examples are apparent during courtship displays when many birds and fish respond to obvious visual characteristics of their partners. Tinbergen referred to a specific environmental feature which elicited a response from an animal as a *sign stimulus* (Tinbergen 1951). This term was not confined to social situations and is descriptive rather than interpretative. The term *releaser* implies more about the relationship between the presumed sensory input and the response. The use of both terms assumes, however, that mechanisms for detecting simple and complex sensory patterns exist in the animals.

For the purposes of this chapter, the behavioural observations which are most relevant are those which indicate that an animal has the ability to detect some environmental characteristic. In order to understand how a recognition system functions, however, it is desirable to consider the overall context of the behavioural or physiological observations. As Marler and Hamilton (1966) put it 'The essential quality of this sign stimulus or stimulus complex is that it is shared by all situations in which the response is appropriate'. The animal must recognise the sign stimulus in all situations where it is meaningful and must not respond to it in inappropriate situations. The question of when a response is, or is not, appropriate is easier to answer if the selection pressures which have acted during the evolution of the mechanism and the past experience and present state of the individual animal are considered. Members of a species which frequently feeds on a particular item would be more likely to possess a sensory mechanism for recognising that item than would members of other species. There would be strong selection for the spread in the population of gene-complexes whose effects facilitated the functioning of such a mechanism. Past experience over a long period may alter the effectiveness of a sign stimulus. An individual which had fed previously on only one food item might fail to respond to an item of food which most members of its species would recognise and eat (see Chapter 6). Immediate past experience may affect the likelihood of a response to a sign stimulus, perhaps by altering sensory functioning. For example, a food item may be ignored if danger from a predator has been detected or if the animal is satiated with food. Alternatively, receptor sensitivity may be increased in certain circumstances. Such moderations in sensitivity are discussed later in this chapter with reference to central effects on sensory systems by means of centrifugal nerve fibres.

All these factors make mechanisms of sensory detection more complex than Lorenz's original idea that sensory reception would inevitably release a specific response. It is very difficult to determine by behavioural observation what aspects of a potential sensory input an animal has detected. The distinction between releasing and motivating effects of sensory input, which Lorenz makes in his early writings and which persists in the writing of Eibl-Eibesfeldt (1967), does not help to explain behaviour control systems. As Hinde (1970) observes: 'Often, however, there is little evidence that releasing and motivating effects depend upon different types of mechanism, and the distinction seems to depend on the nature of the stimuli rather than the way they act: motivating stimuli are often continuously present, whereas releasing stimuli appear more suddenly'.

Behavioural evidence for sensory analysers

Tinbergen's work and related studies

The evidence from Tinbergen's many experiments in which sign stimuli were analysed demonstrates clearly that some specific responses are elicited by a very narrow range of complex sensory patterns whilst others are elicited by a wide range of different sensory patterns. Laboratory studies and subjective impressions had already indicated that this was so but corroboration by Tinbergen's elegant field experiments impressed the scientific world and therefore warrant description. The digger wasp (*Philanthus*) hunts for bees, which it paralyses with its sting, and uses as provisions for its larvae. Tinbergen (1935) found that a hunting *Philanthus* would approach any bee-sized object moving near heather (*Erica*) flowers. It would position itself 5–15 cm downwind of the object but would pounce only if a bee odour was present. The pounce was guided by visual cues and the action of stinging was initiated only if the appropriate smell was detected.

A study of the gaping response of young blackbirds (*Turdus merula*) and song thrushes (*Turdus ericetorum*) (Tinbergen and Kuenen 1939) showed that very young birds would gape when the nest rim was shaken. Gaping was sometimes elicited by strong air currents but never by increasing or decreasing light levels or temperature. Gaping was inhibited when the birds had just been fed and when the parent alarm call was heard. Responsiveness changed as the nestings developed. After the eyes opened at 9–10 days the appearance of certain objects near the nest elicited gaping whilst other objects did not (Figure 19). The fact that a dark circle with any discontinuity in it was sufficient for gaping to occur provides some indication of the sensory mechanism which must exist.

The male grayling butterfly (*Eumenis semele*) will approach and court females which fly near it. Tinbergen, Meeuse, Boerema and Varossieau (1942) demonstrated that it also approached other butterfly species, beetles, bumble bees, flies, dragonflies, grasshoppers, birds, falling leaves, thrown pine cones, and the shadows of any of these objects. Stationary objects were ignored and some moving objects were more likely to be approached than others. Studies with models suspended from fishing rods produced results like those shown in Figures 20 and 21. Approach was dependent upon size, colour and type of movement but not on shape. It seems likely that the disadvantage of wasting energy by approaching the wrong object is small enough to be counteracted by the advantage of not missing any individual which might possibly be a potential mate. The female grayling butterfly will rapidly accept most courting males whose scent scales (androconia) are exposed on their wings but not males whose androconia have been removed and the area of the wing covered with lacquer. These and other studies by Tinbergen demonstrated the feasibility of precise experimental field studies. They improved our understanding of the mechanisms which control behaviour, a contribution to knowledge which resulted in the award of the Nobel Prize for Medicine and Physiology to Tinbergen together with Lorenz and von Frisch in 1973. The work on butterflies has been followed up by a number of other research workers. For example, Hidaka and Yamashita (1976) found that male swallowtail butterflies (*Papilio xuthus*) showed most approach to objects coloured viridian-yellow. If the objects were striped viridian-yellow and black they would show contact behaviour like the close following that they show to females. In their famous study of the herring gull (*Larus argentatus*) Tinbergen and Perdeck (1950) recorded the pecking preferences of young chicks to approaching objects which resembled the head of their parent (Figure 22). A cardboard model painted like a gull's head was sufficient to elicit the response and would elicit some pecking if the bill was painted yellow with a black or white spot instead of the normal red one. A series of experiments with gull heads and bills painted in various ways established that the colour of the head had little effect on the response but that a bill with a contrasting spot was better than a bill with no spot or little contrast. The maximum response was achieved if the spot was red but Tinbergen (1951) emphasised that an object shaped like a gull's bill with contrast and some red coloration seemed to be most effective in eliciting the reaction. Later work showed that a stick which was longer and thinner than a gull's bill and which was coloured red and with white bands near its tip was even more effective than a model herring gull. Hailman (1967), working with laughing gulls (*Larus atricilla*), confirmed that the presence of a contrasting spot elicited pecking and showed that a dark spot on a light background was more effective than a light spot

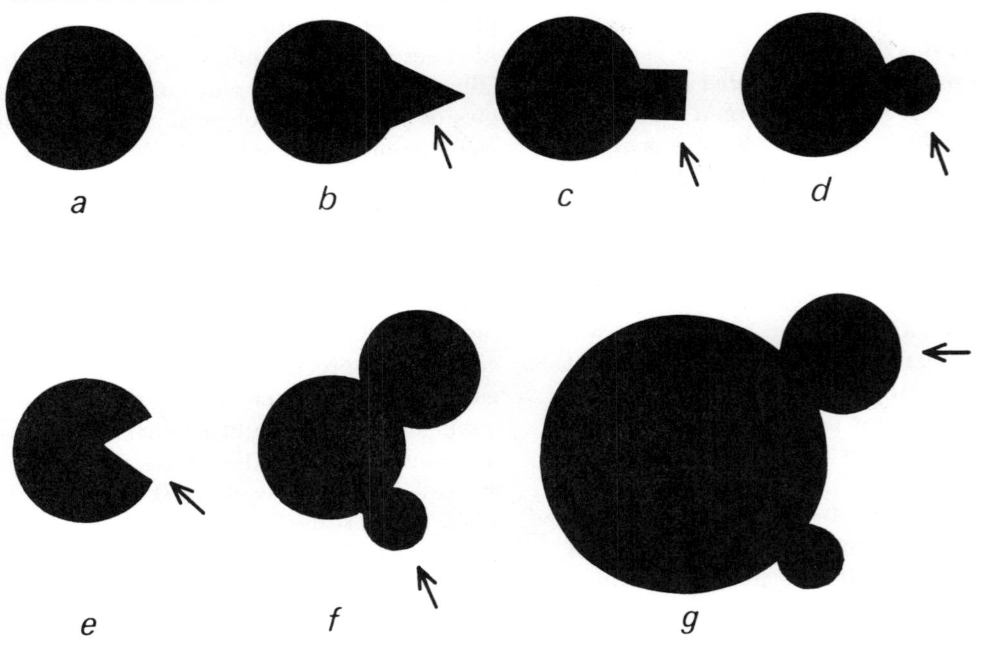

19 The gaping responses of 8–9-day-old nestling blackbirds are directed (arrow) towards black discs with a discontinuity (*b–e*) but not towards *a*. If there are two discontinuities, the size relationship between 'head' and 'body' is important in determining which elicits gaping. Modified after Tinbergen and Kuenen 1939.

on a dark background. He also found that herring gull chicks pecked at any model gull head in which the spot moved a large distance as the head was moved (Figure 23). The optimum head movement was a horizontal, side-to-side movement at a speed of 12 cm s^{-1}, almost the same speed as that normally used by parent gulls. These discoveries provide more precise evidence as to the sensory analysers which must exist in gull chicks. Hailman's evidence for changes in responsiveness with age, such that chicks with more experience of feeding by adult gulls are less likely to respond to crude models, may mean that the visual pathway analysers themselves change. It may mean, however, that older chicks use inputs from more analysers before the decision to peck is made.

20 Percentage of male grayling butterflies which approached four different brown-paper shapes, all 'dancing' in the same way on the end of a fishing-rod. Modified after Tinbergen, Meeuse, Boerema and Varossieau 1942.

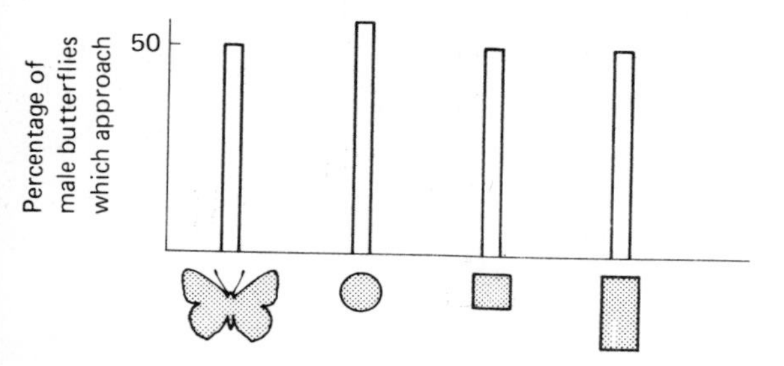

21 Percentage of male grayling butterflies which approach white or brown-paper model butterflies moved smoothly or made to 'dance' (double-headed arrows) at the end of a fishing-rod; experimental conditions slightly different from Figure 20. Modified after Tinbergen *et al.* 1942.

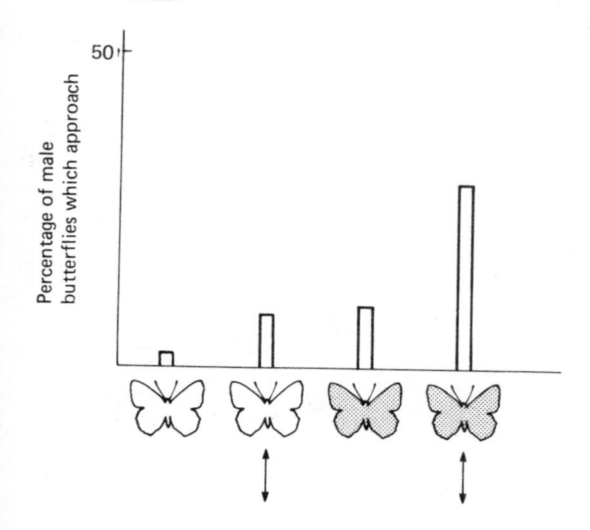

Detecting faces and eyes

Just as the studies of the pecking preferences of gull chicks tell us something about the sensory mechanisms they must possess, so studies of the gaze direction of human infants provide similar information for man. Fantz (1965) recorded the amount of visual fixation to face-shaped targets by human babies less than one week old. (Figure 24). A complex pattern with the configuration of a human face, or with the same components rearranged, elicited more fixation than simpler patterns. A face-shape marked with two eye-sized spots was fixated more than a plain white or grey face. The initial experiments can be explained in terms of relative visual complexity but human infants soon look most at a face with two eyes horizontally arranged and, later still, can distinguish amongst faces. Such experiments demonstrate the minimum visual ability which must exist at each age. If an individual demonstrates by its behaviour that discrimination is possible then it must possess the sensory ability to make the distinction. A lack of discrimination does not

22 The herring-gull chick aims its peck at the moving bill of the parent. It pecks close to the red spot which contrasts with the yellow bill.

23 The number of pecks directed at flat model herring-gull heads. If the red spot (circle) is distant from the pivot point (cross), so that it moves a lot when the head moves, the model elicits many pecks irrespective of the location of the spot relative to the bill. Modified after Hailman 1967 and Tinbergen 1951.

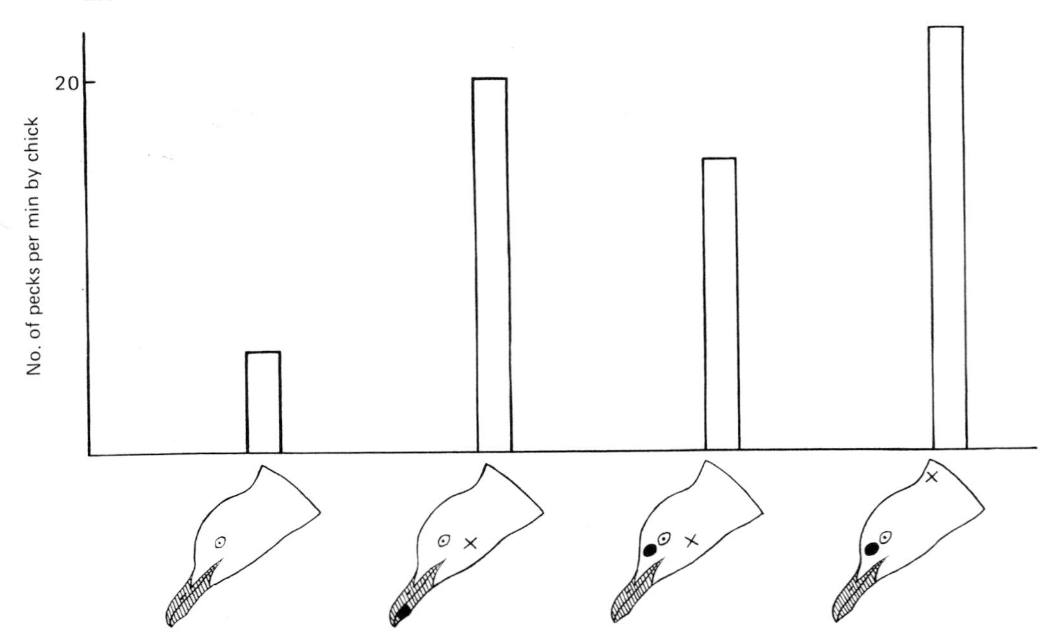

24 These six figures were exposed in front of 0–7-day-old human infants for 30 s each in repeated random sequence. The percentage of total fixation time and the number of infants looking at that target for the longest time is shown. Modified after Fantz 1965.

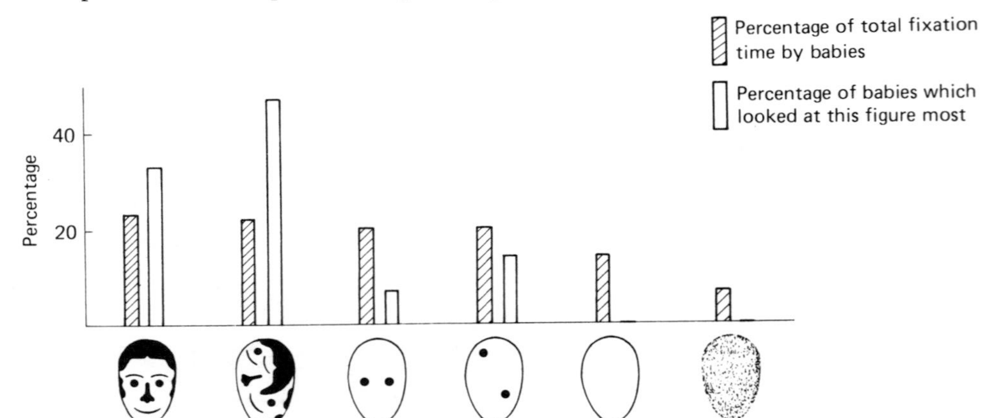

prove, however, that the sensory ability is not present. There is a large literature describing discrimination experiments as evidence for perceptual ability. Such work provides a high proportion of our knowledge about perceptual ability but only a small selection of studies can be mentioned in this chapter.

The features of the human face which are recognised most easily by a human infant are the eyes. Evidence for eye recognition mechanisms exists for a great variety of animal species. Predators often have forward-facing eyes so if an animal detects two eyes which are spaced and oriented as in a facing head, there is often a risk of attack by a predator. An attack might also follow if the eyes are those of a member of its own species. Alternatively some other interaction might follow. It is clearly advantageous to possess a sensory mechanism which facilitates the recognition of eyes. The characteristics of the front of a vertebrate predator's head are the presence of two discoid, horizontally positioned elements, each of which includes some concentrically arranged pattern. In some situations the eyes look bright because they reflect light and can be seen to be three-dimensional. Experiments by Coss (1972), 1978*a*, *b*) showed that mouse lemurs (*Microcebus murinus*) avoided eye-to-eye contact with conspecifics, i.e. with members of their own species. They were also more likely to give alarm barks if confronted with a human face in which both eyes were exposed (Figure 25). Similarly, marmosets (*Callithrix jacchus*) made shorter eye-to-eye contacts with human observers if both eyes were exposed than if one eye was covered and young jewel fish (*Hemichromis bimaculatus*) fled most readily from two-eyed models. Human children, however, looked at a strange adult more if both of the eyes were exposed. Coss also studied autistic children, who make few social contacts with other people and found that they looked at a face most if both of the eyes were covered. It is apparent that eyes are readily recognised, whether as sources of potential interest or potential danger.

The signalling function of eyes can be accentuated by the elaboration of their concentric pattern. The selective advantages of such accentuation have resulted in the evolution of eye exaggerators in many species including man (Figure 26). Many individuals of our species choose to elaborate the concentric pattern even further

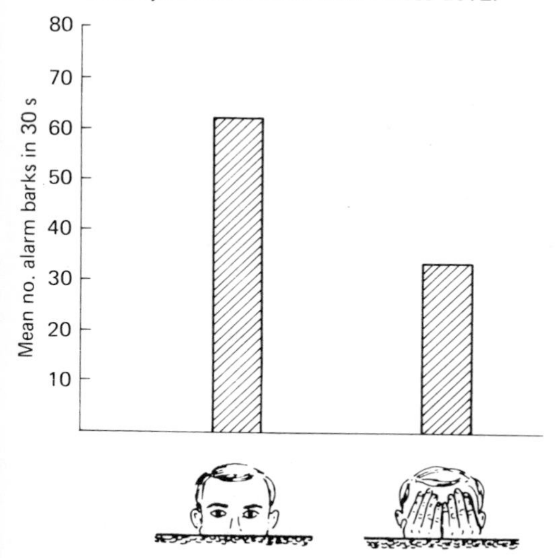

25 Mouse lemurs were more likely to give alarm barks if confronted with a human face with eyes uncovered. After Coss 1972.

26 Many species possess features which exaggerate the concentric nature of eyes. This suggests that there are visual analysers which respond more readily to such exaggerated patterns than to the eye alone. The bird is a grey-breasted white-eye *Zosterops lateralis* and the monkey is a dusky langur *Presbytis entellus*.

by artificial means. Small vertebrates often recognise and respond to the eyes of owls, other birds of prey and ground predators. The rapid avoidance response when such eyes are seen has resulted in the evolution of elaborate eye mimic deception mechanisms in harmless species (review by Wickler 1968). The sensory mechanism which detects the two large eyes of a predator will also respond to butterflies or moths which have eye spots on their wings (Cott 1940) or toads with eyespots on the back of the body (Edmunds 1974). Although small birds will attack small round spots, Blest (1957*a*, *b*) demonstrated that they retreat from butterflies such as *Nymphalis io*, or moths in the Saturnoidea and Sphingidae which have large eye spots. The eye spots in most of these animals are coloured so that they appear to have three-dimensional eyeballs as well as having concentric features. Most of the animals move their wings or bodies in a way which is consistent with the slow movements of potential predators. All these studies provide some insight into the sensory analysers which must exist in the visual system of the animals which are observing and responding to the eyes, or apparent eyes. However, many experiments demonstrate the interrelations of the visual cues with the context in which they occur, such as movement, visual surroundings, or auditory cues such as rustling noises or vocalisations. This emphasises the difficulties of determining the components of the sensory analysis system from purely behavioural evidence.

More complex visual analysers

In order to facilitate comparison of anatomical and physiological evidence with observed behaviour, the simpler rather than the most complex behavioural examples of sensory recognition systems have been mentioned. It is important, however, to emphasise the fact that many elaborate recognition systems exist and that their mechanisms may eventually be elucidated. Much of the best work on perceptual mechanisms has been carried out on human subjects. We are all aware of our ability to distinguish a human face in a complex mass of lines or in an elaborate natural background. We can recognise the faces of individual people in the presence of many other faces and we respond to the sound of our own name or a familiar voice when we are surrounded by many other potential sensory inputs. It is apparent that many other species rapidly distinguish amongst conspecifics by visual, auditory, or olfactory means. Pigeons have been trained to respond to photographs which include people and could do so as well as could the experimenters (Herrnstein and Loveland 1964). Rhesus monkeys can easily distinguish among photographs of other monkeys or, with practice, amongst individual pigs or dogs (Humphrey 1974). When we are reading, we rapidly identify the various letters. The number of symbols recognised is very much larger in China than it is in those countries where the Roman or Cyrillic alphabets are used. The letter B, for example, is distinguished in various orientations and with various minor modifications. A study by Morgan, Fitch, Holman and Lea (1976) demonstrated that free-living pigeons could be trained to approach and respond to certain letters and not to others, whichever of forty different typefaces was used.

When an animal is feeding on a single sort of food item or searching successively for other items which closely resemble one another, recognition is very fast. Complex patterns are recognised faster with practice, as is evident from our poor initial ability to distinguish between people of another race but our rapid improvement when frequently in the company of such people. The indications from many perceptual studies are that man, and presumably other species, identifies components of complex visual patterns such as straight lines, circles, corners, or three-dimensional characteristics, but that sets of such features making up all or part of a complex pattern, are also recognised in their entirety.

Unexpected sensory abilities

Observations of behaviour have sometimes forced the conclusion that the animal concerned has sensory abilities which man does not have or which are surprising because of their extreme specificity or sensitivity. von Frisch demonstrated that honeybees could determine the position of the sun when it was obscured by cloud provided that some clear sky was visible. This ability was explained when, in 1949, he confirmed that the honeybee could detect the plane of polarisation of light. Subsequent work has shown that the plane of polarised light can be detected by many other arthropods, which have compound eyes (Waterman 1961), by *Octopus* (Moody and Parriss 1961), by fish and by birds (Kreithen and Keeton 1973, Delius, Perchard and Emmerton 1976). The insect eye has long been known to be sensitive to a range of colours different from those detectable by man and Lubbock (1882) showed that ants were more likely to move their pupae out of normal sunlight than out of sunlight from which ultra-violet light had been filtered. Daumer (1958) photographed flowers by using an ultra-violet filter and demonstrated that many flowers which bees visit and which appear to us to be uniformly coloured have ultra-violet coloured centres and guidelines. This makes them more conspicuous to bees, as he confirmed in experiments with artificial flowers behind light filters. Bees approach the region which is ultra-violet in colour even if it is at the periphery of the flower. As a result of a series of such experiments, Daumer concluded that bees possess receptors which are maximally sensitive to ultra-violet 360 nm, or blue-violet 440 nm, or green-yellow 588 nm and that they could distinguish the following colour mixtures; bee violet (67% blue-violet + 33% ultra-violet), bee-purple (79% yellow-green + 21% ultra-violet), and blue green (yellow-green + blue violet). Goldsmith (1960) recorded from the bee retina and was able to confirm the existence of such receptors. Figure 27 shows that the wings of female cabbage-butterflies *Pieris rapae* reflect ultra-violet light whilst those of males do not (Obara 1970). This accounts for the ability of males to recognise females rapidly at a distance (Obara and Hidaka 1968): see also Silberglied and Taylor 1978).

The ability of flying bats, even if they are blind, to avoid obstacles has been known since the observations of Spallanzani in 1793 or 1794. Griffin and Galambos (1941) discovered that the bats make and hear ultra sounds of 25–100 KHz in order to navigate by echolocation. Some of the moths which bats hunt can also hear these sounds, by means of sound receptors which are very frequency-specific, and can take evasive action when they hear them (Roeder and Treat 1961). This response has been used to protect crops from the corn borer moth (*Ostrinia nubilalis*); they can be frightened away from crops in which they would otherwise lay eggs, by playing bat sounds in the field (Cherrett, Ford, Herbert and Probert 1971).

Economic exploitation and other deceptions of sensory analysers

In many situations where an individual needs to recognise a distant potential mate or food source, olfactory cues are used. When the female cabbage-root fly (*Delia brassicae*) is looking for a brassica plant beside which it can lay its eggs, it recognises the plant by the smell of mustard oil and flies upwind to the plant (Hawkes 1971). Clearly it has a receptor which detects mustard oils. This fact has been utilised in order to trap the flies in water-traps which contain allyl isothiocyanate, a mustard oil, in brassica fields. The number of flies caught, and consequent reduction in crop infestation by the larvae, is also dependent upon the colour of the trap and its height above the ground. Fluorescent yellow traps set at ground level were maximally efficient (Finch and Skinner 1974).

The members of one sex of a species can advertise their presence and receptivity by producing a volatile chemical which can be detected by the other sex. Male silk moths (*Bombyx mori*) will walk upwind when they detect hexa deca-(10,12)-diene-(1)ol which is produced by the female (Butenandt, Behmann, Stamm and Hecker 1959). Electrophysiological records from the antennal nerve of male silk moths reveal that electrical responses are shown to the natural and synthesised sex attractant. This receptor specificity is associated with remarkable

27 The wings of male and female cabbage butterflies *Pieris rapae* look the same to us in natural light, *a*, but are readily distinguished by the butterflies. When illuminated with ultra-violet light, *b*, the wings of the female are seen to reflect the light whilst those of the male do not. The background is a cabbage leaf. After Obara 1970.

a

Photographed in natural sunlight

b

Photographed through filter transparent to near ultra-violet region only

sensitivity (Table 2). Insects can be controlled using knowledge of such receptors to sex attractants, for example the cotton boll-weevil (*Anthonomus grandis*) is readily caught in traps baited with a chemical, extracted from the male weeviles, which is more attractive to both sexes than is the smell of cotton plants (Tumlinson *et al.* 1970).

Sexual behaviour in mammals also depends upon input from specific receptors. Pig herdsmen can determine whether a gilt or sow is in oestrus because it stands rigid when pressed on the back in the presence of a boar. The boar may be detected visually, by its rhythmic vocalisations, the 'chant de coeur', or by the odour of the secretions from the preputial pouch of the boar (Signoret and du Mesnil du Buisson 1961). The odoriferous substance, 5α-androst-16-ene-3one, has been isolated and synthesised (Patterson 1968). A tape recording of the boar's 'chant de coeur' or the androstenone sprayed by aerosol is certainly detected by gilts or sows in oestrus for they will show the rigidity response and thus allow the herdsman to determine the time of oestrus.

Much of our knowledge of sensory analysers has come from experiments in which sensory systems are deceived by models. As has been mentioned already, we can find out more about the sensory analysers by varying the characteristics of a model and recording which variations still elicit a response. As has already been mentioned when discussing responses to eyes, mimicry of signals to which other species will respond is widespread. Plants use animals to promote their chances of pollination and seed dispersal, sometimes by providing an obvious reward and sometimes by deception. *Arum maculatum* produces a foetid smell which attracts small flies and traps them by means of hairs which prevent their exit from a chamber. When the pollen has matured and the flies have it on their bodies, the hairs wither, the flies depart and are then attracted to other plants whose stigmas are receptive to pollen. The chemical attractant must mimic some food or mate attractant. A bee orchid (Figure 28) resembles a female solitary bee sufficiently closely to elicit mating behaviour from male solitary bees which encounter it. When the bees subsequently visit and attempt to mate with another bee orchid they cross-pollinate it.

The fly *Stomoxys* is able to delude driver ants so that they accept and carry its egg, which will hatch into a larva predatory on the ant

28 Male solitary bees are deceived by the visual characteristics of the bee orchid *Ophrys apifera*, which resembles a female bee, so they attempt to mate with the flower. Photograph: G. C. Bellamy.

brood (Oldroyd 1964). The many brood parasitic birds such as the European cuckoo (*Cuculus canorus*) or the North American cowbird (*Molothrus ater*) delude other species which act as foster parents for their young. Such mimics provide further information about sensory mechanisms and are discussed at length by Wickler (1968). Surprising accidental occurrences also serve as reminders of the fallibility of the recognition systems. A cardinal (*Cardinalis cardinalis*) which built its nest near a goldfish pond, stopped when returning with food for its young and, seeing goldfish opening their brightly coloured mouths near the surface, fed the goldfish instead of its own young.

Visual pattern recognition: the toad as an example

The interpretation of behavioural evidence for mechanisms of sensory pattern recognition in terms of known aspects of the anatomy and physiology of sensory pathways was scarcely possible until recently and is still difficult. In certain sense organs anatomical and biochemical characteristics could be related to function, for example certain visual pigments are bleached most by light of the same wavelength that elicits most behavioural response. The advent of precise methods for recording from cells in sense organs and in the brain has resulted in much improved understanding of complex pattern analysers. The best physiological evidence for the existence of sensory pattern detectors results from studies on vision. The visual system of the toad has been extensively investigated with close reference to behavioural abilities and will now be described in detail.

The toad's behaviour

The common toad (*Bufo bufo*) is 60–80 mm long: it feeds on worms, insect larvae, adult insects, wood-lice or other small animals, provided that they are moving, and is subject to predation by various large birds, mammals and snakes (Smith 1951). When a possible predator is detected the toad may adopt a defensive attitude with head lowered and body inflated, or it may merely crouch away from the enemy, or turn and run. Hinsche (1928) found that small objects elicited prey-catching movements but objects larger than 75 cm² did not. When Hinsche presented toads with the legless lizard *Anguis* less than 23 cm in length, they usually showed prey-catching behaviour. Encounters with larger *Anguis*, rabbits, or guinea-pigs resulted in the adoption of a defensive attitude. Prey-catching behaviour, when the prey is nearby, is readily recognised for it involves a sequence of motor patterns: (1) orienting by turn-

29 A toad confined in a glass cylindrical container shows orienting movements and other prey-catching activity when a worm-shaped object moves past it horizontally. No such movements are shown to the same object, still moving horizontally but in a vertical position. Modified after Ewert 1968.

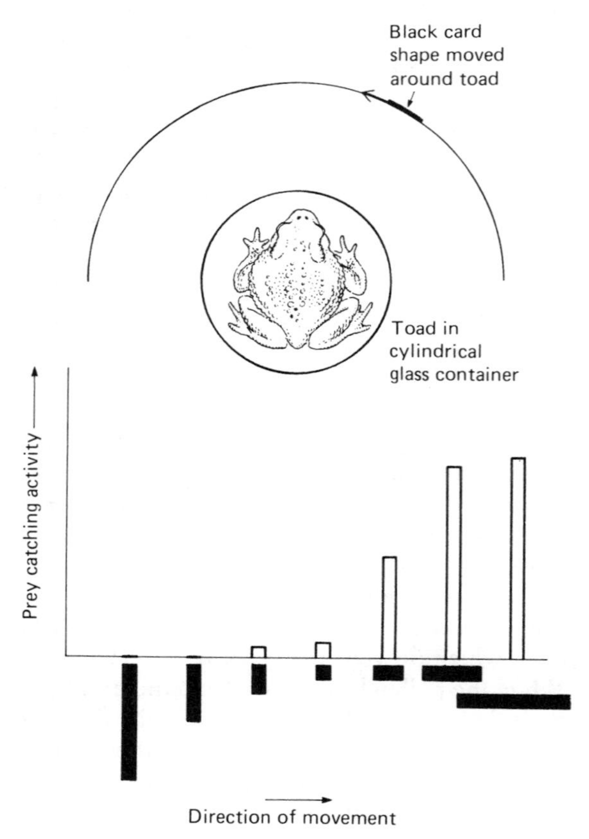

ing head and body; (2) binocular fixation; (3) snapping by extending neck and tongue; (4) gulping; (5) cleaning snout with fore feet (Schneider 1954). Eibl-Eibesfeldt (1951) and Schneider pointed out that objects were treated as prey or enemy by toads only if they were moving and contrasted with their surroundings. Whilst prey objects are usually small and near the ground, enemies are large and loom above the toad. Ewert (1968) showed that a horizontal bar moved horizontally around a toad elicited a prey-catching response (Figure 29). The greatest response occurred when the bar was at least eight times as long as it was wide. If the bar was vertical, but still moved horizontally, prey-catching responses were rare and avoidance behaviour occurred. The likelihood of avoidance depended upon the shape of the moving pattern (Figure 30) (Ewert and Rehn 1969). The toad is able to discriminate between two bars of the same size and shape. A horizontal bar is similar to a worm in orientation and direction of movement and elicits a feeding response. A vertical bar, on the other hand, which moves horizontally is not like a worm and elicits freezing behaviour. How does the toad distinguish between 'worm' and 'anti-worm'? Another observation requiring explanation is that toads presented with two meal-worm larvae (*Tenebrio*) hesitate to attack and the feeding response is inhibited by the presentation of two parallel horizontal moving bars or one bar with dots above one or both ends (Ewert, Speckhardt and Amelang 1970).

Retinal functioning

The anatomy of the eye of the frog was described in detail by Ramon y Cajal (1909) and the description confirmed by Lettvin, Maturana, McCulloch and Pitts (1959), and Maturana, Lettvin, McCulloch and Pitts (1960). The frog eye has two types of rod together with single and double cones. All of these are connected, via bipolar cells etc., to four principal types of ganglion cell (Figure 31). The axons from the ganglion cells make up the optic nerve. In a study which was a landmark in the development of our knowledge of sensory analysis, Lettvin and Maturana and their colleagues recorded from the optic nerve whilst objects were moved

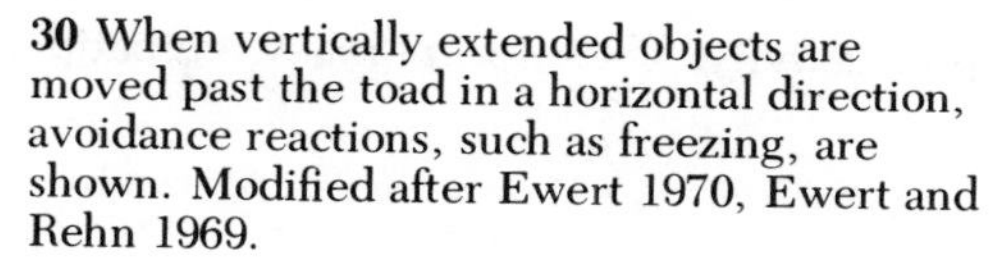

30 When vertically extended objects are moved past the toad in a horizontal direction, avoidance reactions, such as freezing, are shown. Modified after Ewert 1970, Ewert and Rehn 1969.

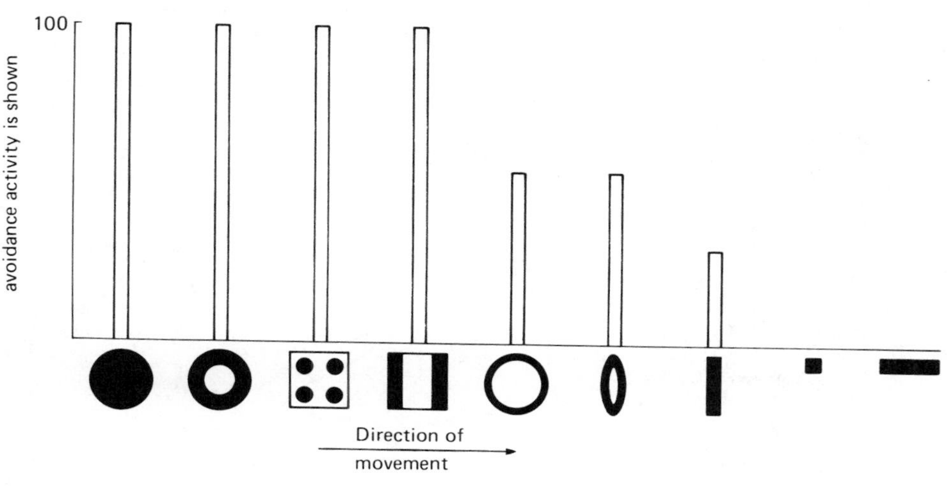

across the visual field of a frog by means of a magnet on the other side of a screen (Figure 32) and found four main types of responses (Table 3). The frequency of finding the four different units in the optic nerve corresponded to the frequency of occurrence of the four types of ganglion cells detected histologically. Table 3 also shows which types of ganglion cell were present in tadpoles (Pomeranz and Chung 1970). The class 2 ganglion cell responds to a dark, convex, moving boundary about 1° across in a field of 3–5° but does not respond when the frog is shown a colour slide of a frog's-eye view of the world around the edge of a pond (sedges, flowers, trees etc.). Such a ganglion cell certainly responds to a small insect like a fly, a major item in the diet of frogs but not in the diet of young tadpoles.

31 Diagram of section of frog retina showing shape and positions of cones and two sorts of rod, bipolar cells, and ganglion cells. The four principal types of ganglion cell are: Class (1) one level constricted field; Class (2) many level E-shaped; Class (3) many level A-shaped, and Class (4) one level extended field. Modified after Ramon y Cajal 1909.

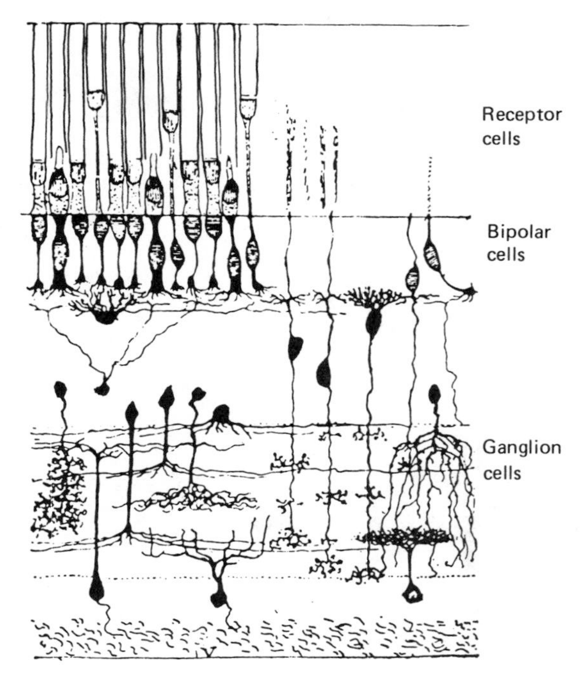

How could these ganglion cells analyse visual input as much as they do? One principle of receptor function which must be involved here is lateral inhibition. This important phenomenon was first described in a study of the eye of the horseshoe crab (*Limulus*) by Hartline, Wagner and Ratliff (1956). The compound eye of *Limulus* is made up of ommatidia, adjacent ones having lateral connections between them. Small spots of light were flashed on individual ommatidia or groups of adjacent ommatidia whilst recordings from the eccentric cell axons were made. Longer bursts of firing were initiated by brighter flashes of light but the number of impulses produced was smaller if adjacent ommatidia were also illuminated. Each ommatidium which fired had an inhibitory effect on firing in adjacent ommatidia, the inhibition being greater for close than for distant neighbours and greater if it was firing a lot than if it was firing a little (Figure 33) (Hartline and Ratliff 1957). When Hartline and co-workers gradually dissected away the interconnecting plexus between responding ommatidia, the extent of inhibition decreased in proportion to the amount removed. This suggests that inhibition is mediated by graded potentials rather than by spike activity. The effect of many ommatidia act-

32 The frog is shown objects, moved by a magnet which it cannot see, whilst the output from the ganglion cells is recorded. Modified after Lettvin *et al.* 1959.

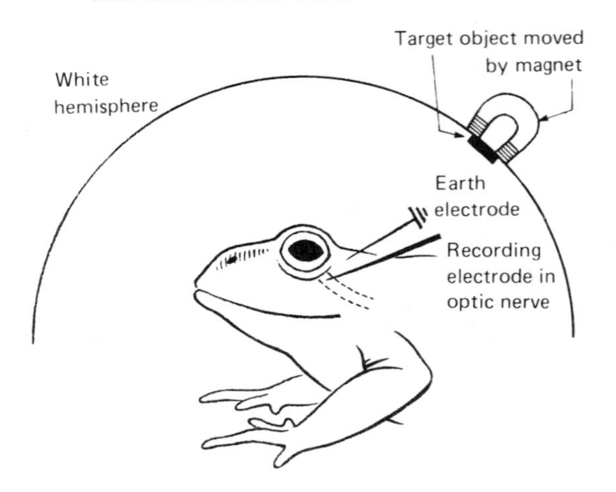

ing together is a cumulative inhibition (Hartline and Ratliff 1958) and has the effect of emphasising any light–dark boundary seen by the eye (Ratliff and Hartline 1959). Figure 34 shows that because of differential inhibition of ommatidia nearer to and further from a light–dark edge, lateral inhibition results in the accentuation of the edge. For such work, Hartline received the Nobel Prize in 1967. The function of class 1 ganglion cells in the frog eye (Table 3), which respond to stationary edges, could be explained by such a mechanism. As a result of work on cats (Kuffler 1953; Rodieck and Stone 1965) it is apparent that lateral interactions in the vertebrate retina can include excitation as well as inhibition. A variety of studies indicate the possibility that neural delays and persistence of response may be an important aspect of lateral interactions in sensory systems.

The toad (*Bufo bufo*) had three types of

Table 3. *Physiology and anatomy of the eye of adult frogs and their tadpoles (data from Lettvin, Maturana, Pitts and McCulloch 1961; Pomeranz and Chung 1970)*

	Adult frog			Tadpole	
	Type of detector	Field size	Ganglion cell shape	Detector	Ganglion cell
Class 1	stationary boundary	2–4°	1-level constricted	absent	absent
Class 2	dark, convex, moving boundary	3–5°	many-level E-shaped	present in centre of field only	
Class 3	moving or changing contrast	7–11°	many-level H-shaped	present	present
Class 4	dimming	12–16°	1-level extended	present	present

33 The rate of firing in ommatidium *A*, in the *Limulus* eye, is affected by the level of illumination and by inhibition from adjacent ommatidia. Inhibition is greater when the inhibiting ommatidium is close and when it is itself firing fast. Data from Hartline and Ratliff 1957.

ganglion cells which project along the brain pathway to the optic tectum. These correspond in function with class 2, class 3 and class 4 in the frog (Ewert and Hock 1972, Table 3). These ganglion cells can transmit information along the visual pathway about the position of an object in the visual field, the angular size of the object, its angular velocity, the degree of contrast between the object and its background, and the overall illumination level. They will not distinguish prey from enemy; a 'worm' is not distinguished from an 'anti-worm'.

From retina to brain in the toad

Ewert and his co-workers have investigated the system for analysis of visual input by recording from various points in the visual pathway in the brain by stimulating electrically in these areas and by cutting the pathway at various points and investigating which aspects of visual discrimination were still possible.

The anatomy of the visual pathway in Amphibia has long been known (see Ariëns Kappers *et al.* 1936). When Ewert and von Wietersheim (1974*a*) removed an eye from a toad, three layers of degeneration, corresponding with the three types of ganglion cells, were found in the contralateral optic tectum, just as Keating and Gaze (1970) had found four layers in the frog tectum. The posterior thalamic pre-tectal region intervenes between the retina and the tectum. Electrodes placed in this region by Ewert (1971) picked up a variety of evoked electrical responses in freely moving or awake but immobilised toads which were shown black targets on a white screen (Figure 35). The electrodes whose tip was 5 μm, or as small as 1–2 μm in

34 A light–dark boundary on the eye is emphasised due to lateral inhibition. Ommatidium 1 is brightly illuminated but is inhibited by other brightly illuminated ommatidia around it. Ommatidium 2 is illuminated as much as 1 but is inhibited less because some of the adjacent ommatidia are in shadow. Ommatidium 4 is dimly illuminated and little inhibited but ommatidium 3 is dimly illuminated and inhibited more because some of the adjacent cells are in light and firing fast. Modified after Ratliff and Hartline 1959.

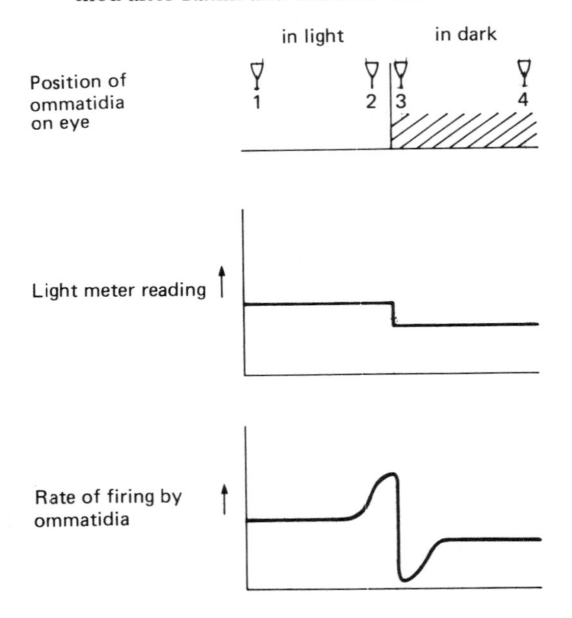

35 Units were found in the posterior thalamus of the toad which fired when a vertically elongated object was moved horizontally through the receptive field. From Ewert 1971.

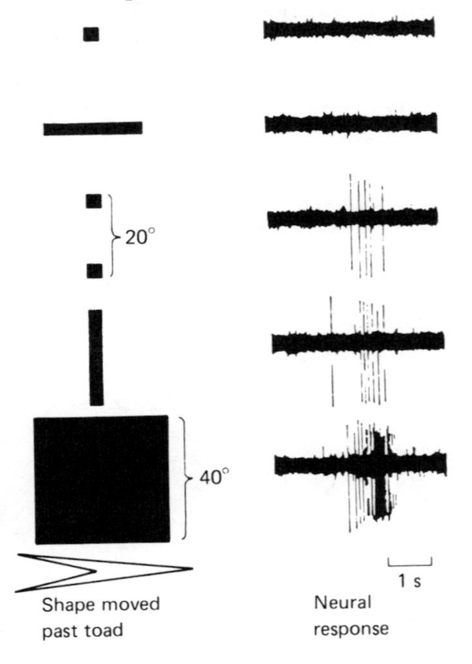

diameter, were recording the activity of single neurons. Some of these units responded to tactile input or body tilting, but most responded to visual input such as dimming, stationary objects, movement of a large object across a small part of the visual field, changing movement across a large part of the visual field, movement towards the animal's eye, or movement in an area which changed in size in different parts of the visual field. Two such units are of particular interest since they would respond to a vertical bar or square moving horizontally (Figure 35) and to an enemy looming towards the toad (Figure 36). The first of these would allow the animal to detect an 'anti-worm'. All thalamic pretectal units were activated better by threatening objects than by prey objects. Electrical stimulation in the region often elicited escape behaviour but not prey-catching (Ewert 1968) but stimulation in the optic tectum did elicit prey-catching motor patterns (Ewert 1967).

Unit-recording in the optic tectum of the toad revealed movement-sensitive neurons with a large field, which would enable the toad to localise large, moving objects. There were also other movement-specific units with a small field (Ewert and Borchers 1971). Detailed study of these small-field units showed that some (tectal type 1 neurons) responded to objects which were moving horizontally and elongated in the direction of movement irrespective of their height. Others (tectal type 2 neurons) responded in the same way except that their response was inhibited in proportion to the height of the object (Ewert and von Wietersheim 1974*b*). These tectal type 2 neurons would respond to an object which looked and moved like a worm but not to a vertically oriented worm or a more obvious enemy. Ewert suggests that the tectal 2 neurons are excited by the tectal 1 neurons and inhibited by the 'enemy detectors' in the thalamic pre-tectal region. This idea is supported by the observation that toads with lesions on one side of their brain in the thalamic pre-tectal region will show prey-catching responses to everything, including vertical bars moving horizontally or even their own feet and stationary black squares which they encounter as they walk (Figure 37) (Ewert 1968, 1969). If the responses of tectal 2 neurons are recorded after lesions have been made in the thalamic pre-tectal region they are found to respond to objects which are vertically and horizontally elongated (Ewert and von Wietersheim 1974*c*); a response which the thalamic pre-tectal units would normally inhibit. The behavioural avoidance of two parallel worm-like objects or of the horizontal bar with a dot over one end, which looks somewhat snake-like, is also abolished by thalamic pre-tectal lesions (Ewert *et al.* 1970).

In summary, the 'worm' and 'anti-worm' or other 'enemy' detectors have been found. The overall recognition mechanisms still require much study, however, for there are subtle differences in the behavioural responses of toads. For example, those to black objects against a white background, to white objects against a black background, to the same objects in different positions in the visual field and to objects at different times of day or year. Further investigation is revealing differences in ana-

36 This unit in the posterior thalamus of the toad fires when a large dark object moves towards the toad. From Ewert 1971.

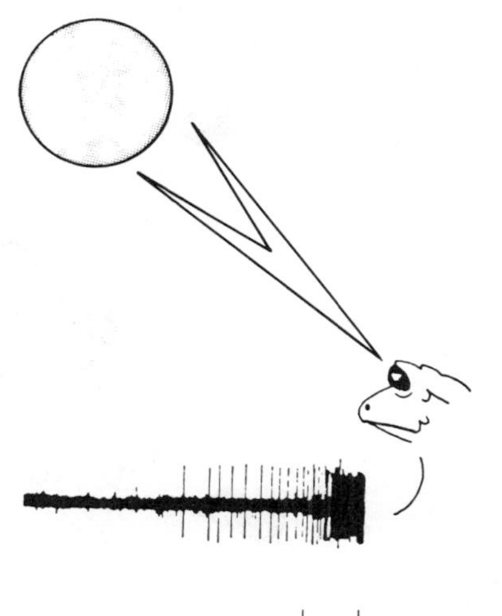

lysers which receive inputs from different parts of the visual field and there must also be centrifugal modulation of the sensory analysis by means of efferent axons: thus other activities in the brain can affect the sensory analysis system. The behavioural responses are subject to habituation which is readily reversed. When toads were presented with successive inaccessible worm-like models, visible to the right eye, their prey-catching responses eventually disappeared. After a brief presentation to the other eye or a touch on the body, however, the response to objects shown to the right eye returned (Ewert and Ingle 1971). Tectal recordings from animals which are satiated with food or which do not readily feed in the laboratory are much less clear when a worm model is presented than are recordings from hungry animals which do feed in the laboratory. Preliminary developmental studies indicate that the behavioural and physiological distinctions between 'worm' and 'anti-worm' are less pronounced in toads which have been fed by hand. They do respond, however, to the experimenter's hand

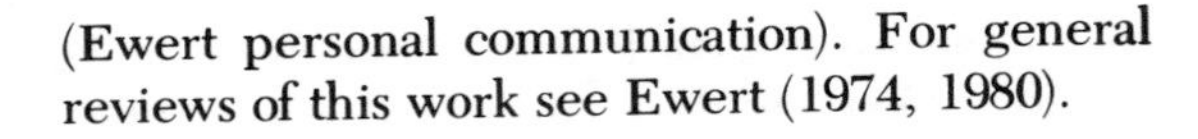

(Ewert personal communication). For general reviews of this work see Ewert (1974, 1980).

Visual pattern recognition mechanisms in mammals

The mammalian visual system has also been extensively investigated but whilst physiological evidence for many pattern detectors has been found, it has proved more difficult than in the toad to relate the variety of behavioural responses to distinguishable analysers in the brain. The visual pathway in mammals (Figure 38) is different from that of amphibia. Most axons from the retinal ganglion cells go to the lateral geniculate nuclei of the thalamus, whence nerves go on to the visual cortex in the telencephalon (see Figure 5). A much smaller proportion of ganglion cell axons goes to the pre-tectal nuclei or on to the optic tectum (here called superior colliculi). The principal sites for recording have thus been the optic nerve, the lateral geniculate body and the visual cortex. The work on frogs and toads shows that complex analysis takes place in their retinal ganglion cells

37 If toads are lesioned on one side of the brain in the thalamic or pre-tectal region, their prey-catching behaviour is abnormal, being directed towards their own feet (*a*) or towards stationary black squares which are encountered while walking (*b–f*). After Ewert 1969.

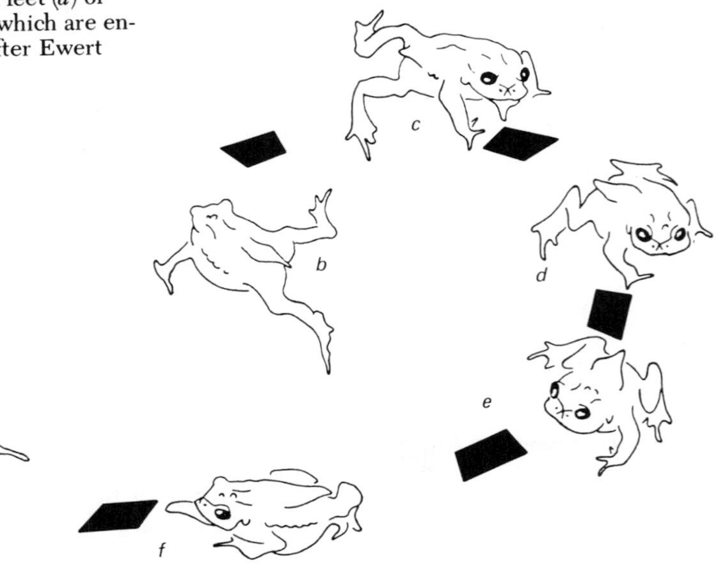

and more analysis occurs at later stages in the pathway. Work with cats and monkeys indicates that the majority of such analysis occurs in their visual cortex (Hubel and Wiesel 1959, 1962, 1963*a*, 1965, 1968). Studies on rabbits (Barlow and Hill 1963) and ground squirrels (Michael 1966*a, b*) show that very much more analysis of visual patterns takes place in their retinal ganglion cells. Work on birds indicates that pigeons, at least, are more like rabbits in this respect (Maturana and Frenk 1963, Maturana 1964). With this rather small sample of species it appears that those which use binocular vision extensively, carry out less analysis in the retina (Hinde 1970) and it may be more efficient for them to analyse at a point where inputs from both eyes are available. These differences emphasise the importance of making clear reference to the species studied when discussing mechanisms of visual analysis.

The development of a tungsten electrode which made possible long periods of recording from single cells in the brain (Hubel 1957) allowed Hubel and Wiesel to initiate an important new field of research into sensory processing in vertebrates. Following up Kuffler's work, in which he shone light spots on to the retina of the cat, they were able to describe the analysers which are present in the lateral geniculate nuclei and in the visual cortex. Most of the ganglion cells of the cat have circular receptive fields which are excited by light shone on to the centre of the field and inhibited by light on the periphery, or vice versa (Kuffler 1953, Rodieck and Stone 1965). Such units could operate by means of a simple combination of excitation and inhibition from the receptors (Figure 39) and a set of on-centre–off-periphery cells would act so as to emphasise light–dark boundaries (Horn 1962) (Figure 40). The cells in the lateral genic-

38 The visual pathway of a mammal showing that the major tracts go to the lateral geniculate nucleus and visual cortex. Modified after Milner 1970.

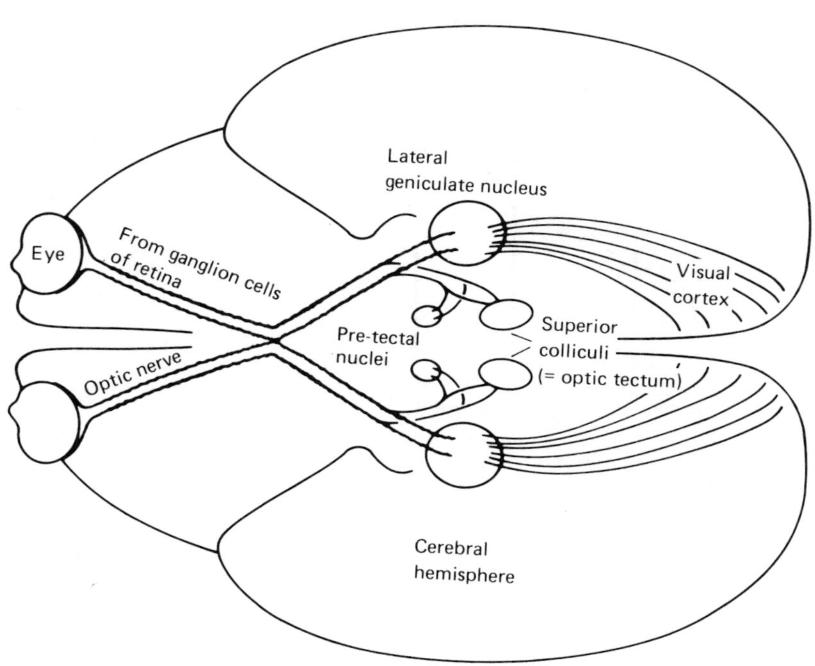

ulate nuclei are more responsive to spatial differences in illumination because of the greater inhibitory effects of the periphery of the field (Hubel and Wiesel 1961). Some cells are present which will respond to black or white objects entering or leaving their receptive fields so they must receive inputs from on-centre and off-centre retinal ganglion cells (Kozak, Rodieck and Bishop 1965).

When a recording electrode is inserted into the brain, not much time is available in which to find the visual patterns which evoke a response from a cell. As a consequence the positive responses discovered must always be regarded as delimiting a minimum range of responsiveness. There is still debate about the range of responsiveness of units in the cat's cortex but it is apparent that there are (1) circular field cells which respond like retinal ganglion cells; (2) simple cells (Hubel and Wiesel's terminology) which respond to edges, bars or slits with a particular orientation and sometimes with an optimum direction of movement in a particular part of the visual field (Figure 41); (3) complex cells which are like (2) but which respond irrespective of location in the visual field;

39 Receptor cells in the eye may be connected via bipolar cells to amacrine cells which produce an inhibitory input to the ganglion cells. Inputs from the centre of the field produce an excitatory ganglion cell input. If all the field is illuminated the two effects cancel out but light on the centre excites and light on the periphery inhibits. Modified after Dowling and Boycott 1966.

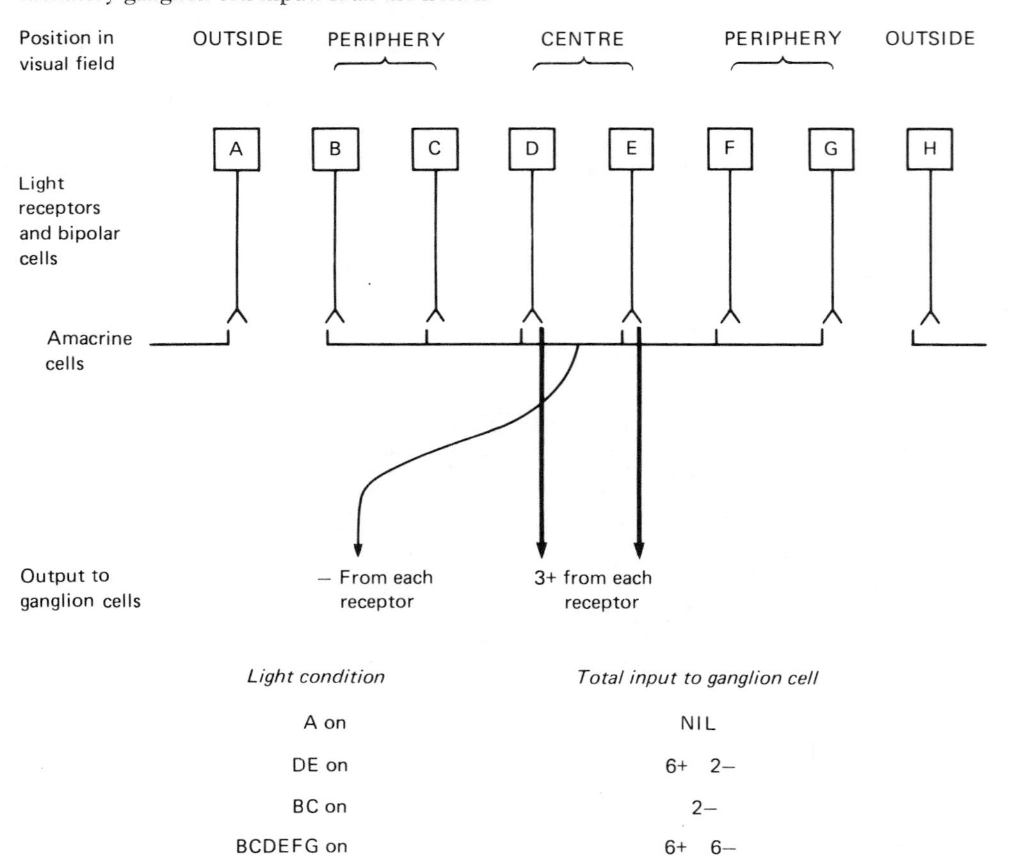

Light condition	*Total input to ganglion cell*
A on	NIL
DE on	6+ 2–
BC on	2–
BCDEFG on	6+ 6–

40 A battery of ganglion cells in the vertebrate eye which fire faster when the centre of their receptive field is illuminated but are inhibited by illumination of the periphery of their receptive field will emphasise a light–dark boundary in much the same way as was described for *Limulus* eye in Figure 34. Ganglion cell *B* will fire at a slower rate than *A* because its periphery is partly illuminated whilst cell *C* will fire faster than cell *D* because it is inhibited less. After Horn 1962.

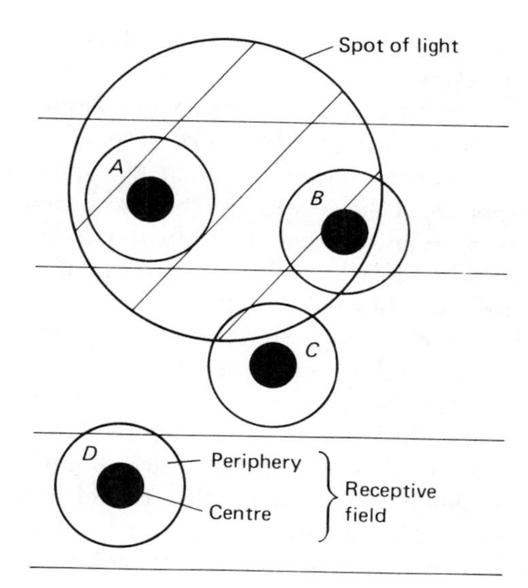

41 By recording from the cat's visual cortex when bars of light were shone onto the eye, units were found which fired only when the bar was in a specific orientation and in a specific part of the visual field. Some units responded only when the bar moved in one direction. After Hubel and Wiesel 1959.

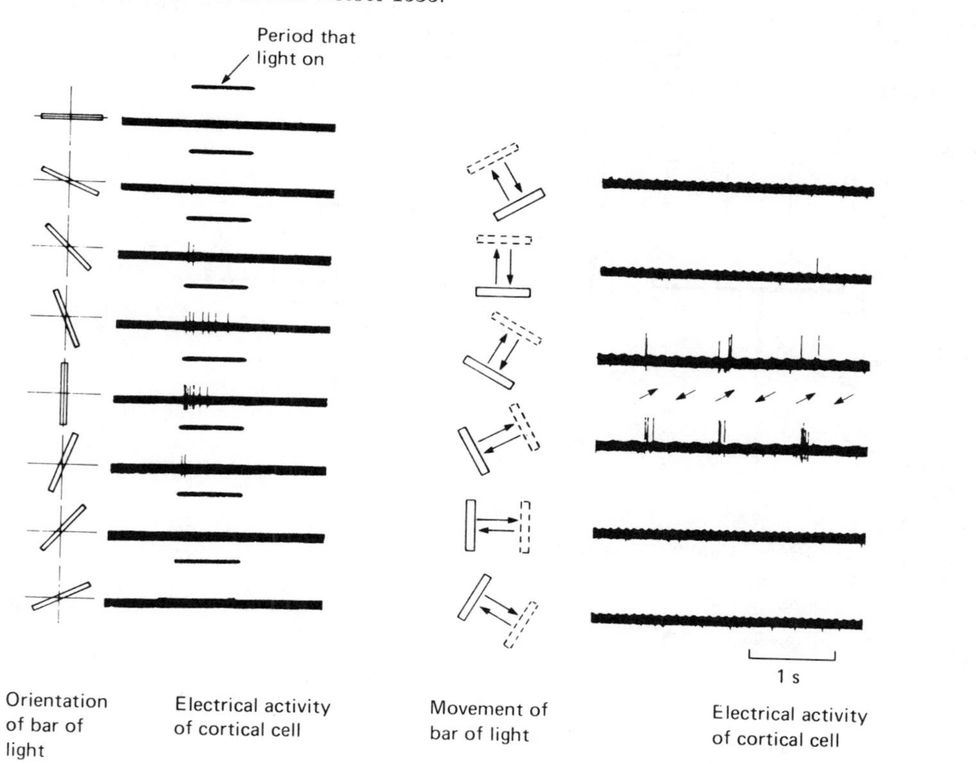

(4) lower-order hypercomplex cells which respond to edges, corners, tongues and angles of particular sizes; (5) higher-order hypercomplex cells which are like (4) but which respond to more than one preferred direction 90° apart. The responses of these cells are each explained in terms of the integration of inputs from lower order cells by Hubel and Wiesel (1965). Many of the cells are binocular in that they receive input from both eyes (Hubel and Wiesel 1962) and some respond to small binocular disparities and might thus be important in depth perception (Barlow, Blakemore and Pettigrew 1967). Various studies show that specific colour boundaries are detected by some cells. Hubel and Wiesel showed in 1963 that, in the visual cortex, cells responding to light bars of particular orientation are often arranged in columns at right angles to the surface of the cortex, an especially interesting observation since Mountcastle (1957) had described functional columns in the somatosensory cortex. There is also evidence for functional specialisation of areas of the visual cortex. Zeki (1978) has shown, for the rhesus monkey, that most orientation-selective cells are in the areas which he calls V2, V3 and V3A. Most colour specific cells are in the area V4, whilst directionally selective cells are concentrated around the superior temporal sulcus in the pre-striate cortex.

The rabbit has some retinal ganglion cells which will respond only to an object moving across the receptive field in a particular direction or at a particular speed (Barlow and Levick 1965). The directional response could be explained by a system such as that shown in Figure 42, in which bipolar cells excite but horizontal cells from one side inhibit, or vice versa. Responses to particular speeds could depend on inhibitory or excitatory inputs with fixed neural

42 Proposed 'wiring circuit' for retinal ganglion cells which fire only when light moves across their receptive field in one direction. Due to the inhibitory connections of the horizontal cells shown above, ganglion cells would receive excitatory input only when light moved from left to right. Modified after Barlow and Levick 1965.

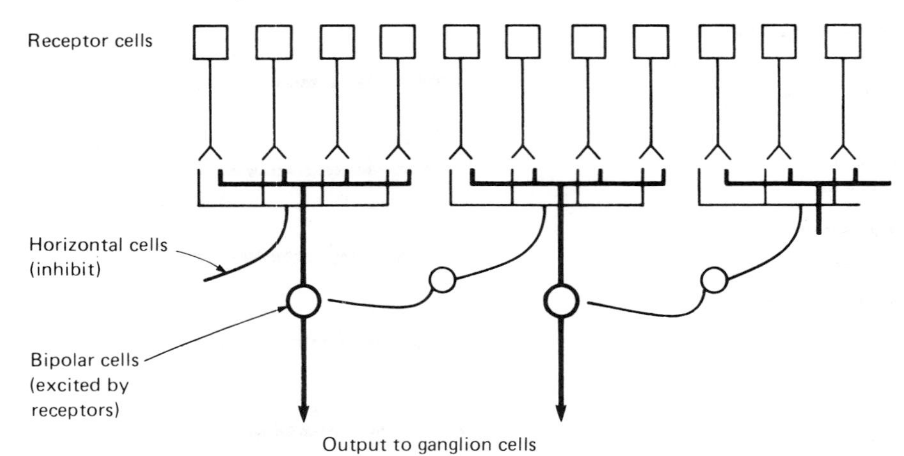

Condition	*Output to ganglion cells*
Continuous light	NIL (excitation = inhibition)
Light moves from right to left	NIL (bipolar output still inhibited)
Light moves from left to right	+ (ganglion cells excited before any inhibition from horizontal cells)

delays. Later work by Levick (1967) revealed the presence of ganglion cells which would respond if an object stopped in their receptive field but not if it moved continuously across the field (Figure 43). The rabbit also has elaborate analysers in the visual cortex, for example Chkikvadze (1975) recorded from units which responded only to one light intensity, not to brighter or dimmer lights, whilst Chelidze (1975) found units which responded only to visual events with a specific time interval between them.

Effects of experience on visual analysis

If mammals are reared in darkness for long periods, the anatomy and biochemistry of their visual systems are altered from those of animals reared under normal conditions. The retinal ganglion cell layers in the eyes of rabbits reared in darkness for 10 weeks were much smaller than were those of light-reared rabbits and the amount of nucleic acid present in the layer was drastically reduced (Brattgård 1951, 1952).

43 The record of response of this ganglion cell from the rabbit's retina shows that it fires if a dark spot moves into its receptive field and stops but shows little response if the spot moves straight through the field. From Levick 1967.

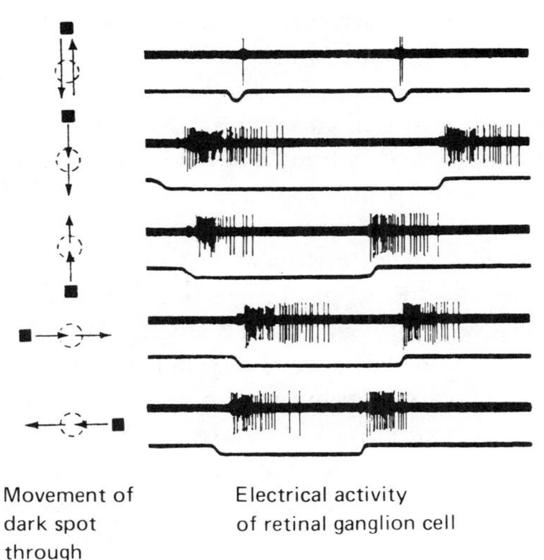

Chow, Riesen and Newell (1957) and Rasch *et al.* (1961) confirmed these results for rats, cats and chimpanzees kept in darkness for several months. Wendell-Smith (1964) found that the optic nerve from a mouse's eye which had been covered for 75 days after birth was 10% smaller than that from the other, uncovered eye; both the axon diameter and sheath were smaller. Retinal ganglion cell size was normal if the animals had one and a half hours in light each day. If input from the eye to the cortex, in a rabbit, was prevented by lesions at birth, after 30 days the pyramidal cells in the cortex had localised zones in which the number of pre-synaptic buttons was greatly reduced (Globus and Scheibel 1966). When rats were kept in darkness for 21 days the pre-synaptic buttons in the visual cortex were very small. The size of pre-synaptic buttons increased in some areas of the visual cortex, however, after only three hours' exposure to light and increased more after longer periods in light (Cragg 1967). Subsequently Cragg (1975) has shown that fewer than 1% of synapses in the visual cortex develop before eye opening. It is apparent that light is essential for the normal development of the visual system but to what extent are its effects non-specific and how much of the development of specific visual analysers depends upon specific visual input?

Physiological function in the visual system can be altered by visual experience. When Wiesel and Hubel (1963) recorded from the visual cortex of kittens reared with one eye covered they found that this eye had made very few connections with cortical neurons. They were almost all monocular instead of mostly binocular. If the two eyes are alternately covered during development they do make connections with the cortex but not normal binocular connections (Blakemore and van Sluyters 1974). It is therefore clear that the cortical neurons form connections, during a certain period of development, with the eye or eyes which are transmitting information at the time. These aspects of visual system development have considerable plasticity but when Hubel and Wiesel (1963*b*) recorded from the visual cortex of kittens whose eyes had not yet opened they found units which responded to specific position in the visual field, contrast, direction of movement, velocity of movement, and orientation. The view that the functioning of the visual analysers in the cortex was not dependent upon visual experience held until the independent studies of Hirsch and Spinelli (1970) and Blakemore and Cooper (1970). Hirsch and Spinelli kept kittens in the dark for the first three weeks of their lives. They then equipped them with goggles (Figure 44) and put them in light until they were 10–12 weeks old. Blakemore and Cooper kept kittens in darkness from 2 to 20 weeks of age, except for the five hours per day which they spent on the platform in the horizontally or vertically striped cylinder. In both experiments the eye of the kitten was exposed to either vertical or horizontal stripes. Hirsch and Spinelli recorded from the visual cortex at 10–12 weeks and found that half of the units had elongated fields corresponding to the experience of that eye, and were activated by only one eye. Blakemore and Cooper's kittens were tested behaviourally at 20 weeks. They had normal pupillary reflexes but showed no visually guided placing with the paws when brought toward a solid object. They did not show a startle response when an object approached them and showed clumsy movements, bumping into things when they moved around. If they had been reared with horizontal stripe experience they did not appear to see a vertical grating but did see a horizontal grating. At 30 weeks, 124 of 125 cortical units were vertical or horizontal according to the subject's experience. Kittens, reared as in Hirsch and Spinelli (1970), were trained by Hirsch (1972) at 11–19 weeks on tasks involving discrimination of patterns with horizontal and vertical components. The experimental animals were markedly poorer at learning such tasks than were the controls but their ability improved after some weeks in more normal surroundings. If the rearing goggles had two orientations of diagonal lines, instead of vertical and horizontal, the cortical units responded to the expected diagonal (a condition not normally found) or to either vertical or horizontal (Leventhal and Hirsch 1975).

44 Kittens were reared so that they could see only vertical stripes with one eye and horizontal stripes with the other. Light, but no patterned vision, entered the sides of the goggles so that the patterns at the end could be seen. The cardboard cone prevents the kitten from dislodging the goggles. Drawing from photograph in Carlson 1977, referred to by Hirsch and Spinelli 1970, 1971.

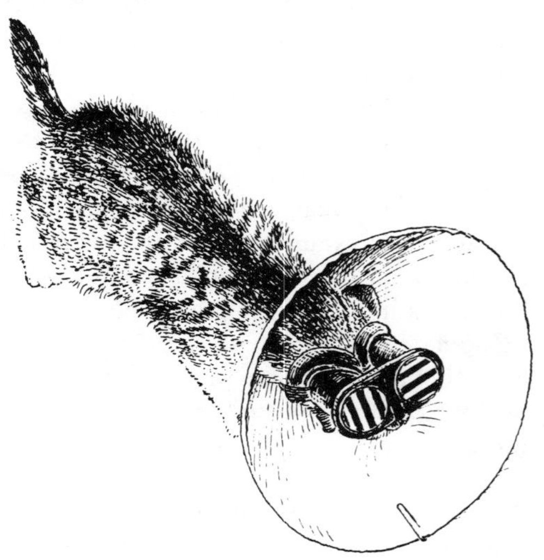

Some workers have been unable to confirm the effects of visual surroundings during rearing (Stryker and Sherk 1975) but this may be because the animals did not look at the stripes or because of variation in the critical developmental age when such effects occur. Blakemore (1974) has found that the effects of the orientation of stripes on cortical functioning did not occur if kittens were exposed before 4 weeks or after 10 weeks of age but did occur if exposure was between 4 and 8 weeks. The normal connections between the retina and the visual cortex are preserved in the cat even when one eye is rotated by 90° before it has opened. Most cells in the cortex are monocular, in such animals, but the animals can use their rotated eye in vertical *versus* horizontal discrimination tests and, with reduced efficiency, in distance perception (Blakemore, van Sluyters, Peck and Hein 1975; Mitchell *et al.* 1976). In summary, it can be seen that visual experience is important in modifying the function of visual analysis units but that certain components of the analysis systems can function without it. It seems likely that in all visual recognition mechanisms, at least the 'fine tuning' depends upon specific visual experience. As Barlow (1975), in a useful review, puts it 'neither an inexperienced animal, nor an inexperienced neurone, has been shown to have the normal powers of resolution'.

Centrifugal effects on sensory function

One method of filtering out biologically important information from the great variety of environmental changes which impinge upon an animal is to have very specific receptor systems. An extreme example of this is the ear of the moth which will respond to the sounds made by hunting bats and to little else (Roeder and Treat 1961). The retinal ganglion cell which detects directional movement in the rabbit (Barlow and Hill 1963) is one of many other examples. In such units most of the incoming information is thrown away. Much of the design of sensory systems has the effect of eliminating most possible input and such a function is essential to the animal but some of the rejected information might have been useful. As Collett (1974) has pointed out, an alternative method of filtering is to determine which sensory pathways and which analysers operate at any given time by means of centrifugal control. As mentioned earlier in this chapter, neurons which are efferent or centrifugal (coming from the central regions of the nervous system) have been described in many sense organs. Miles (1970) stimulated centrifugal fibres which go to the avian retina whilst recording from the retinal ganglion cells and found that the ganglion cells responded to a wider range of visual changes after such stimulation. This occurs because the inhibitory effects in an on-centre–off-periphery unit are reduced, and hence the unit becomes better at detecting that a change has occurred but poorer at discriminating the precise pattern of the change (Miles 1972).

The possibility of determining what sorts of environmental change will elicit a response, by central control of which sensory pathway is switched on at any moment, is discussed further in Chapter 4 and by Broadbent (1958, 1971). Many studies show that the functioning of sensory analysers which are several synapses away from the receptor cells is altered by centrifugal output; for example units in the cat's visual cortex produced a different response to light flashes when simultaneous weak electric shocks were given to the skin (Horn 1963). Visual neurons in the superior colliculus of the rhesus monkey were more likely to fire when a light was seen if they had been trained to respond to it with a saccadic eye movement than if they had not (Goldberg and Wurtz 1972). Another function of centrifugal nerves is to protect sensory systems against overload. The sensitivity of the lateral-line organs of fish is drastically reduced when the animal itself moves, for without this blunting of response the strong pressure waves produced by the fish's movement would render the sense organs functionless for some time after the movement, or damage them (Roberts and Russell 1972). Similarly the functioning of the echolocating bat ear is attenuated at the time of squeak production but recovers in time to hear any echoes (Henson 1965). Mechanisms which allow an animal to distinguish between sensory input from its own movements (i.e. reafference) and sensory input from environmental changes are described in Chapter 3. Interaction between sensory analysis systems and between motor and sensory systems are important fields of research which are clearly relevant to attempts to explain how individuals can recognise and respond to patterns of sensory input, especially those which are spatially or temporally complex.

3
The control of movement

How do mice groom?

As an example of sequences of movements which are readily observed, consider a mouse which is grooming itself. Questions about the behaviour and the mechanisms which control it can then be formulated and discussed. The animal sits and moves its fore paws over its face, often apparently repeating movements several times. The mouth and fore paws are then applied to the belly, back, and tail, again with some apparent repetition of single movements and of pairs of movements (Figure 45). The mouse may then stop grooming, walk to its nest and cease movement. The actions from the beginning to the end of the grooming sequence are the result of a series of muscle contractions and depend upon the mechanics of the motor system. The muscle contractions are initiated by output from some motor control centre in the nervous system. Any part of that output might be affected by input from receptors and the characteristics of the control centre are certainly the result of interactions between genetic and environmental factors during development. Before considering genetics, development and physiological control, more detailed description of the movements is necessary. Are there sequences of muscle contractions which we can usefully refer to as a single motor pattern because the whole sequence is repeated? If so, are some patterns temporally related to others and do the relationships change with time or according to the sensory input received by the mouse central nervous system?

45 This mouse is sitting on its haunches and grooming. A forepaw is being moved along the snout.

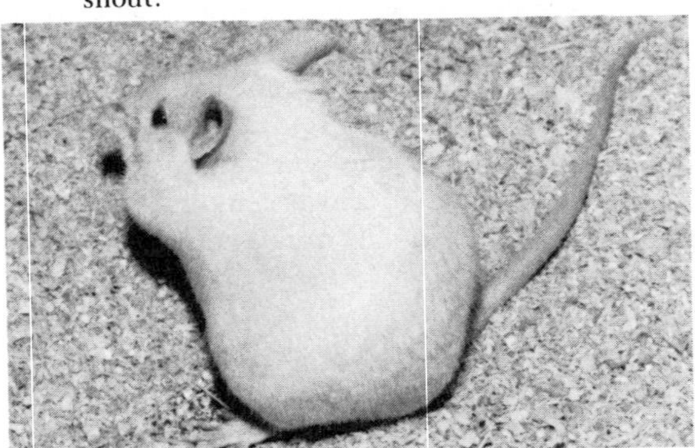

Levels of description

Fentress (1972) filmed mice which were grooming and then, after using single frame analysis, looked for temporal patterns in various behaviours including grooming the face, belly, back or tail. He found that these behaviours were not distributed randomly with respect to one another, nor did they occur in invariable sequences. Mice were likely to oscillate between grooming the belly and grooming the face and were likely to return to a movement after they had interrupted it to carry out another movement. Some of the relationships between activities were altered if the mouse had been disturbed by an object moving overhead. In a detailed study of face grooming, Fentress and Stillwell (1973) described seven distinct *components:*

a rapid vibration of the body (H);
circling movements of the fore limbs below the face (C);
licking the fore paws (L);
synchronous asymmetric overhand movements in which the forepaws move over the head (O);
parallel excursions of both fore limbs around the sides of the snout (P);

single stroke horizontal movements on the sides of the face (S);
and pauses with fore limbs at chest height (N).

Some of the transitions from one of these components to another were commoner than would be expected by chance so there is some order in face grooming. Knowledge of the preceding component allowed prediction of the next with a mean accuracy of one in five or six. The duration of a component was found to affect subsequent behaviour. For example, licks (L) lasting up to 0.3 s were followed on 64% of occasions by overhand (O) and on 12% by circle (C) but licks lasting more than 0.7 s were followed 6% by overhand (O) and 7% by circle (C) (Fentress 1980). As a result of this type of observation Fentress and Stillwell looked for units made up of combinations of components of face grooming. They described five *units* as follows:

1. varying combinations of all components except (S) e.g. O N L C P L C O L;
2. long periods (at least 0.8 s) of licking (L) which may be momentarily interrupted by circling (C) and pauses (N) e.g. L C C C L C L;
3. a series of repeated single strokes (S) interrupted by an occasional brief lick (L) and usually preceded by one or more parallels (P) e.g. P S S S S S S S S:
4. repeated overhands (O) with rare interjections of brief licks (L) e.g. O O O O;
5. varying combinations of all components except circling (C) and body vibration (H) e.g. P O L O L P L O.

Sequences of these units were analysed and were often found to include units in ascending numerical order: 2 then 3 then 4 etc. The predictability of a unit when the preceding unit was known was one in two.

Grooming by mice can be considered at various levels. Impulses in motoneurons initiate changes in muscle end-plates which result in the contraction of *groups of muscle fibres* (e.g. see Aidley 1978). The contraction of one or more muscles causes a simple *movement* of body, limb or head. More complex sequences of muscle contraction and relaxation can produce a *component* of grooming such as (S), the single horizontal stroke of fore limbs on face or an even more elaborate movement such as the overhand (O) or licking (L) components. These components are aggregated into *units* of grooming. *Groups of units* occur in certain sequences. Grooming itself is associated temporally with certain activities more than with others.

The remainder of this chapter explains something of what we know about complex movements and their control. This explanation must inevitably refer to some ideas about motivation but most discussion of the factors which affect whether or not an activity such as grooming or feeding will occur at any moment is in Chapter 4.

Cerebellar and cortical control

A movement like the horizontal single stroke from snout to cheek (S), which the mouse uses during grooming, necessitates commands from the brain to a succession of different muscles. The accuracy of temporal patterning of the commands must be sufficient to avoid conflict between different muscles and to achieve fur smoothing and cleaning. The strokes appear to a human observer to be very similar to one another and they may occur as rapidly as ten times per second (Fentress 1972). The commands for such rapid, complex movements are coded in the cerebellum (see Figure 5, for the anatomy of the brain). They are so fast that modulation of the movement during its execution, as a result of sensory feedback, is most unlikely. The course of events in the motor system of the mammalian brain during such a movement has been described by Eccles (1973). Firing in the pyramidal cells of the cerebral cortex results in excitatory inputs to the mossy fibres and climbing fibres of the cerebellum. These connect to the large Purkinje cells and the granule cells of the cerebellum. Outputs from the cerebellum go via the thalamus back to the pyramidal cells and can affect the commands travelling down the pyramidal tract to motoneurons in the spinal cord. The circuit time for such a loop is probably about 0.001 s. It seems likely that output to the motoneurons occurs after a series of such loops so that the final command takes account of the complex pattern programs in the cerebellum.

The cerebellum also controls posture, balance and some locomotor movements by means of commands produced after input from proprioceptors. These are all rapid movements but slow precise movements seem to be controlled by the basal ganglia (which include the caudate nucleus and amygdala) in the fore brain (Kornhuber 1974). Movements which require much guidance from somatosensory information for their regulation are controlled principally by the pyamidal system in the cerebral cortex (Kornhuber 1974) where the neurons may respond in as little as 25ms to a tactual input (Evarts 1974). Some of the components of a mouse's grooming activity are skilled, ballistic movements which appear to occur in a predictable, unalterable way once they are initiated and hence are controlled principally by the cerebellum. Other components are slower and more variable so they are likely to involve more modifications due to sensory feedback whilst they are occurring.

Sensory feedback

Some information about the role of sensory feedback during mouse grooming has been obtained as a result of experiments by Fentress (1972). The trigeminal nerve receives input from sensory receptors on the face of the mouse. If mice were anaesthetised and their trigeminal nerves lesioned, on recovery from anaesthesia they groomed their faces at a higher rate than usual for the first two days but all the components of grooming occurred. It seems, therefore, that the performance of the various components of face grooming does not depend upon sensory input from the face. Not only were the components unaltered but the grooming patterns shown when mice were put into strange boxes, which Fentress (1972) describes as being 'very stereotyped and prolonged sequences', were also unaltered by the trigeminal nerve lesions. Differences in the proportions of the components, especially an increase in single strokes (S), were observed, however, when the mice were in their home cages. This result emphasises the interrelationships between motivation and motor control for it would appear that the control of the movement is affected by sensory input in the home situation but not in the stressful strange-cage situation. When sensory input from all sources, including feedback from muscles, was prevented by sectioning the dorsal roots in the spinal cord, mice were still able to perform each of the components of face grooming. The combinations of components into units did not occur, however, so sensory input is probably necessary for this more complex level of motor patterning. Although there is still much to discover about the brain mechanisms controlling grooming movements it is clear that motor control programs, somewhat like those for a computer, exist.

Development and genetics

Studies of grooming in developing animals and in different genetic strains are providing further information about the control mechanisms. The ages at which various grooming movements appear in the developing rat have been described by Bolles and Woods (1964). A detailed investigation by Richmond and Sachs (1981) has revealed the interesting fact that the parts of the body which are groomed first during development are those which are groomed first during a sequence of grooming by an adult rat. Figure 46 shows that the adult grooming order: nose, face, ear, belly, hind leg, back, tail is very similar to the order of appearance during development. The developmental changes probably occur in this order because of the way in which the neuromuscular systems mature. The head-grooming movements that appear first during development are the simplest adult movements, which occur in the middle of a sequence of adult head-grooming (Fentress 1981*b*). Another interesting change in mouse grooming during development is what Fentress (1981*b*) calls 'its progressive emancipation from being dictated by sensory cues of the moment'. He noticed that young mice, 6 to 12 days old, sometimes initiated abbreviated grooming sequences when their fore paws passed near the face during walking or swimming movements. Grooming was also initiated on most occasions when the young mice were placed in an upright posture. Older mice and rats did not groom automatically in this way unless their central nervous systems had been disrupted in some way (Golani, Wolgin and Teitelbaum 1979). It seems that the control systems which operate are of a higher order in adults than in young mice (Fentress 1981*a*). Young mice can carry out co-ordinated movements of shoulder and tongue during grooming even if they have been without fore limbs since birth (Fentress 1973). It can be concluded, therefore, that all aspects of sensory feedback are necessary for the development of grooming components, although they may be necessary for complex grooming patterns to occur.

Some mutant mice groom abnormally and have modified brain structures so studies of them provide insight into the brain mechanisms controlling grooming behaviour. The mouse strain 'reeler', which cannot walk or use its limbs normally (Sidman 1968), has disoriented granule and Purkinje cells in the cerebellum. These result from defects in the system controlling the migration of these cells during early development (Sidman 1972, 1974). In another genetic strain of mice, 'nervous', 90% of the Purkinje cells in the cerebellum are atrophied (Sidman and Green 1970). These mice showed recognisable components of grooming behaviour but the overhand component (O) was irregular and sequential linkages between components were significantly more random (Fentress 1972).

46 The ringed figures show the order in which an adult rat grooms the various parts of the body. The other figures indicate the mean age in days at which this part of the body was first groomed during development. Modified after Richmond and Sachs 1981.

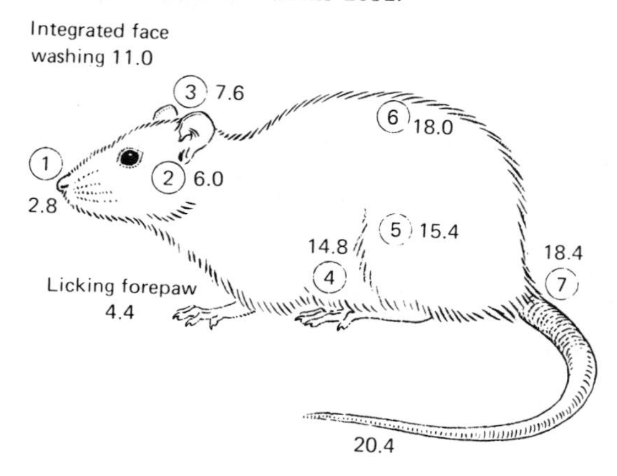

Other grooming and preening studies

Acts and bouts

Further evidence concerning the different levels of motor control has resulted from studies of grooming and preening in animals other than rodents. The general structure of preening movements in birds and the timing of preening in relation to other activities have been described by Andrew (1956), van Iersel and Bol (1958, and Rowell (1961). Andrew and Rowell were concerned principally with the factors which affect the occurrence of preening and other activities but Rowell did find it useful to distinguish between acts (Russell, Mead and Hayes 1954) and bouts. When a chaffinch (*Fringilla coelebs*) was observed 'bending forward, wiping the bill on the perch, and resuming an upright posture' Rowell recorded a single *act*. Groups of acts, separated from other acts by gaps, whose minimum length was defined, were called a *bout*. Rowell arbitrarily defined a gap as '15 seconds of locomotion or inactivity,' but a better criterion is required. Slater (1974*a*,*b*, 1975) described preening and feeding in zebra finches *Poephila castanotis*. The distribution of gaps between preening acts included many very short gaps within bouts and fewer long gaps of divers lengths. The criterion for the gap-length, which Slater used as the minimum inter-bout interval, was determined by identifying the first discontinuity on a log-survivor function of gap-lengths. This method is described in detail by Slater (1975) and alternative methods of defining bouts are discussed by Machlis (1977).

Stochastic analysis

The frequency of transitions between activities such as 'preen' and 'stretch' were calculated by Andrew (1956*a*) for buntings (*Emberiza* spp) but a much more detailed analysis of the interrelationships of body maintenance activities of this kind in skylarks (*Alauda arvensis*) was carried out by Delius (1969). Delius was one of the first to emphasise that analytical methods in which behaviour is treated as if it is the result of deterministic processes are often inadequate. Random variables inevitably influence behaviour so stochastic analysis is often desirable (see also Attneave 1959, Dawkins and Dawkins 1974, Broom 1979). Such analysis involved applying probability rules for if some, at least, of the variation of a process is a consequence of the action of random factors, then it is a *stochastic process*. Delius showed that certain behaviour sequences were much more likely than would be expected by chance, e.g. body-shake followed within one minute by stretching both wings. It had been pointed out previously by Nelson (1964) and Schleidt (1965) that the description of sequences of movements necessitates measurement and analysis of intervals between the actions. Interval distribution plots by Delius indicated that whilst the intervals between some pairs of actions are randomly distributed, others, like the interval between preening and body-shake, are biassed towards shorter intervals so that actions must occur close together in time. Spectral analyses of time series which include preening and other comfort behaviours indicate that there are periodicities in the occurrence of these behaviours and the comparison of spectral-density functions for comfort behaviour and flying shows that these are statistically independent processes.

Fly grooming

The time course of eight grooming and two non-grooming components of fly (*Calliphora erythrocephala*) behaviour were recorded by R. and M. Dawkins (1976). They found many examples of oscillation between two activities. Figure 47 shows that the probability that front-leg grooming (FL) will follow head grooming (HD) is very high and that the probabilities that FL will be the third and fifth activities after HD are also high. These two activities occur alternately up to about ten times each. A sequence of behaviour like this, in which the likelihood of occurrence of an activity depends only upon the occurrence of the preceding activity, is called a first-order *Markov chain*. If the occurrence of the activity depended only upon the activity before last, the sequence would be called a second-order Markov chain. Further analysis of the fly-grooming data showed that recognisable combinations of events occurred at the beginning and end of grooming bouts and that various points where a decision is made about the next event occur within bouts. The clustering in time of grooming components is suggested by R. and M. Dawkins and is emphasised in further analysis of their data by Cane (1978). When the components were divided into three clusters by inspection of a transition matrix and the sequences of clusters were compared to the values predicted for a first-order Markov chain it was apparent that the clusters were real and that they occurred in predictable sequences. Thus

47 The sequence of components of the grooming behaviour of a fly was recorded. The probability that front-leg grooming would occur first, second, third etc. (lag 1, 2, 3, etc) after head grooming was calculated. Modified after R. and M. Dawkins 1976.

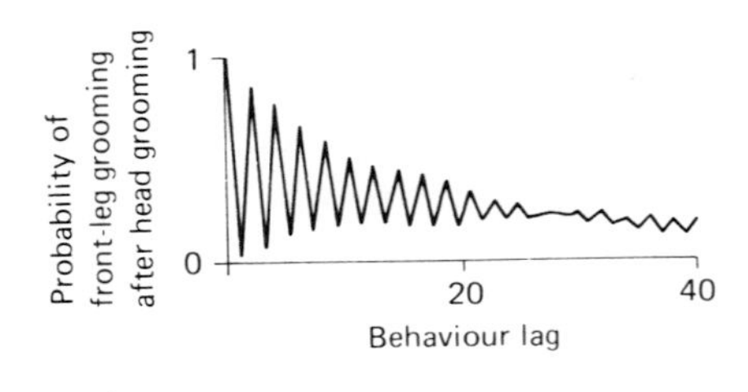

fly grooming appears to be less variable than mouse grooming, perhaps because the control programs depend upon fewer or less-variable sensory inputs.

Mantis grooming

The experiments of Zack (1978*a,b*) on head grooming by the praying mantis (*Sphodromantis lineola*) help to elucidate the control mechanism and the role of sensory feedback. Single-frame analysis of cine film allowed Zack to describe all the movements as flexion, rotation etc. of limb parts, head or mouth parts. Each synchronised combination of movements is referred to as a functional unit of grooming (Figure 48, Zack 1978*b*). The durations of these functional units changed during a bout of grooming and that of one unit, femur brush-cleaning, was different according to how the grooming was elicited. If water, acetic acid of various strengths, ethanol, glycerin, flour or methyl benzoate were applied to a mantis, the substance applied affected the probability of eliciting grooming, the latency to the start of grooming and the number of cycles per grooming episode. The same three units of grooming were readily observed in mantids whose femur brush had been removed and grooming could be elicited just as easily in these as in normal animals (Zack 1978*a*). The number of grooming episodes with a large number of cycles was higher than in normal mantids and the duration of cleaning the femur brush was much shorter. Zack's model for the mechanism controlling grooming therefore assumes that there is a detailed patterning program in the brain but that certain control variables depend upon sensory input. The finding that there is a program for an elaborate sequence of actions which, when switched on, starts at the beginning and then follows a stereotyped pattern has parallels in Fentress's (1972) mouse-grooming study. He found that mild irritants placed on a mouse's back often elicited a grooming sequence which

48 The grooming movements of a mantis are shown plotted against time. Each movement is shown as a line and the sequence is divided into three functional units. Modified after Zack 1978*b*.

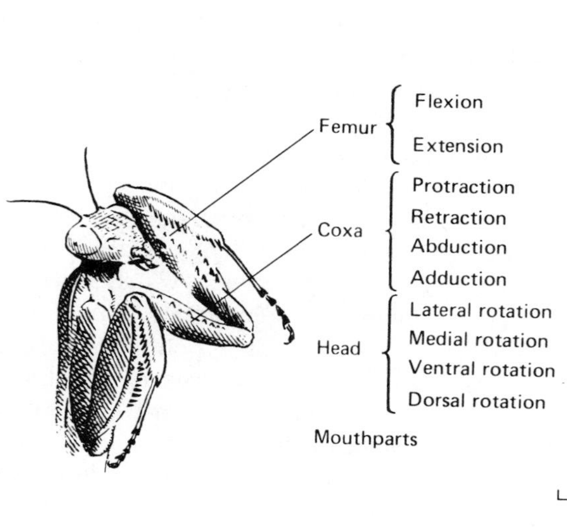

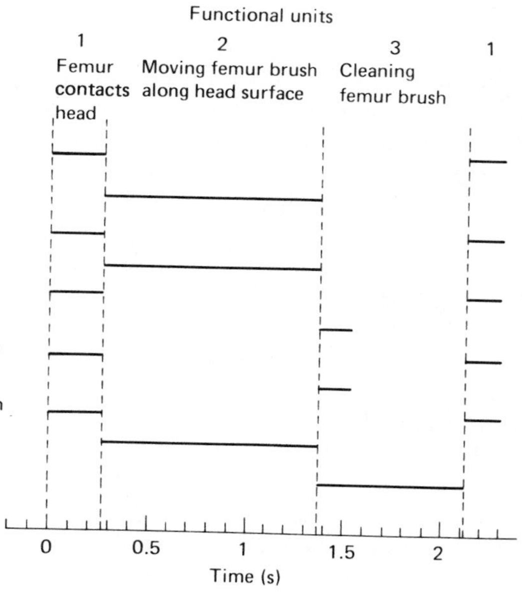

started with face grooming despite the fact that the face was not irritated. When people have dirt on a localised region of their hands they often go through their complete handwashing routine in the process of removing the dirt.

The apparent similarity between grooming control systems in different animals, for example gulls (van Rhijn 1977*b*) and mice (Fentress 1972), has led Baerends (1976) to speculate that such behaviour control mechanisms may be phylogenetically old: 'older than species, families or even phyla'.

General ideas about action patterns

Some general principles about movement patterns in animals are exemplified by the studies on grooming described above. There are several levels of complexity in the movements and the repeatability of actions declines with increasing complexity. A single stroke of the fore limbs over the head involves an elaborate sequence of muscle contractions and relaxations but when we watch a grooming sequence we may be unable to distinguish by eye between movements of the same type. Analysis using cine film, videotape or sound recording might allow us to detect differences between successive acts. Some of these differences might be considerable but the sequence of acts may occur too fast for us to see the differences at normal speed. The different acts would warrant description as distinguishable components of the behaviour pattern, especially if they altered the effect of the act, e.g. in cleaning the face. The smallest differences between successive acts, however, might be due merely to noise in the system, e.g. variability in muscle function. The ultimate criterion for distinguishing between acts must involve consideration of the neural control mechanism.

It is useful to group acts together into components if they occur as a result of the operation of the same neural control circuit. Such information is seldom available but there are some studies where detailed behavioural analysis and neurophysiological studies are combined and these are discussed later in this chapter. Grooming components in mice (Fentress and Stillwell 1973) are not all equally complex but they seem to be consistent in organisation and they can occur irrespective of sensory input. They are altered only in a genetic strain in which the cerebellum, which controls complex movements, is affected. The units of grooming, which are combinations of components, are much more variable in mice or gulls than they are in mantids or flies but they occur with sufficient frequency to be readily recognisable to an observer. Hence it is reasonable to assume that there is some more general neural program for them. Fentress's observation that mice in strange surroundings groom in a much more stereotyped way emphasises that such a program must exist but need not always operate.

Variability of action patterns

The question of how fixed or stereotyped patterns of movement are, has been extensively debated ever since the early writings of Lorenz and Tinbergen (e.g. Lorenz and Tinbergen 1938). Barlow (1968, 1977) and Schleidt (1974) have reviewed these arguments. Barlow points out several examples of behaviour which have been called fixed action patterns but which are modifiable in certain circumstances. These include the quivering, flickering display of orange chromide fish (*Etroplus maculatus*) and, if analysis is sufficiently precise, the strike of a snake. Analysis of films of pythons striking at rats (Frazetta 1966) shows that the python can compensate in the midst of its strike for movements by the rat. Hinde (1970) points out that stereotypy is a relative term and Schleidt (1974) explains that there are three ways in which a behaviour pattern might be stereotyped: firstly in the number of elements in repetitions of the pattern; secondly in the degree of coupling between those elements; and thirdly in the fidelity of repetition of the whole pattern including the size, duration, speed of movement and position of each component. Complex and variable behaviour sequences are composed of more-or-less variable units of behaviour, which are in turn composed of even less variable components. None of these is absolutely fixed but most are partially repeatable and hence include some elements of stereotypy in Schleidt's sense. Therefore the term 'fixed action pattern' does not seem very relevant or useful. Barlow's (1977) 'modal action pattern' is better but *action pattern* is used in this book. The measurement of the degree of variability in number of elements, coupling and pattern, which both Barlow and Schleidt advocate, is clearly a worthwhile method of finding out about behaviour patterns, especially if combined with neurophysiological investigation.

An example of a behavioural study of an action pattern is the work of R. and M. Dawkins on drinking by domestic chicks (R. and M. Dawkins 1973, M. and R. Dawkins 1974). Analysis of 290 videotaped drinking movements showed that the very frequent sequence of movement was as follows: downstroke of head, bill strikes water, bill comes out of the water, end of upstroke. When the variability of the movements was expressed, the greatest was found in the downstroke and the least in the time the bill was in the water. The downstroke must be an action during which the chick is taking decisions about what to do next, hence the uncertainty. Other components are much less variable so, presumably, decisions are seldom taken whilst they occur. This does not mean that decisions cannot be taken then for if the chick was provided with shallow water, the time that the bill was in the water was lengthened greatly.

The existence of very similar behaviour patterns in many of the individuals of a species, or individuals of several species or families, was a point often emphasised by early ethologists. The term 'fixed action pattern' thus included reference to ideas about behaviour development and genetics as well as to invariability of the behaviour in a single individual. Eibl-Eibesfeldt's (1967) introduction to the concept of the fixed action pattern is subtitled 'inborn skills'. He emphasises the lack of effect of environmental factors on the development of the behaviour and the similarity between the behaviour patterns of different individuals. As explained in Chapter 1, all behaviour patterns do necessarily develop as a result of interaction between the genetic material and environmental factors but it is impressive that the genetic programs should interact with such predictable environmental factors that great uniformity amongst individuals can occur. Complex action patterns can, however, develop as a result of the sensory feedback which results from practice. People can learn to ride bicycles: birds such as great tits can learn to remove the metal-foil tops from milk-bottles. Many of the motor patterns which are shown during feeding etc. are developed and perfected as a result of complex correlations of motor output and sensory input.

Motor control and command neurons

The difficulties of defining the systems which control motor patterning are emphasised by Fentress (1976*a*). He discusses the points mentioned earlier about problems of defining and describing the observed behaviour and also points out that it is likely that there is operational overlap between systems initially classified separately on the basis of behavioural output. This overlap could vary in time so that systems which are interconnected on one occasion may be independent on another. A system may be interactive, in that it can be activated by a variety of factors normally considered as extrinsic to the system, but self organising, in that once activated the system generates patterns of activity that are largely independent of extrinsic factors. Fentress is considering complex motor patterning control systems and these have not yet been investigated neurophysiologically in detail. Simpler motor patterns have been investigated and the ideas of control centres and command neurons are clearly relevant to this discussion. Doty (1976) reviews work on the control of simple vertebrate motor patterns and lists six basic questions which are asked when analysing motor control centres.

1. What is the afferent code which triggers activation of the centre?
2. How is the central excitatory state set?
3. How is the spatiotemporal output sequence generated?
4. How is the output of the centre guaranteed effective command of motoneurons?
5. To what extent does afferent feedback modify output?
6. How does the hierarchical organisation linking sequences operate?

Doty's own research on swallowing has helped to answer some of these questions. The sequences of muscle contractions shown by dogs carrying out six different activities are shown in Figure 49. A different control centre is needed for each of these actions and the sequence varies very little from one repetition of the act to another. The basic pattern of swallowing is organised exclusively in the medulla (Doty 1968) and its output is undisturbed by cutting a great variety of afferent or efferent nerves. Its initiation depends upon sensory input which provides the information that there is something to swallow. Although its approximate location in the brain is known, the neuronal circuits cannot be discovered by existing neurophysiological methods.

49 Electromyographic records from a normal, alert dog showing six sequences of muscular activity. Each horizontal line shows the degree of contraction of individual muscles plotted against time during each of the six movements. Modified after Kawasaki, Ogura and Takenouchi 1964.

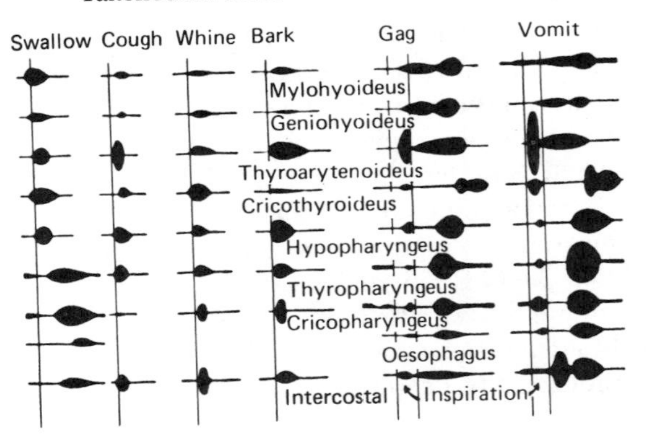

It is only in invertebrate nervous systems, especially those of molluscs whose neurons are very large, that any close approach to a definition of the circuit controlling a behaviour pattern can be made. It is even possible in some cases to find a single neuron which, when stimulated, initiates a complex motor pattern (Bullock 1975). Such neurons have been referred to as *command neurons* but, as Kupfermann and Weiss (1978) point out, in order to be sure that such a neuron does 'command' it is also necessary (i) to record from it and establish that it fires when a stimulus is presented and the behaviour executed and (ii) to find that the function is lost when the neuron is removed. The advantages of the greater simplicity and accessibility of mollusc and insect nervous systems, as compared with those of vertebrates, are considerable when studying the control of motor patterns (Hoyle 1975, 1976). Consequently the two examples described in greatest detail below are the escape response of the mollusc *Tritonia* and the song of the cricket. Other motor systems described briefly are those controlling bird song, lizard display and locomotion in various species.

Escape swimming by *Tritonia*

Some cells in the central nervous system of certain molluscs are large enough to be seen and constant enough in position to be recognised in different individuals by the use of a low-powered microscope. In 1967, Willows reported that electrical stimulation of cells in the brain of the nudibranch mollusc *Tritonia* elicited complex motor patterns. The same motor pattern could be elicited whenever the cell was stimulated. In addition, Willows and Hoyle (1969) were able to record from cells in the brain of *Tritonia gilberti* which fired when an escape-swimming response was initiated.

Detailed investigations of the escape-swimming response in *Tritonia diomedia* have been carried out by Willows, Dorsett and Hoyle (1973*a*,*b*). If an individual of this large nudibranch mollusc is resting on the sea-bed and is touched by any one of several species of mollusc-eating starfish it rapidly responds by starting to swim. The swimming response has four components which are illustrated in Figure 50. The pattern of movement varies in the extent of local withdrawal, the number of swimming movements and the number of termination movements, but the sequence is always the same. It can be elicited in the laboratory in an animal kept in running sea-water at a temperature similar to that of its normal surroundings (9–12°C) by applying salt crystals or soap solution to the oral veil of the *Tritonia*, as well as by touching the animal with a starfish. The brain is composed of cerebral, pedal and pleural ganglia (Figure 51) and it can be exposed and supported on a wax-covered platform moved by a micromanipulator as shown in Figure 52. It is then possible to locate individual cells and record from them with an intracellular micropipette or

50 The sequence of movements during the escape-swimming response of *Tritonia* which follows contact with a starfish, salt or soap solution. Modified after Willows *et al.* 1973*a*.

Starfish touches resting *Tritonia*

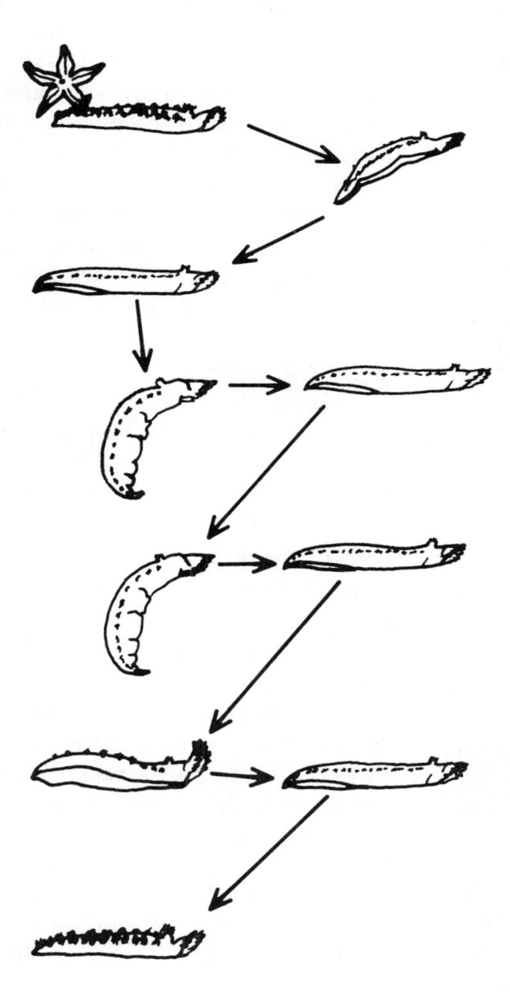

Components of escape swimming response

1 Reflex local withdrawal
region contacted is withdrawn; oral veil, branchial tufts, rhinophores retracted; slight ventral bending.

2 Preparation for swimming
circular muscles in midsection contract resulting in body lengthening and enlargement of oral veil and tail regions producing paddle-like structures.

3 Swimming
contraction of the ventral longitudinal muscles flexes body and pushes animal off substratum; dorsal flexion of equal vigour; 1–8 cycles (mean 5).

4 Termination
up to 5 dorsal flexions of higher frequency and shorter duration than swimming movements; rhinophores and branchial tufts re-extended and animal starts creeping locomotion.

51 Drawing of the brain of the nudibranch mollusc *Tritonia diomedia* showing the three ganglia, the giant cells on the surface of the ganglia and the nerve trunks (numbered). After Willows *et al.* 1973*a*.

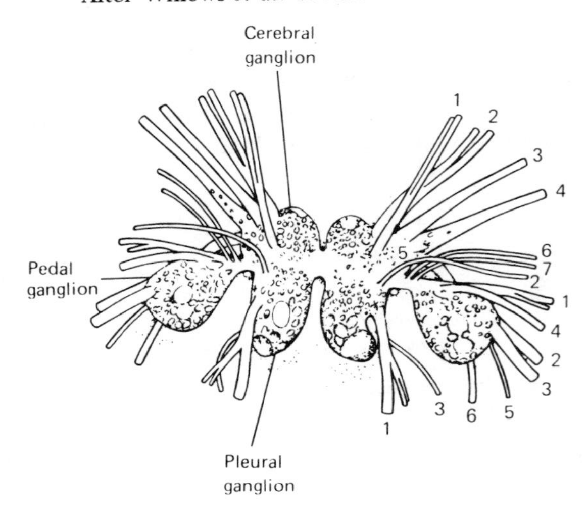

52 In order to record from, or stimulate electrically, the brain of *Tritonia*, the animal is suspended in a tank of running sea-water by a group of small hooks as shown. This allows the animal to move freely but the brain is held still on a wax-covered platform which can be moved by a micromanipulator. After Willows *et al.* 1973*a*.

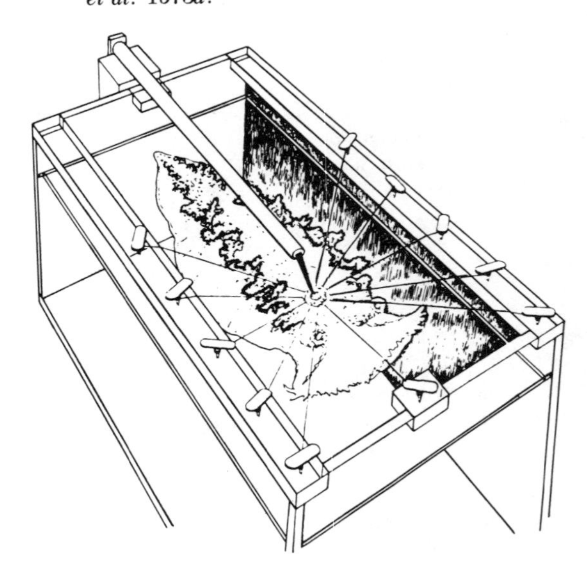

stimulate them by using a polyethylene suction electrode.

The first component of the escape-swimming response, local withdrawal, is too fast for central control and can still occur after the animal's brain has been removed, so it must be controlled by peripheral neuromuscular reflexes. The second component, preparation for swimming, occurs at the same time that cells in two groups on each pleural ganglion show bursts of activity. These cells, which Willows *et al.* call 'trigger group neurons' are coupled together so that activity in each neuron occurs almost synchronously. Stimulation of the neurons initiates components of preparation for swimming but sometimes also elicits the complete response. Other neurons which fire at the same time as the trigger group neurons include the S cells, afferent neurons in the pleural ganglia (Getting 1976). Figure 53 shows the sequence of firing in the five principal neuron types during an escape-swimming sequence like that in Figure 50. The next activity during the preparation phase is in a group of neurons which initiate contraction in the dorsal longitudinal muscles. This produces the dorsal flexion which occurs at the end of the preparation phase.

Immediately prior to the first swimming movement (phase 3) a burst of firing occurs in a pair of symmetrically placed neurons, called CeW, on the cerebral ganglion. Taghert and Willows (1978) suggest that the CeW neurons control the generation of the rhythmic swimming movements. Each swimming cycle involves the sequential firing of three groups of neurons. These are the CeW neurons, the group which initiates ventral flexion movements and the group which initiates dorsal flexion. The final termination stage occurs when the ventral flexion neuron burst ceases and weak bursts continue in the dorsal flexion neurons (Figure 53). The function of the neurons has been worked out as a result of experiments with intact animals. If the brain is removed, the animal can still crawl and all the muscles operate but no feeding, copulating or swimming is possible

(Hoyle and Willows 1973). The isolated brain can be stimulated electrically to mimic the sensory input which would be received when a starfish touches the oral veil. When this is done, recordings from the neurons involved in controlling the motor pattern are just like those obtained when the brain is in the animal so the patterned output for swimming can certainly occur without any necessity for sensory feedback (Dorsett, Willows and Hoyle 1973). The components of the neuronal mechanism controlling *Tritonia* escape-swimming are known as well as that for any action pattern but, as research continues, the mechanism is shown to be more and more complex. The number of participating cells is large and the number of possible interactions between them is even larger. The factors affecting the variability within and between individuals, e.g. in number of swimming cycles, have yet to be thoroughly investigated and developmental or genetic studies are difficult.

Sound production by crickets

The nervous system of even a comparatively large insect like a cricket is composed of cells which are very much smaller than those of *Tritonia*. It is possible, however, to stimulate electrically and record from nerves and localised regions of the brain and other ganglia. The animals can be kept and bred in the laboratory for many generations so genetic and developmental studies can be carried out.

The motor patterns which have been subjected to greatest analysis are those of certain thoracic muscles which result in sound production by crickets. The sounds which are produced can be recorded and displayed on a sonogram which plots frequency of sound against time. The male European field cricket (*Gryllus campestris*) makes its song by raising the tegmina, i.e. the leathery-looking fore wings, and rubbing them together causing a file on one tegmen to rub against a scraper on the other. Each inward movement of the tegmina produces a sound-pulse, Figure 54, whose length is dependent on the number of teeth on the file which make contact. The frequency of the sound depends upon the morphological and mechanical structure of the wings and the speed of wing movement. The patterning of the pulses depends upon the timing of the contraction of the various muscle groups (Huber 1962). The song,

53 Sequence of activity in five neuron types during a typical *Tritonia* escape-swimming sequence like that in Figure 50. One burst from the neurons connected to the dorsal flexion muscles occurs during preparation and weak bursts occur during termination. Each swimming cycle is preceded by a burst from the CeW neurons. Modified after Taghert and Willows 1978.

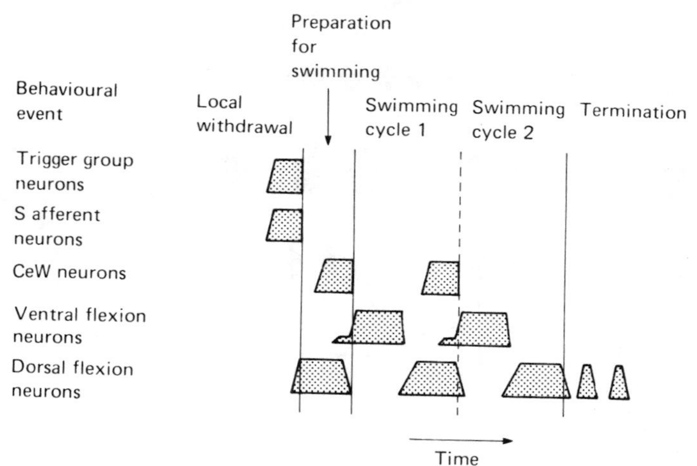

known as stridulation, results from the action of fourteen pairs of muscles in the second thoracic segment. The three functional groups of muscles acting on the fore wing are (a) elevators which raise and hold them in a specific posture, (b) depressors, and (c) adductors which draw them over one another and bring the teeth in contact with the scraper (Huber 1960).

Neurons from the second thoracic ganglion innervate all the muscles involved in stridulation. By recording from muscles or nerves and simultaneously making sound recordings, the sound pattern has been shown to correspond exactly to the pattern of muscular events. This pattern depends upon distinct spatial and temporal interactions of motoneurons within the second thoracic ganglion (Kutsch 1969, Huber 1974). The control of stridulation in the cricket, which never flies, is similar to the control of flying in related insects (Huber 1962, 1978). When Bentley (1969) put micro-electrodes into the second thoracic ganglion he was able to find and record from individual motoneurons which fired when the elevator muscles contracted and others which fired when the depressor muscles contracted. Interneurons which fired either at the time of pulses of sound or at the time of intervals between pulses were also found.

The cricket's second thoracic ganglion is connected to the brain and electrical stimulation of these connectives, or of the brain itself, can elicit any one of the three types of song (Huber 1962, Otto 1971). The pathway from brain to muscle is shown in Figure 55. Stimulation in different sites in the mushroom bodies could inhibit or elicit song or modify the pulse-rate of the song. Stimulation of the central body or of the mushroom bodies could lead to the production of abnormal song if the electrical input was above a certain level. It was clear that the brain had some influence on calling behaviour but

54 A field cricket (*Gryllus campestris*) and the sound pulse patterns of its three types of call. Modified after Huber 1962.

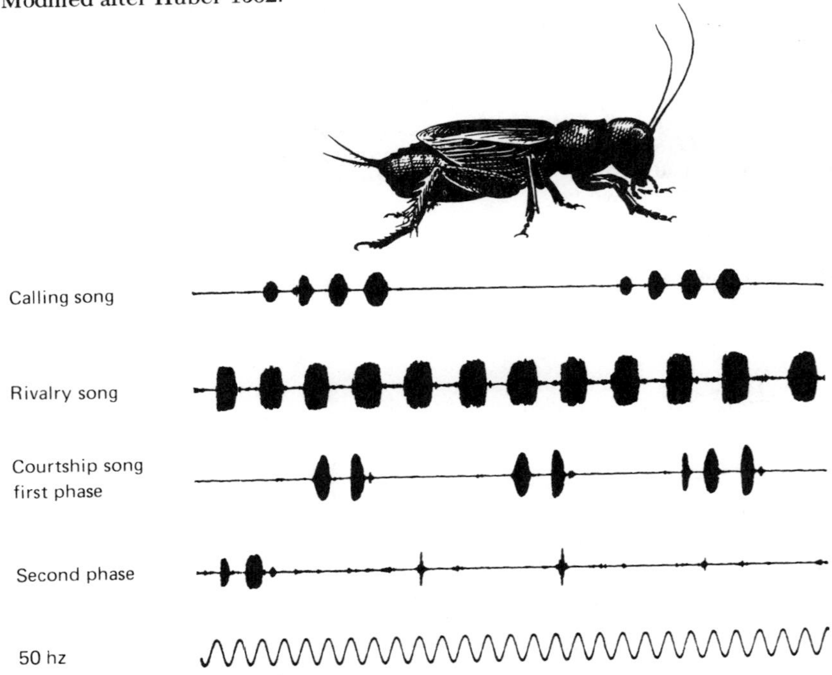

when Otto cut the connectives to the brain a normal calling song could still be produced: so the pattern can be generated within the second thoracic ganglion. The basic motor patterning is coded in the second thoracic ganglion but the brain determines the type of song, the posture and general body movements necessary for song to occur, and other movements involved in related behaviour such as courtship. A nervous system isolated from all phasic input can still produce a normal song pattern (Bentley 1969) but calling and courtship songs require input from the region of the genital apparatus (Huber 1960). The interactions of the brain and the various ganglia in the control of stridulation and other motor patterns are reviewed by Huber (1974, 1978).

Huber (1960) demonstrated the inhibitory function of the brain by showing that male crickets which were lesioned in the dorsal region of the mushroom bodies would often sing continuously. Such inhibition must also be important

55 The brain and second thoracic ganglion of the field cricket are concerned with the control of calling. Efferent fibres to the stridulatory muscles are found in the nerves shown in black. Modified after Huber 1962.

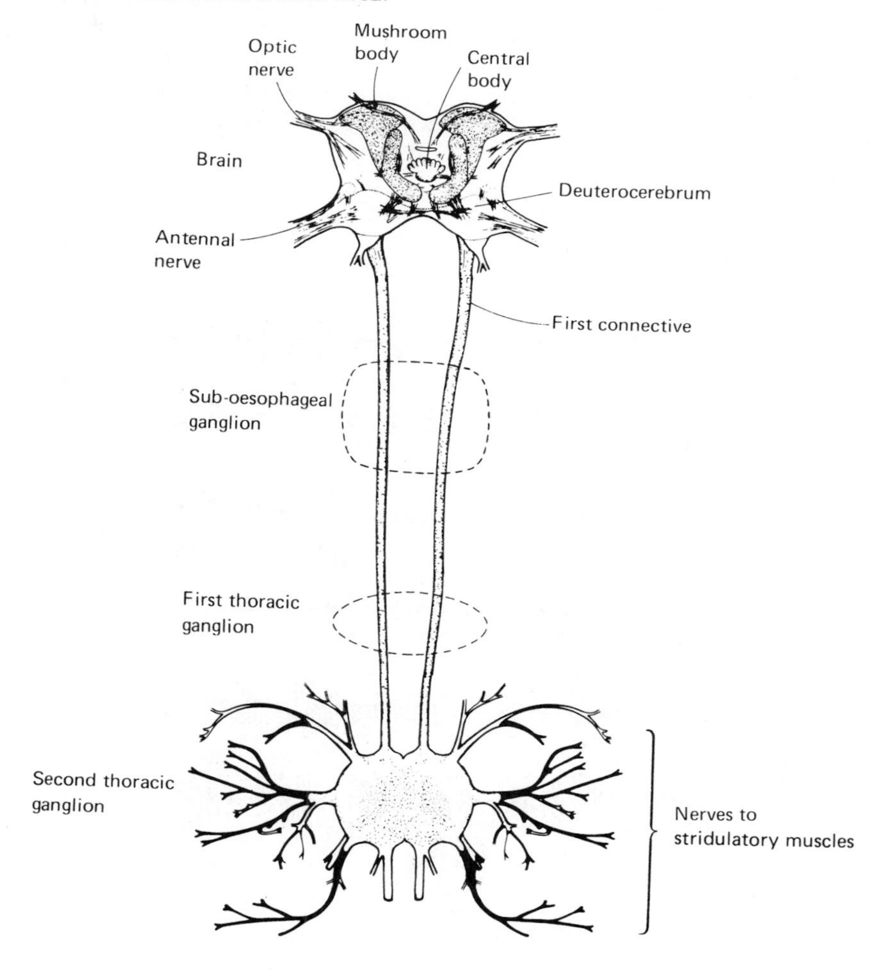

in cricket nymphs. Bentley and Hoy (1970) made similar lesions in nymphs and were able to elicit the motor patterns for singing despite the fact that normal nymphs never perform these movements. This discovery opened up possibilities for a study of the development of the motor pattern. Most individuals of the Australian field cricket (*Teleogryllus commodus*) develop into an adult after ten moults. Bentley and Hoy recorded from the song-producing muscles of brain-lesioned nymphs in the last instar before the last moult. They found that the patterns for the rivalry song, the courtship song and the most structured part of the calling song could be produced in these nymphs. In a related study on the patterning of neuronal impulses for flight movements they found that the first signs of the adult flight pattern appeared during the instar four moults before adulthood. The song pattern within a species is not very variable and was not changed obviously when crickets were reared under different conditions of temperature, diet, light cycle, season or population density (Bentley 1971).

The Australian field cricket (*Teleogryllus commodus*) and the Polynesian field cricket (*T. oceanicus*) have quite distinct songs (Figure 56). It is possible to hybridise these species and the songs of hybrids are intermediate in form between those of the parents. They continue to reflect the patterned neuronal output in the second thoracic ganglion (Bentley 1971). Since male crickets have no Y chromosome, F1 males from reciprocal crosses will be almost identical except for their X chromosomes. Detailed analysis of the intertrill intervals, the number of pulses per trill and the number of trills per phrase shows that some features of the song depend upon genes which are on the X chromosomes whilst others do not. Studies of the fea-

56 Sound pulse patterns in the calling songs of the field crickets *Teleogryllus oceanicus*, *T. commodus* and their hybrids. Each phrase is a chirp followed by a series of trills. Modified after Bentley 1971.

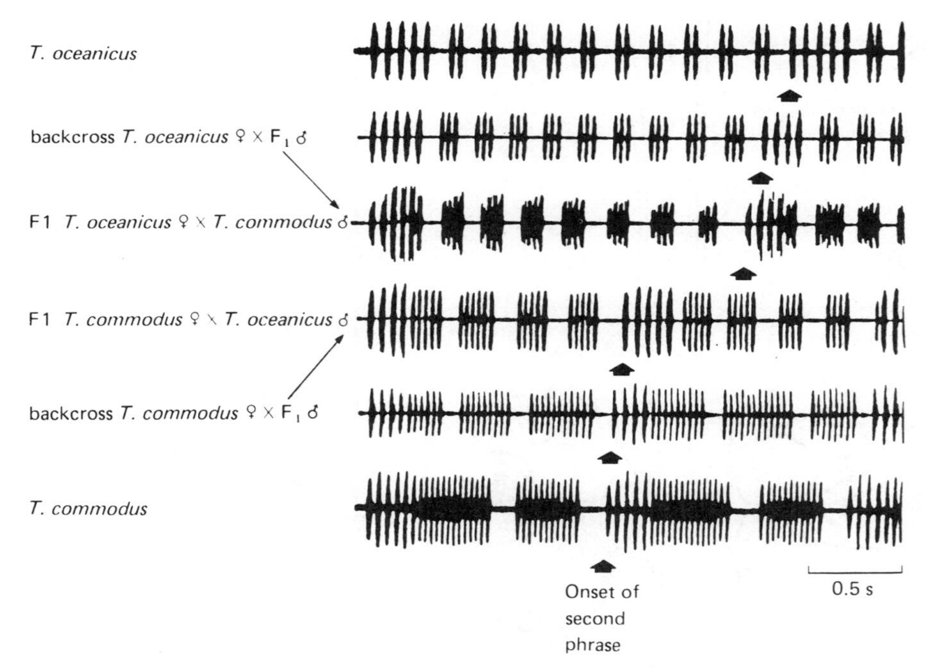

tures of the songs of individuals produced by back-crossing F1 hybrids with the parents show that the genetic information for the neuronal network controlling the motor pattern which results in song is widely distributed through the genome. More detailed analysis of the songs of *Teleogryllus* hybrids confirms that the songs are controlled by a polygenic multichromosomal system. Hybrids between *Gryllus armatus* and *G.rubens* showed at least three different sorts of songs which are produced by quite different neuronal systems (Bentley and Hoy 1972). Further experiments on hybrids, in which recording of muscle action potentials has been carried out, have provided information about the role of command interneurons in pattern production (Bentley and Hoy 1974). Hybridisation experiments have provided a new means of investigating the motor control mechanism (Bentley 1976).

Other examples of action patterns

Various other action patterns have been subjected to detailed analysis by using behavioural or neurophysiological techniques but there is a great need for a combined approach. Valuable information about action patterns could also be obtained by further studies using developmental and genetic methods. The limitations of recording methods have made detailed analysis of short sequences of behaviour difficult. The gap between the time scale of the behaviour record made by the human observer and that of the oscilloscope was difficult to bridge but such difficulties are lessened by computerised recording methods. The equipment which did allow detailed recording and analysis of events lasting a few seconds was that for sound recording. It was the tape recorder and sonograph which made possible the cricket-song work and the extensive work on bird song.

Bird song

Detailed bird-song studies were pioneered by W. H. Thorpe. Sonograms like that in Figure 57 are a representation of the effects of sequential muscle contraction in the bird's vocal apparatus, the syrinx. The variability of the songs of individuals of a species is described by Thorpe (1961) and he recognised the potential of bird song for developmental experiments. More detailed description of vocalisations as action patterns came from the work of Schleidt (1964*a*, *b*, 1965, 1974) on turkey gobbling. The male turkey's gobbling call showed very little variation in loudness or duration. The gobbling rate varied from 1.45 to 8.25 gobbles per minute but the variability, i.e. difference between fifteenth and eighty-fifth percentile, of inter-call intervals within a single record was the same irrespective of the length of the median interval. Such mathematical descriptions of behaviour help in understanding the type of neural control mechanism which might exist. When sequences of strutting and gobbling were analysed, the occurrence of strutting was found to have no effect on that of gobbling except for inhibition during the 4 s after a strutting bout. Gobbling on the other hand, was likely to be followed by strutting within 2 to 8 s.

Another bird whose song has been subjected to detailed analysis is the cardinal (*Cardinalis cardinalis*). Lemon and Chatfield (1971) described syllables, their subsyllables, their durations, the intervals between them, the number of syllables per utterance and the number of utterances per bout of song. Analysis of the probabilities of occurrence of the various song types indicated that prediction of which song type would be sung at any moment was possible if the previous song type was known. Cardinal song can be described, therefore, as a first-order Markov chain (page 60) as can the songs of various American thrushes (Chatfield and Lemon 1970, Dobson and Lemon 1979).

Vocal behaviour, especially that used for personal or territorial advertisement often shows periodicity. This is obvious in some insects and periodicities have been investigated in the songs of various birds such as quail (Guyomarc'h and Thibout 1969, Schleidt and Shalter 1973, Schleidt 1981).

Most bird calls develop normally when birds are reared in isolation, whether or not the birds have been exposed to sounds of other species (Marler and Hamilton 1966). Sensory input does play a part in development, however, since chaffinches deafened at an early age produce abnormal call notes (Nottebohm 1967). Songs, on the other hand, are more complex. Whereas some develop normally after isolation-rearing, in many species, an individual's song depends for much of its fine structure on the sounds which it hears of its own and other songs (Thorpe 1958*a*, 1961), Marler and Tamura 1964, Marler 1976*b* and Figure 57). When Nottebohn (1968) deafened chaffinches at 88 days of age, well before they started to sing, the songs which they produced later were merely variable trills. This was not due to inability to sing when deafened, since a chaffinch which sings a typical, elaborate song, continues to sing it immediately after deafening. If the chaffinch can hear itself but never hears any other bird song, it eventu-

57 Sonograms of chaffinch song after different treatments. Modified after Thorpe 1961 and Nottebohm 1968.

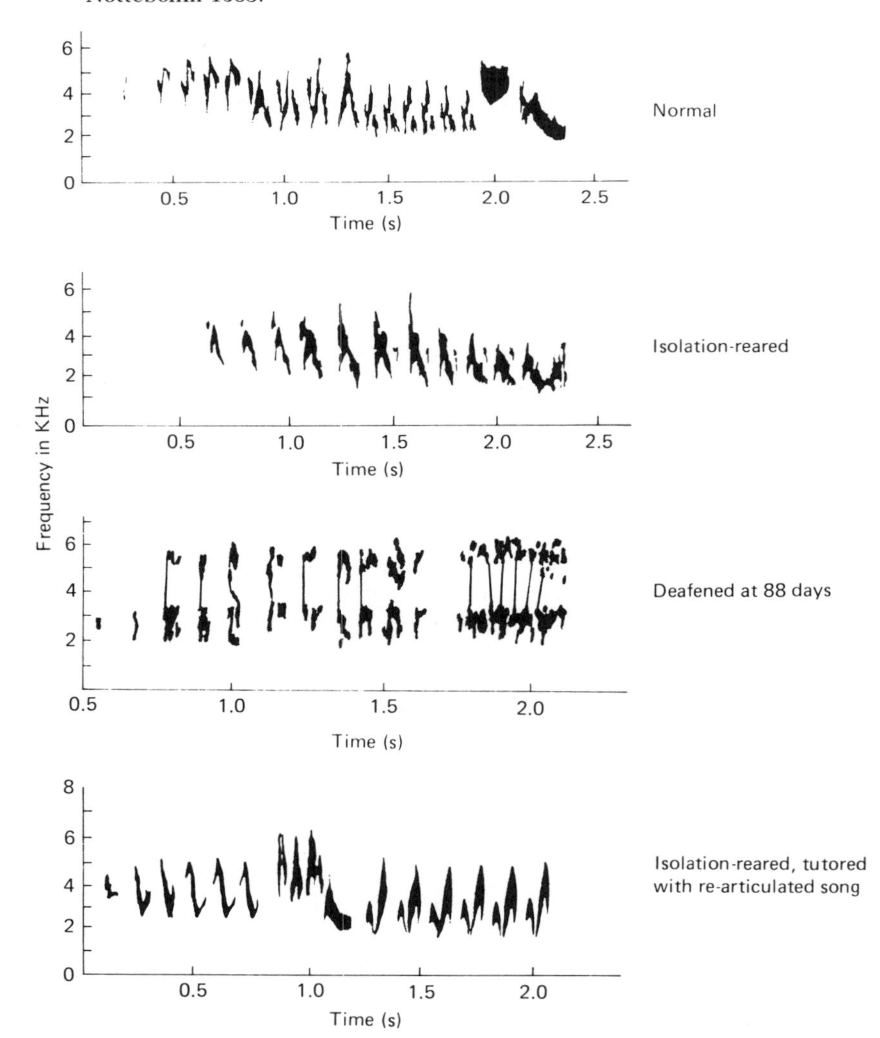

ally produces a song which is more complex than that of a deafened bird but which is much simpler that that of a normal bird (Thorpe 1958*a*). Chaffinches reared with other chaffinches produce a song which varies according to the songs which they hear. In one example, shown in Figure 57, an artificially modified song was copied by a chaffinch (Thorpe 1958*b*). The age at which exposure to song has most effect on song learning and the range of songs which are most likely to be copied has also been studied by Marler and Tamura (1964) in the white-crowned sparrow *Zonotrichia leucophrys* (Figure 58). The readily distinguishable dialects of the white-crowned sparrow in different parts of the North American west coast have been shown to be produced as a result of young males hearing and subsequently copying the songs sung in their own area when they are 2–8 weeks old. They do not copy the songs of closely related species with similar songs. There is much variation amongst bird species in the extent and timing of song copying. The New Zealand saddleback (*Philesturnus carunculatus*) copies the songs of the males which it hears when it settles in a territory and does not copy songs heard previously. Even an established bird may modify its song to match the songs of the birds which arrive in the area (Jenkins 1978).

58 White-crowned sparrow (*Zonotrichia leucophrys*) singing.

Lizard display

Many early ethological studies were inspired by observations of sequences of display behaviour which are relatively constant within and between individuals. An example of an apparently stereotyped display whose structure and variability have been described in detail, is the territorial display of the lizard *Anolis aeneus* (Stamps and Barlow 1973). The components of display by a resident male lizard when an intruder was introduced to its cage were called fan-bob, gorged-throat etc (Figure 59). The fan-bob was composed of (1) the fan, an extension of the dewlap, (2) an increase in height, and (3) the signature-bob. The signature-bob was very constant for any one individual but varied among individuals, hence the name which the authors gave it. If a lizard was attacked during a signature-bob, it finished the movement before returning the attack or retreating. Other components of the display did vary according to the motivational state of the resident, which was affected by intruder distance. The amount of side flattening and dewlap extension during the fan, and the height of the head, could vary and stepping and jerking movements could be included before or after the signature-bob. More prolonged sequences of displays and other movements have been analysed by many authors and described in terms of deterministic models (Nelson 1965*b*), Markov processes (Nelson 1964, Chatfield and Lemon 1970), or renewal processes, i.e. those in which an event depends upon the previous occurrence of that same event but not on intervening events (Schleidt 1964*a*, Hauske 1967, Heiligenberg 1974). These

59 Displays of lizard (*Anolis aeneus*). Modified after Stamps and Barlow 1973.

descriptive terms are discussed by Metz (1974). These successions of motor patterns require discussion of motivational variables for there is overlap between considerations of action patterns and of behaviour sequences (Chapter 4).

Locomotion: pattern, development and genetics

In contrast to most display patterns, locomotor movements are repeated many times and are very useful subjects for neurophysiological and developmental studies. Studies of walking in insects in which sequences of leg movements were described (Wilson 1966) or in which electrical changes in motoneurons and interneurons were recorded (Pearson and Iles 1970, Burrows and Horridge 1974, Horridge and Burrows 1974) clearly demonstrated the existence of central patterns for walking. The role of resistance reflexes is small. There is also a central control mechanism for crayfish swimming (Figure 60). Kennedy, Evoy and Hanawalt (1966) were able to elicit swimming movements by stimulating an interneuron in the nerve cord and rhythmic bursts of activity could be recorded in motoneurons even if the nerve cord was completely isolated from any input (Ikeda and Wiersma 1964, reviewed by Kennedy 1976). Swimming by fish is controlled by a combination of central pattern and sensory feedback (Lissmann 1964*a*,*b*, Roberts 1969), as is walking and swimming by amphibia (Gray and Lissmann 1946). The fact that reflex flexion of one hind leg in mammals initiates a reflex extension of the opposite hind leg and a flexor extension of the diagonal fore limb was discovered by Sherrington (1898). Subsequent work indicates that complex feedback mechanisms interact with central patterning during mammalian walking but the control system is not completely understood (see reviews by Grillner 1975, 1976).

60 Crayfish (*Procambarus clarkii*) showing electrodes positioned to record electrical changes in nerve cord and in motoneurons. After Schrameck 1970.

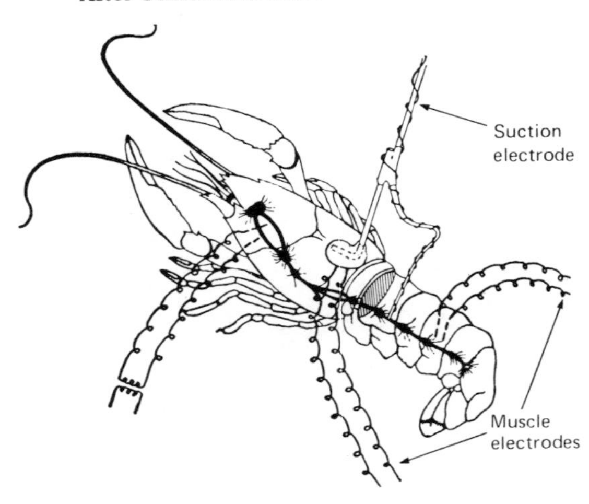

Studies of the development of locomotor movements in vertebrate embryos have provided interesting evidence about the control of movement. Early ideas that all locomotor movements arise as a result of the integration of local reflexes which appear earlier in development (Windle 1944) or, on the contrary, that all movement results from the 'progressive expansion of a perfectly integrated total pattern' (Coghill 1929) are now both seen to be inadequate as general explanatory theories. Carmichael (1926) showed that salamander larvae which had been anaesthetised by chloretone throughout development and hence were unable to move, were able to show apparently normal swimming movements on emergence from anaesthesia. Fromme (1941), however, found that there were some differences in the behaviour of amphibian tadpoles reared in this way. The development, independent of feedback, of co-ordination mechanisms in mammals has been demonstrated by Crain, Bornstein and Peterson (1968). They removed cortical and spinal cord cells from rodent embryos early in development and cultured them *in vitro* in the presence of blocking agents which prevented spontaneous firing. When the blocking agent was removed after development, complex, long-lasting discharges were produced immediately. Hence functional synaptic networks had developed whilst no nerve cells could fire. The electrophysiological development was found to be incomplete, however, in a replica experiment by Leiman, Seil and Kelly 1975).

The view that practice was important in the development of motor patterns was championed by Kuo. For example he proposed that leg movements by chick embryos helped to shape locomotor co-ordination (Kuo 1932). These ideas are difficult to test satisfactorily and most studies of embryo behaviour involve attempts to correlate anatomical, biochemical, physiological and behavioural changes (Hamburger 1968, Abu Gideiri 1971, Chapter 1).

Humphrey (e.g. 1964) studied the behaviour and the extent of the nervous system development of human foetuses born very prematurely by Caesarean section. She concluded that some of the neural control mechanisms which cause some early foetal movement sequences still operate in neonates. The movements of early foetus and neonate are not the same because the motor system has changed but she suggests that 'the neurons which have functioned the longest require the least stimulation to fire them and less oxygen to react so they are discharged more readily. Readily-activated motor-control circuits would be energetically economical but the time of first functioning of such circuits must vary widely among species according to biological needs, especially predator avoidance and respiration. Some embryos need to move at an early stage in development in order to survive whereas others develop to an advanced stage without much movement. Motor control systems would develop in different ways in these two circumstances and such differences account for some of the apparently anomalous results when interspecific comparisons of behavioural development are made. For detailed reviews of the embryology of behaviour see Marler and Hamilton (1966) and Gottlieb (1973).

Comparisons of locomotor patterns in different genetic strains has revealed something about the control of locomotion, e.g. the work of Sidman (1968, 1972, 1974) on the 'reeler' mouse quoted earlier in this chapter, but differences in locomotion might be due to variation in sensory function. Bentley's (1975) study of cricket nymphs which failed to jump when air was puffed at them showed that there were single gene mutants in which certain mechanoreceptors were absent. The mutant nematodes *Caenorhabditis elegans* described by Brenner (1974), which often performed rolling movements, and many of the fruit-fly (*Drosophila*) mutants described by Benzer (1973) may have differed in either sensory or motor abilities. One *Drosophila* mutant called 'hyperkinetic' shows abnormal firing in thoracic motoneurons (Ikeda and Kaplan 1970 *a*). By producing sex mosaics which are part male and part female, thoracic ganglia precursor cells on the surface of the blastula have been shown to be modified in hyperkinetic mutants (Ikeda and Kaplan 1970*b*, Hotta and Benzer 1970, 1972). In the future, such techniques should provide much new information about the neural control of locomotion and other behaviour (Bentley 1976).

Sensory-motor co-ordination

It is apparent from many studies of patterned behaviour that, whilst some central control mechanisms can operate independently of sensory function, most require complex interaction with afferent input from receptors. Considerable constancy of behaviour pattern can still occur if sensory input is required during the execution of the action or if the development of the pattern has necessitated extensive sensory involvement.

Whenever muscles contract as a result of a motor command, there are changes in potential sensory input. As a consequence, problems are posed for sensory analysis since input which is a consequence of an individual's own movements must be distinguished from other input. von Holst and Mittelstaedt (1950) called input due to changes in the external world *exafference* and input produced by the subject's own movement *reafference*. They pointed out that an *efference copy* of the motor command would provide an expected value which would match the reafference if no change in the external world was occurring. The implications for perception are that if the eyes are moved when looking at a stationary train, the train does not appear to move but if the train is moving, the same visual input may occur but movement is perceived. In the first case the visual input is reafference but in the second the input is exafference. Apparent movement is also seen when the eyeball is pushed with the finger for there has been no command to the eye muscles to match the visual input, which is therefore reafference.

Gaze redirection is an example of one of the two principal means of motor control: the *open loop*. The mammalian eye is moved by the eye muscles in rapid, jerky movements which do not depend upon any feedback for their function (Figure 61). Open-loop control systems occur in situations where the movement has to be so fast that there is insufficient time for feedback and correction of motor output, for example, prey catching by arm extension of mantids or cuttlefish, or when the input has disappeared before the movement is completed, such as when a firefly orients to a light flash. Closed loop systems, however, do include feedback from receptors and are much more widespread, e.g. the movement of a mammal's limbs (Figure 61).

An example of sensory-motor co-ordination in a control system which has been studied in detail is prey capture by the mantis (*Parastagmaptera unipunctata*) (Mittelstaedt 1957, 1962). When the mantis detects prey, such as a fly, it turns its head towards it and provided that the fly is within range, rapidly (10–30 ms) extends its fore legs and grasps it. The control system for this operation (Figure 62) requires input from proprioceptive hairs in the neck, as well as from the eyes, and output to the neck muscles as well as to the muscles which extend the fore legs. Experiments by Mittelstaedt showed that the mantis could compensate for a load stuck on to the head but could not strike accurately if the input from the neck hairs was modified or if the head was fixed to the thorax. His proposed control system accounts for the ability to compensate for extra load on the head, such as that which would occur if the mantis had a fly in its mouth (see Figure 60). The output from the proprioceptive hairs (u) is subtracted from the optic output (w) thus producing the neck muscles input (v). The overall output from this proprioceptive loop (x) is modified by head movement until

61 The open-loop control system for eye-position control and closed-loop control system for mammalian limb-position control. After McFarland 1971.

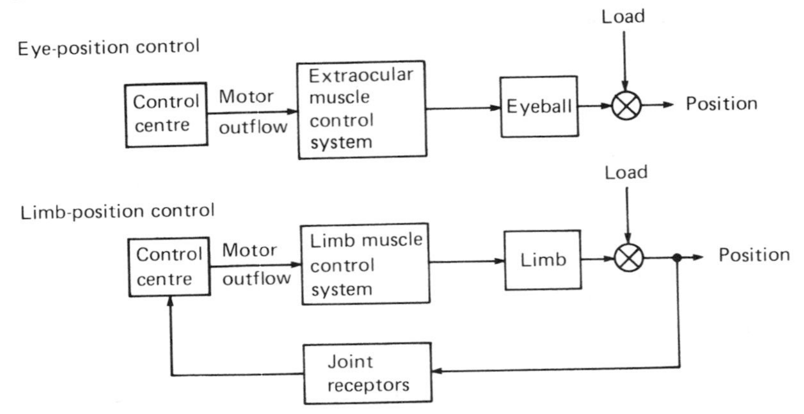

it has a particular relationship to the visual input (y). Once the head has stopped moving, the output from the optic mechanism (w) can be multiplied by a constant factor to give the correct strike angle (s).

The importance of sensory-motor co-ordination in the control of mammalian motor patterns has been mentioned earlier in this chapter and is discussed further by Grossman (1967) and Hinde (1970). One example of work which has produced interesting information on this topic is that of Held. The co-ordination of arm movements and visual input in human subjects was studied in individuals adapting to wearing goggles with prisms in them. Accurate finger placement was poor when distorting prisms were first worn but improved if the subject was allowed to practise moving his arm whilst watching it. During such practice, an efference copy of motor command could be related to visual reafference. If the subject's arm was moved passively, so that there was no motor command to the arm and the visual input was exafference, adaptation to the prisms did not occur (Held and Hein 1958). In another elegant experiment, Held and Hein (1963) reared kittens in darkness except for three hours per day when they were in the arena shown in Figure 63. One of each pair of kittens could walk around and its visual input was a consequence of its own movements. The other kitten had similar visual input but it was moved passively. After about 30 h in the apparatus the active kitten in each pair showed normal visual-motor co-ordination. It blinked when an object approached the eye; it put out its fore paws when it was moved towards a surface; and it avoided the apparent steep drop on the deep side of a visual cliff. At the same stage in the experiment, the passively moved kitten failed to show these types of behaviour, which presum-

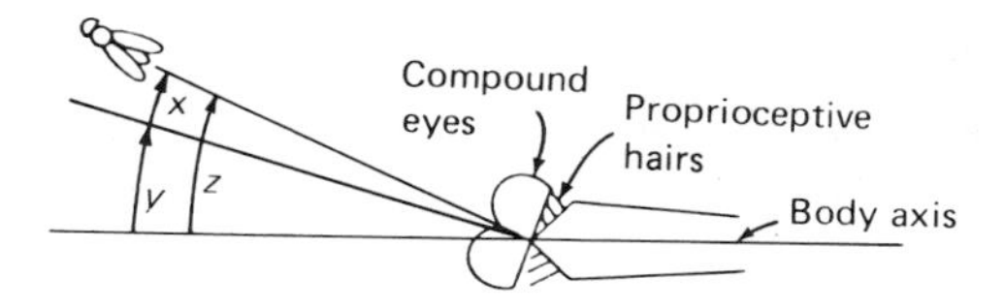

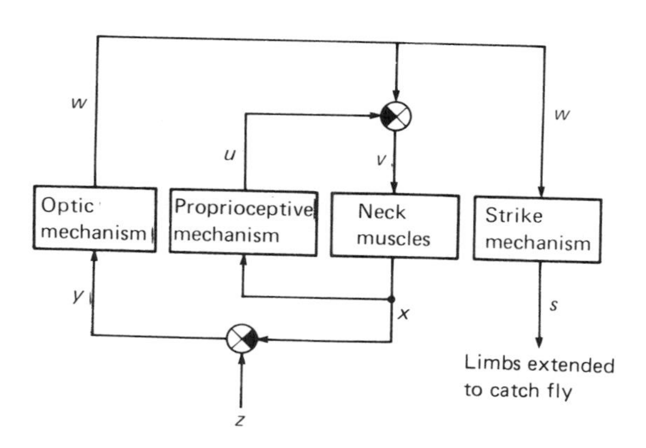

62 Mantid's head and thorax when about to catch a fly and the prey-capture control system. z, angle between prey and mid line of thorax; y, optic input angle; x, angle between head and thorax, w, optic output; v, input to neck muscles; v, output from proprioceptive hairs; s, strike angle. White segments within circles indicate addition, black segments subtraction. Modified after Mittelstaedt 1957.

ably require reafferent input for their development. After a few days living in a normal environment, efficient visual-motor co-ordination was shown by all the kittens.

63 The active kitten moves and the passive kitten is thus moved also so that the two animals receive similar visual inputs. After Held 1965.

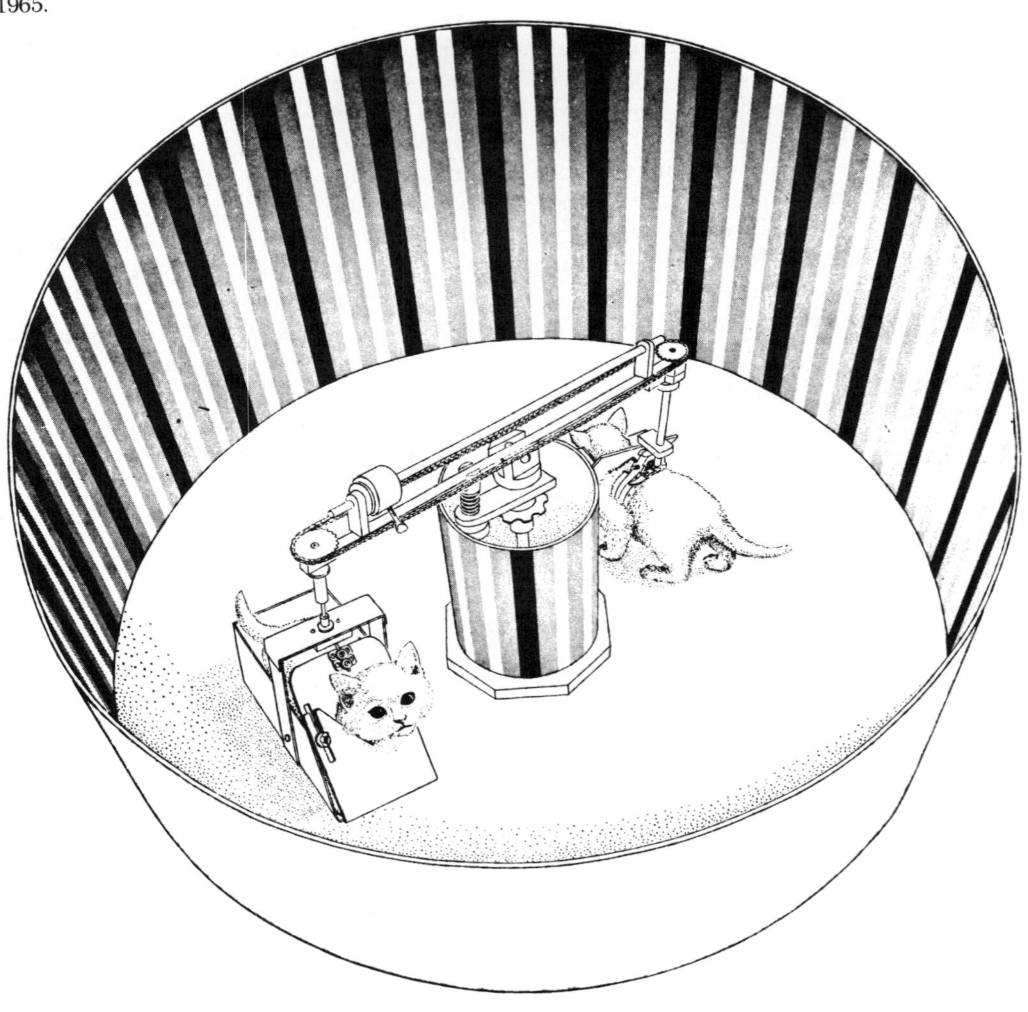

4
The allocation of resources

Each individual possesses motivational mechanisms (see Chapter 1) which result in decisions about the nature and timing of behaviour. At any moment, various functional systems will be operating but the behaviour shown by the individual will be a component of only one or two of those systems. The decisions result in the allocation of resources. One way of measuring resources is in terms of the energetic costs of their accumulation. This method includes time as a resource, for during time, energy is utilised and other resources might have been accumulated.

No behavioural observations can be interpreted usefully without some understanding of motivation and no study of learning and motivation can be complete without some consideration of biological function. As emphasised in Chapter 1, the motivational mechanisms which exist now are the result of the action of natural selection. As a consequence of the competition between alternative genes during evolution, the genes which are present now are those whose expression in the animals best promoted the chances that the gene would survive in the population. In most motivational mechanisms, improved gene survival resulted from increased efficiency of operation of the individual animals bearing the genes. The word efficiency here refers to the ability of the animal to assess biological priorities and act accordingly. McFarland (1977) quotes the example of a herring gull (*Larus argentatus*) which is sitting on its nest and incubating eggs (Figure 64). Whilst doing this it is unable to feed but when its mate returns and takes over the incubation it can leave and find food. If it leaves before the return of the mate the eggs are vulnerable to predation. If it waits and the return of the mate is delayed because of death, desertion, injury or difficulty in finding food the sitting bird becomes more and more food-deprived. At what point should it desert the nest? If it stays too long it may become too weak to find food when in competition with other gulls and hence will not survive to breed another time. Early desertion of the nest, in a situation where the mate would have returned soon, may result in unnecessary loss of the eggs. Genes which increase the chances that individuals bearing them would desert the nest very early or very late would be less likely to survive in the population than those which facilitated an accurate assessment of biological priorities. The fact that the characteristics of the mechanism will depend upon genetic factors does not mean that all individuals will be the same in this respect. There will be individual variation in the decision-making mechanism, since the relevant experience of the gulls will vary.

In order to understand mechanisms for taking optimal decisions it is necessary to consider what alternative courses of action there might be and how biological priorities might be assessed. This can be done only if the general biology of the animal is understood. There is considerable variation during life in the *urgency* with which decisions must be taken. Decisions

64 Herring gull (*Larus argentatus*) incubating its eggs and awaiting the return of its mate.

about maintenance behaviour, feeding and reproductive behaviour after pairing are often not urgent. For example a decision about when to stop resting and start grooming might be taken at any time during a long period. The benefit derived by taking the decision at one moment, rather than another, changes little throughout this period. Urgent decisions and the consequent actions, on the other hand, must occur rapidly in order to avoid the costs associated with injury, death or failure to find a mate. Examples of such decisions include: when to stop feeding and start running away from an approaching leopard; hazard avoidance by an individual faced with a landslide or falling tree; mate acquisition when mates are scarce and rivals are present; food finding when failure to find the food might result in starvation; or even grooming if a mosquito which is carrying yellow fever is about to bite. Some genes must promote speed of decision and action, even at the expense of energetic efficiency, in such situations and hence survive in the population. It is easy to appreciate that a gene which increased response time when a predator approached would not leave many replicates in the next generation. The decision to rest and the decision to flee from immediate danger are two extremes of an urgency continuum and many decisions have an intermediate urgency. Costs and benefits are assessed ultimately in terms of gene survival, but at the less urgent end of the continuum the input and output of energy is a useful estimate of these. In certain situations, which are exemplified above as being of high urgency, costs and benefits must be measured directly. Failure to appreciate that there is an urgency continuum which makes difficult the measurement of costs and benefits, has resulted in some confusion in the literature on motivation.

The next part of this chapter reviews some studies in which biological priorities are assessed. The general question of how to switch from one functional system to another is then discussed. Aspects of this problem and of more minute changes from one activity to another within a system depend on attentional mechanisms and methods of modifying the readiness of motor systems to respond. Input selection is modified according to motivational state and provides an important method of switching from one activity to another so these aspects of attention are considered later in the chapter. The physical difficulties and costs of changing from one activity to another, stereotypies, general ideas about the regulation of motivational state and the control of rhythms are also mentioned.

Assessing biological priorities

Rational decision making

Edwards (1954) wrote about the ideas of economists concerning choice and tried to relate them to studies by psychologists. He referred to the decision maker as 'economic man' who is completely informed, infinitely sensitive and rational. The word *rational* implies that possible combinations of options can be ordered in terms of the probability of their outcome and that choices can be made so as to maximise something, e.g. the benefit or 'utility'. 'Economic man' deals with situations where there is more or less risk and Edwards discusses observations and experiments on what to eat, which items to buy, whether to change jobs and, generally, how much to gamble. As McCleery (1978) has pointed out, this approach is relevant to all species, especially because of the rationality of most decisions. He considers situations where an animal has to choose between a small amount of food in a safe place and a larger amount whose acquisition involves some risk of predation. When these situations are plotted on a graph of risk against available food, contours of equal utility, i.e. an *indifference function*, can be plotted (Figure 65). The animal should be just as likely to go for the food at all points along this line. A food/risk situation like *A* is outside this line so the animal should choose that food in preference to any situation on the line. Above and to the right of the utility contour, the greater the perpendicular distance from it, the more the utility. Feeding should be more likely, therefore, at situation *B* than at *A*. The shape of indifference curves like that in Figure 65 will vary slightly during the life of an animal if the benefit or utility is measured in terms of gene survival. This benefit is, approximately, chances of reproduction minus chances of dying. The relationship between food and risk of predation will be different in breeding and non-breeding situations. A non-breeder may expose itself to higher risks. An example of a rational approach by a parent might be that it is worthwhile to make ten visits to a very good food source where there is a 1 in 50 chance of being caught by a predator, rather than to make 100 visits to a source where the risk is 1 in 1000, if three young can be reared before the death of the parent in the first case but only one young in the second case. In studies of motivation where the factors which modify the shapes of curves of equal utility do not vary much, it is useful to try to discover such indifference functions. Most attempts to do so demonstrate that animals do behave rationally, for example Logan (1965) was

65 Hypothetical indifference curve for an animal choosing food patches of different sizes in places where there is a greater or lesser risk of predation. (Utility) is higher at *B* than at *A*. Modified after McCleery 1978.

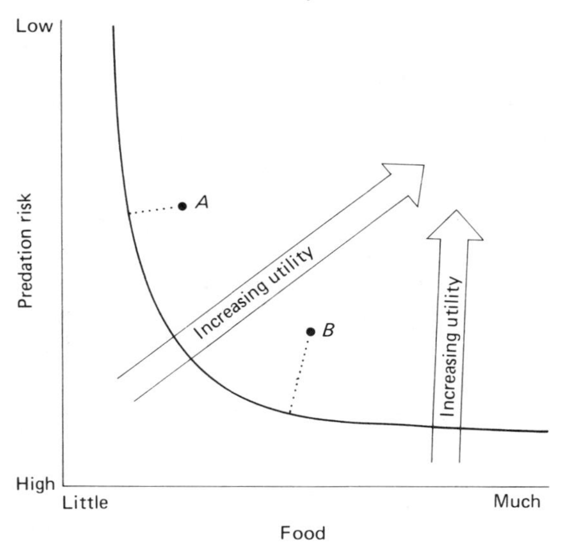

able to plot indifference functions for rats provided with different quantities of food after various delays in the arms of a maze.

Optimality

The biological priorities, which we attempt to assess when trying to understand behaviour and the motivational systems underlying it, are the selection pressures which have acted during evolution. Each gene or gene-complex has survived in the population because it out-competed alternative genes or gene-complexes. Hence we assume that, within the constraints of what is phenotypically possible, the mechanisms controlling sequences of behaviour are efficient. Sometimes that efficiency may come close to an absolute measure of *optimality* but for most mechanisms the criterion for optimality must be related to evolutionary fitness (Sibly and McFarland 1976). If there are several alternative gene-complexes affecting a behaviour sequence then the optimal gene-complex is, by definition, the one which is supplanting, gradually or rapidly, all the alternatives. There will inevitably be much variation in the environmental factors affecting this behaviour sequence, hence individuals which are identical in respect of the relevant genetic characteristic will differ in behaviour. It is still of interest to consider the optimality of the behaviour sequences shown by individuals.

Another exercise is to use optimal control theory to predict the characteristics of the best possible model, i.e. theoretical animal, for this behaviour sequence and then to compare real animals with this model (McCleery 1978). Such theoretical models may be informative but it would still be necessary to estimate real costs for real animals. In practice it is very difficult to do this, although there have been some useful attempts, for some costs are difficult to measure and it is not possible to know that all have been considered. As McFarland (1977) has pointed out it is necessary to investigate interactions between motivational systems in order to understand decision-making in behavioural sequences. He and his collaborators have developed analytical methods in order to calculate optimal paths in situations where there are alternatives.

Costs and optimal paths

In order to try to understand whether or not behaviour is optimal it is necessary to consider the costs of the behaviour to the individual. The term *cost* refers to reduction in fitness and a *cost function* is the relationship between risks and causal variables. For a particular behaviour sequence, a mathematical statement of the cost function can be obtained by direct measurement. In order to do this it is necessary to know about the consequences of each activity in terms of depletion of food reserves, predation and other hazards, etc. Some aspects of costs have to be estimated. The procedure described by Sibly and McFarland (1976) involves integrating the cost (C) of being in state (x) and of selecting behaviour (u) over the time between the beginning (O) and the end (T) of the sequence. A behaviour sequence is optimal if it maximises fitness, i.e. if it minimises:

$$\int_O^T C\ (x,\ u)\mathrm{d}t.$$

One way of trying to find the optimal behaviour sequence which results in a change from particular starting to finishing states (see later for further explanation of 'state') is by continually modifying behaviour so as to maximise a quantity H which is referred to in mathematics as a *Hamiltonian function*. This method was utilised by Sibly and McFarland (1976) and is further explained by McCleery (1978). Essentially, it involves balancing estimates of the cost of being in a certain state against the cost of performing the behaviour required to change the state. If the state at a moment is far from the optimum and hence the cost is high, then it is worthwhile to use high-cost behaviour to change to a state where the cost is lower. As the state approaches the optimum where cost is at its lowest, however, it is less worthwhile to use high-cost behaviour. If the animal could move infinitely fast and could perform all possible activities at once it would be able to reduce all its costs exponentially and simultaneously. In practice there is evidence for exponential changes but the animal has to choose sequences in which different costs are ameliorated at different times. Examples of such studies are discussed below. From these it is apparent that an approach to motivational analysis which incorporates ideas about biological priorities is useful as one possible basis for the explanation of behaviour sequences.

Causal factor space

The occurrence of each behaviour of an individual depends upon a set of causal factors. *Causal factors* are internal variables but some are altered rapidly by external events. For example, the sensing of blood-glucose level or the sight of a potential food source would alter causal factor levels. In contrast with these, a causal factor which depended upon the concentration of a steroid hormone in the blood would change slowly. Some of these factors can be estimated accurately by using physiological measurements whereas others will probably never be measured directly. The relationship between causal factors and behaviour is sometimes investigated by attempting to find out all of the causal factors which affect a particular behaviour and sometimes by describing all the behaviours which are affected by a single causal factor (Hinde 1970). Since all changes in behaviour are the manifestation of the animal's response to changes in causal factors, an important way to study the motivational basis of behaviour sequences is to investigate situations where more than one experimentally modifiable set of causal factors is acting. This is the approach developed by McFarland, who worked initially on interactions between effects of food deprivation and water deprivation (e.g. McFarland 1965*a*), following earlier work by Verplanck and Hayes (1953) and Bolles (1961).

If a dove is deprived of food there will be

inputs, from the various monitors in the gut, blood and other tissues, to the control areas in the brain. These inputs must interact with system parameters so that the levels of one or more causal factors are modified. The way in which the causal-factor level is affected by the input from a physiological monitor will depend upon the previous experience of the animal (Sibly and McFarland 1974). If two causal factors are considered, plotting them against one another shows a two-dimensional causal-factor space. Figure 66 is such a plot for a dove deprived of food and water. If provided with equal opportunities to eat or drink, the dove at point *A* is most likely to drink and the dove at *B* most likely to eat. In experiments in which doves were deprived, firstly of water alone and then,

66 Positions of the motivational state of animals *A*,*B*,*C*, and *D* in two-dimensional causal-factor space. Animal *A* is most likely to drink whereas animal B is most likely to eat. Animals *C* and *D* were initially deprived of water with the result that their position in causal-factor space moved away from the origin *O* upwards and also to the right, because water-deprived animals eat less, to points *C*1 and *D*1. From then on, the animals were also deprived of food until they reached points *C*2 and *D*2. They were then equally likely to choose food or water if offered both. Modified after McFarland 1971.

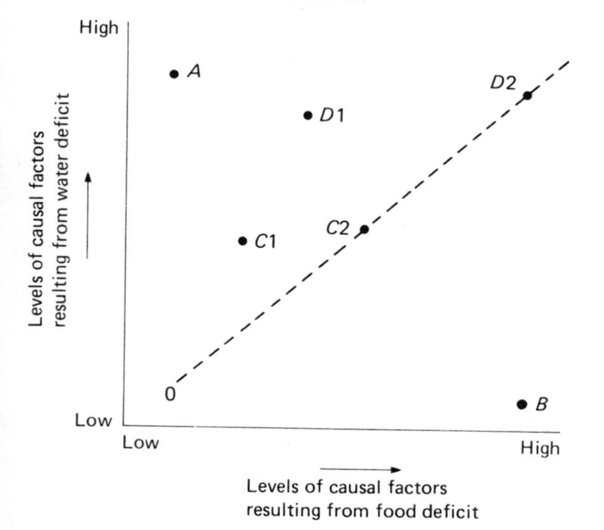

at a specified time later, of food as well, McFarland (1971) was able to find a series of points in the causal factor space at which food and water preferences were equivalent (Figure 66). There is a complex interaction between eating and drinking, for water is used when feeding and digesting. This is at least part of the reason why doves deprived of food drink less water (McFarland 1965*a*) and doves deprived of water reduce food intake (McFarland and Wright 1969).

Most experiments on the interactions of motivational variables involve the experimental manipulation of only two sets of causal factors. These are often associated with different functional systems. It is important to emphasise, however, that there will be many causal factors affecting decisions in a behavioural sequence. Causal-factor space is, in reality, multidimensional. At any instant the levels of any of a large array of causal factors might affect behaviour. The dove which has the option of eating or drinking might preen, or sleep, or flee from a predator (see the discussion of the dog's behaviour, in Chapter 1). If three causal factors are considered, it is possible to draw a three-dimensional diagram of causal-factor space but if four or many causal factors are considered it becomes difficult to conceptualise the space and the movement of the motivational state of the individual within that space. Throughout this book the term *motivational state* is used to mean the position of the individual, at that moment, in multidimensional causal-factor space. The state is thus defined in terms of the levels of each possible causal factor.

When the levels of all causal factors have been assessed, one of several candidates for the behavioural final common path (McFarland and Sibly 1975) will be selected. This is the activity which we as observers have to try to measure and interpret as an indicator of the motivational state. By experimental manipulation, the course of the motivational state of an individual through two-dimensional causal-factor space can be followed. Consider a dove which has been trained to peck at different keys for food and for water. It is deprived of both food and water and then provided with both. Such a dove often spends some time pecking the key associated with food, then switches to the water key and then switches back and forth before being satiated with both. The water intake and food intake can be taken as indicators of the food and water deficit. The intakes can thus be plotted against one another and the course through this causal-factor space depicted (Figure 67). The position of the boundary between motivational states where drinking is most likely and those where eating is most likely can be estimated by means of a 'time-out' test after each bout of pecking a particular key (see Chapter 3 for discussion of the term *bout*). In experiments by McFarland and Lloyd (1973), this 'time-out' was merely a period of about 10s during which no food or water could be obtained. It served to minimise the effects of other activities on the feeding–drinking sequence. The line which could then be drawn between feed–drink transitions and drink–feed transitions, as shown in Figure 67, was called a *dominance boundary* by McFarland and Lloyd. The technique is discussed further by Sibly and McCleery (1976) who found that for doves fed in similar ways prior to the experimental period and tested in the same way, the slope of the dominance boundary did not vary much. There was a tendency for the boundary to be steeper when a bird started with a long dominant feeding bout and shallower when it started with a long drinking bout but the slope was not altered by supplying some water through an oesophageal fistula during the test (McFarland 1974*b*, Sibly and McCleery 1976).

The slope of the dominance boundary can be altered by changing the availability of food. When Sibly (1975) increased the rate of food intake the dominance boundary became steeper (Figure 68). Another study in which the slope of the dominance boundary was altered was that of Larkin and McFarland (1978) in which the cost of changing from one activity to another was varied. The doves used in this work were fed with wheat grains buried under sand on one side of a room and were provided with water in metal-foil containers on the other side of the room. The 'time-out' was one minute with the lights off in the experimental room. When the doves had to fly up to a higher level or over a partition in order to change from feeding to drinking or

67 Changes in motivational state plotted in two-dimensional causal-factor space. Causal-factor levels approximate to measured intake of water and food after deprivation. The results of a series of choice tests (Feed or Drink) indicate whether feeding or drinking was dominant at that point in causal-factor space. Modified after Sibly 1975.

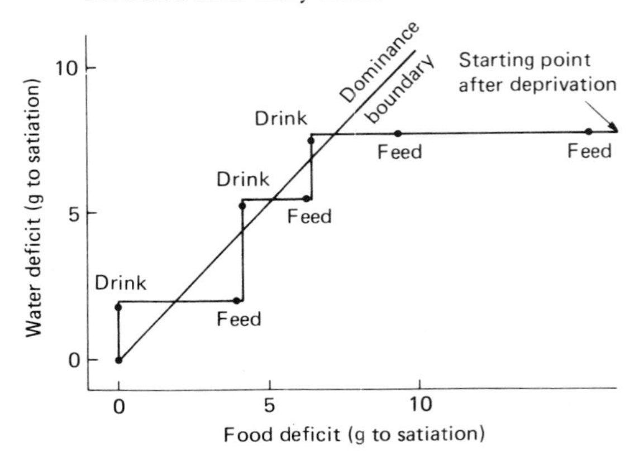

68 This experiment is similar to that illustrated in Figure 67 but the food-reward rate is increased at the change-point shown, with the result that the dominance boundary is steeper. Modified after Sibly 1975.

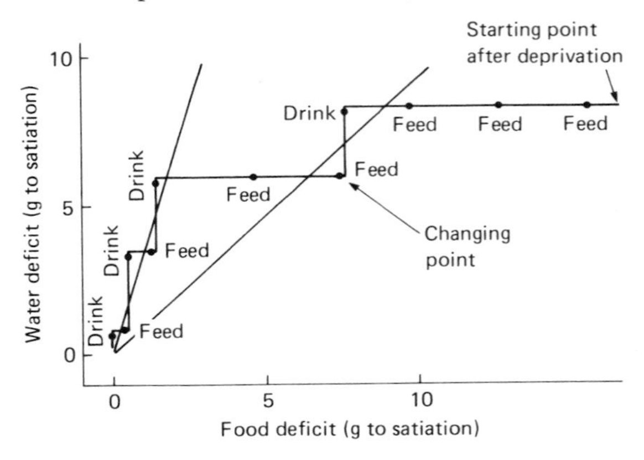

vice versa, they prolonged the ongoing activity before changing. As a consequence of this, Larkin and McFarland used additional lines on the causal-factor space plot, namely the line outside which feeding only occurred, and the line outside which drinking only occurred. When the food was raised above floor level, so that the bird had to fly up in order to change from drinking to feeding, the angle subtended by the feed-only boundary was substantially reduced and that subtended by the drink-only boundary was slightly reduced. If the water was raised above floor level, the angle of the feed-only boundary was decreased a little and that of the drink-only boundary was increased a little. If the bird had to fly over a partition in both directions there was an increase in both angles, especially that subtended by the drink-only boundary. The results demonstrate that the cost of changing from one activity to another is allocated to the activity to which the dove is changing.

Motivation terminology

It may be useful at this point to relate current ideas about motivational mechanisms to terms which have been widely used in the motivation literature such as drive, reinforcer, activation, arousal and attention. Early attempts to account for the timing and nature of switches from one activity to another often invoked instincts as hypothetical controlling factors. As late as 1923, McDougall wrote about the 'feeding-instinct' operating through particular 'motor habits'. The term instinct implied to many people that the mechanism developed without the influence of environmental factors so it was soon dropped from the literature on motivation and was replaced, in the writings of many authors, by *drive*. This term which was introduced into common usage by Woodworth (1918), soon came to be used to describe those internal variables which affect the likelihood of activities, such as eating, and which are themselves affected by short-term experience such as food deprivation. The most frequent usage of drive was by those who conducted experimental studies of learning, usually in situations where feeding occurred during the experiment, but there was no uniformity in the meaning ascribed to it. Skinner (1953) stated that a drive is not a physiological or psychic state but merely a set of operations which have an effect on the rate at which a particular activity occurs. Hence withholding food from an experimental subject for a certain number of hours was called a drive. Most authors, however, did refer to the internal state itself, rather than to the factors which affected it. Tolman (1951) talked about *needs* which had to be fulfilled by the effects of behaviour and Hull (1952) postulated that the likelihood of showing a particular behaviour depended upon an interaction between drive strength and the *habit strength* of that behaviour. Hilgard (1958) provided definitions of need and drive: 'need as the physiological state of deprivation or of tissue injury, and drive as the psychological consequence of this state'.

Many ideas about drive were based on observations made during a narrow range of experimental procedures and depended upon what events acted as *reinforcers* for certain behaviours. A frequent experiment involved a rat which was placed in a situation where it could perform a behaviour which could be quantified easily, such as pressing a lever. If the rate of lever-pressing by a food-deprived rat increased when the presentation of food occurred immediately after it had pressed the lever but at no other time, food was said to be a reinforcer for that behaviour. The concept of drive depended jointly upon the previous deprivation treatment and on the reinforcing properties of food. The fact that an event did act as a reinforcer provided some evidence about the nature of the drive. It was assumed by most theorists that the drive had quantitative characteristics, the quantity being greater after longer periods of deprivation and being reduced by the reinforcer.

Ideas about drives

One important present-day idea which has a long history in behavioural research is that motivational systems function in a way which maintains an animal's internal stability. Drives have often been thought of as components of homeostatic systems (see Chapter 1) in which the result of the behaviour which they facilitate is to reduce deviations from the optimum internal state. It was assumed that the drive level was reduced by the effects of the behaviour (e.g. Miller 1959). Lorenz (1937) and Tinbergen (1951), considering the function of essential biological mechanisms, referred to 'action specific energy' rather than drive and suggested that this was reduced consequent upon the performance of particular activities. The extent to which Lorenz and Tinbergen viewed drive in quantitative terms is revealed by their analogies with reservoirs of water which discharged water and electrical capacitors which discharged current via one, or a range of, motor activities. Tinbergen and his group often thought of drives as being closely related to specific activities, e.g. Bastock, Morris and Moynihan (1953) referred to 'behaviour belonging to drives' and 'the executive motor patterns of one of the . . . drives'. If drives are considered as being very specific to certain behaviours then it is necessary to postulate the existence of very many of them. As soon as the drive concept is applied to sequences of behaviour in the everyday life of an individual, instead of a very small range of experimental situations, its inadequacies become apparent.

Some of the different ways in which the term drive has been used were listed by Hinde (1959) and explained further by Hinde (1970). He pointed out some of the problems associated with its use in a very general or a specific sense and has been responsible for a widespread reduction in its usage. He considered (Hinde 1970) that 'drive concepts can be useful only if defined independently of the variations in behaviour they are supposed to explain' and he suggested that it is conceptionally uneconomical to discuss thirst drive if the only measurable variables are hours of water deprivation and amount of water drunk. Tolman's (1932) concept of the thirst drive as an *intervening variable* would, however, be of greater use if it intervened between several factors which affected thirst, i.e. variables which are independent of drinking behaviour and several measurable behaviours associated with water intake, i.e. dependent variables. Hinde uses Miller's (1959) studies on drinking by rats to illustrate the idea of drive as an intervening variable. The amount of water drunk by rats can be increased if they have been (1) deprived of water, (2) fed on dry food or (3) injected with hypertonic saline. Water deprivation has effects on (*A*) the amount of water the rat drinks in a given time, (*B*) the rate at which the rat will press a bar if water presentation is contingent upon bar pressing, and (*C*) the concentration of quinine which it will tolerate in its drinking water. The three independent variables (1), (2) and (3) can be said to affect the intervening variable, thirst, which in turn affects the dependent variables (*A*), (*B*) and (*C*) (Figure 69). Hinde goes on to comment on the extent to which the three methods of assessing thirst (*A*), (*B*) and (*C*) are correlated for different durations of water deprivation. He concludes that lack of agreement makes it unlikely that a single intervening variable is involved. This may be true (see Fitzsimons 1968) but as Houston and McFarland (1976) point out, an ordinal scale was used for the measures (*A*), (*B*) and (*C*) and the agreement is much better if the results are expressed on an interval scale, i.e. ranks rather than numbers.

As explained earlier in the chapter, motivational state is used here to mean position in multidimensional causal-factor space. This definition acknowledges that the probabilities of occurrence of activities are affected by a wide variety of causal factors. Hinde (1959) criticised the idea of unitary drives and his paper was the beginning, for many people studying behaviour, of a considerable broadening of ideas about the number of variables which had to be considered in order to have any chance of understanding motivation. Although we now appreciate that the problems are much greater than had previously been thought, we still do not know much more about the steps between the array of causal factors and the behavioural final common path (McFarland and Sibly 1975). A problem which now exists in motivation research is how many of the possible causal factors should be considered in any investigation? How many dimensions of multidimensional state space are needed in order to describe motivational state? McFarland and Sibly (1972) tried to find out when the problem could usefully be simplified by coupling different causal factors. If feeding is dependent upon causal factors related to levels of protein, carbohydrate and fat, under certain circumstances two of these factors might vary together with sufficient frequency for them to be treated as one. An animal which has adapted to its surroundings, including sources of food, must have reduced the number of dimensions. As an animal acclimatises to an environment it maintains stability by reaching a 'space of minimum dimensionality'.

General drive, activation and arousal

Current views on motivation have some antecedents in the multiple drive ideas of Guthrie (1952), Estes (1958) and Miller (1959), as well as those of Lorenz (1937) Tinbergen (1951) and Hinde (1959, 1970) which are more biologically oriented. The alternative view, that changes which affect an individual all modify a general drive state which determines behaviour, was proposed by Hull (1943) and supported by Spence (1951) and Brown (1961). Such a general drive seems logically unlikely, for every causal factor might then affect every behaviour, and the idea has been criticised by many authors (e.g. Bolles 1958, 1975, Hinde 1970). Bolles (1975) concludes that the general drive factor is of minor importance for 'behaviour is predominantly determined by the specific drive conditions, specific stimulus situations, and specific habit structures that characterise an individual at any given time'. Nevertheless the idea has persisted because it has been equated with the concepts of activation and general arousal (Hebb 1955).

One of the preludes to Duffy's (1962) concept of *level of activation* was her previous conclusion (Duffy 1941) that different amounts of energy were likely to be utilised in the brain and body according to the 'emotional' state of the individual. Other important findings were

69 The relationships between each of three independent and three dependent variables (see text) can be simplified if an intervening variable is considered. This diagram refers to experiments on the maintenance of water balance in a rat. Modified after Miller 1959.

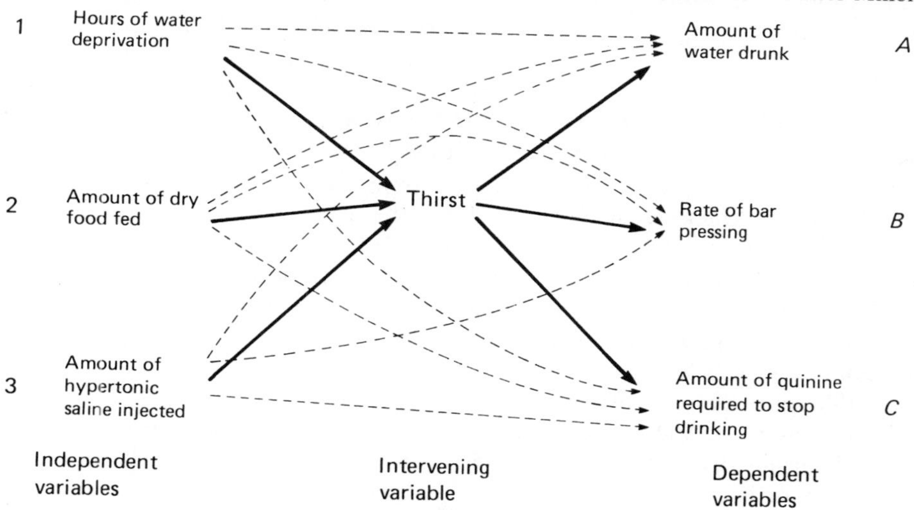

those of Moruzzi and Magoun (1949) and Lindsley (1951) that variations in levels of electrical activity in the reticular formation and the cortex reflected some changes in general measures of degree of behavioural activity and responsiveness. As a result of such work, the terms *cortical arousal* and *behavioural arousal* came to be widely used. Duffy (1962) and Bindra (1959) combined them in describing various levels of activation and arousal. When fluctuations in the various possible measures of arousal or activation were compared, however, they often did not correspond, sometimes going in opposite directions from that predicted (Lacey, Kagan, Lacey and Moss 1963, Malmo 1959). Hence it is best to regard arousal as a name for a collection of variables rather than as a process in the brain (Berlyne 1967, Hinde 1970). Delius (1970) suggested that arousal level could be equated with the rate at which decisions are being taken within the central nervous system, thus even the reticular activity seems likely to be the result of simultaneous activity in a variety of pathways in the brain. The work on arousal has drawn attention to the existence of causal factors, such as those consequent upon general light intensity or temperature level, which affect a wide range of behaviours. Behaviours may fail to occur, irrespective of the fluctuations of other causal factors, unless the levels of these factors are within certain limits.

Reinforcers

The discovery that an event can act as a *reinforcer* for some behaviour provides evidence about the motivational state of the individual and about motivation mechanisms in general. After many years of experiment in which only food, water or electric shock were used as reinforcers, there has been a period when a large variety of potential reinforcers has been described. These reinforcers include:

1. those which fulfill some obvious physiological need, such as food, water, heat, cold, oxygen and various drugs, in appropriately deprived animals;
2. the opportunity to engage in various species-specific behaviours such as pecking, gnawing, grooming, sand-digging, nest-building, offspring-retrieving, courting, copulating, fighting and running;
3. the opportunity to explore;
4. biologically relevant stimuli such as the song of a conspecific bird, sweet tastes or a parent or sibling-like object;
5. electrical stimulation of the brain; or
6. any sort of stimulus change.

Such studies have been reviewed by Bolles (1975) and details of some of them are summarised by Hogan and Roper (1978). As a result of finding these reinforcers, we know that there must be causal factors whose levels can be altered by the effects of each of the reinforcing events. The existence of such causal factors could have been predicted from observation of the behaviour of undisturbed animals but the experimental results do provide clear evidence for the importance of a variety of causal factors in the control of behaviour. Ideas about possible mechanisms by which reinforcers might act were greatly modified by the discovery by Olds and Milner (1954) that rats with electrodes implanted in various parts of their brains would press a lever which resulted in the passage of a small current through the brain. This self-stimulation would sometimes continue for long periods, the behaviour being preferred to any of a variety of alternatives. The frequency of occurrence of self-stimulation can differ according to the levels of many other causal factors; Hogan and Roper (1978) give a review. The technique, therefore, has provided useful information about the control of motivational state. The ubiquity of sites for self-stimulation, however, has precluded the formulation of precise ideas about the localisation of motivational mechanisms in the brain.

Attention

The term *attention* is used in two rather different ways in the scientific literature, as well as in other ways colloquially. In one sense of the word, attention is equated with alertness or vigilance and concerns the 'state which affects general receptivity to input information' (Posner 1975). To say that an individual is in a state of high alertness means that the likelihood of response to any sort of detectable environmental change is high. This concept is thus very close to that of arousal or activation, which is discussed above. It is best to confine the use of the word *attention* to the mechanisms which limit and discriminate amongst channels of input of sensory information. This definition includes *selective attention*, a concept based partly on experiments in which human subjects receive more than one simultaneous input (e.g. Broadbent 1958, 1971, Treisman 1964). Another example of such studies is the assessment of a response which necessitates the utilisation of one type of perceptual input and of the distractibility of the individual when an alternative input is provided (Andrew 1976). Both types of experiment involve attempts to assess the functioning of input control mechanisms by measuring responsiveness. The mechanisms of selective attention have also been studied by recording from sensory pathways in the brain and monitoring the electrical changes, i.e. evoked potentials, which occurred when appropriate physical changes were presented to receptors. Early studies (Hernández-Peón, Scherrer and Jouvet 1956) seemed to show that output from the reticular formation might modulate input to cortical analysers and that this could be the mechanism by which input was controlled in attention. The experimental procedure in this work has been shown, subsequently, to have been inadequately controlled (Horn 1965, summary by Milner 1970). There is, however, some work on human subjects which indicates that an instruction to attend to input to one ear rather than the other can result in differences in evoked potentials (Hillyard, Hink, Schwent and Picton 1973). The work on the functioning of efferent neurons in receptors (Chapter 2) also shows that central control of sensory input is possible. Most work on selective attention, however, indicates that there are important attentional mechanisms which act after most sensory analysis has been carried out.

The control of attention is an important tool of motivational mechanisms. Many changes in behaviour must be preceded by changes in attention. In order that an individual can change from walking to a prolonged bout of grooming, it is necessary that a suitable resting site be found. The attentional changes which occur preparatory to such a behavioural change will result in the alteration of some causal factors associated with the two activities. This will, in itself, facilitate the change but all decisions about activity changes must take into account a variety of causal factors. Modifications of attention are initiated as a result of changes in motivational state which may or may not be readily attributable to sensory input. When there is a detectable change in the environment of the animal, the first response is often an *orientation reaction* which is not specific to the input (Pavlov 1924, Sokolov 1960). Orientation reactions involve general preparations for response such as alterations of heart rate, ventilation rate, muscular tonus, etc. as well as orientation to the source of the environmental change. Such orientation reactions are rapidly followed by appropriate modification of attention.

The assessment of motivational state

We cannot determine motivational state directly but estimates can be obtained by observing current behaviour, observing sequences of behaviour in surroundings which change little, assessing responsiveness to detectable environmental changes including food presentation etc., monitoring physiological variables such as blood-sugar or hormone levels, recording from those parts of the brain likely to be concerned with commands and motor output, and manipulating motivational state by deprivation or brain stimulation. A better assessment is obtained by the use of several methods. Rapid changes in motivational state are difficult to monitor by any means other than behaviour measurement or brain recording but, as McFarland and Sibly (1972) have pointed out, such measurements may be inadequate or misleading. Various commands, resulting from quite different motivational states, might lead to the same initial consequences for behaviour. The observation that an individual stands up and starts walking, or recording the corresponding electrical events in the appropriate motor area in the brain, does not provide precise information about motivational state. It is also possible that different brain mechanisms and behaviours might be alternative means of achieving the same end and might reflect the same motivational state. There are some activities which are likely to be more precise indicators of motivational state than others. For example, it is easier to estimate the levels of particular causal factors if the animal is feeding or copulating than if it is walking or standing alert. Even the observation that the individual is feeding, however, does not provide information about the majority of causal factors which affect behaviour. It merely shows that none of the collections of causal factors which promote other activities is at a sufficiently high level for another activity to take over from feeding.

Information about current behaviour as an estimator of motivational state can be obtained by using other estimation methods at the same time. What happens when different behaviours are interrupted? Fentress (1968) recorded the behaviour of two species of voles, *Microtus agrestis* and *Clethrionomys britannicus*, when an object was moved overhead. If the vole had been walking at the time of interruption its response was to flee, but if it had been grooming, it froze. In similar experiments using domestic chicks Forrester and Broom (1980) described the different responses shown when ten ongoing behaviours were interrupted in the same way and Culshaw and Broom (1980) assessed the effects of interruptions at different points in bouts of behaviour.

Differences in response could be attributed to differences in motivational state at the time that the interruption occurred. The interruption of behavioural sequences has also been used as an analytical technique by McFarland (1974), Cohen and McFarland (1979) and others. In experiments with sticklebacks, the effects of trapping a male en route from courtship to nest visiting, or vice versa, were studied (Figure 70). If the male was prevented from proceeding to the other end of the tank during a changeover in behaviour, would it continue after a one-minute interruption or would it return to the previous activity? A similar question was asked by Larkin and McFarland (1978) who used periods of darkness in situations where doves were changing to or from feeding and drinking. The results of such experiments provide evidence about mechanisms for changeover between behaviours (see later).

The widely used techniques of investigating motivational state by measuring responses to particular detectable environmental changes and by depriving the individual of necessities are discussed elsewhere in this book and by Marler and Hamilton (1966), Milner (1970) and Thompson (1975). Brain-recording studies have increased enormously our knowledge of sensory analysis (Chapter 2) and have also provided

70 Water tank in which a male stickleback can be trapped temporarily in the middle section whilst changing from courtship to nest visiting or vice versa. Modified after McFarland 1974.

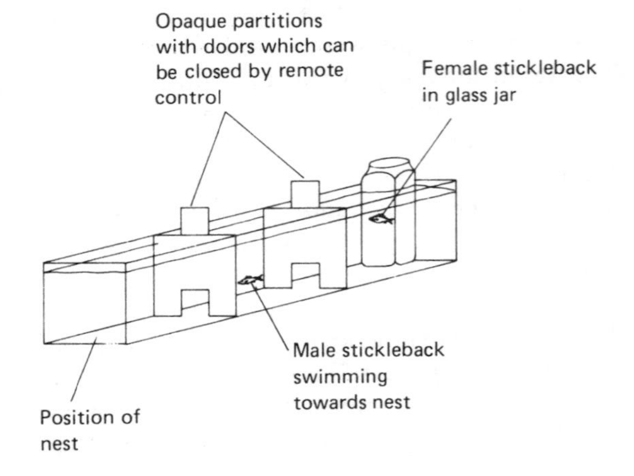

much information about motor control (Chapter 3). They have not, as yet, proved particularly valuable in the estimation of motivational state but it is likely that there will soon be considerable progress in this research area. The fact that an area in the brain is electrically active when an activity occurs, or when it is inhibited, provides only a small amount of information about the decision-making mechanism. Examples of such studies include the discovery by Oomura *et al.* (1964), with cats, that a variety of treatments which inhibit eating, e.g. glucose-feeding or stomach-distension, result in increased firing by many cells in the ventro-medial hypothalamus (for brain anatomy diagram, see figure 3). This part of the brain had already been implicated in the control of eating by the use of brain ablation and stimulation studies. In the same way, the lateral hypothalamus of female cats seemed to include pathways involved in sexual behaviour; electrical activity in this region was increased when the cat's vagina was probed (Porter, Cavanaugh, Critchlow and Sawyer 1957, Barraclough and Cross 1963). These results were the more significant because the electrical response was confined to cats in oestrus, or to those which had been treated recently with progesterone.

Much more information about motivational mechanisms has been obtained by means of experiments in which the brain is manipulated by electrical stimulation, chemical treatment or ablation. One way of approaching the decision-making mechanisms is by following motor pathways back from the hind brain. Phillips and Youngren (1976) stimulated chicks at various points in the mid brain and hind brain which they had previously discovered to be part of the pathway for control of calling. They recorded the patterned output from the motor nerves and also from sites in the medulla posterior to the pons. Anterior to this there was no patterned output. In the somewhat different system of the canary, pathways from the mid brain and from the archistriatum have been explored (Nottebohm, Stokes and Leonard 1976). Landmarks in the utilisation of brain stimulation to find out about motivation have included Andersson and Wyrwicka's (1957) stimulation of goat's hypothalamus, leading to drinking, von Holst and von St Paul's (1960) report of interaction between brain stimulation at different sites and between stimulation and sensory input in determining fowl behaviour, and the work of Roberts (1958) who elicted 'alarm' and 'escape' responses from cats stimulated in the lateral hypothalamus. Other examples of brain stimulation etc. are described in Chapters 5, 6, 8 and 9.

Describing behaviour sequences

Before considering further how individuals change from one behaviour to another it will be useful to consider how sequences of behaviour can be described. The combination of muscle contractions into motor patterns, of motor patterns into components (e.g. the overhand grooming movement of a mouse), of components into units and of units into bouts of a certain type of behaviour, is described in Chapter 3. Even during the organisation of motor patterns into components, decisions about alternative movements have to be made. This is emphasised by the work of R. and M. Dawkins (1976) on drinking by chicks and that of Stamps and Barlow (1973) on lizard display. Fluctuations in causal factors must be considered when studying most aspects of patterning, both within one category of behaviour and in sequences of different categories.

Feeding by finches involves several types of pecking behaviour, according to the food item which is picked up, but most human observers can easily limit the use of the term feeding to pecks aimed at potential food items. The pecks are obviously organised in bouts for there are intervals with pecking and intervals without pecking. How should a bout of feeding be described? This question has been treated in detail by Slater (1974*a*, 1975) (Chapter 3). The timing of events within a bout of pecking by domestic chicks has been precisely described and models have been generated which behave in approximately the same way as does a chick. Such studies do not yet tell us, however, how patterning of components, such as pecks, is organised (Dawkins 1969*a*,*b*, Machlis 1977). The temporal organisation of behaviour sequences has often been described at the level of changes from a bout of one behaviour to a bout of another. An early detailed study was that of Nelson (1964) on fish courtship. He described the frequency of *transitions* between different behaviours and also the distribution of intervals between events. If a particular transition between behaviour *A* and behaviour *B*, in a single animal, is especially frequent we assume that *A* and *B* depend upon the levels of the same causal factors, or that the occurrence of *A* affects the causal factors upon which *B* depends. Heiligenberg (1974) also studied fish display and measured transition frequencies. He was able to make suggestions about the control of display and the interrelations between the different components of display and other behaviours. Some of the difficulties of interpreting transition analysis are pointed out by Slater and Ollason (1971).

An early consequence of the analysis of transitions between behaviours was that experimenters noticed that some of the variation in the data could be explained if it was assumed that *Markov processes* (see Chapter 3) were operating. For example, fly-grooming components could be predicted quite well by knowing the preceding component (Cane 1978). Altmann (1965) found that much of the social behaviour of monkeys depends upon the immediately preceding event, so the sequence is a Markov chain. Other examples of the applications of Markovian analysis are explained by Gottman and Notarius (1978) who emphasise that the method can be used for events two or three before the present event and can also be combined with other analytical methods. Although the Markov chain is used to describe behaviour it is frequently found to be bad at describing longer sequences. Staddon (1972) recorded pecks, wing flaps, position, etc. of pigeons in a conditioning box and found that a Markov chain did not explain the transition frequencies. Metz (1974) considers that Markov-chain analysis is useful for short sequences but not for overall patterns of display or other behaviour. He also emphasises the importance of renewal processes in which an event depends only upon the previous occurrences of that event.

Another important method of describing sequences of behaviour is in terms of periodic occurrences of events (see Chapter 1). Short-term cycles of activity are very widespread in behaviour, the period ranging from a few seconds to a few hours, and *time series* analysis is readily used to describe them (Delius 1969, Slater 1975, Lehmann 1976, Broom 1980). This topic is considered later in this chapter.

All these methods of describing behaviour sequences have implications for ideas about motivation. If the occurrence of a behaviour seems to depend upon the previous activity, or the last occurrence of that same activity, or a clock which initiates the behaviour at regular intervals, then we must look for causal factors and interactive mechanisms which might account for such dependence.

Mechanisms for switching from one behaviour to another

What is happening to causal factors when a behaviour starts, then stops and is followed by another? Behaviours which form part of different functional systems (Chapter 1) will be considered first. For example a person arrives at a blackberry bush, picks and eats blackberries for some time and then stops to urinate. As in the example of the dog described in Chapter 1, it is immediately apparent that fluctuations in a large range of causal factors may affect which transition in behaviour occurs and when. The levels of some of these causal factors, such as inputs from some gustatory and alimentary receptors and certain timing mechanisms will affect whether or not the food items are eaten. At the same time, there will be input to the decision-making mechanism from monitors of hormone levels, muscular fatigue and various internal clocks. These, together with input from sensory analysers, will all be causal factors which might affect behaviour. The causal factors most likely to affect the occurrence of urination will be, firstly, input from receptors affected by bladder pressure and, secondly, a complex set of factors from systems which assess the availability of urination sites taking account of social and anti-predator constraints.

The levels of some causal factors must differ when individuals are engaged in different behaviours. There is indirect evidence for this from the many studies of the effects of manipulating causal factors by deprivation or by stimulus presentation. Direct evidence is provided by the findings of Fentress (1968) and Forrester and Broom (1980) that responses to a particular environmental change differ according to the behaviour which is occurring at the time of the change. When Forrester and Broom interrupted one of ten activities of young domestic chicks, by illuminating a small light-bulb on the wall of the home-pen, the behaviour after interruption was different according to the ongoing behaviour (Table 4). Using the same type of experiment it is possible to look for fluctuations in the levels of causal factors during a bout of a single behaviour. It seems like that some of the causal factors which promote the ongoing behaviour must wane as it continues, for example when the person eating blackberries fills his stomach or when a grooming act removes a source of irritation. Culshaw and Broom (1980) watched the behaviour of chicks whose behaviour was interrupted at the beginning, or at a time predicted as being near the end, of a bout of preening or feeding (Table 5). At the end of a bout, the response to interruption was stronger, the ongoing behaviour persisted less and it reappeared later than if the interruption was at the beginning of a bout. It seems possible that the levels of causal factors which promote a behaviour remain high at the beginning of a bout of behaviour due to a positive feedback system which 'locks on' the behaviour (Wiepkema 1971, Toates and Archer 1978, Forrester and Broom

Table 4. *Behaviour shown after interrupting various ongoing behaviours of chicks. (Data from Forrester and Broom 1980)*

Ongoing behaviour of 6-day-old domestic chicks	Behaviour during first minute after illumination of small light-bulb on wall of home pen
Crouching	Fixate light-bulb, immobile for 47% of minute, 17% of chicks peep, 6% of chicks jump
Preening	Stop preening, fixate, immobile for 39% of minute, 31% of chicks peep, 5% of chicks jump
Walking	Fixate, immobile for 39% of minute, 57% of chicks peep, 39% of chicks jump
Feeding	Stop feeding, fixate, immobile for 22% of minute, 58% of chicks peep, 23% of chicks jump

Table 5. *Behaviour shown after interrupting at the beginning or end of a bout of behaviour. (Data from Culshaw and Broom 1980)*

Ongoing behaviour of 6-day-old domestic chicks		Behaviour during first five minutes after illumination of small light-bulb in home pen
Preening	(1) interruption at beginning of bout	Stop preening after a few seconds, fixate for 20% and immobile for 29% of first two minutes
	(2) interruption at end of bout	Stop preening immediately, fixate for 50% and immobile for 36% of first two minutes
Feeding	(1) interruption at beginning of bout	Stop feeding for two minutes and then resume at low level, fixate for 55% and immobile for 20% of first two minutes
	(2) interruption at end of bout	Stop feeding for more than five minutes, fixate for 73% and immobile for 51% of first two minutes

1980). This would be a beneficial mechanism because it would reduce energetically expensive vacillation amongst behaviours if there were rapid changes in causal factor levels. As the levels of some causal factors wane due to the effects of the behaviour, levels of other, unrelated, causal factors must be increasing. At some point during these changes the motivational system probably prepares for a change in behaviour by altering attentional functioning. The operation of such a mechanism would have the consequence that the response would be stronger when an activity is interrupted at the end of a bout, as described above.

Competition and disinhibition

The multidimensional nature of causal-factor space makes difficult any theorising about how transitions in behaviour might be initiated. The simplification which is usually adopted in discussions of this topic, like those in the section on causal factor space in this chapter, is to consider two variables. These might be single causal factors which are the major ones promoting the occurrence of two different behaviours, or they may be collections of causal factors which have these effects. Such an argument referring to two variables must often be an oversimplified description of changes in motivational state but it is a useful start to motivation analysis. In the discussions of the chick nearing the end of a bout of feeding, or of the person who stops picking blackberries and urinates, the following question arises. Are these behaviour transitions affected principally by the fluctuations in causal factors which promote feeding or those which promote the activity which takes over from it? If the onset of urination is determined entirely by the increase in causal factors which promote this behaviour then this is an example of *competition* (McFarland 1969, Figure 71). If, on the other hand, urination becomes possible because there is a decline in causal factors which promote eating, this is known as behavioural *disinhibition* (van Iersel and Bol 1958). Competition and disinhibition are illustrated diagrammatically in Figure 71. This figure is drawn so that a crossover in levels of causal factors determines a transition in behaviour. McFarland (1969) says that 'the time of occurrence of a disinhibited activity is independent of the level of causal factors relevant to that activity'. If this strict definition is adhered to, a situation which must often arise is that in which

71 Transitions in behaviour may be due to the level of causal factors promoting new activity *B* or to disinhibition of *B* by a decline in the level of causal factors promoting ongoing behaviour *A*. Two starting levels are shown in examples 1 and 2 and the same changes in *B* are shown in examples 3 and 4. The decline in *A* may not be sufficient to disinhibit *B* but competition may then be possible (example 5). The lines in this figure are drawn straight for simplicity but it is likely that the causal factors which promote an activity will normally decline during the time that the activity occurs. Modified after McFarland 1969.

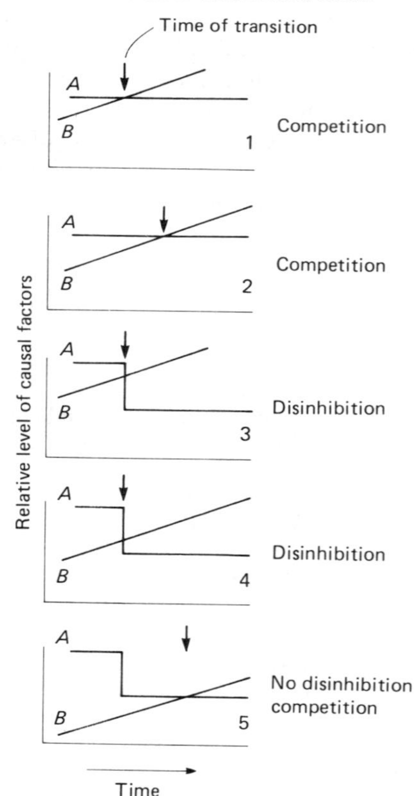

there is a decline in the causal factors promoting the ongoing behaviour but the low levels of the causal factors promoting the other behaviour are not reached (Figure 68). Disinhibition would not occur at all in this situation but a transition due to competition might occur later. It is possible that the levels of causal factors which promote a behaviour may be at or near their maxima but the behaviour is not shown because that behaviour has a low priority at that time. Both eating blackberries and urinating would be inhibited by the approach of a leopard.

Motivational dominance

The idea that a causal factor promoting one activity might be dominant over others, perhaps permanently, has been proposed by McFarland (1974*b*). The dominant behaviour starts when its causal factors rise above a threshold level, irrespective of what else is happening and ends when its causal factors decline again, thus disinhibiting some other activity. McFarland uses the term 'time sharing' for such a mechanism, thus making an analogy with the sharing of computer time on a foreground–background basis in which the foreground programs always have precedence. 'Time sharing' can easily be taken to have a much less specific meaning which would not imply that dominance determines the sharing so it is better to use the term *motivational dominance*. An example quoted by McFarland (1974*b*) and extended by Cohen and McFarland (1979) is of the stickleback in the apparatus shown in Figure 70. In McFarland's preliminary studies it appeared that trapping the stickleback in the middle compartment, after it had been courting, was followed by continuation to the nest if the interruption was short but was always followed by courtship if the interruption was long. Trapping en route from nest attendance to courting was always followed by courtship, whatever the duration of the interruption. Courtship was thus said to be dominant to nest attendance, for the latter occurred only when disinhibited by a drop in causal factors promoting courtship. Subsequent work (Cohen and McFarland 1979) showed that either of these two behaviours could be dominant in different fish or in the same fish at different times. When snails were dropped on to a male stickleback's nest, some males changed from courtship dominance to nest dominance. Such nest disruption altered the timing of behaviour only during nest-dominant sessions.

Cohen and McFarland concluded from their study that, at any time, one behaviour is dominant to all others and that the timing of its occurrence is determined solely by fluctuations in the causal factors which promote it. They do, however, acknowledge that dominance does change and subsequent work by McFarland (personal communication) indicates that there can be a period during which the timing of behaviour is influenced by other causal factors. When McFarland interrupted transitions between feeding and air-breathing in the blue gourami (*Trichopteris*), an obligate air-breathing fish, feeding was dominant, in that long interruptions were always followed by feeding. After successive interruptions of air-breathing attempts the fish still returned to feed. The feeding bout, however, was very short for the fish soon swam up to breathe. This observation raises the question of whether dominance switches might be preceded, sometimes, by an intermediate state in which there is no clear dominance. In many behaviour sequences, causal factors are changing fast and dominance may change so rapidly that the concept is of little use in explaining behaviour. An important question which this motivational dominance hypothesis poses but does not answer is: what determines dominance? There would be an answer based on the theory of natural selection, for genes which favoured efficient mechanisms would survive, but how is dominance determined in a single animal during a behaviour sequence? Perhaps some of the ranking of functional systems remains throughout most of life, e.g. feeding may normally be dominant to body maintenance, but there may be times when body maintenance is so important for ensuring adequate predator-avoidance ability that it becomes dominant to feeding. It seems most likely that any simple dominance order for functional systems should be subject to over-ride mechanisms if the causal factors from any system are very high. Feeding might be dominant to temperature regulation, osmoregulation etc. on most occasions but in extreme conditions of temperature or osmotic state the corrective behaviour must occur very rapidly. The motivational dominance approach is not very useful in such circumstances but may be useful in situations where most of the changes in causal factors are a consequence of the actions of the animal itself.

The newt's problem

Another example of a situation in which switches from one functional system to another have been studied in detail is Halliday's work on the courtship of the newt (*Triturus vulgaris*) (Halliday 1976, 1977, Halliday and Sweatman 1976). The male newt is active during courtship for he has to perform elaborate displays in order to encourage the female to accept his spermatophore (see Chapter 9). This activity results in oxygen depletion so the male newt needs to ascend to the surface of the water to breathe air. If he does this too early in the courtship sequence, or too often, the female's receptivity declines and successful spermatophore transfer is less likely. 'The newt's problem', as Halliday and Sweatman put it, is to court as efficiently as possible before running out of air. He delays the breath for longer if a female, held in a straight-jacket by the experimenters, is present and for longer still if she makes movement towards him. If the gas above the water in his tank is nitrogen he speeds up his courtship and breathes earlier than if it is air. If the gas is oxygen he shows a slightly slower display sequence and more resting time. The causal factors promoting breathing and those promoting courtship have an effect on the timing of behaviour. A descriptive model of newt courtship has been suggested by Houston, Halliday and McFarland (1977) and a model which describes what the newt may be optimising is described by McCleery (1978). Such models are often useful because they draw together the important variables in a behaviour control system but each model describes only one of many possible ways in which those variables might be interrelated.

Physical limitations

When one is trying to understand the mechanisms which determine the occurrence of one particular behavioural change rather than another, it is important to consider physical and attentional limitations. Some transitions from one activity to another are not possible without an intervening activity, e.g. jumping could not follow lying and vocalisation is inhibited until a mouthful of food has been swallowed. Physical difficulties may inhibit certain transitions or may be represented by variable causal factors related to the energetic cost of the transition. As mentioned earlier in this chapter, the probability that a transition from feeding to drinking would occur at any moment is altered if an energetically costly flight is necessary before drinking can occur (Larkin and McFarland 1978). Another limitation is the capacity of the brain to process information (Augenstein and Quastler 1967). It is likely that there are limits to the number of input channels to decision makers, or at least that an increase in the number of channels results in a decrease in the complexity of input on any channel which can be processed. There are probably similar limits to the number of output channels which can be operated simultaneously. As Delius (1970) emphasises, however, input capacity varies. When the range of operational input channels increases, more energy is needed for their function and the cells in those pathways of the brain cannot fulfill metabolic functions at that time. There are physical limitations and energetic optima for brain processing mechanisms just as there are for effector mechanisms. When an orientation reaction occurs, or when a behavioural change is imminent, input and output capacity can be expanded. If little processing is likely to be needed in the immediate future, or if there is the possibility of information overload, there can be a reduction of input and output capacity which conserves resources and hence increases fitness.

Attentional mechanisms

There is much evidence, such as that from the work of Broadbent (1958), concerning the number of inputs which can be processed and the likelihood of response to a particular input channel. The work of Andrew and his collaborators suggests strongly that motivational mechanisms can operate by modifying attentional function. The hormone testosterone has a variety of effects on behaviour and some of these effects are mediated by attentional changes. When testosterone is injected there is a change in the levels of some causal factors and this change in motivational state increases the persistence of attention to one type of stimulus. One series of experiments showed that chicks treated with testosterone continued to peck at one type of food grain longer than did the controls, before they switched to another type of grain (Andrew 1976). In another experiment Archer (1974) found that testosterone-treated chicks which ran towards a food dish were less likely to be distracted by a change in the walls of the runway than were the controls. Similar results were obtained with male rats and Birke, Andrew and Best (1979) found that female rats were distracted more readily in a runway at dioestrus than at oestrus. A third type of experiment showed that responses to a novel object in the home cage persisted more in testosterone-treated chicks (Andrew 1976). These experiments and many others emphasise that whilst behavioural changes are often attributable to changes in input, the threshold of responsiveness to any type of input varies greatly and is under motivational control.

Positive feedback control

The changing inputs to decision-making mechanisms in the brain are causal factors which affect motivational state and hence can affect behaviour. An individual can alter its motivational state by initiating changes in behaviour or in attentional function which are likely to result in changed causal factors. This positive feedback system might operate in one of the following ways.

1. A motor command initiates a behavioural change which alters the position of receptors and hence the input from them. This might be merely a head orientation movement or it might be a complex series of motor patterns which restricts sensory input to a predictable series of changes. When a man starts pacing up and down a room there are inputs from proprioceptors and also repeated, similar, inputs from eyes, ears etc. The motivational state may become comparatively constant due to the predictable input and its resulting causal factors. The sensory consequences of the behavioural change may result in an increase or decrease in the causal factors which promote further behaviour change.
2. An output from a decision centre via an efferent neuron modifies the functioning of a sense organ by acting at a peripheral level. Such efferent control of input is described in Chapter 2. The consequent reduction or enhancement in input may again alter motivational state. This same effect might sometimes be achieved by altering the functioning of some stage in the sensory analytical pathway but there is little good evidence for such control mechanisms.
3. A command from a decision centre may initiate or remove a gating mechanism which prevents input from one type of sensory analyser. Such mechanisms are very important and widespread.

It is likely that all three types of mechanism sometimes occur together. An individual may prepare for greater behavioural change by such methods of increasing sensory input. Alternatively, as discussed in the next section, behavioural and attentional modification may be used to reduce the likelihood that major sensory or behavioural change will occur. It may also be possible to avoid specific kinds of sensory input.

Displacement activities

A final point in this discussion concerns the use of the term 'displacement activity'. The idea originated with Kortlandt (1940*b*) and Tinbergen (1940) and has been extensively discussed by Bastock *et al.* (1953), van Iersel and Bol (1958), Bindra (1959), McFarland (1965*b*, 1966), Delius (1970) and others. It has been used to describe activities whose occurrence was surprising to the investigators because they could not account for it in motivational terms.

An example from van Iersel and Bol's study (see also Chapter 3) is of a tern (*Sterna*) landing near its nest after being alarmed and preening before reaching the nest. Preening was called displacement activity. The behaviours which were considered to be most likely by the observers were incubation and antipredator behaviour so the occurrence of preening was incongruous. Preening has been reported by others in situations where its functional importance was not clear to the observers. The observation by van Iersel and Bol, however, that terns with wet feathers preened more often, does suggest that plumage maintenance cannot be ignored when interpreting the sequence of events.

In recent studies of behaviour the concept of 'displacement activity' has fallen into disuse because it is not possible to distinguish such activities from any others. The fact that it is difficult to ascribe a function to an activity is not sufficient justification for classifying that activity as different from another. The research which has been carried out on this subject has been valuable, however, for it has made an important contribution to our understanding of behaviour sequences and their control.

Stereotypies

Caged animals in zoos often spend much of their time repeating one sequence of movements, particularly when visitors are present. Farm animals in cages or small pens, pets kept in restricted quarters and people who are obliged to wait in an unfamiliar place also show such motor stereotypies. As mentioned in Chapter 3, stereotypy is a relative term and there will be some variation in the movement patterns but the similarity of successive sequences, which may be complex and protracted, is often very striking.

Effects of environmental complexity

Before speculating on the motivational basis of stereotypies, the effects of environmental conditions and previous experience on the occurrence of stereotypies will be considered. Hediger (1950) described a variety of locomotor and other stereotypies in zoo animals housed in quite small cages but their incidence was less in larger cages (Meyer-Holzapfel 1968). Levy (1944) reported that stereotyped head shaking in battery hens was directly related to the amount of floor space in the cages. Smaller cages have less variety of potential sensory input and there is often no opportunity for the animals to hide if the cage is approached by man. The occurrence of stereotypies in the absence of varied sensory stimulation is stressed by many authors. For example Levy (1944) observed more stereotypies amongst children in hospital if toys were removed and Fox (1968) found that dogs which were frequently put in new cages spent less time engaging in stereotypies. A person in a waiting room is more likely to walk up and down jangling keys on a key-ring if the room is empty than if it is full of distractions. Keiper (1970) found that stereotyped route tracing by canaries in a cage was reduced by the introduction of plastic beads, a mirror, or the opportunity to look at another canary in an adjacent cage. Individuals reared in less varied surroundings are more likely to develop stereotypies than are those whose rearing conditions have included possibilities for diverse sensory inputs and activities. Rhesus monkeys spent little time showing stereotypies if reared and kept with mother and peers but showed progressively more stereotypies if reared without peers (Meyer, Novak, Bowman and Harlow 1975) or with a stationary rather than a moving surrogate mother (Mason and Berkson 1975).

Effects of disturbance

Whilst there is no doubt that stereotypies are shown by undisturbed animals in surroundings which vary little, many of the observations of stereotypies are made by observers who are themselves disturbing the subjects. Since there is much evidence for increases in the occurrence of stereotypies when subjects are disturbed, any objective study of the incidence of stereotypy should take account of all aspects of the observation situation including any recent alteration in the subject's surroundings. Berkson, Mason and Saxon (1963) found that the incidence of stereotyped behaviour in chimpanzees reared in isolation was increased if novel objects were put into their home cages or if the chimpanzees were moved to a novel cage. Hutt and Hutt (1965) watched stereotyped rocking and movements of hand, shoulder and head by autistic children in an empty waiting room. These stereotypies were increased if a box of wooden bricks was placed in the room and increased further if an adult sat passively in one corner of the room. Increasing stereotypy of behaviour and increasing speed of stereotyped movement was reported by Fentress (1976*a*) for two mammalian species. A Cape hunting dog (*Lycaon pictus*) in a zoo cage had developed a stereotyped figure-of-eight pattern of locomotion in its cage (Figure 72). When a chain was introduced in its path, the animal jumped over it if undisturbed but moved faster and stumbled over the chain if disturbed. After the chain had been there for several months it was removed. The stereotyped movement pattern now included chain-jumping but whilst pacing at the undisturbed rate, the hunting dog soon adapted to the absence of the chain. If disturbed by loud noises, however, the speed of locomotion increased and the jump was again included in the locomotor pattern. Similarly, Fentress reported that caged voles *Microtus agrestis* or *Clethrionomys britannicus*, developed stereotyped movements around the spouts of their water bottles. They soon compensated when the spout was removed but if they were startled, spout avoidance movements were still included in the motor pattern. All of these examples demonstrate that disturbance can elicit well-established stereotyped behaviour.

72 Caged animals often develop stereotyped locomotor movements. This cape hunting dog *Lycaon pictus* repeatedly walked in a figure-of-eight pattern.

Functions of stereotypies

All individuals show stereotyped behaviour in certain circumstances. The person standing in a corridor awaiting a summons to a job interview or a court appearance may rediscover childhood stereotypies. This is a situation which combines the two factors which seem to elicit stereotyped behaviour: low rate of change of sensory input and high unpredictability of important future events. The sensory input during stereotyped behaviour patterns may be very complex but is also very predictable. It seems that the greatest degree of stereotypy, and hence the most predictable sensory input, occurs when the individual is disturbed. That disturbance involves unpredictable external changes and the magnitude of a particular disturbance will be greater to an individual whose rearing condition has been simple than to one whose rearing condition has been complex. Fluctuations in motivational state might be large if there is high unpredictability of events which will require a response and this might result in decreased efficiency of action. Stereotyped behaviour might therefore be used as a regulator of motivational state. A precursor of this idea was advanced by Hutt and Hutt (1965) who suggested that stereotypies by autistic children 'may serve to maintain arousal within acceptable limits'. The general arousal concept has been criticised earlier in this chapter but there is no doubt that the performance of stereotyped behaviour would increase the mean predictability of sensory input more than would remaining immobile. The idea that it is useful to consider behavioural and some physiological changes as the means by which an individual controls its motivational state is mentioned by McFarland (1977). Forrester (1980) has developed the idea that behaviour might often be used as a regulator of motivational state and he stresses that the levels of various causal factors could be adjusted toward a more favourable position by the repetition of a single behaviour pattern. This behaviour might be simple, as in the bill-shake movement of a domestic chick, or complex, as in the route-tracing movements of zoo animals or expectant human fathers.

The control of rhythms

As mentioned earlier in this chapter, internal clocks with a periodic output (see Chapter 1 for definitions) determine the levels of some of the causal factors which contribute to motivational state. If we try to explain the occurrence of particular behaviour transitions, such causal factors are often important. Some activity rhythms may be determined entirely by fluctuations in external factors but there is extensive evidence for the existence of pacemakers within the body. Some of these are involved in motor control. For example, the pacemaker in the spinal cord of a cat results in periodic bursts of firing in a hind-limb motoneuron at a wavelength of 1.8 s. This occurs even when the spinal cord is severed behind the brain and all afferent input is prevented by the administration of curare (Edgerton, Grillner, Sjöstrom and Zangger 1976). The existence of other rhythms is deduced from experiments such as those of Renner (1957) on honeybees. Bees which visited feeders at particular times of day were placed at the bottom of a salt-mine in conditions of constant light, temperature, cosmic radiation etc., without any alteration in the circadian rhythm. In another experiment the time at which bees visited feeders was not altered by taking the bees from their rearing area in Paris to New York, despite the shift in the time of dawn, the time of dusk and the height of the sun at noon.

Little is known about the mechanisms controlling such pacemakers but some periodic activities can be delayed by cooling the animal. Other periodicities can be lengthened by substituting the heavy isotope of hydrogen, deuterium, for normal hydrogen in the water, so that 'heavy water' is consumed by the animal (Pittendrigh, Caldarola and Cosbey 1973). Circadian pacemakers seem to be localised within the brain. Stephan and Zucker (1972) have reported that ablation of the suprachiasmatic nucleus from rodents often results in the loss of circadian rhythms in constant conditions. The pineal gland appears to be involved in the control of circadian rhythms of birds for Binkley, Kluth and Menaker (1971) found that removal of the pineal gland from a house sparrow (*Passer domesticus*) resulted in its losing circadian rhythm in constant conditions. Reimplantation of the pineal was followed by recovery. As emphasised by Rusak and Zucker (1975) it is not known whether these areas of the brain serve as a master clock, a central coupler of rhythms, or as one complex oscillator in a multi-oscillator system.

When activity levels are recorded at intervals of a few minutes, behavioural periodicities much shorter than 24 h are often detected. Periodicities of 20–30 min have been reported for the feeding behaviour of zebra finches (*Poephila castanotis*) (Slater 1974*b*) and for locomotor activity of domestic chicks with hens (Guyomarc'h 1975) or isolated in cages (Broom 1980). Periodicities of 2–3 h were shown by the chicks in cages and also by voles (*Microtus agrestis*) with a running wheel in their cage (Lehmann 1976*a*). Most of these short-term rhythms are superimposed on circadian rhythms. When Lehmann kept voles in conditions of constant temperature and sound during spring and autumn, however, they emerged from the nest to run in a wheel for periods lasting about 14 min and rested for about 120 min so the periodicity was about 134 min. Such results raise the question of whether there is a periodicity based on a pacemaker or whether stochastic processes are involved in the patterning of the behaviour (Lehmann 1976*b*). A session of feeding which is limited by gut capacity, followed by a session of digestion and energy utilisation, will produce a more-or-less-regular rhythm of activity. It is not necessary to postulate the existence of a pacemaker to ex-

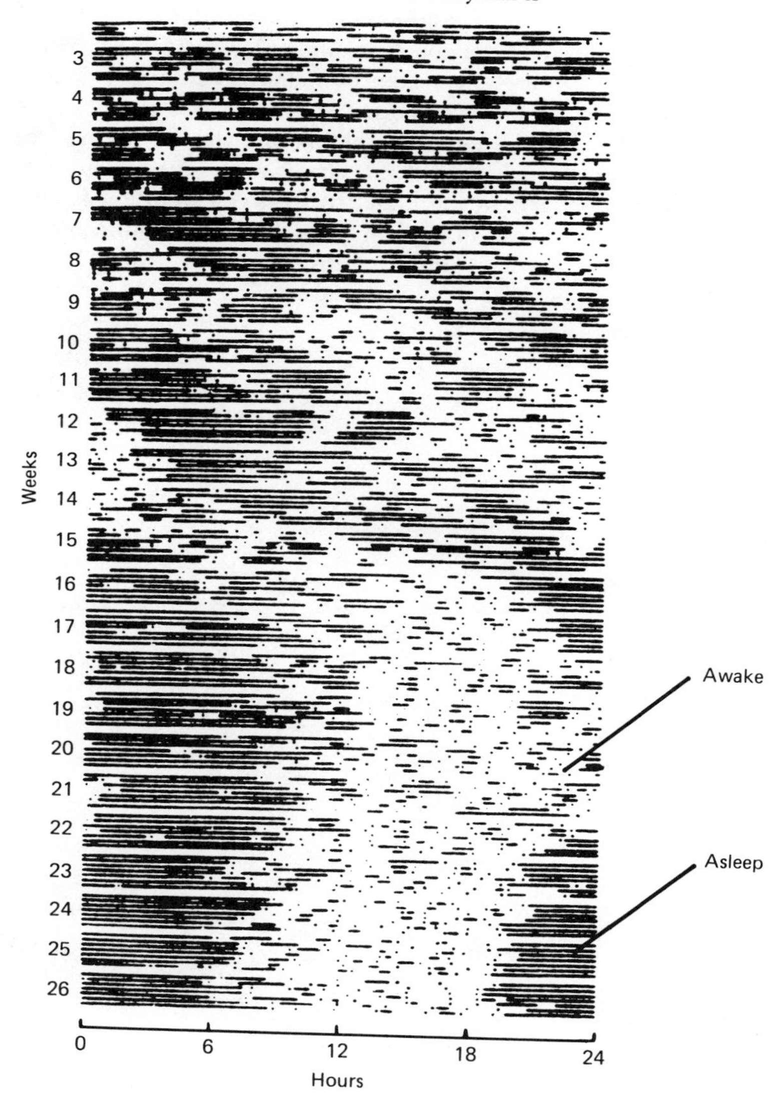

73 Patterns of sleep and wakefulness on each day during the first 26 weeks of life of a human baby. After week 16 a circadian rhythm is apparent. Modified after Kleitman and Engelmann 1953.

plain this rhythm for the system as described has parameters which result in the occurrence of events which are far from random in time. It is possible, however, that some pacemaker might develop in the individual so that the feeding rhythm persists even if environmental factors change. If the pacemaker then acts as a causal factor which interacts with others in determining behaviour the result will be a fluctuating rhythm. It is then very difficult to decide whether or not a pacemaker is involved.

Does a pacemaker control the timing of sleep and wakefulness? Initial analysis of Kleitman and Engelmann's (1953) data on human babies (Figure 73) suggested that a rhythm of wavelength 4–6 h during the first few weeks was succeeded by a circadian rhythm after week 16. Lehmann (1976*b*) points out that the duration of successive periods of wakefulness remains fairly constant at about 2 h but the duration of sleep-bouts changes. There are no serial correlations, i.e. those between the duration of wakefulness and the duration of the succeeding bout of sleep, during the first four months. This suggests that the patterning during this time results from stochastic processes and that the duration of each sleep-bout is controlled separately by physiological mechanisms. After the sixteenth week, significant serial correlations are detectable and it seems likely that a circadian oscillator is involved in the control. For further information on this subject see Enright (1980). Such considerations are relevant to discussions of behaviour transitions for it is clearly necessary to consider pacemakers as causal factors on some occasions. On other occasions, however, the interaction of causal factors resulting from such physiological mechanisms as recovery from fatigue or limits to food consumption is sufficient to explain the existence of quite regular behavioural rhythms.

5

Body regulation, maintenance and hazard avoidance behaviour

The functional systems described in Chapter 1 maintain the body of an animal in a viable state, reduce the chances of predation or other harmful events and facilitate reproduction. Feeding, predator avoidance and reproduction are discussed in Chapters 6, 7 and 9 whilst the role of social behaviour in a variety of functional systems forms the subject of Chapters 8 and 10. Most of the other functional systems in which behaviour plays a part are discussed in this chapter. The systems have behavioural and physiological components, for example, in thermoregulation an animal which is overheating may go to a cooler place, or sweat, or both. Behavioural components are discussed at greater length than physiological components but it is essential to be aware of the existence of both when one is trying to understand either behaviour or physiology. The evidence concerning the brain mechanisms which are involved in these functional systems is reviewed briefly in each section. For further information about physiological mechanisms see Schmidt-Nielsen (1979) or Wilson (1979) and for more details of brain function see physiological psychology texts such as Milner (1970) or Thompson (1975). The first four sections of this chapter deal with breathing, regulating body-water, thermoregulation and body-surface maintenance. The next section, on resting and sleeping, concerns the characteristics and functions of these behaviours. The avoidance of physical and chemical hazards is then considered with especial reference to the responses of pests to the chemicals with which we try to control them. The final section refers to the effects of overloading regulatory mechanisms and to the concepts of pain and stress.

Obtaining oxygen

Movements which result in air or aerated water coming into contact with respiratory surfaces continue throughout the lives of all animals except the very small or the very inactive, in which diffusion suffices. There is, however, considerable variation within individuals in the rates at which air is inspired and expired, or at which gills, buccal cavities etc. are moved. Such variation is apparent in any detailed study of behaviour and may provide useful information about the motivational state of the individual. Hence the mechanism in the brain which controls breathing is obviously not automatic and unmodifiable. The basic patterning of breathing movements in mammals results from interaction between groups of neurons in the hind brain but breathing is affected by very many factors via the cortex (Guz 1975). The neurons in the medulla (for brain anatomy diagram see Figure 3) which fire on inspiration or on expiration are not aggregated into well-defined centres. Their activity is modulated, firstly, by centres in the pons which control the amplitude of ventilation and secondly, by direct input from various carbon-dioxide-sensitive receptors and from stretch receptors in the lungs. If the carbon-dioxide level in the blood rises there is a change in output from the receptors in the carotid body. There are also other carbon-dioxide receptors, for example on the inner surface of the medulla, which is in contact with the cerebrospinal fluid (Dejours 1962). This mechanism for the regulation of breathing can operate in individuals which have no fore brain or mid brain. There is much evidence, however, for the influence of the fore brain on breathing. Changes in breathing which occur during an orientation reaction, talking, sneezing, or coughing require input from the cerebral cortex. An example of a very specific effect on breathing is the change in breathing rate shown by a new-born human baby when presented with a novel odour. The response wanes with repeated presentation (Engen and Lipsitt 1965).

Brief pauses and changes in rates of breathing may be observed in animals which spend time under water. Diving mammals, birds, Amphibia, insects, etc. must come to the surface more often if they use up their oxygen at a high rate, for example if they chase prey vigorously. Spurway and Haldane (1953) described the occurrence of breathing in a newt, *Triturus cristatus*. A reduction in the proportion of oxygen in the gas above the water in which they were swimming resulted in an increase in breathing frequency. The same results have been obtained with smooth newts (*T. vulgaris*) and the timing of breathing has been shown (Chapter 4) to be heavily dependent upon the possibilities for courtship (Halliday and Sweatman 1976, Halliday 1977).

Regulating body-water levels

Aquatic animals

Osmoregulation, the problem of maintaining the volume of body fluids within tolerable limits, is solved, largely, by the functioning of the kidney in most freshwater and marine animals. Most ionic regulation is also carried out by the kidney. Marine teleost fish, however, have body fluids which are more dilute than sea water. In order to maintain this state they drink sea water, pass out divalent salts through the gut and secrete sodium chloride from the gills. Marine reptiles, birds and mammals are not subject to much water loss by evaporation from the body surface or lungs. It seems that they do not drink, but obtain water from their food, which is often teleost fish. Marine mammals seem to be able to eliminate excess salt via the kidneys but sea-birds also have nasal salt glands.

Water conservation on land

Terrestrial animals, which are vulnerable to desiccation, must drink or absorb water by some means and must conserve water. The methods used for taking in water and reducing water loss vary according to water availability, the anatomy of the animal and its mobility. The structure and way of life of animals is determined, partly, by the necessity to avoid dehydration. Many animals are restricted to damp places, such as the soil, and possess behavioural mechanisms for avoiding dry places. An earthworm will turn away from dry soil or dry air and a wood-louse will keep walking if conditions are dry but stops if they are damp. Experiments by Hughes (1966, 1967) showed that wood-lice (*Porcellio scaber*) left warm, dry, brightly illuminated areas by walking away from them in straight lines. When obstacles, such as the walls of a maze, forced the animals to turn, they corrected with another turn when this became possible. The efficient departure from the unfavourable conditions necessitated complex sensory and motor ability as well as memory during the interval between the forced turn and the opportunity to correct. Wood-lice spend much of the daytime, especially when conditions are dry, in crevices amongst stones or in wood. Dehydration is minimised by many shore animals in a similar way. Those which are mobile, hide in cracks in rock, in weed, or under stones; barnacles and limpets remain still, keeping moisture within their shells; serpulid worms retreat into their calcareous tubes and close the opening; and sea anemones on rocks withdraw their tentacles and minimise their surface area to volume ratio. Animals living in deserts often reduce their thermoregulation and water-economy problems by remaining underground during the daytime and emerging at night. The wood-louse *Hemilepistus* lives 30cm underground in a burrow during the hot part of the day. Even birds may spend the day in burrows, e.g. the burrowing owl (*Speotyto cunicularia*) does this. Diurnal insect hunters like shrikes

(*Lanius*) reduce overheating and water loss by sitting inside bushes instead of on top of them, their characteristic position in cooler conditions.

Very small animals and those with skin which is not waterproof such as Amphibia, can take up water which is in contact with the general body surface (Bentley 1966, Bentley and Yorio 1979). Schmajuk (personal communication) has suggested that toads have osmoreceptors on their skin surface. Most of the larger terrestrial animals drink. Some wood-lice are known to drink water or to take up water through the anus (Edney 1954) and drinking has been observed in the land crab (*Birgus*) and in many insects. Travelling from the feeding or resting site to a place where drinking is possible may be energetically costly and may involve predation hazards but it is essential for many animals living in dry habitats. Daily drinking flights are made by birds such as budgerigars (*Melopsittacus*) and sandgrouse (Pteroclididae), and sheep on dry ranges may have to walk long distances each day to drink (Squires 1974). Differences amongst species in their abilities to conserve water result in great variation in water requirements. The Zebu cattle (*Bos indicus*) conserve water better in hot conditions than do the closely related European cattle (*Bos taurus*). Compared with *B. indicus* at 28°C *B. taurus* drank 30% more water per unit dry matter ingested, but at 38°C, 100% more water was drunk (Winchester and Morris 1956).

Drinking behaviour

The movements carried out when drinking and the rate at which water is ingested, vary among species. Most birds take water into the buccal cavity and throw the head back in a characteristic way prior to swallowing but pigeons can suck up water without raising the head. The drinking patterns of mammalian predators are different from those of mammals which are constantly vulnerable to predation when drinking. Cattle and antelopes suck up water very rapidly and scan around with eyes and ears whilst drinking. Members of the dog and cat families lap water with their tongues and take much longer than an antelope of similar size to drink the same amount of water. An unusual drinking method is that of the elephant which sucks up water with its trunk before squirting it into its mouth. The chimpanzees which Goodall (1964) observed, obtained water from small pools by dipping handfuls of leaves into the water and then squeezing the water into their mouths.

Some drinking methods in man are mentioned in Chapter 1 and the detailed analyses of mammalian swallowing (Doty 1968) and chick-drinking movements (R. and M. Dawkins 1973) are described in Chapter 3. Rats lap 6–7 times per second (Stellar and Hill 1952) and these movements are aggregated into bouts. In the deer-mouse (*Peromyscus*) bouts lasted for about 3 s and intervals without any drinking lasted from 5 to 40 min (Kavanau 1963). The lengths of inter-bout intervals will be affected by many factors but variability within individuals is less for bout-length and less still for the patterning of the component sequences of movements. Drinking movements are, however, likely to be modified by experience in different drinking sites and, like the chicks studied by R. and M. Dawkins, the sequence of movement will be affected by factors such as water depth. The frequency of alternative drinking movements and the location chosen for drinking will depend upon what the individual has learned about the many aspects of its environment.

Water is ingested with food, sometimes in large quantity and some animals obtain all of their water requirements without drinking. This applies to some desert species (Wright 1976) but in many animals whose food has a lower water content, water is necessary for ingestion. The swallowing of food is facilitated by drinking during a meal or by salivary secretion. Digestive enzymes in each region of the gut are secreted as solutions and the removal of faeces inevitably involves water loss. Hence drinking must occur at a higher rate if such animals feed than if they do not. Fitzsimons and Le Magnen (1969) showed that if water was provided *ad libitum*, the quantity drunk was largely determined by food intake. Similarly, McFarland (1971) showed that the consumption of water and food by doves was positively correlated both in timing and in quantity. Many studies of cattle have demonstrated positive correlations between water intake, dry matter intake, growth rate and milk production, see, for example, Bond, Ittner and Kelly (1954). The occurrence of drinking is thus interrelated with that of feeding in many species. The use of water evaporation as a cooling method and the relation between temperature and rate of water loss, both result in interactions between the systems regulating body temperature and water level. As stressed in Chapter 4, such interactions must be considered when we try to understand how transitions in behaviour sequences are determined.

The regulation of water level in the body

The mechanisms for regulating water level in the body can be represented by a control-system diagram. That shown in Figure 74 refers to changes over a long time scale in birds and mammals. It has been showed to be valid for doves (McFarland 1965*b*) and is probably widely relevant. There are two feedback loops which affect the net loss. Ingestion behaviour is modulated and terminated by satiation mechanisms. In the other feedback loop, conservation mechanisms in the kidneys play an important part. Their conservative function is regulated via the secretion of the hormone vasopressin in the hypothalamus. The kidneys can also excrete excess body water. Other conservation mechanisms are behavioural, e.g. walking to a place where less water loss will occur.

Whilst the amount of water drunk during a period of several days is explicable in general terms by a simple model, such as that shown in Figure 74, in the short term, drinking is discontinuous and is clearly regulated by the more than one feedback mechanism. When an individual drinks, he receives input from various sources which might be used to terminate a drinking bout. Receptors in the mouth, stretch receptors in the crop or stomach, those measuring blood concentration or blood volume, cells throughout the body whose degree of hydration is increased after drinking, or monitors of kidney activity could all be used. Evidence for the role of oral receptors includes the results of studies in which rats were allowed to lick at a jet of cool air. This behaviour must result in some sensory input which mimics that received during drinking, for it was much more frequent in water-deprived than in satiated rats. A period of air-jet licking reduced the consumption of water presented immediately afterwards (Hendry and Rasche 1961). Oral receptors can terminate drinking bouts. Bellows (1939) put an oesophageal fistula into dogs, so that anything

74 Control system for the regulation of body-water balance. The net water loss is modified (black segments in input circle) by conservation mechanisms and by the operation of the ingestion–satiation feedback loop. After McFarland 1971.

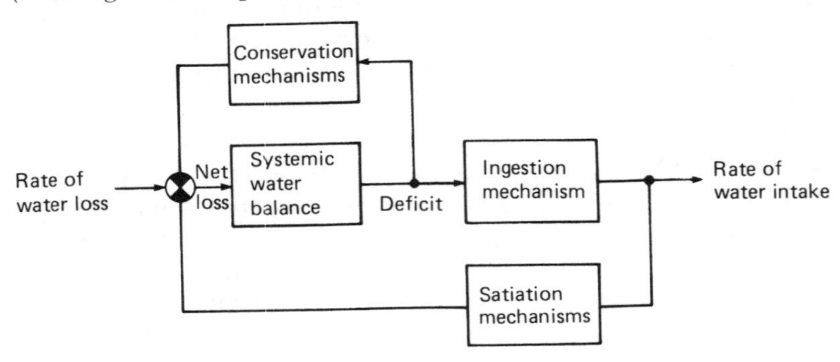

swallowed passed out of the oesophagus and down the tube instead of going into the stomach. When these dogs drank, after being deprived of water, they did stop after a while, although they drank for longer than did normal dogs. If the dogs were deprived and then water was inserted directly into the stomach, or a balloon was inflated in the stomach, they drank a much smaller amount of water than did equally deprived controls (Towbin 1949). Hence input from receptors in the stomach can also influence the timing of the termination of drinking. If the osmolarity of the blood was increased by the injection of hypertonic sodium chloride solution, drinking was very likely to occur but if the same concentration of a potassium salt was injected there was no such effect (Falk 1961). The potassium passes out of the blood rapidly so blood osmolarity is altered only briefly. The effect of increased blood concentration is limited in extent and in man, an experimental increase in blood osmolarity did not increase drinking (McCance 1936).

The work of Andersson (1953) suggested the existence of osmoreceptors in the brain for he found that the injection of 0.1 ml of 1.5–2% sodium chloride solution into the hypothalamus of conscious goats elicited drinking. Again there is evidence that this is only part of the control mechanism, for Novin (1962) showed that less was drunk in such experiments than if the same salt concentration in the brain was initiated by water deprivation. Andersson still emphasises (1978) that 'cerebral sodium-sensitive receptors are essential for the maintenance of water balance'. The possibility that extra-cellular fluid volume in the body might be monitored as part of the water-level regulatory system was raised by Fitzsimons (1961) who increased the incidence of drinking in rats by inserting a colloid, which sequesters body fluids with their dissolved salts, into a body cavity. This technique thus reduces the volume of extra-cellular body fluid, i.e. it produces hypovolaemia, but does not change its osmolarity. Later work by Fitzsimons and Simons (1968) and Fitzsimons (1968, 1969) showed that if the volume of blood reaching the kidney is reduced, the peptide hormone angiotensin is secreted by the kidney and the angiotensin acts on the hypothalamus so as to increase drinking. A drop in blood pressure also results in increased firing in the baroreceptors in the heart and their output promotes water conservation in the kidney.

The role of the hypothalamus in the control of drinking was indicated by Andersson (1953) and confirmed by the use of brain stimulation. When the anteromedial hypothalamus of a goat was stimulated, the goat drank (Andersson and McCann 1955) and a conditioned response reinforced by drinking could also be elicited by such stimulation (Andersson and Wyrwicka 1957). Lesions in the anterior hypothalamus and preoptic area of goats are sometimes fatal but if not, they prevent drinking. The injection of the cholinergic stimulant carbachol into the septal area, dorsomedial hippocampus, preoptic area, lateral hypothalamus, anterior hypothalamus, or amygdala (Figure 5) elicited drinking in satiated rats (Fisher and Coury 1962). The injection of carbachol into the amygdala elicited drinking in water-deprived but not in satiated rats (S. and L. Grossman 1963) so this area of the brain may receive input from water satiation receptors. The injection of angiotensin into the brain is followed by drinking if the septal, preoptic or anterior hypothalamic regions are injected (Epstein, Fitzsimmons and Rolls 1970). Drinking is elicited very readily if angiotensin is injected into the brain ventricles and there may be receptor cells in the optic recess of the third brain ventricle (Buggy *et al.* 1975).

In summary, it seems that drinking is initiated by the hypothalamus when angiotensin is released from the kidney and, perhaps, also when salt concentration increases in the brain cells. The major body change which results in the commencement of drinking is thus a decrease in extra-cellular blood volume. Drinking may be terminated following input from oropharyngeal and gastric receptors but the command which initiates drinking is likely to be produced for a limited time irrespective of any feedback. These mechanisms are reviewed by Fitzsimons (1972, 1979) and Andersson (1978).

Thermoregulation

Many animals expend little or no energy on the regulation of their body temperature, either because they live in conditions where the temperature varies little, for example in much of the sea, or because they can tolerate wide fluctuations in internal temperatures. Some animals can survive repeated freezing and thawing, for example the larvae of *Chironomus* midges (Scholander, Flagg, Hock and Irving 1953). The ability to resist or survive freezing is shared by many animals which live in polar regions, or on shores and in other damp conditions during cold winters. At the other extreme, some insects are physiologically adapted to withstand temperatures which are almost high enough to denature proteins. No activity is possible until the temperature of the animal has risen from very low temperatures or fallen from very high temperatures to within a certain range, but animals which have survived extremes of temperature can then show quite complex behaviour. When an animal commences muscular activity, heat is generated; all animals can thus increase body temperature, or lower it by reducing activity. The effects are brief in very small animals, whose surface-area to volume ratio is large, but all animals have some capacity for *endothermy*, i.e. the regulation of body temperature by means of internal heat production. The alternative method of heat regulation, heating or cooling the body by use of external conditions, is called *ectothermy*. Most animals which regulate body temperature use both methods. There are some species which always maintain their body-core temperature within a narrow range, e.g. man, cold-water fish; others which allow their temperatures to change during prolonged torpid states, e.g. a squirrel which hibernates; others which maintain a fairly constant temperature during warm days but become cool and torpid at night, e.g. many lizards; and many others which have some thermoregulatory ability so that lethal levels can be avoided, usually, but which are very dependent on external temperatures.

Changes in body temperature are detected by a wide range of receptors (see Hensel 1974). *Cold receptors*, whose rate of nerve-impulse production is increased on cooling, and *warm receptors* have been found in mammalian skin, mucous membranes e.g. the nasal area, spinal cord and many organs deep in the body. Receptors to cold only have been found in many places including pigeons' beaks and cockroach or locust antennae. Evidence for receptors in the hypothalamus and other parts of the brain will be discussed at the end of this section.

Temperature regulatory mechanisms can operate by *feedback*, in which the regulatory response is determined when a drop in internal temperature has occurred, or by *feedforward*, in which a change in peripheral receptors initiates responses which will alter body temperature (Figure 75). Examples of feedforward effects in man include the effects of seeing snow and ice or of feeling cold air on the face, on mus-

75 Body temperature can be regulated by feedback mechanisms and by feedforward in which a predicted change in body temperature is counteracted by regulatory responses. After McFarland 1971.

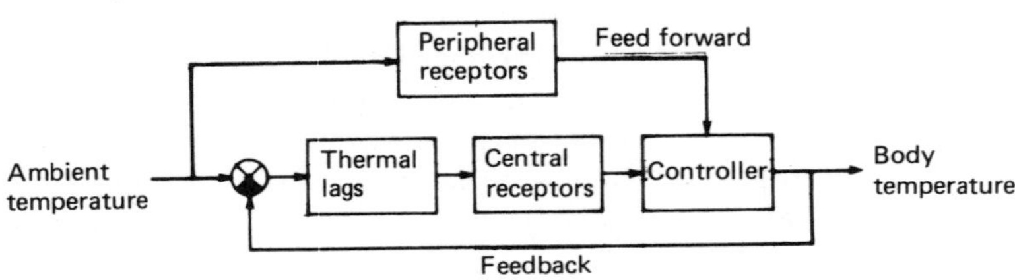

cle tone, vasoconstriction and the likelihood of actions preventing cooling such as donning warm clothing.

Long-term responses to unfavourable temperatures include nocturnal rather than diurnal activity, hibernation, migration and modification of breeding cycles. These may be universal adaptations of all members of a species or they may be utilised by individuals in extreme conditions only. Many desert animals are largely nocturnal and animals which are physiologically adapted to temperate climates may become nocturnal when they are in hot places. Seath and Miller (1946) found that dairy cattle at 27°C spent 37% of the daytime grazing but at 29.5°C only 11%. Payne, Laing and Raivoka (1951) found the dairy cattle in Fiji did 67% of all grazing at night when the day temperature was 27°C whereas cattle in Britain are largely diurnal grazers.

Physiological and behavioural body-temperature modifiers

Changing metabolic rate

A long-term, physiological response to temperature change is the initiation of increased metabolic rate after body cooling. This occurs following production of thyrotrophic hormone from the pituitary and hence increased production of thyroxine in the thyroid gland. It is, however, an energetically expensive means of increasing body temperature. As Johnson (1977) points out, mammals which will be exposed to low temperature usually grow longer coats that improve insulation. This minimises the necessity for inefficient increases in metabolism at a time when food may be scarce.

Piloerection

Piloerection is an effective response to lowered temperatures because the erection of hair or feathers increases the depth of the insulating layer of still air around the skin. McFarland and Budgell (1970) were able to show that feather erection in doves is closely related to ambient temperature and hypothalamic temperature.

Sweating

When body-temperature rises, one physiological response is sweating, which cools the body by evaporation. Most of the sweat glands in man are controlled by the hypothalamus via cholinergic sympathetic neurons but those on the palms and the soles of the feet are under cortical control. Man can produce up to 1.5l of sweat per hour but many animals, including dogs, cats and all birds, cannot use this cooling method.

Panting

A temperature regulatory mechanism which is obviously behavioural is panting. Schmidt-Nielsen, Bretz and Taylor (1970) demonstrated that when a dog pants, the air is drawn in through the nose where heat is lost to it and then the warmed air is rapidly blown out through the mouth. When a cheetah runs at about $100 kmh^{-1}$, it accumulates heat and its body temperature rises to a near lethal 40.5°C within 60 s. It must then stop and lose heat by panting so its prey must be caught within that time (Taylor and Rountree 1973). Panting is an effective means of reducing body temperature when the body is considerably warmer than the ambient temperature. Breeds of cattle from temperate regions breathe about 20 min^{-1} at 5–17°C but increase this rate to 100 min^{-1} at 32°C (Findlay and Beakley 1954). At high temperatures the heat production due to panting and the cooling effect counterbalance one another and there is a maximum panting rate which is 130 min^{-1} for adult Jersey cows and 110 min^{-1} for Jersey-Zebu crossbreeds (Schein, McDowell, Lee and Hyde 1957).

Gular fluttering

Another example of an activity which results in the movement of air over a body surface is gular-fluttering in birds. The gular membrane, which is the floor of the mouth, is highly vascularised in certain nightjars, owls, doves, herons, cormorants, pelicans and boobies. It can be fluttered rapidly at a resonant frequency with little energetic cost so this method of heat dissipation is used whenever the birds have to sit in hot conditions. One North American nightjar species, the poor-will (*Phalaenoptilus nuttallii*) has to sit, camouflaged, in the open during the daytime. In hot weather the inconspicuous gular-fluttering movements are thus invaluable. The effectiveness of gular-fluttering is much greater than is that of panting by the cardinal (*Cardinalis cardinalis*) at high temperatures (Figure 76, Bartholomew 1964).

Saliva spreading

The cooling effect which is produced when water evaporates is utilised when panting and also when cats and other mammals lick their fur. Salivation and the spreading of saliva over the fur, is an important means of reducing temperature in several species including the opossum *Didelphis marsupialis* (Higginbotham and Koon 1955).

Adopting a posture or orientation

Another type of behavioural response to temperature is the adoption of a particular posture or orientation in certain conditions. When they feel cold, cats, dogs, people and many other species will curl up so that the extremities are folded against the body as much as possible. When they are warm, however, the surface area exposed to the cooler air or ground is maximised by spreading the limbs. After a cool night, lizards such as *Phrynosoma* (Bellairs 1957) and butterflies of many species bask in the sun. The lizard turns its body so that it is exposed maximally to the sun and the butterfly spreads its wings and holds them at 90° to the direction of the sun's rays. In the day, when lizard and butterfly might become overheated, they turn their bodies so as to reduce the area exposed to the sun. The butterfly does this by folding its wings over its back and pointing them at the sun. It seems likely that the fossil mammal-like reptiles with a large fan on their backs, such as *Dimetrodon*, used these fans for thermoregulation in the same way. Young Laysan albatrosses (*Diomedea immutabilis*) which have to sit in the sun for long periods, also orient with respect to the sun. They turn so that their large feet are in the shade of the body and rock back so that the feet are elevated and can be cooled by any air movement (Figure 77, Howell and Bartholomew 1961).

76 Active cooling methods produce heat as well as causing heat loss. The net effectiveness of gular-fluttering (see text) in a nightjar, the poor-will (full line), is much greater than is that of panting in the cardinal (interrupted line), especially at high temperatures. Modified after Bartholomew 1964.

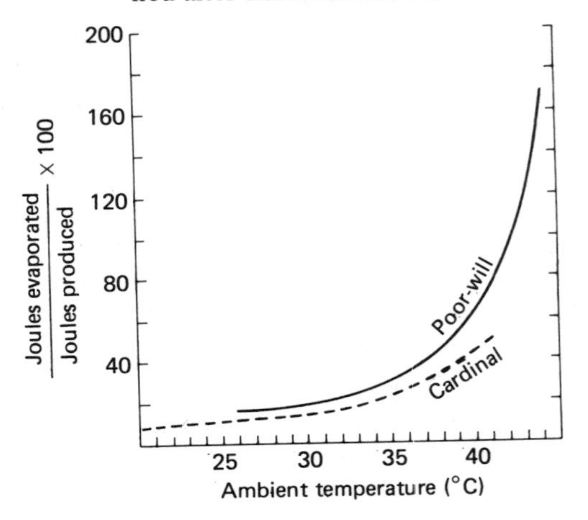

77 Juvenile Laysan albatross resting on its heels with its feet in the air so as to facilitate heat loss. The bird keeps its feet in the shade of the body. Drawn from photograph in Howell and Bartholomew 1961.

Locomotion

Many other active means of temperature regulation involve animals in moving from one place to another. Arctic sculpin fish (*Myox ocephalus*) were taken from their home tank, which was kept at 5°C, placed in tanks at 20°, 16°, or 12°C and given the opportunity to swim down tubes to a tank at 5°C. Whatever the external temperature the fish remained in the tank until their body temperatures reached 8°C and then they swam down the tube. Such swimming was delayed if the fore brain was cooled (Hammel, Strømme and Myhre 1969). Similarly, blue-tongued lizards (*Tiliqua scincoides*) which could stay in either of two containers at 15°C or 45°C, moved from one to the other in such a way that their body temperatures were maintained at 30–37°C. Their movements were delayed by cooling or warming the brain (Hammel, Caldwell and Abrams 1967). Locomotion is used to modify body temperature by pigs, sheep, cattle and many other species which walk into shade on sunny days or into shelter from cooling winds on cold days. Inevitably, such behaviour occurs at different temperatures according to species and breed. Seath and Miller (1946, 1947) found that, on average, Jersey cows sought shade at 27°C when their body temperature was at 39.9°C, 0.4°C above normal. Holstein cows sought shade at lower temperatures and were more likely to wallow in mud or water. Wallowing behaviour, which helps evaporative cooling as well as bringing the animal in contact with cooler mud underneath that on the surface, is frequently shown by pigs, water buffalo and other large ungulates.

Using burrows and nests

The poor heat-conducting qualities of soil result in considerable thermal protection possibilities for those animals which burrow or use caves. Many invertebrate animals, as well as vertebrates, burrow when overheated. The antelope ground squirrel (*Citellus leucurus*) lives in desert conditions where it soon overheats on the surface in the daytime. It then enters its burrow and flattens itself against the cool floor with the result that it loses heat to the ground by conduction as well as to the walls by radiation (Hudson 1962). As mentioned in more detail in Chapter 8, social behaviour can also be important in thermoregulation, for example hive fanning by bees and clustering in bees, birds and mammals. The construction of artefacts, such as nests, has thermoregulatory benefits when carried out by individuals but has most effect when co-ordinated in social groups, for example in social insects (see p. 178–9) and man. Individual nests with insulatory properties are built by many small rodents. Nest-building by rat is initiated by peripheral cooling. It is inhibited in air temperatures of 27°C in all but females with young (Kinder 1927).

Temperature change as a reinforcer

The investigation of active methods of thermoregulation and of the preferred thermal ranges of animals have been facilitated by experiments in which animals are provided with an opportunity to control their own temperature. The two-chamber experiments of Hammel *et al.* (1967, 1969) are examples of such experiments. It is advantageous for all individuals which possess behavioural thermoregulatory methods to learn how best to use them. The return of body temperature to the optimum range after it has been experimentally modified, has been used as a reinforcer in various experiments. When young domestic chicks were given the opportunity to run from a cool box to a source of heat they soon learned to do so and the cooler the box, the faster they ran (Zolman and Martin 1967). Rats (Weiss and Laties 1961), sheep (Baldwin 1972), pigs (Baldwin and Lipton 1973) and doves (McFarland 1971) could be trained to press a lever or to place their noses in an aperture if the consequence was a period during which the ambient temperature was raised. Hogan and Roper (1978) reviewed such operant experiments and concluded that several mammal and bird species would use behavioural means for temperature regulation where these were easy but would start to use physiological means when the behavioural regulation became difficult. This order of choice would be reversed in those situations where the behaviour made the individual more vulnerable to predation.

Thermoregulation in young animals

The thermoregulatory problems of young animals are often quite different from those of adults. The pups of small mouse species produce little metabolic heat until they are about 7 days old and, although larger rodents produce metabolic heat earlier, the brown rat in the laboratory does not reach its adult level of endothermy until it is 25 days old. The pups die if exposed for long periods to the ambient temperature at which most adult rodents live. They depend for their survival on warmth from their parents and on heat conservation, which is facilitated by remaining in a nest with their siblings. Separation from the nest, and the consequent cooling, results in ultrasonic calling (30–60 KHz) by many murid and cricetid rodent species (Zippelius and Schleidt 1956, Sewell 1968). This has a thermoregulatory function. The responses of parents to these sounds include orientation towards the sound-source and nest-building (Allin and Banks 1972, Noirot 1974). When a pup is found, the parent retrieves it to the nest. Noirot and Pye (1969) suggested that the likelihood of eliciting parental behaviour might depend on the total sound energy produced by cooled pups. When Mongolian gerbil (*Meriones unguiculatus*) pups were cooled, their ultrasonic calls reached a peak, at 4–6 days of age, in sound level, rate of calling, duration of calls, complexity of calls and lengths of calling bouts (Broom *et al.* 1977). These changes with age are positively correlated with data on the thermoregulatory ability of the pups (McManus 1971). The major thermoregulatory behaviour in these rodent pups is thus ultrasonic calling.

Newly hatched domestic chicks also vocalise when cold for, although they can pant, they do not shiver and their ability to regulate body temperature is poor. There is slow improvement during the first three weeks of life (Randall 1943, Moreng and Shaffner 1951). Mother hens respond to the loud peep calls of chicks by approaching the chicks and brooding them. The chicks do, however, show active temperature regulation when the mother is present. Sherry (1981), using a stuffed hen with brood-patches whose temperature could be varied, showed that chicks moved close to or further from the hen according to air temperature, brood-patch temperature and their own body temperature. Their behaviour was quite complex for if air temperature was high they would tolerate a greater drop in body temperature before going to the warm brood-patch.

Brain mechanisms

The involvement of the hypothalamus in temperature regulation was initially discovered by brain surgeons who found that sweating and panting occurred when a hot cauteriser was moved near this part of the brain. This was verified experimentally with cats (Magoun, Harrison, Brobeck and Ranson 1938), and cooling the hypothalamus was found to elicit vasoconstriction and shivering in dogs (Hammel, Hardy and Fusco 1960). It is now known that many neurons in the preoptic region, the anterior hypothalamus and in parts of the mid brain of mammals will fire when they are warmed or cooled (review by Hensel 1974). Stimulation of the same area in goats was found to elicit panting and cutaneous vasodilation (Andersson, Grant and Larsson 1956). The involvement of chemical mediators, presumably transmitters, in such control mechanisms was emphasised by the discovery that perfusate from the anterior hypothalamic region of a monkey, when injected into the same region in another monkey's brain, would elicit shivering in the recipient if the donor was cold, but sweating if the donor was hot (Myers and Sharpe 1968). It is possible that the transmitters adrenaline and serotonin are differentially involved in warming and cooling mechanisms (Feldberg and Myers 1963). Physiological thermoregulatory mechanisms can act locally for if a hand is cooled, vasoconstriction occurs in that hand only. The brain control of local and general physiological responses to adverse temperature conditions has been extensively investigated but the control of behavioural components of temperature regulatory mechanisms has received less attention. If the preoptic region of a pig's brain is cooled the pig will carry out an operant response, putting its nose where this results in an infra-red heater being switched on (Baldwin and Lipton 1973). Similarly, warming or cooling the anterior hypothalamus of a sheep modifies the frequency of the same response (Baldwin and Yates 1977). The work of Hammel *et al.* (1967, 1969) which has already been mentioned, also demonstrates brain control of a behavioural thermoregulatory response but there is much scope for more work in this area.

Body surface maintenance

Rapid movement may be impaired if foreign or recently moulted material is adhering to the surface of mammalian skin or fur, bird feathers or insect cuticle. Such material, or mere disarray of feathers or fur, might also interfere with camouflage, thermoregulation, or efficient courtship. The body surface could also harbour disease-causing organisms. Grooming, preening and related movements remove unwanted material and restore the body surface to its optimum state, so in many species such movements occupy a substantial part of the individual's time. The temporal patterning of movements during preening and grooming in flies, mantids, mice and various birds has been extensively discussed in Chapter 3 and the allocation of time to such maintenance behaviour has been discussed in Chapter 4. Some of the ways in which body-surface maintenance is achieved will now be considered.

Grooming

If a kangaroo rat (*Dipodomys merriami*) is deprived of sand for ten days, the pelage becomes matted and greasy (Borchelt, Griswold and Branchek 1976). When sand is provided, sandbathing and grooming movements are used to clean the coat. The first process is the secretion of lipid on to the fur from superficial epidermal and sebaceous glands. Next sand-bathing, brushing strokes with the limbs, scratching and nibbling movements are used to remove excess lipid. Finally, further grooming and shaking movements remove sand from the fur and re-align the body hairs. Most of these movements occur more after long periods of sand deprivation, for the final condition of the fur is similar at the end of most undisturbed periods of maintenance behaviour. Many other mammals and birds use dust or sand as an aid when cleaning their pelage and some birds apply ants to their feathers and skin, presumably as a aid to cleansing. When cattle groom, they lick all parts of the body that they can reach and scratch themselves against trees or fence posts (Brownlee 1950). Calves have been found to spend 52 min per day grooming (Roy, Shillam and Palmer 1955) and adult beef cattle showed 180 bouts of grooming during a day (Schake and Riggs 1966). Allogrooming, in which two or more individuals groom one another, also occurs in cattle (Wood 1977), primates (Simpson 1973) and other mammals and birds but it is seldom clear to an observer how useful it is for body maintenance.

Preening

Bird flight feathers do not function properly and body insulation by contour feathers is less efficient if the barbules are not properly linked on the feathers. Flight necessitates free movement of feathers and is adversely affected by any extra load. Preening and other feather maintenance behaviours are, therefore, very important to birds. Figure 78 shows such behaviours, called comfort behaviours by Delius (1969), in the skylark (*Alauda arvensis*). Evidence for organised sequences of bathing, head shaking, breast preening, shoulder preening and back preening in two tern (*Sterna*) species came from the work of van Iersel and Bol (1958). This work has been followed by other detailed studies (McKinney 1965, Delius 1969, Ainley

1974). The more elaborate body-care sequences in herring gulls (*Larus argentatus*) (Figure 79, van Rhijn 1977*a*), start with bathing which is followed by shaking, oiling and preening, as shown in Figure 80, together with the component movements. The oil comes from the preen gland on the rump by the base of the tail. Most birds have a preen gland and it is especially well developed in aquatic birds. The principal function of oiling seems to be to help to prevent the feathers from breaking. Whilst waterproofing has also been suggested as a function, van Rhijn's (1977*b*) studies on herring-gull feathers show that the removal of oil does not make properly aligned feathers more likely to be wetted. Oiling might aid waterproofing in other species.

Developmental studies on grooming and preening include Sach's study of rat grooming (see pages 58–9) and work on adélie penguins (*Pygoscelis adéliae*) by Bekoff, Ainley and Bekoff (1979). Penguin chicks are covered with fluffy feathers at hatching and they show no preening movements until day 7. Stretching, yawning,

78 Some body-surface maintenance behaviours of the skylark (*a*) sun bathing, (*b*) body-shake, (*c*) preening wing, (*d*) bill and head wiping, (*e*) rain bathing, (*f*) wing and leg stretch, (*g*) dust bathing, (*h*) scratching, (*i*) both wing stretch. Redrawn after Delius 1969.

79 Herring gull (*Larus argentatus*) preening its tail feathers.

80 The mean distribution of bathing, preening and shaking during twelve observations in which herring gulls were observed to apply oil to their feathers. Modified after van Rhijn 1977*a*.

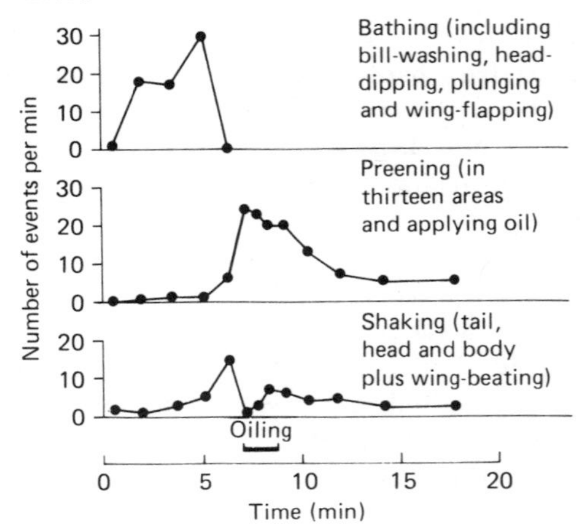

head shaking and wing flapping occur before this time. From day 7 to day 21, 24 body maintenance movements appear. The preen gland starts secreting oil between day 30 and day 33 and on day 35 the first oiling movements are seen. The duration of sequences of preening which do not include oiling are the same from day 13 to adulthood but chicks perform fewer acts per minute than do adults. Analysis of sequences show that those of chicks resemble those of adults in a number of ways by day 28.

Resting and sleeping

Types of rest

There are two sorts of resting, one of which follows necessarily after vigorous exercise and another which need not be preceded by a period of high activity. If a motoneuron is stimulated electrically at a sufficiently rapid rate, the muscle will cease to contract when it becomes *fatigued*. This fatigue may result from an increase in the concentrations of toxic substances, probably not including lactic acid, or from a decrease in available adenosine triphosphate (ATP) or creatine phosphate in the muscles. It may be a consequence, however, of the cessation of impulse transmission across the synapse at a neuromuscular junction. Since the circulatory system removes toxic substances and replenishes energy sources, synaptic fatigue is the most likely explanation of muscle failure except in some laboratory preparations. After a brief rest, normal muscular function can be resumed. Vigorous muscular activity may also lead to depletion of oxygen levels in the blood. This can be remedied by rapid breathing and reduced activity. Some periods of rest are, therefore, restorative in function but the time needed for such restoration is very short in comparison with the total rest-time shown by most animals.

Other possible functions of resting include increasing the amount of available energy which can be allocated to a particular system such as digestion or grooming, avoiding wastage of energy on activities which are unlikely to be productive at that time and reducing the risk of predation. It seems likely that most resting is a behaviour which is chosen, rather than an inevitable consequence of previous high-energy activities. The location at which resting occurs usually allows some protection from predation, for example trees are selected by ground-feeding birds, crevices or holes are selected by mice and mother is selected by a young rhesus monkey (Simpson 1979).

A definition of sleep

An extreme form of resting is sleep. Definitions of sleep are difficult to formulate and depend upon which species and which states the definer wishes to include. Meddis (1975) lists four characteristics of sleep which are common to a wide variety of animal species: (1) prolonged periods of inactivity which often occur with (2) circadian or tidal periodicity, (3) reduced response thresholds, and (4) a characteristic posture in a characteristic site. He lists examples of very many mammals, birds, reptiles, amphibians, fish, molluscs and insects which show this behaviour. If, in addition, characteristics of the electroencephalogram, defined on the basis of mammals, are included in the definition then only birds are sufficiently similar to mammals to be included. The amount of time taken up by sleep varies greatly among species and is also quite variable within species. Table 6 shows the results of studies of the duration of sleep in a variety of mammals. In addition to the porpoise and the shrew there are birds, such as swifts and albatrosses, which spend the night on the wing and do not sleep, according to the definition given above.

The functions of sleep

There has been discussion for many years about the functions of sleep. The argument that sleep occurs in order to rectify muscular fatigue seems unlikely to be correct and those individuals who engage in much physical exercise do not necessarily sleep longer than relatively inactive individuals. Many people have assumed that mental fatigue necessitates sleep, hence statements like 'sleep basically has a restitutive or restorative function' (Hartmann 1973). This may be correct but the evidence for it is inconclusive. The timing and duration of sleep can be altered in individuals and often seems to be determined by the optimisation of feeding or predator-avoidance behaviour rather than by any recuperative consideration. Toutain and Ruckebusch (1973) found that the duration of sleep by cattle which were fed cut grass *ad libitum* was 30% longer than that of the same animals when grazing at pasture. Horses, cows, pigs and sheep slept for less time if kept in less comfortable quarters, if disturbed at infrequent intervals by farm staff, or if moved to new quarters (Ruckebusch 1975). Sleep times are often adapted to alternate with necessary activities, for example sleep is split into two sessions in species which are active at dawn and dusk. Many other examples of the sleeping–waking cycle and its control are given by Enright (1980).

Table 6. *Number of hours asleep per 24 h for various mammals. (Modified from Meddis 1975)*

Species	Hours
Two-toed sloth[a]	20
Armadillo, opossum, bat	19
Lemur, tree-shrew	16
Hamster, squirrel, mountain beaver	14
Rat, cat, mouse, pig, phalanger	13
Chinchilla, spiny ant-eater	12
Jaguar	11
Hedgehog, chimpanzee, rabbit, mole-rat	10
Man, mole	8
Guinea-pig, cow	7
Tapir[a], sheep	6
Okapi, horse, bottle-nosed dolphin[a], pilot-whale[a]	5
Giraffe[a], elephant[a]	4
Dall's porpoise[a], shrew	0

[a] no EEG data

The differences amongst species in the duration of sleep during 24 h has lead several authors to consider the idea that sleep might have evolved as a mechanism for ensuring immobility at certain times, rather than as a restorative mechanism (Milner 1970). Hediger (1969) pointed out that antelopes, which are subject to predation in all but the darkest conditions, sleep for a shorter time than does a predator such as a bear which can rest in the comparative safety of its den. This difference persists in quiet zoo conditions. Allison and Van Twyver (1970) also referred to the inverse relationship between safety from attack and duration of sleep. Webb (1974) emphasised that sleep as an adaptation to life style would therefore vary from species to species. Meddis (1975) proposed that immobilisation is the principal function of sleep and that it occurs whenever survival is promoted by having such periods of immobility. He points out that if restorative functions were of major importance, those individual people who sleep for a few minutes or for one or two hours, in each 24 h, would be expected to suffer in some way. There would also be much greater effects of sleep deprivation than are apparent from experimental studies.

Brain activity and sleep

The development of brain recording methods, starting with the electroencephalogram produced from electrodes on the skull surface, has made possible the classification of sleep into slow-wave sleep, active sleep which is often associated with rapid eye movements (REM), and the intermediate quiet sleep (Figure 81). Stimulation of various parts of the mammalian brain can elicit sleep and lesions of the Raphé nucleus in the centre of the brain stem drastically reduce sleep duration (Jouvet 1967). Many of the neurons in the Raphé nucleus are sensitive to the transmitter serotonin. Jouvet showed that various treatments which increased serotonin levels in the brains of cats also increased the amount of slow-wave sleep and reduced REM sleep. The locus coeruleus, which is between the mid brain and the pons (see Figure 5 for brain anatomy), includes many noradrenaline-sensitive neurons. When Jouvet removed it, transitions from slow-wave sleep to REM sleep were disturbed by violent twitches and the duration of REM sleep was reduced.

During REM sleep, neural activity and the rate of blood-flow through the brain is greater than at almost any other time. If individuals are deprived of REM sleep by awakening them as soon as it occurs, they do not show serious adverse effects. When they next sleep, however, they start rapid eye movements much earlier than usual. Akindele, Evans and Oswald (1970) used drugs which deprived individuals of REM sleep for many months with no detectable ill effects. Meddis follows Snyder (1966) in arguing that REM sleep is the relic of an alarm mechanism which is still present in many mammals and birds. He suggests that it serves to awaken, at intervals, those at risk but no longer awakens those not at risk. Kleitman (1939) considered that periods of light sleep might be a relic of the different sleep periodicities of the newborn (see pages 101–2). Neither of these arguments is very convincing but the elaborate slow-wave–REM-sleep control mechanism is likely to have some function.

81 Electroencephalogram of human subjects who are awake (active, relaxed or drowsy) and asleep (with and without periods of active sleep). Modified after Penfield and Jasper 1954.

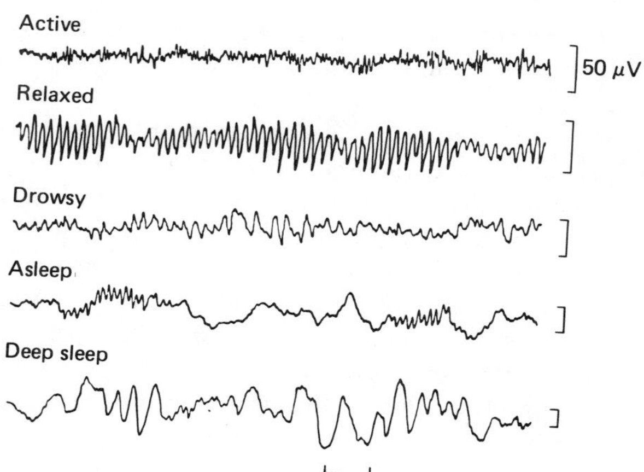

Sleep occurs in young birds (Schadé, Corner and Peters 1965) as well as young mammals and shows different patterning at different ages so the investigation of its development may shed some light on theories about its function. The origins of sleep behaviour might well have been a response to predation pressure and the consequent advantages of immobility when resting, but it is not possible to discount the suggestion that sleep allows some restorative process to occur. It is certain that physiological activities in body and brain are different during resting and during active periods. Some of the processes which occur in sleep control centres may be part of various regulatory mechanisms.

Avoiding physical and chemical hazards

Collisions and cliffs

Mechanical damage to the body, due to falls or other collisions with large or small objects, must be minimised by all individuals. The avoidance of damage caused by predators or competitors is discussed in Chapters 7 and 9 but other physical hazards necessitate complex behavioural responses, especially in large and fast-moving animals. Fish avoid swimming into objects, being hit by moving objects, being squashed, or being buried by responding to input from the lateral-line organs, the eyes or electric receptors. Terrestrial animals are subject to the additional hazard of falling down holes or cliffs and must use vision and hearing to detect all dangers. Flying animals travel faster than most of those which run or swim. They are therefore more vulnerable to injury if they collide with objects in the air and they may also fall to the ground. Vision is especially well developed in birds and in those insects which are largely aerial. Nocturnal flight without high risk of collision is facilitated in bats and in oil-birds (*Steatornis*) by the use of sonar. They produce high frequency sounds and listen to the sound reflected from solid objects (Griffin and Galambos 1941).

The ability to modify locomotor movements or to take evasive action when there is a danger of falling or of collision is improved by practice. The skills of people practised in sport or in some circus acts provide clear evidence for the effects of training. Young individuals must develop their perceptual abilities and with them their sensory-motor co-ordination. They must also improve their ability to decide whether or not to carry out activities in which they might be subject to physical hazard, for example to climb a tree or descend a cliff. Many experiments have been carried out in which the development of cliff avoidance in young animals has been studied. Most of these have utilised the 'visual cliff' developed by Gibson and Walk (1960). If adult mammals and birds are placed on a board with a shallow drop on one side and with glass, illuminated from below, over a deep drop on the other side, the 'deep' side is avoided. Young chicks show such avoidance as early as 3 h after hatching (Tallarico 1961) but young altricial mammals whose eyes have opened recently do not avoid the deep side. Kittens reared in darkness and tested on a visual cliff at 27 days do not avoid the deep side, as light-reared kittens do, and require a week of visual experience before they do avoid it (Walk and Gibson 1961). As mentioned in Chapter 3, Held and Hein (1963) showed that cliff avoidance and other behaviours requiring sensory-motor co-ordination, were impaired unless relevant reafferent input has occurred. The lack of behavioural avoidance of cliffs is often attributed, with reason, to inadequate perceptual ability but chicks or rats reared with a deep drop under the glass floor do not avoid the deep side of a visual cliff (Tallarico and Farrell 1964, Carr and McGuigan 1965). It is likely that these animals have adequate depth perception but that their decisions about avoiding the 'deep' side are different from those of normally-reared individuals.

Gibson and Walk reported that turtles do not avoid the deep side of a visual cliff but its similarity to a bank with water below it makes such a response seem adaptive. For most terrestrial animals, however, it is important to avoid falling into water and completely aquatic animals usually have behavioural mechanisms which reduce the likelihood that they will be stranded on land. A familiar environment reduces the dangers mentioned in this section; individuals are less likely to walk into solid walls if they have previously discovered by exploration that the wall is there. Exploration, in animals of any age and of any species, thus seems to reduce physical hazards as well as to improve feeding efficiency and reduce predation risk.

Avoiding contact with toxins – insect repellents

If an aquatic animal detects a toxic substance in the water around it, it takes evasive action. The direction for evasion is often determined by the use of information about a gradient of concentration of the substance and the animal usually has additional cues about its possible source. Toxic substances are encountered less frequently on land, except in the diet (Chapter 6) or as part of animal defence mechanisms (Chapter 7). Plant defences against insects and other herbivores do include the production of noxious chemicals and the odours of some of these must be detected before the animal touches the plant.

Evidence concerning the behavioural responses of insect pests to chemicals is now being accumulated and may prove very important in many methods of pest control. It has long been known that insects which bite man are deterred by smoke and also by camphor and plant substances such as cinnamon and citronella oil. A search for chemicals which would deter mosquitos and other insects from landing on human skin and biting it, produced first dimethyl phthalate and subsequently *N-N*-diethyl-m-toluamide. The latter substance, if spread on the skin, does not kill insects which land on it, but it does inhibit them from biting for several hours. Wright (1957) has suggested that the insects avoid the substance because it affects their humidity receptors in such a way as to create in the insect an illusion that the humidity in that area is dangerously high. Other insect repellents such as naphthalene and *p*-dichlorobenzene, which are used to protect clothes against clothes moths e.g. *Tineola*, are insecticidal but probably irritate the moth rather than killing it.

Responses of insects to insecticides

Observations on houseflies (*Musca domestica*) and on the malaria-carrying mosquito (*Anopheles*) have shown that some will not settle for more than a few seconds or minutes on DDT-treated surfaces and some will not enter DDT-treated buildings (Baranyovits 1951, Woods 1974). Such insects are behaviourally resistant to DDT because they never remain in contact with the insecticide for a time which is sufficiently long to ensure that they absorb a lethal dose. It is important to distinguish between behavioural and physiological resistance to pesticides for they must be dealt with in different ways in order to control the pest. Several clear examples of the importance of recognising behavioural resistance has come from work on insect pests of stored products, where insecticide resistance is a very serious problem (Champ and Dyte 1976). The susceptibility of different species, or different strains of the same species, to an insecticide which is encountered by the insect as it moves around may differ according to the general level of activity of the insect. Surtees (1966) showed that a strain of the grain weevil, (*Sitophilus granarius*) which was resistant to the insecticide pyrethrum, was less active in a test chamber than susceptible strains so, presumably, it encountered less insecticide per unit time. Similarly, Tyler and Binns (1973) found that a malathion-resistant strain of the grain beetle (*Oryzaephilus surinamensis*) was less active than susceptible strains. Further studies of the two species mentioned above and of another important pest of stored cereals, the beetle *Tribolium castaneum*, have shown that they actively avoid surfaces treated with the insecticides such as malathion (Figure 82). They

82 Half of each arena was treated with the malathion wettable powder as shown. When flour beetles (*Tribolium castaneum*) were put into the arenas they soon showed clear avoidance of the insecticide. Photograph, K. B. Wildey.

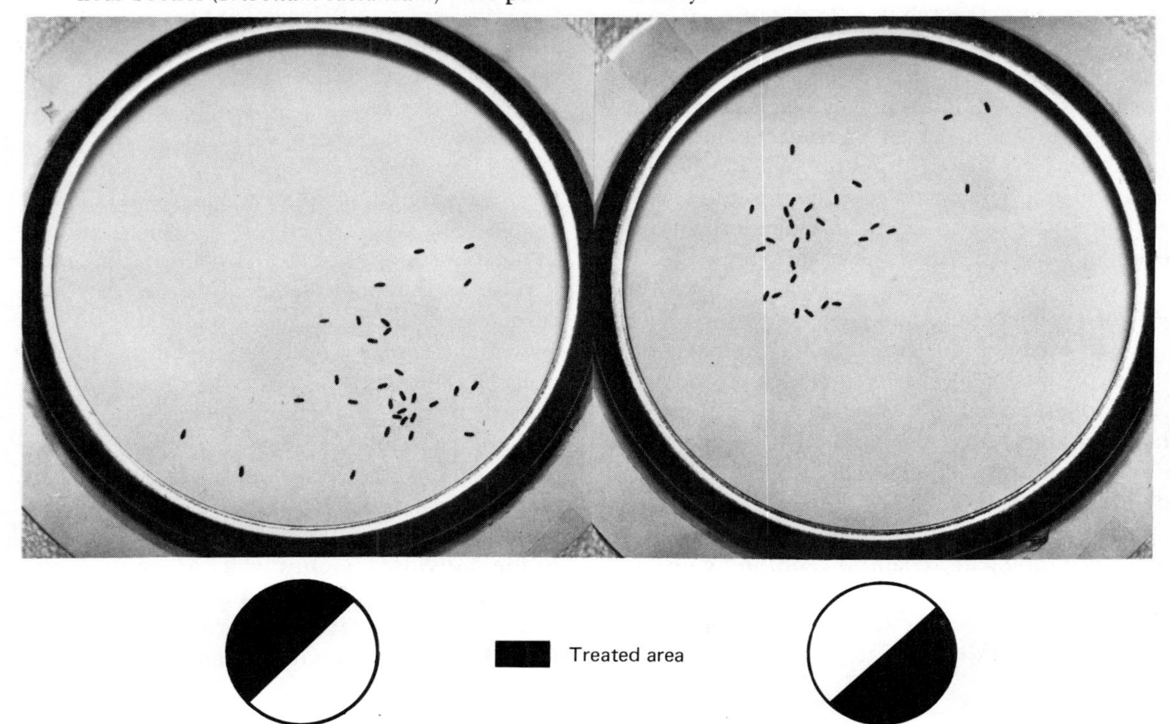

move away from the insecticide and, if possible, creep into crevices which the spray or powder has not reached (Pinniger 1974, 1975, Wildey 1977). Laboratory experiments showed that the beetles were more likely to enter small aluminium cylinders, provided as refuges, if insecticide was present on the surface of the arena outside the refuge. Strains of beetles which had been found to be resistant to the insecticide were more likely to enter the refuge and less likely to leave it whilst the insecticide was present outside. These results explain some of the reasons why the eradication of such beetles has been very much more difficult in some granaries than in others. Those with crevices in the walls are never cleared of the pest so crevices should be filled as much as possible in existing granaries and avoided when new granaries are constructed. Studies of the behavioural responses of the pests to toxic chemicals are helping in the control of these pests.

Very little is known of the responses of other stored product pests or of crop pests when different insecticides are encountered. This is an area where research is urgently needed. Responses to other chemicals also require investigation. The white fly (*Trialurodes vaporariorum*) is a greenhouse pest which is controlled by the use of the parasitoid *Encarsia formosa*. Irving and Wyatt (1973) found that the parasitoid will not remain on a surface treated with certain fungicides so its use in biological control is impaired by the presence of the fungicides.

Overload, pain and stress

In each of the functional systems mentioned in this chapter, if regulatory physiological and behavioural processes are failing to maintain the state of the animal within tolerable limits, additional action must be taken if the individual is to survive. The rate of change of state must be monitored by the use of sensory receptors, including pain receptors, and, in many situations, this information must be related to the rate at which resources are being utilised in order to carry out the regulatory processes. If the most effective regulatory process is operating but is still not sufficient, the operation of a mechanism for increasing energy availability and shutting down activities which are not vital at this time must be initiated. Some regulatory processes may be so costly, in terms of energy or increased predation risk, that emergency mechanisms start to operate whenever these processes start. The principal effectors of these emergency mechanisms are the hormones produced by the adrenal glands.

The concept of stress

The effects of adrenaline (also called epinephrine), which is produced in the adrenal medulla, were described by Cannon (1915). The idea of stress as a bodily response to various extreme situations was elaborated by Selye (1950), whose work concerned mainly the adrenal cortex. The supposedly non-specific reactions which Selye described are shown in response to competitors and predators as well as to oxygen lack, adverse temperature, physical damage etc. They may also be exacerbated by disease. The work of Mason (1975) and others has emphasised that different situations lead to different responses, so there is no single stress response, but responses in many situations share components.

The word *stress* appears in the literature on a wide range of subjects and there is confusion as to its use. Should it be used as in physics to mean force applied to an organism, or should it refer to the state of the organism, or to the response of the organism? Archer (1979*b*) uses it in all three senses and in his definition he calls it an 'inability' but he refers to stress most frequently as being something to which individuals respond. The topics discussed in his review are, however, linked in that they all refer to systems, with adrenal components, for coping with situations which involve actual or potential danger to the individual. Genes which promote the efficiency of such systems would clearly spread in most populations. High levels of adrenal activity must be very frequent in many individuals of many species; indeed Reynolds (1978) goes so far as to state that 'stress is the usual existential state'. Hence other systems must be able to operate in the presence of adrenal hormones and it is very difficult to decide at what point stress should be regarded as bad for the individual. There are certainly large individual differences, depending on previous experience and genotype, in the responses of the adrenals and in responses to levels of adrenal hormones.

Pain and analgesia

In addition to normal receptor function, pain is often an important prelude to the initiation of adrenal activity. Sensory overload, such as that produced by a very bright light or loud sound, is usually classified as pain and there are receptors in the skin which respond to painful changes. In mammalian skin the C-fibres respond to burning pain and adapt rapidly whilst the A-delta fibres respond to prickling pain and adapt slowly. It has been suggested that when tissue change leads to the production of proteolytic enzymes, these produce peptides such as neurokinin and bradykinin which depolarise the receptor cells. The pathway from these C and A-delta fibres terminates in the thalamus. The sensation of pain reported by human subjects varies according to circumstances for it is modulated by the secretion of analgesics which are similar in effect to opiates. Peptides which are produced in the brain and which have this effect were described by Hughes *et al.* (1975), but later work has shown that they affect different brain receptors from those which morphine affects (Chang and Cuatrecasas 1979) so there may be several sorts of analgesics produced in the brain. Such a pain-control system is useful, for whilst adrenal response to pain is initially advantageous to the individual, it is not so advantageous if the pain is prolonged or if there are very strong reasons for reducing metabolic rate.

Adrenal function

Adrenaline and noradrenaline are catecholamines which are produced in the adrenal medulla and released into the blood when the adrenals receive neural input from the brain. They are released when parts of the hypothalamus mid brain or cortex are stimulated (von Euler 1967). Noradrenaline is a neurotransmitter but its other effects on the body are similar to those of adrenaline but less pronounced. Adrenaline facilitates the mobilisation of energy sources. When adrenaline levels in the blood rise, the food reserve glycogen is broken down to sugars in liver and muscle and fats are broken down so that plasma fatty-acid levels rise. Adrenaline increases the rate of action-potential production in heart pacemaker cells and the strength of contraction in the heart and some other muscles. This, together with the vasoconstriction effect on peripheral blood vessels, increases the rate at which blood is pumped around the body and hence the availability of oxygen and energy sources for muscular activity. Arenaline also immobilises the gut. In some situations, at the same time as the initiation of adrenaline production, adrenocorticotrophic hormone (ACTH) is released from the adenohypophysis (formerly called anterior pituitary) in the brain. The hormone passes around the body in the blood stream and has various effects, the principal ones being on the adrenal cortex. The production of the glucocorticoid hormones, cortisone and corticosterone, is initiated and growth of the adrenal cortex may be stimulated if the ACTH level remains high for long enough. The glucocorticoids increase available glucose in the blood, for example by facilitating its production from protein. They also shut down non-vital activities so that after some time at high ACTH level the thymus becomes atrophied and the number of lymphocytes, plasma cells and eosinophils in the blood is reduced (Wolford and Ringer 1962). The sum total of these adrenal hormone effects is to aid vigorous activity.

Effects of aversive situations

Some of the situations which elicit either output from the brain to the adrenal medulla or ACTH secretion, involve the risk of predation or social disadvantage and are discussed in Chapters 7 and 10. Such responses are also elicited when any of the regulatory mechanisms described in this chapter is, or will soon be, inadequate. Examples include situations where there is too high a carbon dioxide level in the blood, no opportunity for rest, too high or too low a blood concentration, too high or too low a body temperature, unremoveable foreign matter on the body surface, danger of collision or falling, pain from burns, lesions, electric shock or chemical damage, sensory overload, severe restriction of movement, or high risk of the withdrawal of any of several essential resources if some complex task is not completed. In man, situations in which the risk is physical rather than social may result in the production of only adrenaline from the medulla, for example parachute jumping and dental surgery (Taggart, Hedworth-Whitty, Carruthers and Gordon 1976). Those in which the risk is complex, for example working on a machine-regulated factory production line in which physical and social risks exist, result in adrenaline and noradrenaline production (Johansson and Lundberg 1978). Situations in which competition between individuals is severe, for example driving in a motor race, result in high proportions of noradrenaline in adrenal medulla secretions (Carruthers 1978). The effects of these hormones can be reduced artificially by the administration of drugs which are 'beta blockers'.

The measurement of the levels of glucocorticoids in the blood is one of the easiest ways to assess adrenal response. Levine *et al.* (1967) and Zarrow, Gandelman and Denenberg (1967) found that newborn rats released corticosterone when subjected to electric shock, heat, histamine treatment, or partial etherisation. Glucocorticoids are also increased when people enter hospital, are etherised, undergo surgery, take examinations or fly aircraft (Mason 1968); when pigs are subjected to low temperatures for one hour (Baldwin and Stephens 1973); and when domestic animals are restrained or transported (Wood-Gush, Duncan and Fraser 1975). For example, glucocorticoids are produced during the first 30 min after calves are confined in small ($0.7m^2$) cubicles (Stephens and Toner 1975). A problem with many experiments of this kind is that the act of taking a blood sample elicits ACTH release. Stephens and Toner found that even sampling from previously implanted cannulae in calves increased levels of glucocorticoids.

If the situation which has resulted in an adrenal response, continues despite that response, secretion from the adrenal medulla declines but glucocorticoid levels often remain high for much longer. A recurrence of the situation will usually elicit a smaller medullary response than the initial one. One experimental situation which has been widely studied is that in which electric shocks are applied to the feet of a rat via the metal floor of its cage and opportunities for the avoidance of this shock are provided. The rat may be in a shuttle box, in which it can avoid shock by moving to another compartment, or it may be able to avoid shock by pressing a lever. Adrenal responses occur when a rat is shocked or when a shock is anticipated. Rats modify their behaviour to avoid shocks, for example by moving to another compartment in a shuttle box, by pressing a lever at appropriate times or rates, or by lying on their backs with their electrically insulating fur rather than their

83 The mean daily urinary adrenaline levels in rhesus monkeys before (Control), during and after 72-hour shock avoidance trials. From Archer 1979*b*.

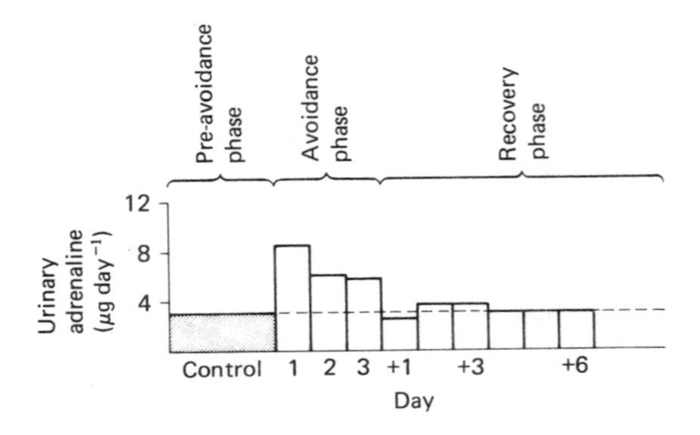

feet on the floor (see review by Fantino and Logan 1979). When Mason *et al.* (1968) kept rhesus monkeys in a situation where, for three days, they had to press a lever in order to avoid a shock, their adrenaline production was high on the first day but declined on the second and third day and returned to normal after the tests (Figure 83). Substantial increases in levels of glucocorticoids and some changes in levels of various other hormones were also recorded during the test period. The ability to learn shock avoidance does not depend upon adrenal function for it can occur after adrenalectomy. The injection of ACTH increases the persistence of avoidance responses (Levine and Jones 1965) and retarded their extinction (Murphy and Miller 1955) even in adrenalectomised animals (Miller and Ogawa 1962). Such effects are presumably beneficial to the individuals, as are the immediate effects of adrenal products.

Prolonged high levels of ACTH and adrenal products must result in inefficiency in certain body systems. If the production of various blood cells or the functioning of the gut are impaired for long periods, susceptibility to disease may be increased (Freeman 1971), ulceration of the stomach may ensue and digestive efficiency may be reduced. There is also some evidence for increased arteriosclerosis and consequently greater risk of heart failure after prolonged adrenal activity in man. These effects and the factors which affect the susceptibility of individuals to them, are the subject of much current medical research.

6
Feeding

Decisions, diets and the optimality approach

Feeding is an important functional system because it provides the source of the energy which is used by animals. The system includes brain mechanisms, with their many behavioural consequences, which result in food intake, and physiological components which involve, firstly, digestion and secondly, the production, storage and translocation of energy-rich compounds. All of these components, together with the nutrient requirements of the body, must be considered in order to understand the allocation of energy to feeding as a functional system (Chapter 4) and to each energy-requiring process within the system.

The description of anatomical and behavioural adaptations for feeding has been a major activity of biologists for many years. Such descriptions, for a wide variety of animals, are available in scientific textbooks such as Fretter and Graham (1976) or Welty (1962) as well as in many general natural history books. Psychologists have used food as a reinforcer in a high proportion of experimental studies on learning. Such work has provided much information about the timing of feeding, the quantities of food taken and the amount of work which will be performed for artificial diet rewards in a laboratory. Information about the brain mechanisms controlling feeding behaviour has come from studies of the effects of the stimulation, ablation or chemical treatment of the brain. Other areas of research on feeding have included studies of human nutrition, the relationships between food intake and production in domestic animals, and the feeding behaviour of insect and rodent pests. Recent ideas in quantitative and behavioural ecology are encouraging interactions between research workers with these different approaches to the study of the feeding functional system.

Decisions about feeding

An individual whose motivational state is such that behaviour will be directed towards feeding (see Chapter 4) must take a series of decisions before and during feeding. This series of decisions (Figure 84) and the set of mechanisms which makes them possible, is the subject of this chapter. The interrelations between feeding and other functional systems, especially predator avoidance (Chapter 7), are discussed in Chapter 4. After a brief discussion of dietary differences and optimality the sections in the chapter refer, in temporal order, to these decisions.

Where the word *foraging* is used it refers to the behaviour of animals when they are moving around in such a way that they are likely to encounter and acquire food for themselves or their offspring. Thus it includes sandpipers finding crustaceans on a beach, sheep grazing in a field, hyaenas hunting wildebeeste, and cabbage white butterflies finding and ovipositing on cabbages. As shown in Figure 84, decisions during foraging will depend on potential energy and nutrient returns and on costs. These costs will include those of searching, pursuit, handling and eating (Schoener 1971); those associated with digestion (Westoby 1974); those of detoxifying poisonous substances in the food; and deleterious effects of substances in the food. As Altmann and Wagner (1978) put it 'given the foods that are available to an animal, how much of each should it consume so as to be above its minimum for each nutrient, below the maximum for every toxin, and concurrently, either to minimise some cost function, such as time expended in foraging or exposure to predators while feeding, or else to maximise some benefit obtained from feeding such as protein or caloric intake'. Foraging will be affected by previous experience in similar situations so the effects of experience are emphasised in relation to many of the examples quoted.

The ecological approach to the study of feeding behaviour is complemented by consideration of physiological mechanisms. These are mentioned in several sections and the section on the control of food intake deals with receptors and brain mechanisms. Most of the studies mentioned in the early sections of this chapter refer to only one aspect of feeding, so two examples of species in which most aspects of the feeding system have been studied, the hummingbird feeding on nectar and the cow grazing on grass, are discussed at its end. The influences

84 Some decisions about feeding and some variables affecting them.

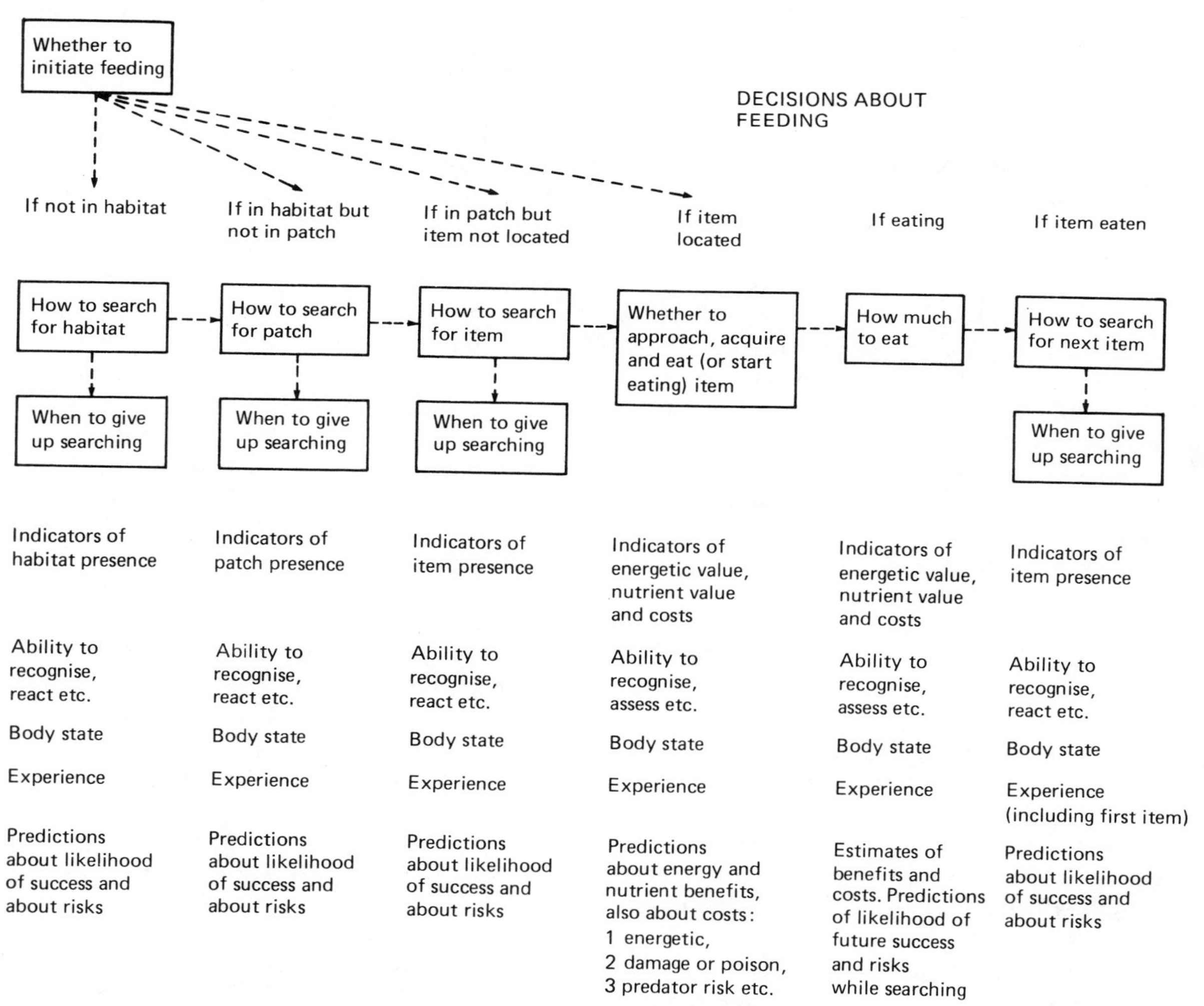

of social factors on feeding strategies are mentioned in this chapter and are discussed further in Chapter 8 and Chapter 10.

Diets and arms races

Feeding behaviour is often divided into separate categories according to whether the food material comes from living animals, living plants or dead organisms. Animal morphology is, to a large extent, a result of adaptations for feeding so there are considerable differences between some animal-feeders and some plant-feeders, but the above categorisation of feeding behaviour is not always useful. Animals which are suspension feeders may accept animals, plants or organic detritus as food. A warbler may hunt for berries or for insect larvae by methods with similar components. Limpets grazing on rocks eat algal sporelings and barnacle larvae, whilst a nudibranch mollusc which is grazing on hydroid colonies will show behavioural similarity in its feeding methods to an insect or mammal grazing on plants. Different digestive mechanisms are required for some components of animal and plant food but there is much variation within vegetable and animal diets and some plant-eaters actually absorb the products of commensal animals (Protozoa) or bacteria.

Although there are often similarities in the feeding behaviours, digestive processes, etc. of animals eating food of diverse origins, some aspects of feeding are specific to one type of diet. The digestive process in grass-eating ruminants has no parallel amongst species in which the diet is composed of animal material. Likewise, no plant-eater needs to chase prey actively so none has need of the adaptations for efficiency of capture possessed by a wolf or a peregrine. Some herbivores are very different in their feeding from some predators but the terms are less useful when considering the many animals with intermediate feeding behaviour and physiology. The most important behaviour which makes feeding possible for parasites is usually host seeking. The behaviour of a parasite looking for a host should perhaps be considered as a form of habitat selection but some of the behavioural mechanisms are similar to those shown by nonparasitic animals when hunting for food.

Dawkins and Krebs (1979) have pointed out that the evolution of predator and prey species can be regarded as an 'arms race'. Modifications of genotype continually result in an improvement in the chance that the average individual fox will catch a rabbit or that the average rabbit will be able to evade foxes. A similar 'arms race' exists between herbivores and plants, for plants can produce toxic or unpleasant tasting substances and can bear spines, hairs and other deterrents to would-be consumers. Animals can modify their digestive systems and feeding methods to avoid the adverse effects of these plant weapons. Plants may also be able to modify their growth form and their ability to grow in places which are inaccessible to the animals. This arms race is a very slow one which takes many generations and the characteristics of a potential consumer, or a potential food, represent only a few of the many factors which might influence evolutionary changes. The evolutionary competition between lettuces and rabbits is, however, just as real as that between rabbits and foxes. Animals which eat dead matter compete with other organisms which are trying to consume their food, rather than with the food itself. (See also page 158).

Specialist and generalist feeders

A major factor which affects feeding behaviour is the distribution of food. Distribution must be assessed in relation to the size and locomotor ability of the animal. Some animals are surrounded by readily accessible food whereas others may live for long periods without encountering a food item. Before reviewing behavioural methods for finding and acquiring food it will be useful to consider the degree of dietary specialisation or generalisation which is shown by animals. The extent of specialisation which is possible depends upon the abundance and the spatial and temporal distribution of possible food items. Most specialists eat food which is locally or generally abundant but generalists can eat foods which they encounter rarely (Emlen 1973). Animals which eat only one type of food are called *monophagous*. Those which eat a limited range of foods are referred to as *oligophagous* whilst those which will eat a wide variety of foods are *polyphagous*. Examples of species which are usually monophagous include the giant panda (*Ailuropoda melanoleuca*) which eats bamboo shoots, the snail kite (*Rostramus sociabilis*) which eats the freshwater snail *Pomacea*, the grasshopper (*Gesonula punctifrons*) which eats the water hyacinth (*Eichhornia crassipes*), the chalk-hill blue butterfly caterpillar (*Lysandra coridon*) which eats the leaves of the horse-shoe vetch (*Hippocrepis comosa*), and many parasitic species. Animals which are oligophagous include the koala (*Phascolarctos cinereus*) which eats leaves from about five species of *Eucalyptus* trees, the brent goose (or brant) (*Branta bernicla*) which seldom eats anything other than species of eel grass (*Zostera*), the grasshopper (*Chorthippus parallelus*) which eats grasses, and the larvae of butterflies in the family Pieridae which eat plants of the family Cruciferae with mustard oils in their leaves. Many animals are polyphagous. Examples of species of similar sizes and from the same groups of animals as those mentioned above are bears, rats, man, the jay (*Garrulus glandifarius*), the great bustard (*Otis tarda*), the locust *Schistocerca gregaria*, and the moth *Spodoptera littoralis*.

It seems likely that the ancestral forms of most animal groups were polyphagous. Emlen (1973) points out that if the members of a species encountered one food much more often and therefore ate it more frequently than others which provided the same benefit in terms of fitness, the average ability to find, ingest and digest that food would probably improve. Such dietary selection would, in a few generations, result in this food being of greater benefit than other foods. Genes which promoted a preference for this food would then spread in the population and the first step towards monophagy would have been taken. The digestion of one food requires a smaller range of enzymes than is required by a polyphagous animal. The anatomy and the behavioural repertoire can become more specialised and hence more efficient if only one food is eaten. Levins and MacArthur (1969) emphasised that monophagy might be advantageous, in terms of the number of young produced, because there is less risk that an inadequate food source would be chosen. A major problem for monophagous animals is what to do when the food source is scarce. Scarcity is a relative term here for it refers to food which cannot be found in quantities adequate for survival by individuals of that species. Some species live through periods during which the food upon which they specialise is absent, by accumulating fat reserves, hibernating, aestivating, encysting, producing long-lived eggs or by hoarding the food when it is abundant. Food is hoarded by squirrels, rats, jays, acorn woodpeckers (*Melanerpes formicivorus*), ants and man. Specialist feeders thus behave in a way which minimises the variance in the availability of their food when they require it. Animals living in unstable habitats cannot afford to be specialists and thus require a more complex repertoire of feeding behaviour.

Polyphagous animals can exploit food sources which are available briefly and they risk less if they move away from a food source because of predator activity or competition. Monophagous animals can often be found easily by predators which merely search for the food source but polyphagous animals are less readily found in this way. Another advantage for some polyphagous animals is that they do not have to range so far in order to find food so they can live in a home area for long periods. Both polyphagy and monophagy have advantages and the effectiveness of one as a strategy in a habitat will depend upon the diversity and stability of population of potential food organisms and on the strategies adopted by competitors in that habitat.

The optimality approach to studies of feeding

An assumption which is made by many of those who speculate upon the mechanisms of feeding behaviour is that, since the genotype is the result of a long period of evolution by natural selection, the behaviour of individuals today may be close to the optimum for the circumstance (Schoener 1971, Pulliam 1974, Pyke *et al.* 1977). As explained in Chapter 1, a consequence of this assumption is that the energetic costs and benefits of feeding behaviour have been measured in many studies. The result has been a great improvement in our understanding of feeding mechanisms. The optimal foraging approach advocated by MacArthur and Pianka (1966) and Emlen (1966) has been vindicated by combined studies of feeding ecology, physiology and behaviour such as those discussed in detail later in this chapter.

Any attempt at cost-benefit analysis of feeding must take account of physical constraints, digestion costs and the necessity for acquiring the complete range of essential nutrients as well as the energetic costs of the feeding behaviour and the energetic returns from the food. As stressed in Chapter 1, however, it is not sufficient to assess optimality in terms of energy obtained in relation to energy expended. The ultimate measure of optimality must refer ideally to the spread of genes in the population. In the absence of altruistic behaviour towards individuals other than offspring, the long-term reproductive potential of the animals under consideration is the factor which is optimised. Statements about optimality of feeding strategies are not useful unless anti-predator behaviour, avoidance of other hazards, the regulation of body temperature etc. are also considered. This same argument applies to statements about whether a feeding strategy is evolutionarily stable.

Finding food

Members of most animal species have a dispersive stage in their life history so that they need to have the ability to distinguish between areas where food might occur and those where it will not. Once such an area is found, other behavioural mechanisms which increase the chance of encountering food items and make possible their acquisition or rejection, come into operation (Figure 84). The area can then be assessed, the effects of food intake monitored, and further behaviour modified accordingly. The feeding behaviour of some animals includes all of the aspects mentioned in Figure 84, whereas that of others is much simpler. Monophagous animals which spend most of their life eating one individual plant or animal may still have to take decisions about which part of the food organism to eat but their feeding behaviour is less complex than is that of a polyphagous animal which can feed in various habitats.

Finding an area where food might occur

Patches and habitats

The distribution of the food of most animals is clumped (see Chapter 8) so that the individual must find areas of local concentration amidst areas of low concentration or absence. Many people call such an area of local food concentration a *patch*. An environment which includes such patches is called *patchy*. Before finding a patch, the individual must find the sort of habitat where sources of its food might occur. The importance of food in determining what can be a suitable habitat is emphasised by Hassell and Southwood's (1978) description of a *habitat* as 'a collection of patches'. For most animals the habitat must have other characteristics as well. Animals often live in a particular area long enough to learn many of its characteristics. Such familiarity is beneficial when evading predators and reducing physical hazards, as well as when seeking food.

Examples of patch-finding include a squirrel finding a nut tree, a mountain sheep finding an area of grass on a rocky mountainside, a goshawk finding a flock of pigeons in a field, or a ladybird beetle finding an area on a plant with an aggregation of aphids on it. The amount of food which can be regarded as a patch for a large animal may be a lifetime's supply for a small one. One fruit may sustain a moth caterpillar for the weeks during which its larval development occurs but may be consumed in seconds by a monkey. A patch for the monkey might be, for example, one, several or many ripe fruits on a tree. Species like fruit-eating monkeys often have to travel for some time between patches whereas leaf-eaters exploit much larger patches and travel less. For example in a study of the spider monkey (*Ateles geoffroyi*) the animals were shown to spend 28% of daylight in travelling from one forest tree bearing ripe fruit to another (Richard 1970). In contrast, the similar-sized black and white colobus monkey (*Colobus guereza*) ate leaves during 77% of all observed feeding time and travelled for only 5% of the day (Oates 1977). These figures are typical of a general relationship for primate species between time spent travelling and proportion of foliage in the diet (Clutton-Brock and Harvey 1977*b*).

Although there is variation amongst species in the amount of time spent searching for patches, this behaviour is very important for almost all animals. Efficient searching will result in finding better quality patches and hence improve the reproductive potential of the individual.

Habitat selection experiments

Casual observations of habitat selection have been reported frequently and some detailed experimental studies have been carried out. Wecker (1963) found that two subspecies of the deer-mouse (*Peromyscus maniculatus*) occurred in two habitats, grassland and mixed oak and hickory woodland. Members of the grassland subspecies kept in an enclosure with grassland on one side and woodland on the other showed a clear preference for the grassland: they are, presumably, better adapted for feeding etc. in grassland. When Wecker reared some mice in the grassland habitat and others in laboratory cages the individuals reared in grassland showed a stronger preference for it. If the mice were reared in the woodland habitat, they also showed a preference, albeit weaker, for the grassland habitat.

Observations by Gibb (1957, 1960) showed that coal tits (*Parus ater*), which feed most frequently in coniferous trees, preferred coniferous branches in laboratory experiments whereas blue tits (*Parus caeruleus*), which feed in deciduous trees, preferred deciduous branches (Figure 85). Partridge (1974) reared tits of both species in a laboratory environment where they had no experience of any vegetation. When full-grown, the coal tits showed a preference for pine branches whilst the blue tits preferred oak branches (Figure 85). Although the preference was not as marked as that of birds caught in the wild, Partridge's experiment demonstrates that the mechanism which determines the preference can develop in the absence of any experience of the preferred trees. Partridge's conclusion that the preference is 'genetically determined' could be taken to imply that environmental factors are not involved in the development of the mechanisms controlling the behaviour. As emphasised in Chapter 1, all

85 The proportion of time spent in oak or Scots pine branches by blue tits and coal tits given a simultaneous choice. *a*, wild birds; *b*, hand-reared birds. The birds prefer the tree species in which they are found most often in the wild. After Partridge 1978.

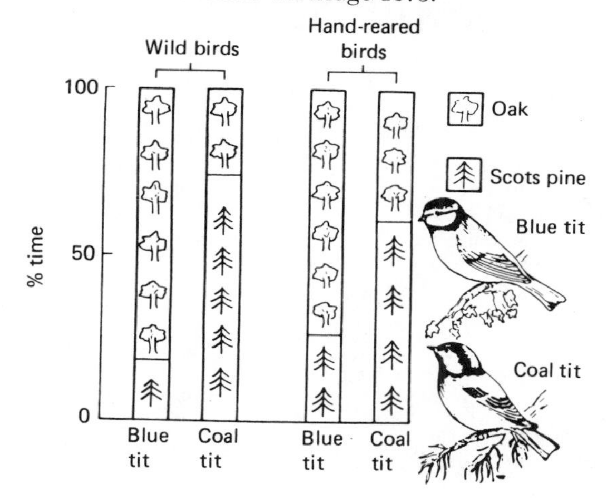

behaviour depends on genetic and environmental factors but Partridge's results do show that input which results from contact with the trees is not amongst the environmental factors which affect the development of this preference. Further experiments and observations of the skills associated with feeding in coniferous or deciduous trees showed that the tits preferred the trees for which their skills were best adapted (Partridge 1978). Studies on a variety of species show that some aspects of habitat preference are modified by experience whilst others are not (Klopfer 1969, Anderson 1973).

Host-finding behaviour

Suitable habitats, and consequently food sources, can be found by moving around and responding to visual and olfactory cues; sometimes at a considerable distance from the habitat. The birds and mammals mentioned above respond to a complex set of sensory cues but an *Anopheles* mosquito which emerges as an adult after its aquatic early life can find human dwellings by a comparatively simple strategy (Figure 86). It flies across wind until it encounters the odour from human dwellings and then flies upwind. Such behaviour is similar to that of active aquatic predators, such as sharks, which will swim up concentration gradients towards bodies which are emitting blood and the starfish (*Asterias*) which will move up-stream readily when the body fluids of their molluscan prey are present in the water (Castilla 1972).

86 The mosquito *Anopheles gambiae* finds human dwellings by flying across wind and then upwind when human odour is detected.

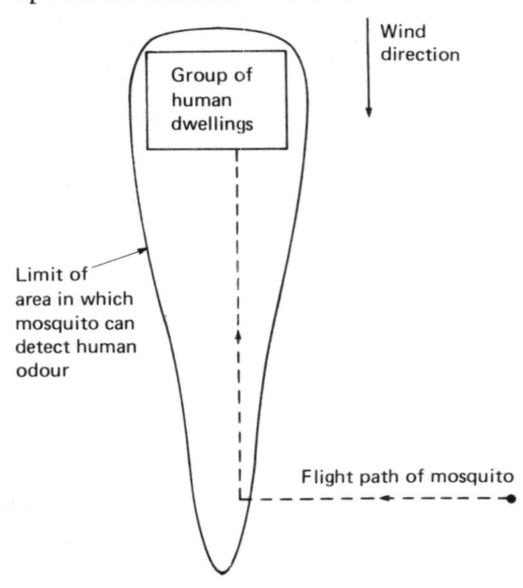

Animals which are not sufficiently mobile to find the required food source have to modify their own behaviour, or modify their surroundings, in such a way as to maximise the chance that they will encounter their food. The habitat of the sheep tick (*Ixodes ricinus*) is a pasture with sheep in it and within this habitat, a patch is a sheep. Sheep ticks increase the chance of encountering a sheep which comes near them by climbing to the tops of the stems and other vegetation. Provided that they do not lose too much water they remain at the tops of the stems until they are disturbed by something whose characteristics are, ideally for the tick, that it is moving, warm and woolly (Lees 1948, Lees and Milne 1951).

Internal parasites are extreme examples of animals which have problems in finding the next feeding site. For a review of host-finding behaviour by miracidia larvae of flatworms see Saladin (1979). There is a large literature on the complexity of parasite life cycles but amongst the most interesting species are those which hi-jack one host or just hitch-hike in order to reach the next. Many parasites have a general debilitatory effect on one host which is then more vulnerable to attack by the parasite's next host. Some parasites, however, modify their environment, that is to say the host, in such a way that the host's behaviour is changed. The acanthocephalan parasite *Polymorphus paradoxus* lives in the freshwater amphipod crustacean *Gammarus* and has to get to its next host, the mallard duck (*Anas platyrhynchos*). Bethel and Holmes (1973) found that, whereas 97% of uninfected *Gammarus* were found in dark areas in an experimental apparatus, only 20% of infected *Gammarus* remained in the dark. The behaviour of the other 80% was altered by the presence of the parasite so that they swam in the light and clung to objects near the water surface. Here they are much more likely to be eaten by ducks than are individuals which remain in the dark areas near the bottom of a pond or stream. A study of the effects of the eyefluke *Diplostomum spathaceum* on the behaviour of fish showed that heavily parasitised dace and trout spent more time swimming near the surface of the water than did those with few parasites (Crowden and Broom 1980). Here they are more likely to be eaten by gulls which are the next host of the flukes.

Finding patches by observing conspecifics

The importance of responding to other individuals, often in the same social group, which move in directions which might lead to a food patch is described in Chapter 8. These ideas were initiated by observations of small birds joining flocks which move towards patches, but there are also examples of large birds in the air keeping visual contact with one another so that when one descends, although it might prefer not to share its find, the others fly towards it. This patch-finding technique is used by vultures finding carcases and by seabirds finding concentrations of fish near the surface of the sea. Even if the individuals moving towards a patch are not seen, the patch itself may be rendered much more conspicuous by the presence of an aggregation of conspecifics or individuals of other species which may feed in similar patches. The presence of these individuals is an indicator of environmental quality (Kiester and Slatkin 1974). In experimental studies, Krebs (1974) found that great blue herons (*Ardea herodias*) were more likely to land at possible feeding sites at which he had put models of herons (see Chapter 8).

Finding food items

Aggregation in patches

The efficiency with which patch-finding mechanisms operate in a variety of species is attested to by the aggregation of animals in regions of high food density. Such aggregation occurs in species in which individuals find their way to the patch separately as well as in species whose members normally move around in groups. Some examples of animals whose distribution is much affected by that of patches of food, are shown in Figure 87. The occurrence of the aggregation is due in part to the use of similar patch-finding mechanisms by many individuals but is accentuated by the greater likelihood that individuals will stay in an area of high food abundance. The advantages of staying in the patch may be counteracted, however, by the disadvantages which result from competition with the other individuals in the patch. The criteria used when deciding whether or not to remain in a patch are discussed later in this section whilst competititon for food amongst members of social groups is discussed in Chapter 10.

Scanning, recognition & reactive distance

Many animals enhance their chances of finding food by *scanning* their surroundings. A monkey looking for a fruit, a mantis looking for a fly, or an owl sitting on a perch at night listening for the rustle caused by a mouse, all adjust the head position so that the receptors can detect and locate the food. Such scanning, whether or not accompanied by locomotion, may account for a high proportion of foraging time, for example in the thrushes observed by Smith (1974*a*). It may be used during patch finding as well as during the detection of items within a patch and is an important factor in food finding by many species (Treisman 1975*a*).

The *recognition* of food items depends upon the functioning of sense organs, together with sensory analysers in the brain (Chapter 2, see also discussion of search images on p. 137), and on the operation of attentional mechanisms (Chapter 4). The efficiency with which sensory analysers function is a factor which limits the distance at which food items can be recognised. The ease of recognition will also depend upon the background against which the item must be detected. Most research on this topic refers to detection by means of vision or olfaction but some animals use other senses. They may have to modify the environment in some way, e.g. by digging into the ground or by moving vegetation, before they encounter a food item.

Precise evidence about the distances at which food is perceived by different species is sparse although we as individuals know the approximate distances at which we recognise mushrooms, berries, shellfish or other items for which we are hunting. Estimates of *reactive distance* have sometimes been made in studies of feeding. Beukema (1968) recorded the distance at which sticklebacks (*Gasterosteus aculeatus*) turned towards or accelerated towards food items. When the food was the small red worm *Tubifex* the reactive distance was 25cm. For another small fish, the dace (*Leuciscus leuciscus*) (Figure 88), the reactive distance to the amphipod crustacean *Gammarus* varied from 20 to 50cm according to the numbers of eyeflukes (*Diplostomum*) in the eyes of the fish (Crowden and Broom 1980).

The effects of experience on the reactive distance is considerable in situations where the

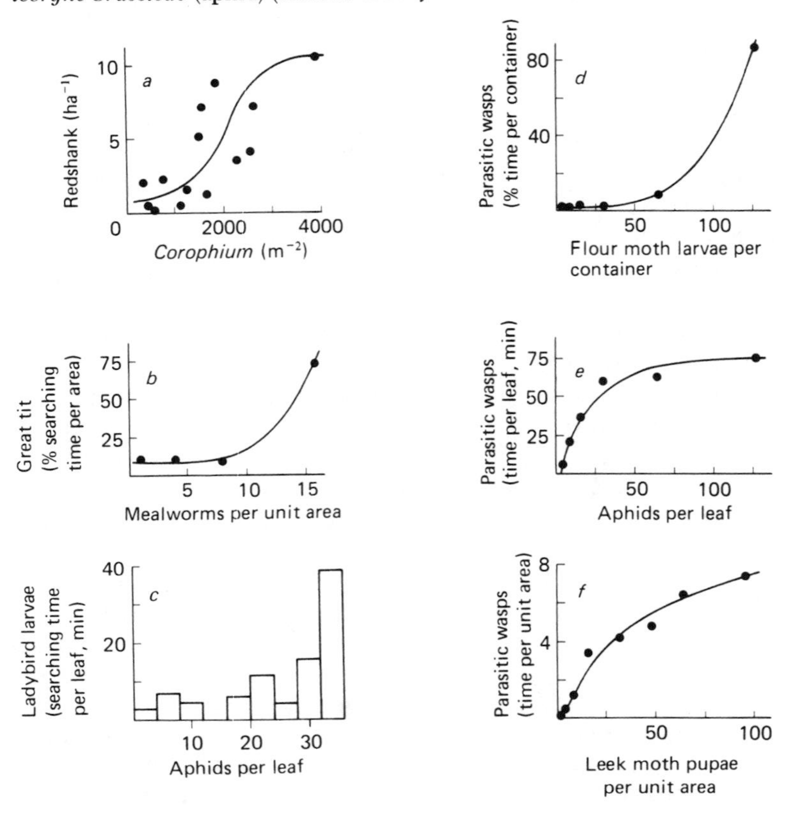

87 Foraging strategies often result in the aggregation of animals at food patches. Examples shown here are (a) *Tringa totanus* (redshank) finding the small crustacean *Corophium volutator* (Goss-Custard 1970), (b) *Parus major* (great tit) finding *Tenebrio mollitor* (meal worm) (Smith and Dawkins 1971), (c) *Coccinella septempunctata* (ladybird) finding *Brevicoryne brassicae* (aphid) (Hassell & May 1974), (d) *Venturia canescens* (parasitic wasp) finding *Ephestia cautella* (flour moth) (Hassell 1971), (e) *Diaeretiella rapae* (parasitic wasp) finding *Brevicoryne brassicae* (aphid) (Akinlosotu 1973), and (f) *Diadromus pulchellus* (parasitic wasp) finding *Acrolepiopis assectella* (leek moth) (Noyes 1974). Modified after Hassell and May 1974.

hunter does not know, initially, the precise characteristics of the prey. When Beukema started to provide sticklebacks with the larvae of the small fly *Drosophila,* they failed to react to the prey at 10cm distance on most of the first twenty presentations. By the fifth set of twenty presentations, however, they reacted to almost all larvae at that distance and to some at 20cm (Figure 89). No stickleback reacted to *Drosophila* larvae at 30cm but some did react at this distance to the larger, red, wriggling *Tubifex*. The reactive distance to a food item, after extensive experience of feeding on that food alone, is likely to be limited by sensory ability but attentional mechanisms are more likely to determine reactive distances in other situations.

88 Dace (*Leuciscus leuciscus*) about to catch the small crustacean *Gammarus*.

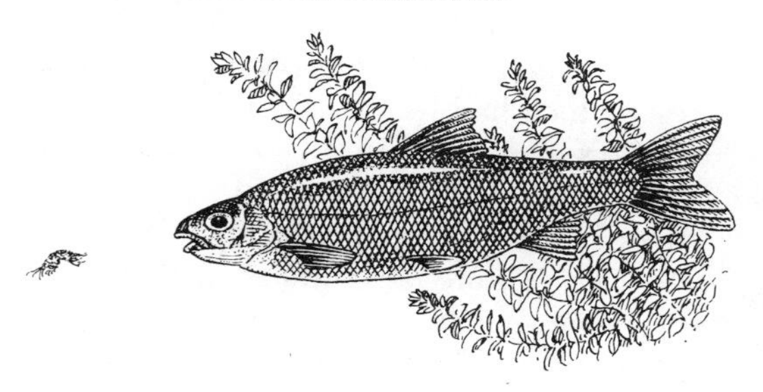

89 Sticklebacks do not react often to *Drosophila* larvae during the first block of 20 trials but the mean reactive distance is less than 10cm by the third block of trials and is almost 20cm by the fifth block of trials. Data from Beukema 1968.

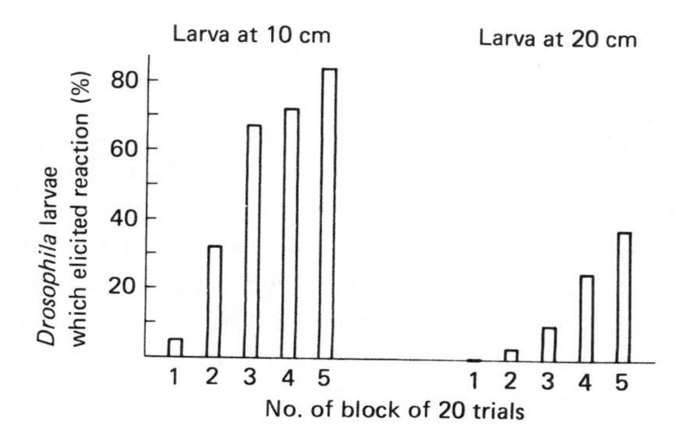

Search paths

The movements of individual animals searching for food within a patch have been studied in detail for several species of birds and insects. The search is unlikely to be random for this would involve repeated search in some areas and would be inefficient unless the food resource was renewed very rapidly (Krebs 1978). Ornithologists have long known that owls systematically 'quarter' fields when hunting for small mammals and that feeding flocks of pigeons or starlings move around fields in such a way that they do not recross their own paths. The direction of movement of an owl is, approximately, a straight line except at the edge of the field where the owl swings around so as to return across the field on a path parallel to the first. The distance between successive traverses of a field must depend upon the distance from which the food item can be recognised and obtained by the owl. The area around a prey species which is a circle whose radius is this distance, has been called the 'danger zone' (de Ruiter 1956). A similar term, but one which takes into account the presence of other prey individuals, is the 'domain of danger' (Hamilton 1971, Chapter 7). These observations of owls and other hunters also indicate that predator species control their movements precisely, use short-term memory of the topography when computing courses and make decisions about what constitutes a boundary of the hunting area. Taking into account all of the factors mentioned above it is possible to formulate models of the optimal searching paths for different hunters in different situations and several authors have attempted to do this (e.g. Cody 1971, Smith 1974*b*, Pyke, Pulliam and Charnov 1977, Pyke 1978*a*,*b*).

Detailed descriptions of searching paths were made by Smith (1974*a*,*b*) who watched blackbirds (*Turdus merula*) and song thrushes (*T. philomelos*) hunting for earthworms in a meadow. The birds moved forward for about 0.5 s and then paused to scan for a mean of 4.8 s before moving again. The mean move length was 340mm for male blackbirds and 450 mm for song thrushes. If no worm was found, the succession of moves often included alternate right and left turns and the beeline direction was approximately straight. If food was found, the thrush often made two or more turns in the same direction and the beeline distance of the twelve moves after capture was shorter than that during the ten moves before capture (Figure 90). In some experiments artificial food, in the form of pastry caterpillars, was provided. Smith found that the search path included more turns and less alternation of turns when thrushes, which had been moving through an area of low food density, encountered an area of high food density. He called this behaviour *area concentrated search*, a more precise description than 'area restricted search' which other authors have used.

Drent and Tinbergen (personal communication) and Tinbergen and Drent (1980) obtained a similar result from observation of starlings hunting leatherjackets, larvae of crane-flies in the family Tipulidae, in grassland. When starlings were hunting for food to take to their nestlings, they were observed to search in comparatively straight lines until leatherjackets were found, then to show area concentrated search. When they had obtained a supply of food they flew off to their nest and then returned to the spot where the high concentration of food had been found. Not only do the birds remember the previous rate at which food had been encountered, when deciding whether or not to adopt area concentrated searching, but they remember localities in open grassland with great precision.

90 Movements of a blackbird hunting for worms. The beeline distance during five moves before capture is longer than that after capture. Modified after Smith 1974a.

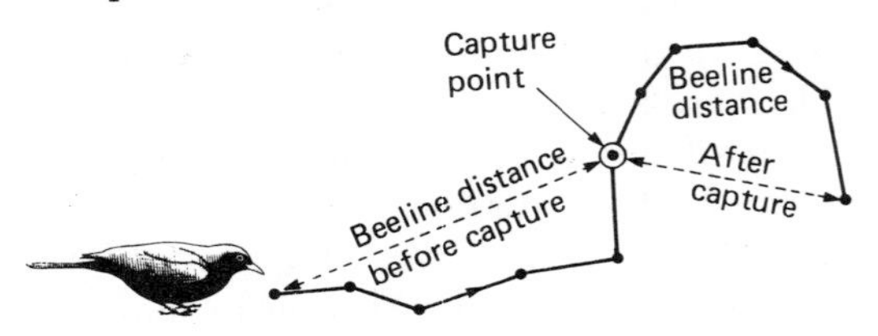

The search paths shown by captive ovenbirds (*Seiurus aurocapillus*) hunting for mealworms, *Tenebrio* beetle larvae, in a 6m × 6m grassy arena were also affected by present and previous food density (Zach and Falls 1976*a,b,c*). When most of the area of the arena was subdivided into four paths which were provisioned with mealworms at densities of 0.9, 1.8, 3.6 or 7.1m^{-2}, the number of visits to a patch and length of search path in a patch were much higher in the highest density patch. The logarithms of the number of visits and of the search-path length were proportional to food density. The length of path per visit and the meander ratio (actual path length/beeline distance) were proportional to food density. The overall result of this searching behaviour was that the highest proportion of prey was taken from the highest density patches. When ovenbirds which had fed in an arena provisioned in this way were returned to it on the following day they immediately started searching in the area which had been most profitable on the previous day (Figure 91). If there was no food there they soon searched elsewhere and if there had been two profitable patches they visited both. Zach and Falls (1976*c*) also provide evidence for the avoidance by ovenbirds of areas which have been searched thoroughly, whilst Thomas (1974) describes the avoidance by sticklebacks of areas in which possible food items had been found but rejected.

Complex search paths are also found amongst invertebrates, for example the bumble bees *Bombus* studied by Pyke (1978*a*). The bees collect nectar and fly from flower to flower in a field on comparatively straight paths but with alternation of left and right turns. In accounting for all aspects of the search paths shown, Pyke concluded that the bumble bee must remember its arrival direction at a flower, its change of direction at the previous inflorescence and the amount of nectar obtained from the flower just visited.

Foraging in relation to food quality and density

Item size

In most of the examples of feeding behaviour quoted so far in this chapter, the ways in which animals deal with situations where there is only one type of food item have been considered. Alternative food items may vary in size, and hence in energy return when the item is ingested. Just as there are components in feeding behaviour which increase the probability of obtaining more food items per unit of energy expended, optimal foraging theory predicts that if items give a greater net energy return they are more likely to be selected.

Where animals of various species are offered the opportunity to search for and to acquire food items of different sizes, such that the larger items are more energetically rewarding, the proportion of larger items taken is larger than that available in the population (see Krebs 1978), especially at high food densities. Examples include bluegill sunfish (*Lepomis macrochirus*) eating the crustacean *Daphnia* (Werner and Hall, 1974), great tits eating mealworm pieces (Krebs, Erickson, Webber and Charnov 1977), redshank (*Tringa totanus*) eating *Nereis* worms (Goss-Custard 1977*a*) and shore crabs (*Carcinus maenas*) eating mussels (*Mytilus edulis*) (Elner and Hughes 1978).

Nutrient quality

If the potential food items which might be encountered differ in quality as well as in size, animals use behavioural mechanisms to maximise the chance that they will ingest items which provide the most energetic and nutrient benefit. In polyphagous species a mixture of different types of food is often needed. The mechanisms used to obtain the food items of the best size

91 The search paths of a typical ovenbird *Seiurus aurocapillus* on day 1, when introduced into an arena with one of the nine squares provisioned with mealworms, and on day 2 when no food was present. The bird returned to the area which had been profitable but left after searching there. Modified after Zach and Falls 1976*b*.

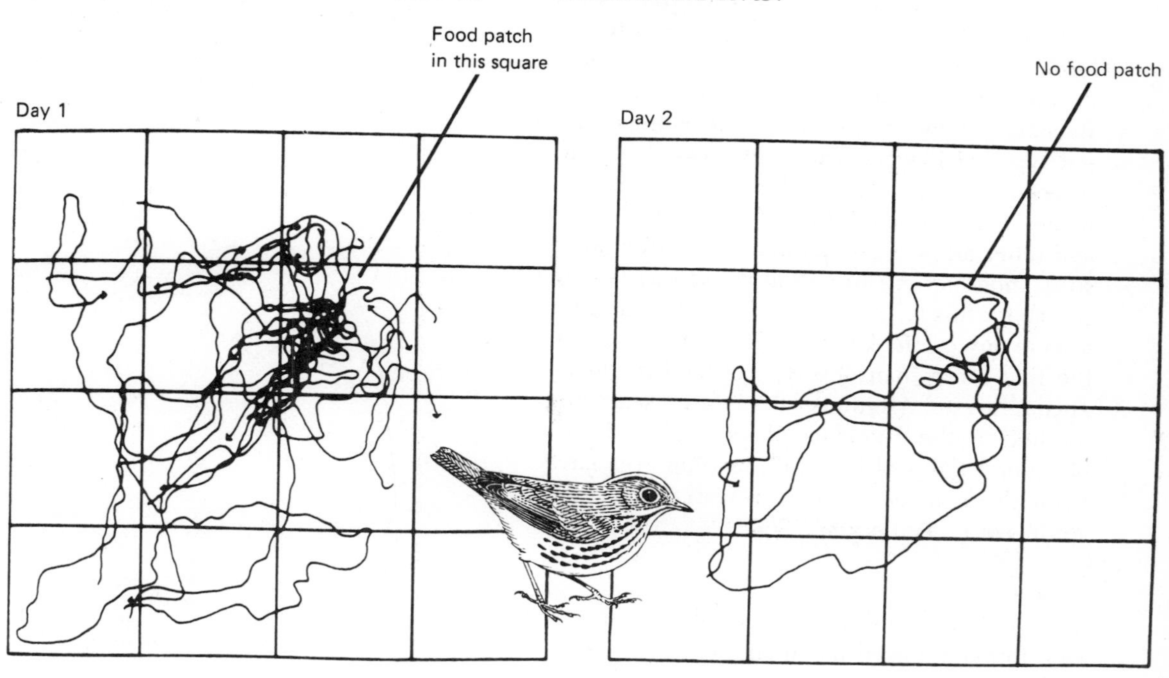

and quality often involve taking decisions about searching behaviour and about whether or not to acquire an item once it is located (see next section). Sometimes changes in searching behaviour are obviously dependent on previous experience. The starlings studied by Tinbergen and Drent (1980) fed their young on a mixture of foods of which the majority (55–80%) were leatherjackets but there was always a proportion of caterpillars of the moth *Cerapterix*, even when leatherjackets were plentiful. A starling which had been collecting leatherjackets from grassland and bringing them back to its nest would fly off in a different direction to a specific area of the salt-marsh where the *Cerapterix* larvae were to be found.

Other examples of observed changes in food searching come from primate studies, for example Chivers (1977) found that the siamang *Symphalangus syndactylus*, an ape which he observed in Malaysia, would climb to the ends of branches in the early part of the day and eat fruit. At some point in the morning it switched to the central trunk region and ate leaves, thus obtaining a variety of nutrients each day. Much more complex sequences of movements to different feeding sites are observed in species of primates which eat a greater variety of foods. Baboons will move from site to site eating different types of flowers, fruits, grasses, rhizomes, insects, birds and mammals (Altmann and Altmann, 1970). Many of the sites visited by these and other animals are familiar to the individual so it knows where to search as well as how to search for this type of food. Whilst searching for food in any particular area, however, it is possible that more than one type of food might be encountered. Attempts to relate searching methods to the frequency of occurrence of each of several types of food in a feeding area have been the subject of much research since the studies of L. Tinbergen (1960).

Item density

The relationships between the number of food items acquired by an animal and the density of occurrence of these items were described as type 1, type 2 and type 3 functional responses by Holling (1959). In his *type 1 functional response* the rate of acquisition is directly proportional to density up to a maximum acquisition rate. He describes the occurrence of such a response in animals which are filter feeders (Holling 1965). The other two types of functional response occur as a result of behavioural mechanisms and are explained in this section. An example of the parabolic *type 2 functional response* is shown in Figure 92 whilst the sigmoid *type 3 functional response* is shown in Figures 93 and 94. In the example shown in Figure 92, the rate of pecking at newly sown cereal grains by woodpigeons (*Columba palumbus*) varied according to grain density, but no pecking occurred at densities of less than $2m^{-2}$ and no further increase in pecking rate occurred at densities of over $30m^{-2}$. Murton, Isaacson and Westwood (1963) suggested that the lower limit was due to the low nutrient return per unit of energy expended at this density whilst the upper limit was a consequence of the time taken to

92 The rate of pecking at cereal grains by woodpigeons feeding in a field has minimum and maximum values as shown. Modified after Murton *et al.* 1963.

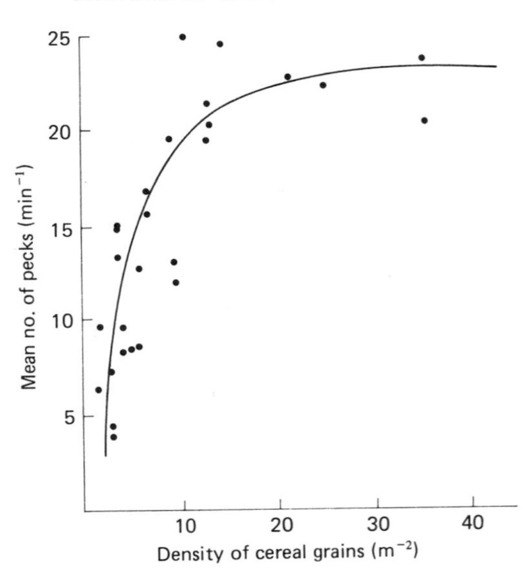

acquire the grain. The importance of this *handling time* was further emphasised by Holling (1959) in experiments with human and insect hunters.

Tinbergen recorded the frequency with which small insectivorous birds, especially the great tit (*Parus major*), brought insects of different species to their young. He related these results to the frequency of occurrence of these insects in the pine-forest habitat of the birds. Most rare insects were taken by the birds less often than chance would lead one to expect and common ones more often (see Figure 93). This result is very similar to that obtained by Holling (1959) who studied feeding by small mammals on sawfly larvae in grassland (Figure 94) and in laboratory experiments. Most of the food items, in these two sets of experiments, were cryptically coloured and Tinbergen proposed that they were not recognised as food when the rate at which they were encountered was low. Once the encounter rate rose above a certain level, the items were taken by the hunter, which then adopted a *specific search image* for this type of food. This idea of a modification in perceptual mechanisms has been widely criticised, for example by Dawkins (1971) and by Royama (1970) who reported that tits often did not collect the same type of food items on successive visits to the same area.

Despite the existence of data which do not fit Tinbergen's theory, there is no doubt that individual hunters do sometimes show long sequences of taking one type of food when another energetically equivalent food is readily available. Murton (1971*a*) conducted a field experiment with wood pigeons feeding on cereal stubble or clover pasture. He provided maize, tic beans, maple peas or green peas all of which had been treated with α-chloralose, which stupefies pigeons. After some time he collected the pigeons and examined the contents of their crops. When equal amounts of tic beans or maple peas were evenly scattered in a field, one pigeon ate only tic beans whilst others ate only maple peas, and the majority showed a clear preference for one or the other food. Murton points out, however, that due to social facilitation within the flock, his data do not fit with all the ideas produced by Tinbergen. Discussions about methods of obtaining food are based on a variety of causal factors so a complete understanding of what happens during foraging necessitates consideration of the effects of experience, motivational mechanisms and the attentional mechanisms which they modify. Tinbergen's search-image concept is clearly an oversimplification of

93 Birds brought few of the larvae of the sawfly *Acantholyda* to the young at low sawfly density but more than expected when the density was high. After Krebs 1973 from the data of Tinbergen 1960.

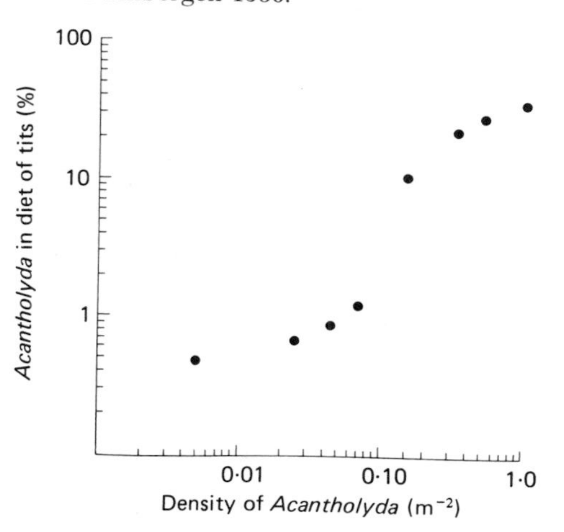

94 The relationship between the number of food items (sawfly cocoons) taken by the deer-mouse *Peromyscus* (circles) and the shrew *Sorex* (triangles); the density of food items shows a type 3 functional response. Modified after Holling 1959.

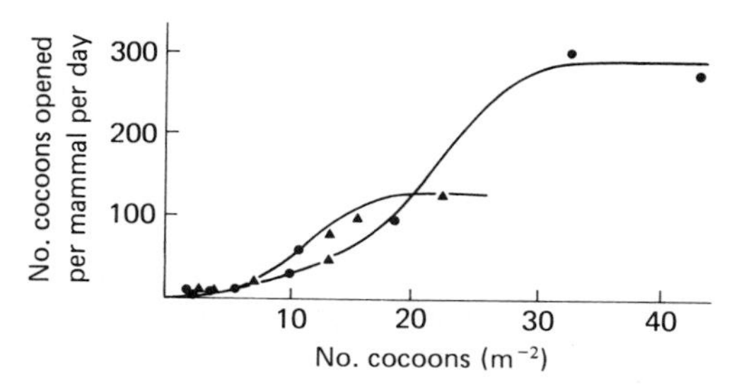

hunting strategy in general. If interpreted in terms of attentional rather than sensory mechanisms, however, it helps in the understanding of some aspects of feeding behaviour. For a review of literature on food finding see Krebs (1973) and Curio (1976).

When to give up foraging in an area

If food is not found in an area the animal must have a mechanism for deciding when to give up searching there (Figure 84). Such a mechanism would also be used after some food had been consumed but no further food was being found. The *giving up time*, during which no food is found, was studied by Krebs, Ryan and Charnov (1974). The likelihood of giving up should increase as environmental quality declines (Hainsworth and Wolf 1979). Charnov (1976) emphasised that the forager must relate intake in the patch to average intake possible elsewhere. It is also necessary for the forager to consider the costs of getting to another patch and the variability of food supplies. This problem is considered further in the section on hummingbird feeding, by Pyke *et al.* (1977) and by Pyke (1978*b*) as well as by the other abovementioned authors.

Eating: energy, nutrients and dangers

Once a potential food item has been found by an individual, the decision as to whether or not to attempt to acquire it may depend upon a variety of factors (see Figure 84). The item will provide benefit in terms of energy and nutrients but the acquisition and digestion of it will involve energetic costs, the time spent on it cannot be spent upon some other item, there could be costs consequent upon resistance to capture by animal prey, there might be costs associated with the physical or chemical characteristics of the food, and there could be costs due to risk of predation or other hazard during the acquisition and digestion of the food. All of these factors, which are detailed below, must be considered at this time, just as they must during food-finding behaviour (Emlen 1966).

Assessing energetic and nutrient value

A redshank (Figure 95) which has arrived at a point on a mud flat where it can see and reach two holes inhabited by *Nereis* is likely to choose the larger worm. Such a worm provides more energy, once digested, and is not much more energetically costly to catch (Goss-Custard 1977*a*). Similarly, a monkey will choose the larger of two fruits and a dog will pick up the larger of two pieces of meat. If the redshank sees a *Nereis* burrow or the burrow of the amphipod crustacean *Corophium* it choses the *Corophium* despite the fact that the net energy return from catching and eating *Corophium* is considerably less than that from *Nereis* (Goss-Custard 1977*b*). It must be presumed that a *Corophium* diet is more efficient than a *Nereis* diet as a means of obtaining an adequate balance of energy and nutrients but there is very little direct evidence concerning the mechanisms of such dietary balancing. A similar example is Smith's (1977) finding that howler monkeys (*Alouatta palliata*) can assimilate much more energy from fruit than from leaves but they eat both. Again we assume that they can thereby obtain a balanced diet.

The digestive functioning of animal species has evolved at the same time as have the mechanisms for food preference and the ability to learn about dietary requirements. Monopha-

95 Redshank (*Tringa totanus*), feeding on a mud-flat.

gous animals can obtain all necessary nutrients from their single food source but polyphagous animals, because of their digestive specialisations, usually need a variety of foods. For example, man and the rat must obtain, respectively, nine and ten amino acids from their food in order to survive. There are mechanisms for ensuring that adequate water (Chapter 5) and energy-giving food (see later in this chapter) are ingested but the only substances which mammals can detect directly in food and ingest when they are deficient in the body, are water and sodium (Richter 1936, Rozin 1976). There is some evidence for a specific calcium appetite in birds. All other balancing of nutrients appears to occur as a result of feeding preferences which are modified by experience.

Animals can learn about diets during development but must also be able to modify food-selection behaviour according to later experience. Young yellow baboons (*Papio cynocephalus*) will sample food which other baboons are eating and will approach individuals which have food in their mouths and sniff the food (S. A. Altmann personal communication). The food choices of the baboon are thus influenced by the choices of the other members of their group and this provides a method of avoiding the necessity to sample harmful foods as well as ensuring that the individual is aware of the existence of enough foods to make a balanced diet possible. It does not, however, result inevitably in the consumption of a diet which is optimal in its composition. There must be much variation amongst individuals in diet, that is to say two individuals may not come to the same decision when confronted with a potential food item.

If an individual is suffering from a dietary deficiency it can modify its behaviour in two ways in order to remedy this situation. Firstly it can refuse to eat the diet which has resulted in the malaise which it presumably feels. Rozin (1967) found that rats fed on a thiamine deficient diet refused the diet after a few days, i.e. they treated it like a poison and took a new diet preferentially. Secondly, the individual can learn to eat diets which remedy deficiency symptoms. Garcia, Ervin, Yorke and Koelling (1967) found that preference for diets containing thiamine was enhanced when the diet was consumed by thiamine-deficient rats. The result of such mechanisms is, as Richter (1942–3) demonstrated with rats, that individuals which have a variety of foods available to them, have the ability to compensate for the variation in nutrient needs which results from pregnancy, lactation, thyroidectomy etc.

Assessing costs of acquisition and digestion

The energy required to acquire food once it has been found may be considerable and in some species it is very much greater than the energetic costs of searching. A cheetah which sees an antelope, or a baboon which sees a coconut or ostrich's egg, still has much to do before it gets a meal. The oyster-catcher (*Haematopus ostralegus*) obtains the flesh of the bivalve mollusc *Scrobicularia plana* by probing down the hole in the mud made by the molluscs' siphons, grasping the shell, walking in a circle whilst pulling and then opening the shell valves (Hughes 1970).

In studies of wolves preying upon ungulates, reviewed by Mech (1970), the majority of the animals killed were found to be young, or old, or to have some disability. The most likely explanation for this is that these are the easiest animals to catch. In Royama's (1970) studies of feeding by great tits the birds were found to concentrate on slow-moving species of spiders, flies and sawflies. Prolonged chases of active prey are energetically expensive and often may be fruitless. This is something which many young animals learn during their development. The experience gained from previous attempts to acquire food items is a major factor affecting many decisions about whether or not to try to acquire a particular item. There have been very many studies of the improvement of various skills when food is a reinforcer, for example, rats or pigeons on complex schedules in operant conditioning experiments (Gollub 1977), finches pulling strings (Vince 1961) and chimpanzees using tools (Goodall 1964). The selection of leaf rather than stem or dead matter by grazing animals is probably related to costs of chewing and digestion as well as to nutrient content (see later in this Chapter). It is likely that digestion costs are a factor which influences decisions about whether to acquire any particular food item but there is little clear evidence for this. The evolution of the mechanisms which result in food preferences must have been affected by digestion costs so that those items which are preferred are usually readily digestible by the individual. Indeed the digestive mechanisms themselves have evolved so that some animals, such as baboons, require low-bulk–high-energy food, whereas others thrive on a high-bulk–low-energy food, eg. colobus monkeys.

In a choice situation where two items of equal energetic value and with equal acquisition costs are available, optimal foraging hypotheses predict that the item whose digestive costs are least should be chosen. Even if the energy costs of digestion or acquisition are similar for two items, if one item takes longer to acquire or digest it may be avoided. The importance of time as a resource was emphasised by Schoener (1971) who pointed out that some animals might be time minimisers rather than energy maximisers in their feeding strategies. An animal might have to wait for a long time before a bivalve mollusc opened widely enough to allow its capture. In this case the animal might decide to leave that mollusc and rather than wait, attempt to find other food. If a type of food takes a long time to digest, energy return in a given time might be reduced because of the long period when the gut is full.

Avoiding damage or poisoning

Damage

The possibility of being damaged by active prey must be a considerable deterrent to a lion deciding whether or not to attack a group of buffalo or a falcon deciding whether to stoop on a crow. Examples of defensive behaviour by prey species are given in Chapter 7. Plant defences include the thorns of acacias or brambles and the irritant chemicals produced by nettles (*Urtica*) or poison ivy (*Rhus toxicodendron*) and its relatives. Animals which might eat the leaves, shoots and bark of such plants are more likely to choose another food source if there is a risk of bodily damage. Small animals also face hazards when feeding on plants. The mite shown in Figure 96 must avoid the hooked hairs on the surface of the leaf and may decide to feed in a different part of the plant or on a different plant if the risk of being trapped is large.

Poisons and food preferences

Decisions about whether or not to eat a particular food will also be affected by previous experience of any poisons or other harmful substances which may be present in such food. As mentioned earlier in this chapter the chemical defence systems of plants and of some poisonous animals have interacted continuously during their evolution with the feeding mechanisms of

96 Scanning electron micrograph of a mite on the surface of a leaf. The mite may decide not to feed in an area where there is a high risk of being trapped by the hooked leaf hairs.

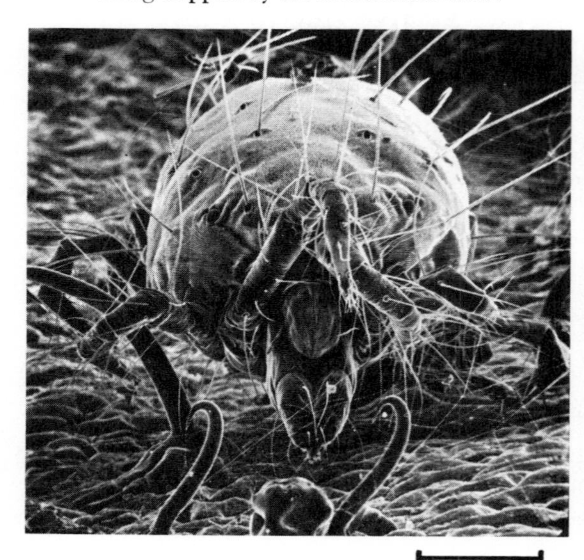

animals. Levin (1976) describes many plants containing phenolic substances, polymeric phenols such as tannins, terpenes, cyanogenic glucosides etc. which, if eaten, impair digestion and damage tissues up to the point of being lethal. Such chemicals may result in the proliferation of some plant species in areas where the growth of others is severely restricted by animal consumption. The locust *Locusta migratoria* is deterred strongly from eating the leaves of the cereal *Sorghum bicolor* by the combined effect of hydrogen cyanide, produced from cyanogenic glucosides, and a mixture of phenolic acids (Woodhead and Bernays 1978). Such *secondary plant compounds* are so widespread in plants that feeding behaviour is very much affected by their presence. Since insects are the principal consumers of plants, plant genes which promoted the production of substances toxic to insects have spread in many plant populations.

Although most of the study of secondary plant compounds has been carried out by those who wish to use the information in order to reduce the depredations of insect pests, there have also been some studies of mammals, especially on farms. Westoby (1974) points out that the presence of poisons which kill cannot be learned if a lethal dose is ingested. Freeland and Janzen (1974) describe biochemical detoxification mechanisms, antibody response and other internal means of combating poisons. Animals combine the development of enzyme systems and behavioural preferences in counteracting the plant defences. Freeland and Janzen list, as desirable behavioural characteristics, that animals should (1) consume only small quantities of new food, (2) have a good memory for different food characteristics, (3) be able to seek out specific foods, (4) sample foods whilst eating staple foods, (5) prefer familiar foods, (6) prefer foods with small amounts of toxic compounds, and (7) have a searching strategy which compromises between maximising variety and maximising intake. Some of these characteristics had previously been described in studies of the responses of rats to food poisoned with arsenic (Rzóska 1953). Tame rats often consumed poisoned bait on first encounter although wild rats avoid traps or other objects as well as novel food which they sample briefly (Shorten 1954). After the adverse effects of a small quantity of poison the rats avoid that food completely. This result has been extended greatly by the work of Garcia (see review by Rozin 1976).

Effects of experience on dietary preferences

The ability to modify food preferences as a result of experience of noxious substances, or after the development of specific digestive mechanisms, is important to many animals. Garcia, Ervin and Koelling (1966) showed that rats learned to avoid a food whose adverse gastro-intestinal consequences occurred more than one hour after ingestion. Subsequent experiments showed that delays in effects of up to twelve hours led to subsequent avoidance of that food (Smith and Roll 1967) and that if such adverse effects occurred after ingesting new and familar foods, the new foods were subsequently avoided (Rozin 1968). Various laboratory experiments indicate that animals associate particular cues obtained from a food, gustatory in rats but visual in birds, with the consequences that the food produces (Rozin and Kalat 1971, Rozin 1976). Such mechanisms affect decisions about whether or not to search for and eat certain items of food in such a way that poisons are avoided and dietary deficiencies are remedied.

Foods which are toxic if consumed in large quantities may be important sources of energy, or of essential chemicals, so decision making processes during feeding are very complex in polyphagous animals. Altmann (personal communication) found that yellow baboons would eat legumes which probably contain anti-trypsin and grasses which contain phytic acid but that they regulated their total intake of poisons so that the adverse effects were not severe. The consumption of such foods depends upon the availability of other foods but some of them were preferred by the baboons to such an extent that they usually formed part of the diet. Toxic foods are also consumed in small quantities by man but most are processed before consumption, for example cyanide is removed from cassava by prolonged boiling. In laboratory studies, rhesus monkeys, like rats, sampled new foods, such as blackcurrant juice, but consumed very little on the first few days (Weiskrantz and Cowey 1963). These monkeys accepted food more readily if other monkeys were observed eating it. Similar imitation by baboons has been described by Altmann and in Itani's (1958) work with Japanese macaques *Macaca fuscata*. Young macaques were more likely than older animals to sample new foods and to develop new feeding techniques.

Studies of rats, monkeys and man show that foods which were consumed without ill effects, whilst the animals were young, are subsequently preferred to novel foods. This method of avoiding the dangers of poison and of wasting energy by consuming unsuitable food, is also used by insects. Jermy, Hanson and Dethier (1968) studied food preferences in caterpillars of the tobacco hornworm *Manduca sexta*, which is oligophagous and eats various plants in the family Solanaceae. The caterpillars were reared from hatching on an artificial diet, or tomato (*Lycopersicon esculentum*), or tobacco (*Nicotiana tabacum*), or Jerusalem cherry (*Solanum pseudocapsicum*). When the caterpillars reached the fifth instar, the last before pupation, they were tested in a choice apparatus which included four leaf discs from each of the three plants mentioned above. The caterpillars crawled around the edge of the container and ate from several of the discs until 50% of leaf from one of the plant species was consumed. The amount consumed at this time was measured and whilst there was no clear preference amongst caterpillars reared on artificial diet or on *Nicotiana*, those reared on *Solanum* or *Lycopersicon* showed a clear preference for these species (Figure 97). Preferences which were present in one larval instar persisted through moulting to the next larval instar. The extent of possible modification of preferences was limited. Exposure to non-host plants by incorporating extracts from them in an artificial diet did not lead to preference for those plants. Similar results were obtained by Cassidy (1978), who found that stick-insects (*Carausius morosus*) were less likely to change their preference away from the plant on which they had been reared if the rearing had continued up to the seventh instar. The advantages of such effects of experience on food preferences are emphasised by the reduced efficiency of food conversion by several species of moth larvae which were forced to change from one food plant to another (Schoonhoven and Meerman 1978). The development of food preferences by insect larvae is not always affected by rearing experience however (Claridge and Wilson 1978) and the food plants on which larvae develop are often determined by the preferences of the adult prior to oviposition (e.g. Finch 1978).

Responses to systemic insecticides

Another area of research in which insects' decisions about whether or not to eat must be studied by entomologists concerned with agricultural problems is where a systemic insecticide has been applied to the plant. When an aphid, a caterpillar or a locust encounters a plant which has been treated with insecticide it may detect the poison before ingesting any of it and then depart. It may ingest a small quantity of plant material, detect the insecticide and leave before acquiring a lethal dose; or it may continue to eat and subsequently die. These behavioural differences are of great importance and more observational studies of insects encountering such situations are required.

Avoiding other hazards whilst eating

The role, in all decision-making situations, of causal factors pertaining to the avoidance of predation and other hazards is stressed in Chapters 4 and 7. A decision about whether or not to acquire and eat a food item may be different in two situations where the item is on flat ground or on a cliff edge or in situations where the probabilities of detection by a predator are different. Another factor which may affect decisions about whether to embark upon the consumption of a food item is the proximity of conspecifics who might take the food, perhaps with accompanying physical violence. An individual monkey which was frequently robbed might benefit by concealing the discovery of a food item until no potential robbers were present. All of these factors may affect both the decision about whether or not to eat the item when it is found and the rate of eating once consumption has started.

The control of food intake

Once food consumption has commenced, how much will be eaten before the individual stops and initiates another activity? How much in total will be eaten in a day or in a year? Some of the factors which affect the amount eaten have been discussed in the previous sections of this chapter (see Figure 82). Any feeding control system must include mechanisms for the avoidance of predation, other hazards including poisons, energetic inefficiency whilst acquiring food, and specific nutrient deficiency. Nevertheless, most attempts to set up models which explain the timing and quantity of food consumption refer to energy input and output but do not take account of the necessity for override or modification when one of the above essential mechanisms is in operation. As emphasized in Chapter 4, since no functional system operates

97 Caterpillars of the tobacco hornworm *Manduca sexta* were reared on one of four diets and allowed to feed, in the apparatus shown, on leaf discs from three different species of possible food plant. Those reared on *Solanum* or *Lycopersicon* preferred these species. Modified after Jermy *et al.* 1968.

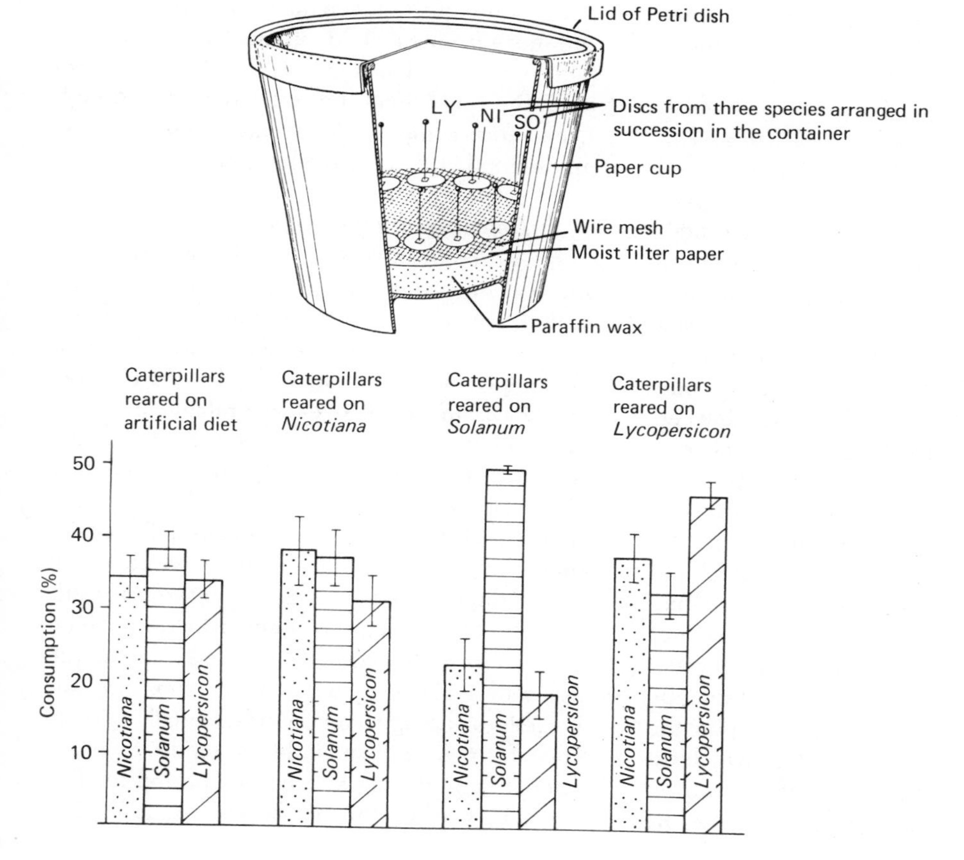

in isolation, the web of interactions with other functional systems will result in any simple explanatory model failing in some circumstances. As Schilstra (1978) emphasises, since the occurrence and duration of feeding is affected by random variables, stochastic models (see Chapter 3) will explain feeding patterns better than deterministic models, especially when day-to-day or hour-to-hour variation is considered.

Meal size and inter-meal interval

Feeding is discontinuous in most animals but it is often difficult to decide when intervals between feeding movements should be considered to be within and when between feeding bouts. Le Magnen and Tallon (1966) chose an arbitrary criterion of 40 min as the minimum inter-meal interval in their study of the laboratory rat whilst Wiepkema (1968) and Slater (1974*a*) used the log survivor function of the distribution of intervals to define bout-length (see Chapter 3). The duration and pattern of occurrences of feeding bouts in zebra finches *Poephila castanotis* (Slater 1974*a*), in domestic fowl (Duncan, Horne, Hughes and Wood-Gush 1970) and in cattle housed separately (Metz 1975) were found to vary considerably amongst individuals in very similar conditions. Like Le Magnen and Tallon, however, they found that the size of the meal was often positively correlated with the succeeding inter-meal interval when food was available continuously. This result has been obtained for some species and situations but not for others, according to the type of food, lighting, effort required to obtain food, age of subjects etc. (Panksepp 1978). The positive correlation between size of meal and succeeding interval, however, has been found much more frequently than any correlation between size of meal and preceding interval in ad-libitum feeding situations. Le Magnen (1971) concludes, therefore, that the timing of the initiation of feeding often depends upon the size of the last meal but the duration of a meal does not depend upon the delay since the last meal. The factors which increase the chances that a meal will be started are different from those which influence the timing of the end of the meal. A further general conclusion from the work of Le Magnen and others is that food intake over a long period is regulated more by frequency than by meal size.

It has often been assumed that the relationship between the size of the meal and the interval before the next meal is a direct consequence of the rates at which food is passed through the gut, absorbed and metabolised. It is possible, however, that animals learn that the satiating effects of ingested food persist for certain times and this may explain why the correlation with succeeding inter-meal intervals is absent in young rats but develops as they reach maturity (de Castro and Balagura 1975). The study of the patterning of feeding when food is provided *ad libitum*, as some of those working in this area now admit (Panksepp 1978), has not made explanation of the regulation of energy balance possible but it has provided information about some of the mechanisms involved. When the more frequent field situation is considered, where various foods are intermittently available, the problem of the regulation of feeding is even more complex. Many meals end because all the food at that source has been consumed.

Deprivation effects and storage

Although meal size is not affected by the interval since the last meal in an ad-libitum situation, it is increased by prolonged food deprivation. As Tugendhat (1960) and Beukema (1968) have shown for sticklebacks, the food deficiency is made up over several meals. If the physiological state of the individual remains within certain tolerable limits, the optimal meal size may stay the same, but if the food deprivation is such that physiological state falls outside these limits, the risk avoidance component of the feeding strategy may result in the consumption of a larger meal. Whenever food intake occurs, some of the food may be used immediately as a source of energy whereas some may be converted to storage compounds. Small animals, such as rats and small birds, need to ingest enough food during the feeding period to allow them to survive the non-feeding period. Rats eat more at night and store food as fat in adipose tissue which breaks down gradually during the day so that the rat does not need to eat as much in daytime (Le Magnen and Devos 1970). Sparrows (Kendeigh, Kontogiannis, Malzac and Roth 1969) and hummingbirds (Hainsworth 1978) store food in the daytime to different extents according to the overnight energy expenditure. These are examples of feed-forward control.

Body monitors and feeding control

Mogenson and Calaresu (1978) list the signals which have been reported as being involved in the control of the timing and quantity of food intake (Table 7) and summarise the reg-

Table 7. *Signals reported to be involved in the regulation of feeding behaviour. (Modified after Mogenson and Calaresu 1978)*

	Initiate	Modulate	Terminate
External			
Olfactory	–	x	–
Visual	x	–	–
Temperature (ambient)	–	x	–
Social and cultural factors	–	x	–
Internal			
Pre-absorptive	–	–	–
Mechanical	–	–	–
Oropharyngeal (taste, texture)	x	x	–
Gastrointestinal tract	–	–	–
stomach contractions	x	–	–
stomach distension	–	–	x
intestinal distension	–	–	x
Hormonal factors	–	–	–
insulin	x	x	–
prostaglandins	–	x	x
cholecystokinin	–	–	x
other gut hormones	–	x	–
Humoral factors	–	x	–
Post-absorptive	–	–	–
plasma glucose	x	–	–
plasma free fatty acids	–	x	–
plasma amino acids	–	x	–
lipid mobilising substances	–	x	x
plasma glycerol	–	x	–
Long term feeding controls	–	–	–
fat stores	x	x	–
hormones	–	–	–
insulin	–	x	–
growth hormone	–	x	–
sex hormones	–	x	–
thyroxine	–	x	–
glucagon	–	x	–
corticosteroids	–	x	–

ulatory mechanism (Figure 98). None of the models which assume the regulation of levels of glucose, fat, amino acids, temperature, or combinations of these factors is very satisfactory in explaining feeding patterns (Russek 1971). The evidence for the importance of input from *hepatic glucoammonium receptors* presented by Russek is convincing, however. Liver metabolism has a crucial role in Friedman and Stricker's (1976) net availability of energy theory. Another general model which is consistent with many observations of feeding in a laboratory rat is that of Toates and Booth (1974) and Booth (1978*a*) which is based on the idea that the propensity to feed appears when the *flow of energy from absorption* becomes too small and disappears again when absorptive flow becomes adequate. None of these models, however, is adequate to explain the feed-forward effects mentioned in the last paragraph, or many aspects of learning whilst feeding, or the influence of most other functional systems on feeding.

The search for receptors and regulatory mechanisms in the brain has emphasised the complexity of feeding control. The idea that the *level of glucose in the blood* might be measured by some receptor in the brain, or elsewhere, and that this would be the key signal which is used in the regulation of feeding, is complicated by studies of diabetics. Insulin facilitates the absorption of glucose from the blood. Diabetics, who are deficient in insulin, have very high levels of blood glucose which they cannot use. Despite these high levels, they will start to feed if food is available. Mayer (1955) suggested, therefore, that receptors would have to measure the rate at which glucose could be utilised rather than the amount present in the blood.

There are neurons in the hypothalamus which respond to glucose level (Desiraju, Banerjee and Anand 1968) but there is no good evidence for neurons which respond to glucose availability. (For brain anatomy diagram see Figure 5.) If the lateral hypothalamus of a rat is lesioned the animal does not eat, but stimulation in the area can lead to eating (Anand and Brobeck 1951): this finding led to the assumption that feeding was initiated by the lateral hypothalamus as a consequence of signals, about blood glucose levels, which probably originated from receptors situated there. Later work shows that this assumption is unlikely, since lesioned animals can be fed on very palatable, moist food until they resume normal eating (Teitelbaum

98 Diagram illustrating some of the signals and feedback loops which may be involved in the regulation of food intake. +, promoting effect; −, inhibiting effect. Modified after Mogenson and Calaresu 1978.

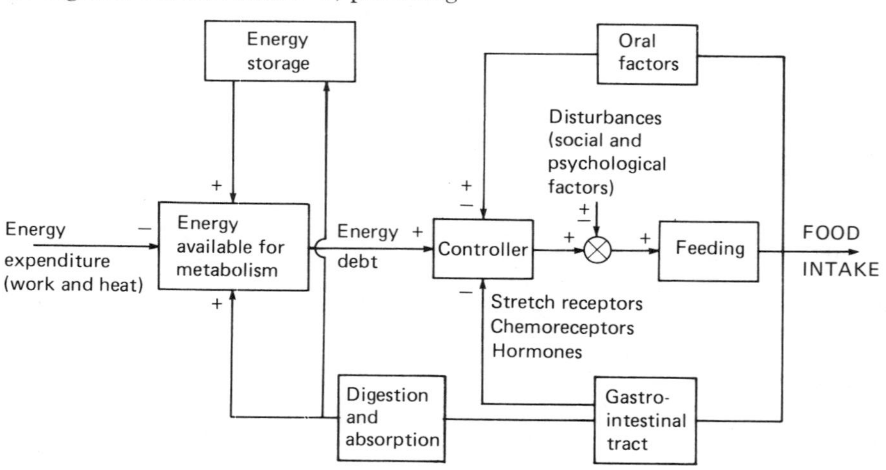

and Epstein 1962) and a lesioned rat whose tail is pinched gently will resume eating (Antelman, Rowland and Fisher 1976). The possibility that there might be, in the ventromedial hypothalamus, satiety receptors which detect high levels of blood glucose and terminate eating was suggested after Brecher and Waxler (1949) had found that injections of gold thioglucose into this area in mice lead to over-eating and obesity. The effect was not produced by gold thiomalate or gold thiogalactose. It was suggested, therefore, that the glucose was taken up by specific receptors but that the cells were then killed by the gold (Mayer and Marshall 1956). Other poisons have the same effect as gold thioglucose (Caffyn 1972), however, and mice with the ventromedial hypothalamus damaged do not become diabetic. It seems likely, therefore, that glucoreceptors, if they exist, are located elsewhere.

The evidence against the idea of the existence of appetite and satiety centres in the hypothalamus is sufficient for some authors to deny their existence altogether (Booth 1978*b*). There is no doubt that the hypothalamus is involved in the control of feeding but it must form a part only of the control system. Many factors affect this system, some modifying short-term aspects of intake and others affecting long-term regulation, but we are at present unable to explain how these factors interact. For reviews of the control of feeding in the pigeon and in the blowfly see Zeigler (1976) and Dethier (1969).

Feeding by hummingbirds

Energy costs and benefits

The feeding strategies of hummingbirds are especially suitable subjects for the optimal foraging approach because the major food is nectar and the feeding behaviour can be precisely described in terms of time and energy utilised. Hummingbirds do eat insects but the energetic input for individuals of most species is derived almost entirely from nectar which is easy to characterise in energetic terms (1ml of a molar sucrose solution yields 5641 Joules). The hummingbirds assimilate all of the nectar, within the range of sugar concentrations found in plant nectar (0.25–1.0M sucrose). For nectar, the sugar composition, the concentration, the amount present and the renewal rate can be measured (Hainsworth and Wolf 1972, Hainsworth 1974, 1978). Hummingbirds will feed in the laboratory where their intake can be measured by the amount taken from a feeder or by weighing the hummingbird when it settles on a perch which is attached to a balance (Gass 1978).

The energetic costs of hummingbird feeding can be measured directly because hummingbirds hover in front of flowers whilst feeding and the oxygen consumption by a hummingbird hovering in a metabolic chamber can be measured. Lasiewski (1963) found that a costa's hummingbird (*Calypte costae*), which weighed 3g, used 42.4ml O_2 $g^{-1}h^{-1}$ whilst *Eulampis jugularis* hummingbirds, which weighed 6–9g, used 43.4ml O_2 $g^{-1}h^{-1}$ whilst hovering (Wolf and Hainsworth 1971). These oxygen consumptions are equivalent to an energy usage of about 890 J $g^{-1}h^{-1}$, the highest rates per unit weight measured for any vertebrate (Wolf and Hainsworth 1978). Energy consumption whilst resting has also been measured accurately and that whilst in forward flight has been estimated. These values for energy consumption are affected by temperature whilst the relative costs of hovering and resting differ according to body size (Brown, Calder and Kodric-Brown 1978, Hainsworth and Wolf 1979).

The efficiency with which nectar can be obtained varies according to the shape of the hummingbird's bill and the shape of the corolla of the plant. The adaptations of the plant promote the chance that one hummingbird will visit and pollinate many flowers of the same species (Grant and Grant 1968, Wolf, Hainsworth and Stiles 1972, Snow and Snow 1972, Wolf, Hainsworth and Gill 1975, Stiles 1978). Any study of hummingbird foraging must take account of body size, degree of adaptation of the bird species to the available nectar sources, etc.

but the measurement of energy costs and benefits is much more precise than is possible for most animals.

Foraging behaviour

Food finding by hummingbirds is facilitated by the advertisement of the food source by the plants. The flowers are rendered conspicuous to the hummingbirds by their large size and long wavelength coloration. Goldsmith (*Science*, *207*, 786, 1980) has demonstrated that hummingbirds can see at wavelengths which we call ultraviolet because objects reflecting these wavelengths cannot be seen by man. Flowers usually occur in patches because they are on a large plant or are part of a collection of small plants flowering synchronously. Nectar production often continues for several days. Patch finding might be energetically costly in some species, such as those which live in the mountainous areas, but hummingbirds must find the majority of patches within a small home range or territory. Territorial defence in hummingbirds and other nectar feeders is discussed by Carpenter (1978) and Wolf (1978).

The importance of learning the location of patches and the rate of renewal of the nectar supply is greater than that of actually finding the food at most times in the lives of hummingbirds. Foraging behaviour in these birds involves a series of decisions about when to leave a patch and when to revisit a patch. Time spent in a patch of inferior quality could often have been spent in a good quality patch. Before considering such decisions it is necessary to review the information about food intake.

Hummingbirds lick nectar from plant nectaries with their grooved tongues and pass it to the crop (Hainsworth 1973). It seems likely that the initiation of feeding depends upon the emptiness of the crop (Wolf and Hainsworth 1977) but many other factors must be involved. The size of the crop is proportional to the weight of the bird (Hainsworth and Wolf 1972). The X-ray studies by these authors show that the food is passed to the rest of the digestive system from the crop, a full crop emptying in 30–40 min in the case of *Eugenes fulgens*. Hummingbirds in the laboratory or in the field usually fly to the food source, be it a feeder or a collection of flowers, ingest enough nectar to half-fill the crop and then return to a perch (Hainsworth 1977, Wolf and Hainsworth 1977). This behaviour is repeated with the result that energy is gained at a relatively constant rate during the day (e.g. see Figure 99).

Feeding strategies

Possible feeding strategies which could be adopted by a hummingbird are discussed by de Benedictis *et al*. (1978). The birds could attempt to minimise feeding time but this would involve filling the crop whenever possible, not half-

99 Hummingbird species of two different sizes showed approximately constant rates of net energy gain during a day. They were fed on 0.5M sucrose and were kept at 23°C on a 14-h light, 10-h dark regime. Modified after Hainsworth and Wolf 1979.

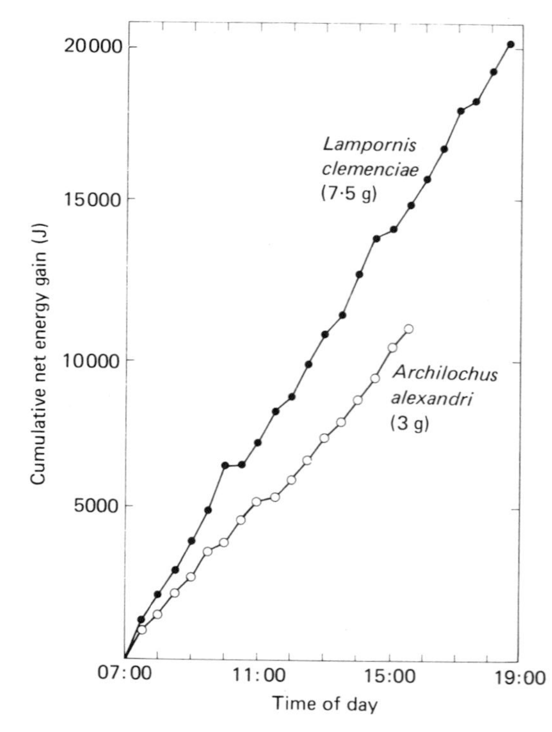

filling it as is observed. The same prediction follows from the strategy of maximising net energy gain per bout. The model of feeding strategy which fitted the observed feeding data best was that in which the rate of net energy gain during the total period from the beginning of one feeding bout to the beginning of the next was maximised. This model took into account the facts that the energetic cost of hovering increases as nectar is ingested, thus increasing body weight, the return flight requires more energy than the outward flight, and the inter-bout period on the perch also involves energetic cost. Hainsworth (1978) has extended this model to incorporate overnight storage effects on energetics. A summary of the system is shown in Figure 100. The information required for this model includes body mass, duration of light and dark periods, body temperature, ambient night temperature, rate of energy expenditure in relation to body mass and temperature differential, and the energy lost when converting sugars to fat and back again. An estimate of the amount of energy required for storage can thus be obtained. This information can be combined with calculation of energy expenditure throughout the feeding period and energy obtained from the food.

Hainsworth's model predicted that maintenance costs during the day would have a major effect on feeding frequency and this was shown to be correct in laboratory studies by reducing temperature, which increased feeding frequency. The model also predicted that increased costs overnight would be remedied by increasing meal size on the following day and this was shown to occur. The function of these mechanisms must be deduced from consideration of the general biology of the hummingbird. The regulatory mechanism in the daytime is that which is the most efficient for maximising the rate of energy gain from the available resources. If the time available for feeding is restricted, such as during preparation for a long night or a migratory journey, the feeding pattern is modified. The meals are larger as part of a feed-forward mechanism which anticipates the greater bodily demands. In situations where the risk is high that food supplies adequate for requirements will not be found, the bird may fill its crop completely.

Another decision which must be taken by a foraging hummingbird is when to give up feeding in a patch. The marginal value theorem of Krebs *et al.* (1974) and Charnov (1976) states that the animal will leave a patch when its net rate of energy intake has dropped to the average net rate of energy intake in the entire habitat. Pyke (1978*b*) suggests that in order to do this a hummingbird should leave a patch when its estimate of the nectar volume obtainable from the next flower, divided by the average time required to move to the next flower and remove the nectar, is less than its overall rate of nectar intake in the patch. Observations suggest that hummingbirds use information about the number of flowers in the patch, the number probed and the amount of nectar in the present and the last, or last two, flowers in deciding whether to depart. Although Pyke's model explains his data, it would seem likely that where patches are widely scattered, an estimate of the energy required to move to another patch might be a factor affecting the threshold level of intake at which departure occurs.

Most of the studies of hummingbirds foraging which have been mentioned are of situa-

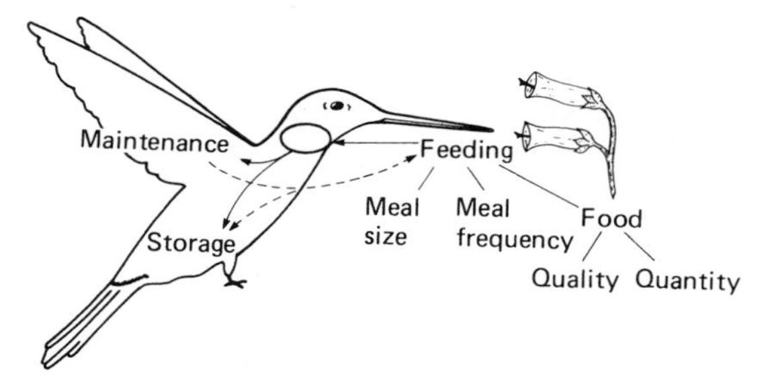

100 Hummingbirds consume nectar which is collected from flowers at intervals and stored briefly in the crop before being used for body maintenance or converted to a food reserve, usually fat, for overnight maintenance. Modified after Hainsworth 1978.

tions where nectar is abundant and feeding behaviour is not much affected by predation risk or by competition for food, nest-sites or mates. Studies of hummingbirds living in mountainous areas where nectar supplies are often sparse and the hummingbird assumes a torpid state during the cold nights, show that vigorous defence of feeding territories occurs and community structure is complex (Carpenter 1978). In such competitive situations, foraging behaviour can be inextricably interwoven with other functional systems (Wolf 1978) so that discussions of optimal foraging alone do not go far toward explaining the behaviour which is observed.

Grazing by cattle

The majority of studies of animal feeding in relation to energy utilisation have been carried out on farm animals and some ideas about feeding strategies have arisen as a result of this work. Other ideas have resulted from work on laboratory animals by physiologists or psychologists and work on wild animals by behavioural ecologists (e.g., the study of moose by Belovsky (1978) and Belovsky and Jordan (1978)). The interchange of such ideas is desirable since many of the same fundamental mechanisms operate on the farm, in the laboratory or in the wild. Cattle are ruminants which regurgitate their food from their multicompartmental stomach and chew it during the resting phase after eating is completed. Much of our knowledge of ruminant nutrition comes from indoor feeding studies but field studies are of great importance for world food supplies. Less than 5% of the world's domestic ruminants spent a significant period indoors in 1970 (F.A.O. 1970). The foraging behaviour of cattle and sheep, for example the detailed studies of the late T. H. Stobbs (e.g. Chacon and Stobbs 1976) and of G. W. Arnold (e.g. Arnold and Dudziński 1978), can be related to information about energetic efficiency. Such studies, therefore, aid in the management of farm animals as well as providing an example of feeding cost–benefit analysis for behavioural ecologists and physiologists.

For many animals, the problems of *food-finding* limit energy input but for ruminants, *food-processing* takes a long time and is often the limiting factor. The gut usually has food in it so digestion, aided by symbiotic bacteria and protozoans, is a continuous process. The upper limit of the overall rate of processing is determined by the cross-sectional area of the gut and the meal must end once the rumen is full. A consequence of the need for large quantities of food and of slow food processing is increased duration of feeding time and of digestion time. A high proportion of the life of the animal, therefore, is devoted to these activities. Wild ungulates, however, also have to find patches of grass or other vegetation upon which to graze or browse. Most wild and domestic ungulates *compete with conspecifics* for the food which is available. The best way of competing is, usually, to eat as fast as possible when food is found and this behavioural strategy is at least as important to cattle whose grass allowance is carefully calculated, as it is to wild ungulates. Another important objective of the feeding strategy in large generalist herbivores is to *obtain the best mix of nutrients* within a fixed total intake (Westoby 1974).

Decisions whilst grazing

The existence of such behavioural components in the feeding strategy is obvious to the behavioural ecologist but is sometimes overlooked by those concerned with animal husbandry. Cows (Figure 101) do not eat randomly all green material which they encounter, as is emphasised in the next section. They have to make a series of decisions, when grazing, which will differ according to the characteristics of the pasture upon which they are feeding. These decisions include whether to lower the head and bite the herbage, how large a bite to take, at what rate to bite, whether or not to stop biting and chew or otherwise manipulate the grass in the mouth, whether to swing the head to the side and how far, whether to take one or more steps forward and whether to raise the head and carry out some other behaviour. The long-term consequences of such decisions are that total daily grazing duration is 4–14h at bite rates of 40–80 min^{-1} and with bite sizes of 0.05–0.7g. Each day, the food intake may be up to 12.5kg dry matter and the cow will ruminate for 4–9h (Hafez and Dyer 1969, Stobbs 1974*b*, Chacon and Stobbs 1976). In temperate countries cattle feed most in the daytime. Visual food selection is then possible but diurnal feeding may be a relic of anti-predator behaviour for animals can feed and keep looking for predators in the daytime but it is safer to remain immobile at night.

101 Before pulling and breaking the grass, the grazing cow wraps its tongue around long grass but it holds short grass with the teeth in its lower jaw and a pad in its upper jaw.

Selection during grazing

The fact that cattle show feeding preferences is apparent when a grazed field is examined. The sward is less uniform than it was before grazing. Some plant strains and species, usually the more digestible, are preferred to others (Ivins 1952, Hughes, Milner and Dale 1964) and leaves are eaten in preference to stems (Stobbs 1973*a*). Some preference studies indicate avoidance of high concentrations of secondary plant chemicals, e.g. tannins, coumarin, alkaloids etc. Donelly (1954) found preferences by cattle, among 1206 strains of *Sericea lespedeza*, for low tannin content and fine stems. A variety of other studies has shown clear relationships between alkaloid levels and food preferences of sheep (review by Arnold and Dudziński 1978). Some ruminants, however, have digestive adaptations which allow them to eat plants with alkaloids or other poisons (see Freeland and Janzen 1974).

Many other studies have related nutritional properties of plants to the extent to which they are eaten by ruminants (see Marten 1969). There is, as yet, no evidence for long-delay learning of nutrient quality by ruminants (Westoby 1974) but the list of dietary requirements of cattle shown in Table 8 (A.R.C. 1965) emphasises the importance of cattle's feeding preferences and the likelihood of occurrence of subtle modifications of diet with experience. It is known that prolonged grazing of grass low in cobalt leads to extreme reduction of intake (Underwood 1956), which would presumably be associated with movement to a new feeding area if this were possible.

Man limits the choice of food plant for many cattle but plants at different stages of development and those contaminated with dung, which is avoided, are still encountered. The composition of pastures is altered by the feeding activities of cattle, by their dung and by the fact that they tread heavily on plants as they move around the pasture. Plants which die when grazed or trodden become rarer, whereas those which are avoided become commoner. When the shoot of a grass plant is eaten to within 5cm of the ground the next shoot is more likely to grow along the ground, so grazing results in a smaller proportion of upright shoots and a larger proportion of horizontal runners and procumbent leaves. Plants near dung pats can survive with a more upright growth form than those in other areas.

Evolutionary changes in anatomy, physiology and behaviour of grazing animals have been a response to the characteristics of grasses but, in response to the grazers, changes in grasses have also occurred. The occurrence of secondary plant chemicals has been mentioned above but the growth form and the mechanical properties of grass must also have changed (see Hickey 1961, Kydd 1966). The arrangement of fibres in grass is such that cows have to work hard in order to break off the leaves.

Table 8. *Daily dietary requirements of cattle. (Data from A.R.C. 1965)*

Nutrient	Requirement for 400kg heifer growing at 0.5kg day^{-1}
Calcium	25.8g
Phosphorus	23.7g
Magnesium	6.9g
Sodium	7.5g
Potassium	(usually adequate in diets)
Chloride	10.9g
Iron	0.24g[a]
Copper	0.08g[a]
Cobalt	0.0008g[a]
Zinc	0.4g[a]
Manganese	0.32g[a]
Iodine	0.001g[a]
Molybdenum	(trace)
β carotene	0.032g
Vitamin D	1000 international units
Protein	210g if low fibre diet
Protein	360g if high fibre diet

[a] Assumes 8kg dry matter intake.

Measuring energy input and output

Intake by grazing animals can be determined by observing behaviour, recording movements automatically, using an oesophageal fistula, estimating dilution of an indigestible substance, or measuring grass before and after grazing (Stobbs and Cowper 1972, Young and Corbett 1972*b*, Jamieson 1975).

The potential energy obtained from plant material can be determined by calorimetry. If a known amount of food is consumed by a cow, the metabolisable energy of the food can be calculated by subtracting from the gross energy obtained from it, the energy obtainable from the faeces, urine and fermentation gases. The metabolisable energy provided by pasture is 7–14 KJ g^{-1} dry organic matter (A.R.C. 1965, Greenhalgh 1975). The metabolisable energy may be used for body maintenance, growth or milk production and some is lost as heat. On a high quality diet, 76% of metabolisable energy can be used for body maintenance, assuming that there is no growth or lactation. Additional food above that for maintenance can be used with efficiencies of 62% for growth and 64% for lactation (A.R.C. 1965). Interindividual comparisons of energy utilisation must take account of body weight. The logarithm of metabolic rate when fasting divided by the logarithm of the body weight is usually close to 0.75 so the scale factor for body weight which is used in comparative energy studies is weight$^{0.75}$ (Kleiber 1965).

Energy expenditure can be estimated by measuring respiratory gas exchange. In the laboratory this can be carried out by enclosing the animal in a respiratory chamber and measuring oxygen consumption and carbon dioxide production (Armsby 1928, Graham 1962, 1965). Alternatively, inspired air flow can be monitored and expired gases can be collected by means of a tube connected to the trachea of the animal (Blaxter and Joyce 1963, Colovos *et al.* 1970, Young and Corbett 1972*a*). Neither method is very satisfactory for the metabolic processes may be altered by the treatment, but both of them allow some estimation of the energetic costs of different activities (Table 9). Indirect estimates of daily energy consumption, by continuous infusion of radioactively labelled sodium bicarbonate and sampling of blood or urine, are similar to those obtained by gas exchange methods. Young (1970) found that young (12–14 month) grazing heifers used 706 KJ $kg^{-0.75} 24h^{-1}$. The results from such studies give values for maintenance energy 20–70% higher for grazers than for animals housed indoors (Baile and Forbes 1974). The total daily energy expenditure estimates are accurate enough to allow the evaluation of particular foods and predictions about requirements. The assessment of the efficiency of feeding strategies, although well worthwhile, is more difficult because results such as those shown in Table 9 are variable. External conditions affect costs, e.g. a 10 km h^{-1} wind increased oxygen consumption of sheep by 40% (Joyce and Blaxter 1965), but sophisticated energy monitoring methods will soon make it possible to assess the energetic costs of the various components of feeding strategies.

The control of grass intake

The physiological mechanisms controlling food intake by ruminants are reviewed by Arnold (1964) and by Baile and Forbes (1974). The initiation of grazing by cattle often occurs at particular times during the 24 h period and the behaviour of individuals is affected by that of others in the social group. It is likely that information from visual, olfactory and gut receptors and from fat depot monitors are factors which affect the start of grazing. The termination of a meal is not normally limited by oropharyngeal factors, for animals whose ingested food is removed via oesophageal fistulas consume much more than they do during normal feeding (Campling and Balch 1961). When Campling and Balch filled the rumen with ingesta from another animal, however, the meal was terminated early. Water or air-filled balloons had the same effect. Removal of rumen contents extended the meal length unless the intake rate was low because the pasture had already been well grazed (Chacon *et al.* 1976). As a consequence of such experiments, most authors agree that when the volume of rumen contents reaches a certain level, receptors detect the distension and the meal is terminated (Blaxter 1962, Baile and Forbes 1974).

Since blood glucose levels decline during feeding and injection of glucose into the body does not reduce the rate of feeding unless the dose levels are very high, it is unlikely that glucose levels are of much importance in terminating feeding. Stimulation of the lateral or ventromedial hypothalamus has similar effects in ungulates to those obtained in rats, so these areas are probably involved in the control of

Table 9. *Energetic costs of different activities by cattle and sheep*

Activity	Cost	Reference
Cattle		
Lying (after 10 min +)	422 KJ $kg^{-0.75}$ $24h^{-1}$	Colovos *et al.* 1970
Standing	490 KJ $kg^{-0.75}$ $24h^{-1}$	Colovos *et al.* 1970
Lying → standing	0.66 KJ $kg^{-0.75}$	Colovos *et al.* 1970
Standing → lying	0.36 KJ $kg^{-0.75}$	Colovos *et al.* 1970
Walking	2.01 $kg^{-1}m^{-1}$	Hall and Brody 1934
Sheep		
Grazing	2.25 J $kg^{-1}h^{-1}$[a]	Graham 1965
Ruminating	1.00 J $kg^{-1}h^{-1}$[a]	Graham 1965

[a] Excess cost over that for lying.

feeding, but no gold thioglucose lesions can be produced in ungulates. Baile and Forbes (1974) consider that input from fat-depot monitors may be important in the control of feeding but the evidence is not clear. They also propose that input from monitors of intestinal content would be more efficient in long-term control of eating than would rumen distension monitors since the latter would provide a poorer estimate of true intake.

Grazing behaviour

The grazing strategies of cattle must include modifications in grazing and ruminating behaviour which depend upon gut contents, upon the quantity and quality of herbage available now and upon that predicted for the future. Domestic cattle have to contend with open range situations, where patches of adequate grazing are widely separated, set stocking, where a herd of cattle is kept in a field whose grass renewal rate is sufficient to supply their needs, and rotational grazing, where they graze on a paddock or strip of pasture for 1–10 days and are then moved to another. The amount of food available may be barely sufficient for body maintenance or may be enough to allow the production of large quantities of milk. The grass diet may be supplemented with high quality concentrates. These extremes may be encountered within a period of a few days by animals on a rotational grazing system.

When a group of six cows were put into a *Setaria* pasture in Australia for 14 days by Chacon and Stobbs (1976) they showed initially a very clear preference for leaves but as the total amount of herbage available declined they ate more stem and dead material (Figure 102). The duration of grazing was quite high (over 9h) at the beginning of the 14 days but increased to almost 11h on days 3–6, and then declined (Figure 103). In other studies, grazing time has been shown to increase considerably from lower initial levels as the amount of available herbage declines (Halley 1953, Arnold 1964, Stobbs 1974*b*). In Chacon and Stobbs' experiments the rate of biting increased during the first eight days but the bite size declined throughout the 14 days. The estimated intake dropped a little between the first two and the second two days, despite behavioural compensation, and dropped considerably after that. These results show that grazing behaviour is modified in a way which is energetically costly but which increases intake when the quality of the food is good. The most costly grazing behaviour is not used, however, when the quality has declined. Such effects of pasture quality on intake are emphasised by Hodgson, Rodrigues Capriles and Fenlon (1977) and by Baker and Barker (1978). The previous experience of cows in rotational grazing experiments must alter the likelihood that they will expend a lot of energy when the grass has been

102 On day 1 in a *Setaria* pasture, six cows ate mostly leaf (data from fistula samples). The amount of leaf available declined but most of their intake was still leaf on the thirteenth day in the pasture. The amount of stem consumed increased during this time but little dead herbage was taken. Modified after Chacon and Stobbs 1976.

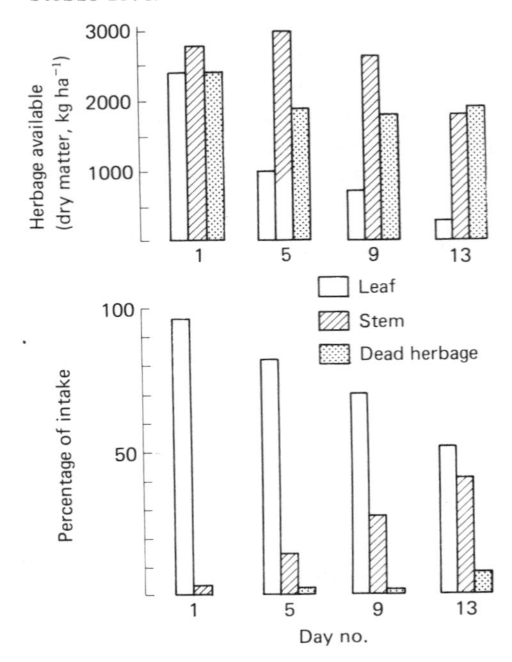

grazed down. Many cows are quite capable of training farmers to move them to a new paddock when available herbage is low. The cue which they train the farmers to use is the sight and sound of a row of cows standing by the fence and bellowing.

In addition to the rate of biting and the bite size, cows can vary manipulatory movements, the amount of walking and the amount of lateral head movement which they carry out during a feeding bout. A comparison of the same animals grazing on long grass, i.e. mean length of longest part of shoot 30cm, and short grass, i.e. mean length 13cm, showed that total grazing time was 7.9h per day on short grass but 6.9h on long grass. The mean times spent walking were 56 min and 30 min respectively. When the cows were videotaped whilst grazing, the rate of biting decreased by 8% and the distance walked whilst grazing decreased by 24%. There was an increase of 27% in the number of manipulatory chews. The bite size of cows which had fasted for as little as 16h was about 10% larger than that of cows which had fasted for only 2h (Chacon and Stobbs 1977) but most variation in bite size and rate of biting is related, principally, to the fact that cows will search for leaves and will take them individually in poor pastures (Stobbs 1973*b*, 1974*b*, 1975, Chacon and Stobbs 1977).

103 For the same experiment as that described in Figure 102 the changes are shown over the 14 days in grazing time (data from vibracorders), bite size, rate of biting (data from automatic recorder on jaw) and total intake (estimated from above). Modified after Chacon and Stobbs 1976.

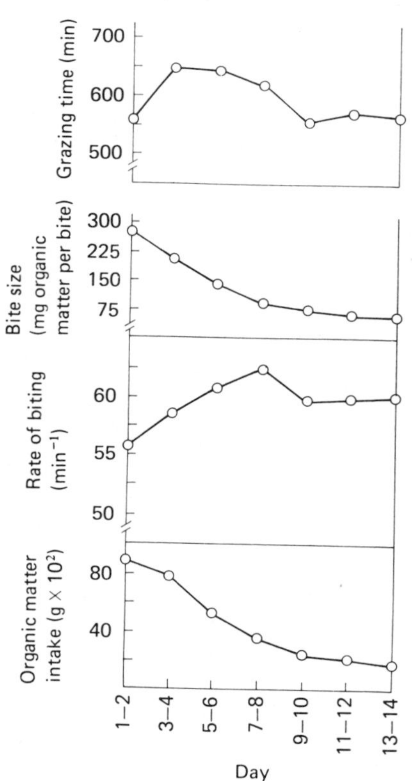

Another factor which affects grazing behaviour is the presence of faeces on the pasture. The dung from one cow, without decomposition, may cover as much as 200m² in a grazing season (Marsh and Campling 1970) and this may affect 10–15% of the pasture required in a season (Greenhalgh 1975, Leaver 1975). Cattle will avoid the area close to dung-pats and will preferentially eat clean herbage rather than herbage treated seven weeks earlier with slurry from a cowshed (Broom, Pain and Leaver 1975). On slurry-treated pasture cows ate less unless they did not have to eat the grass down to the slurry. If slurry was present they were more likely to stop grazing and walk a few steps and were involved in competitive encounters more frequently than on clean pasture (Pain, Leaver and Broom 1974, Pain and Broom 1978).

Grazing behaviour is also affected by weather and by disturbance. Grazing ceases in driving rain or if a large predator is heard or smelled. The sight of a dog or the occurrence of any unfamiliar activity on the farm may result in an interruption of grazing behaviour, probably as part of an anti-predator response. Calves are more likely to be disturbed by any particular event, perhaps because cows have had much previous experience of such events which had no adverse effect on them.

104 Three-year-old sheep which have had no experience of grazing (open circles) are less efficient grazers than are sheep which have grazed (filled circles). Their performance improves after ten weeks' practice. After Arnold and Dudziński 1978.

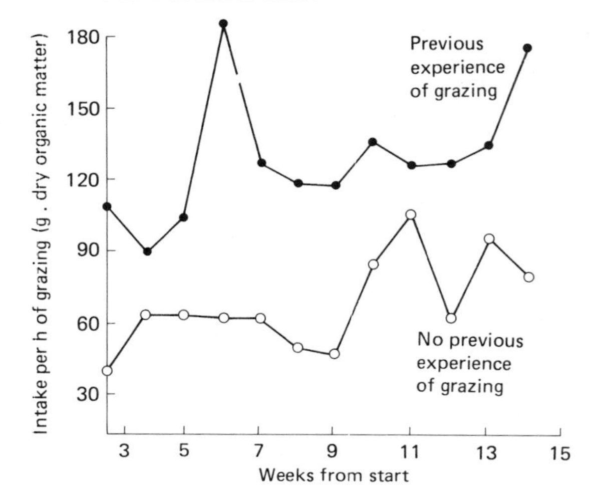

The development of grazing in calves involves an increase in grazing time during the first four months as rumen function develops (Chambers 1959, Hutchison *et al.* 1962). By one year of age, calves graze for an hour a day more than heifers and for 1.6h per day more than 3½ year-old cows (Hodgson and Wilkinson 1969). This may be because calves are more selective (Hafez and Dyer 1969) or it may be because increased mouth size or experience of grazing has improved the feeding efficiency of the older animals. Arnold and Maller (1977) and Arnold and Dudziński (1978) showed that sheep reared for three years with no grazing experience grazed much less efficiently than experienced sheep (Figure 104) and also reported that preferences for pasture plants were much affected by previous experience.

7
Anti-predator behaviour

Defensive mechanisms and their evolution

Anti-predator behaviour is obvious to the field naturalist but is sometimes overlooked by those who carry out experimental studies of behaviour in the laboratory or who try to understand human behaviour. When individuals are making decisions about their next behaviour, those activities which reduce any immediate danger of major bodily damage take precedence over all others. The danger may be that of attack by a predator or competitor, or may result from a physical hazard. Sometimes animals show similar responses to more than one type of danger. Hiding or active defence may be shown when a predator or competitior is detected; fleeing may be elicited by either of these or by a physical hazard such as falling rocks.

The initial physiological responses which prepare the individual for flight from a predator are similar to those which precede other sorts of vigorous action (Chapter 5). In emergency situations, however, there are special responses such as adrenaline release. Physiological and behavioural responses which help to counteract dangers, such as that from predators, may be elicited when there is merely a possibility of danger. It is easy to understand why systems which provide early warning of danger have evolved. Their existence in man and in our domestic animals has important consequences for medicine and for animal husbandry. People, especially children, may show elements of anti-predator behaviour when they predict that there may be danger because it is dark or because they believe there to be ghosts or demons in the vicinity. Some of the responses of domestic animals to man or to unexpected events can also be considered to be components of anti-predator behaviour. Before considering such general mechanisms and the role of experience in their development, the range of different sorts of anti-predator behaviour will be reviewed.

Some defensive mechanisms operate throughout much of the life of an individual whereas others are used only when a predator is detected or when a predator attacks the individual. Edmunds (1974) refers to two types of mechanism, primary and secondary. Following Kruuk (1972) Edmunds defines *primary defence mechanisms* as those which operate regardless of whether or not there is a predator in the vicinity. The behavioural components of such mechanisms may occur before any predator is detected. The mechanisms include (1) hiding in holes; (2) the use of crypsis; (3) mimicking inedible objects; (4) exhibiting a warning of danger to predators; (5) mimicking individuals in category (4); (6) timing activities so as to minimise the chance of detection by a predator; (7) remaining in a situation where any predator attack is likely to be unsuccessful because of possibilities for secondary defence; (8) maintaining vigilance so as to maximise the chance of detecting the advent of a predator. All of these mechanisms decrease the chance that a predator attack will occur. Many of them have a considerable influence upon habitat selection. For some species which are very vulnerable to predation, the avoidance of predators limits the distribution of the animals even more than do factors related to body maintenance and feeding (Chapters 5 and 6). The selection of a suitable habitat may be accomplished by the recognition of a factor which is not in itself advantageous but which is an indicator of the presence of advantageous features, such as those which make possible feeding or predator avoidance. For example, a bird, such as a nightjar, which is camouflaged when amongst dead wood might fly over open downland towards a dark green mass of woodland because suitable hiding places are likely to be found around the edge of the wood.

Secondary defence mechanisms are defined by Edmunds as those which operate during an encounter with a predator. The encounter may just involve the supposed or actual detection of a predator by the prey or it may involve an attack by a predator. Some secondary defences are entirely passive. The poison-spined sea urchin *Diadema* or the thick-skinned rhinoceros are adequately protected from attack. No behavioural specialisation is needed. Active forms of secondary defence include (1) exaggerating primary defence, e.g. a camouflaged animal remaining motionless; (2) withdrawal to a safe retreat; (3) flight; (4) use of a display which deters attack; (5) feigning death; (6) behaviour which deflects the attack to the least vulnerable part of the body, to an inanimate object, or to another individual; and (7) retaliation.

Both primary and secondary defence mechanisms are used by most animals and different mechanisms are often used for different predators. In an encounter with a predator, the prey animal may be able to use several alternative anti-predator behaviours. Decisions must be taken about, for example, whether to remain motionless, to flee, to display, or to fight. If one method is tried, without successful avoidance of attack, others may be tried. There are many possible strategies which involve alternative courses of action depending upon the characteristics and behaviour of the predator.

The evolution of anatomical and behavioural defence mechanisms has been affected by the evolution of the anatomy and behavioural strategies of predators. Dawkins and Krebs (1979) have called this interaction during evolution an arms race (see Chapter 6). The process is very complex. Whilst it is of interest to consider, as Dawkins and Krebs do, the successive modifications to genotype in rabbits and foxes, it is also necessary to consider other relevant evolutionary interactions. For example, those of rabbits with other predators, rabbits with parasites, rabbits with food plants, and foxes with other prey species. A genotypic modification of rabbits which helps them in their arms race with foxes might not spread in the population because it has an adverse affect on the rabbit's chances against stoats or on the rabbit's ability to carry out behaviour in some other functional system, for example the ability to dig burrows.

Since predator attack is a possibility throughout the lives of most animals they need to show exploratory behaviour and to be able to respond and mobilise their resources rapidly. The behaviours and physiological mechanisms which are discussed in the section on 'Exploration and fear responses' later in this chapter are relevant to man and to domestic animals as well as to other species. The occurrence and the forms of defensive mechanisms are usually subject to modification according to the individual's experience of possible danger. For example, an animal which has seen a shadow or an actual predator many times may be less likely to show defence responses than would an individual which sees the shadow or predator for the first time. Such learning mechanisms, and those which occur during development, are also discussed later in this chapter.

Defence behaviour before predator detection

Hiding in holes

Many small animals on land and in water spend much of their life inside holes in the substratum. Some of these animals, such as the lugworm *Arenicola* or the mole *Talpa*, feed whilst under the surface and almost all of their behaviour is adapted for life underground. Others, like the ragworm *Nereis* or the rabbit, use the hole as a retreat where they spend part of each day but from which they emerge to feed. Many coral-reef fish hide in holes in the reef, or in holes excavated in sand, when they are not swimming around (Eibl-Eibesfeldt 1967). Although holes do provide protection against predators they may also serve other functions such as in maintenance of body temperature or body fluid levels (Chapter 5). More frequently, however, they interfere with other functions. If ragworms or rabbits did not have to hide in their burrows for long periods they could spend more time feeding. Some worms which live in holes or tubes are restricted to feeding methods like suspension feeding which do not necessitate dangerous departures from the tube. Reproductive mechanisms also may be less efficient if much of the animal's life is spent in a hole. The construction of a burrow, tube, or other retreat may involve complex behaviour and much time may be spent in this or in searching for suitable hiding places.

The use of crypsis

Animals which are camouflaged so that predators do not distinguish them from their background are said to be *cryptically coloured*. Examples of such camouflage will be familiar to all readers and are extensively reviewed by Cott (1940) and Edmunds (1974). Green grasshoppers or caterpillars in grass, bark-patterned moths on tree-trunks, transparent plankton at the surface of the sea, dark red plankton in deep sea (Hardy 1956), counter-shaded fish, or zebras in tall grass are all hard for predators to see. The behaviour of these animals is just as important as the coloration in providing defence against predation. The camouflage is effective only if the right resting place is chosen and if the animals are stationary for much of their time. The animal must rest if the surroundings match its own coloration and explore further if they do not. Curio (1966) described a moth *Erinnyis ello* whose caterpillars could be green, brown or grey. The resting places which the caterpillars chose were sites on branches or leaves where they were best camouflaged. Cryptic coloration sometimes depends upon the selection of the right orientation in relation to the background. A counter-shaded fish is very obvious if it is upside down. The striped moths studied by Sargeant (1969) need to align themselves with stripes on the bark of trees if they are to be inconspicuous to predators. Figure 105 shows the results of experiments with two striped species. The moths respond to the pattern on the artificial background, just as they do to that on a tree-trunk.

The limitation in the number of possible resting places and the necessity to keep still for long periods in order to maintain crypsis result in a great reduction in the time available for other activities. This disadvantage is reduced if the animal can change its camouflage to match its background. Animals which can change their body colour include the chameleon, many flat-

105 Moths of two species were found to rest on lined surfaces so that the lines on their wings were in the same direction as those on the surface. In control experiments the lines were covered. Modified after Edmunds 1974.

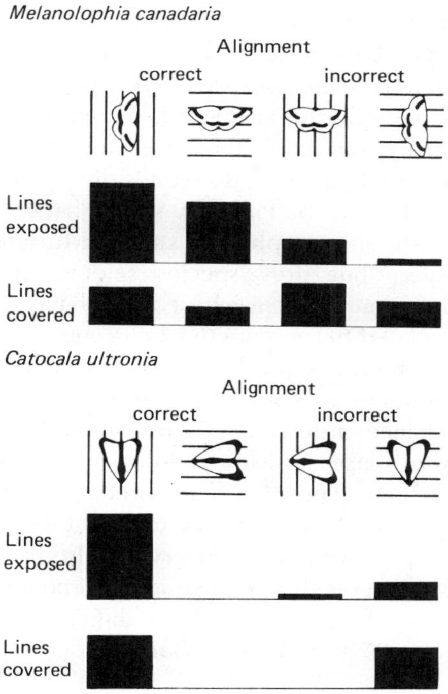

fish and most cephalopod molluscs. The chromatophores in the skin of a cuttlefish or octopus can be expanded or contracted so as to produce different colour patterns, most of which are cryptic when the animal is resting (Holmes 1940, Packard and Sanders 1971). Those animals which use materials from their surroundings to conceal themselves can become camouflaged in various backgrounds. Caddis larvae build cases around themselves with locally available materials (Hansell 1968). Some spider crabs hold algae, sponges and other organisms in their chelae or plant them on their backs (Warner 1977). A crab which moves to a new area could change the materials which it holds or which grow on it.

Hunters will use all of their senses when hunting but to many the sense of smell is important. It is likely, therefore, that *olfactory crypsis* occurs frequently amongst prey species. Our sensory world is so dominated by vision that we have not investigated such mechanisms. It would seem likely that the production of readily detected odours would be disadvantageous in the same way as the exhibition of bright bodily colours. Grooming and preening behaviour might have some anti-predatory function in that decaying, odoriferous material is removed from the body surface before predators detect it. Nest sanitation must also reduce the chance that the nest might be found by smell. Large carnivores are often unpleasant smelling throughout the year but prey species seldom produce strong odours except where these have a function related to reproductive behaviour. Perhaps they sometimes produce odours which mimic environmental odours. Prey animals can reduce the chance of detection by predators if they remain in places which are downwind of the expected arrival position of any predator. Such behaviour also helps the prey to detect the arrival of the predator. Auditory crypsis also occurs in that prey animals minimise sound production or produce alarm calls which are difficult to locate (see p. 195 and Marler 1955).

Mimicry of inedible objects

Cryptically coloured animals are not detectably different from their background. Some animals, however, are obvious but are ignored by predators because they look like inedible objects. Mantids, stick-insects and the caterpillars of moths in the family Geometridae resemble sticks. Even if they are not surrounded by sticks they may be ignored because small sticks are objects which are encountered frequently. The mimicry depends for its effectiveness on lack of movement. This is also true for those caterpillars and spiders which mimic bird droppings. The fish *Lobotes surinamensis* lives amongst mangroves when it is young and it rests on its side near the surface of the water. In size, colour and posture it bears a close resemblance to a dead leaf (Figure 106). When mildly disturbed it swims using its transparent fins, in such a way that it drifts with a zig-zag motion like a sinking leaf (Breder 1949, Edmunds 1974).

106 When young, the fish *Lobotes surinamensis* looks and moves like a leaf floating in the water. Modified after Edmunds 1974.

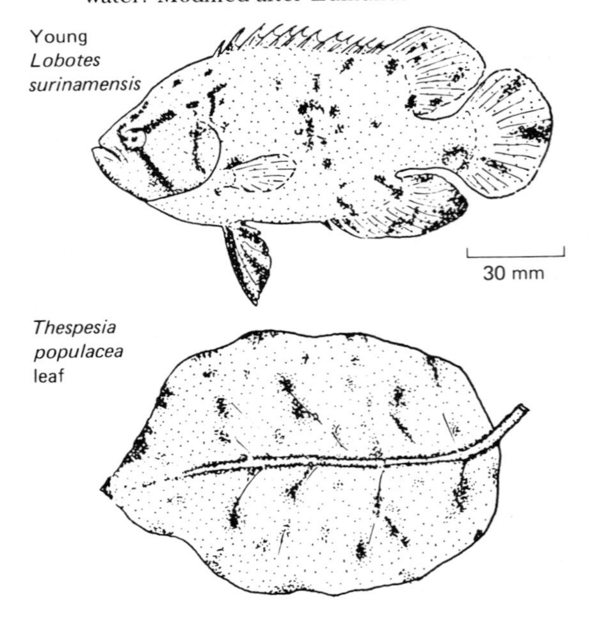

Defence by warning of danger to predators

Animals which are poisonous if eaten or which have poisonous stings or bites are often brightly coloured. The advertisement of such attributes by coloration, sound or other means is called aposematism. Examples include wasps e.g. *Vespa*, ladybird beetles e.g. *Adalia*, coral snakes *Micrurus*, the Gila monster *Heloderma*, and skunks e.g. *Mephitis*. The colours of these animals are yellow, red, pink or white and black. Predators tend to avoid animals coloured in this way, especially after some experience of attacking one. The similarity of coloration and sometimes behaviour, of different distasteful or dangerous species for example, wasps, bees and the yellow and black caterpillars of the cinnabar moth *Callimorpha jacobeae*, is known as Müllerian mimicry. The behaviour required for an aposematic animal is that which makes it conspicuous, for a brightly coloured animal which is hiding in the shadows might be attacked. If the individuals are small, they may be more conspicuous to a predator if they aggregate. A further possible advantage of aggregation was pointed out by Fisher (1958). If the predator has to taste one individual before it avoids others, genes which promote aggregation by close relatives would spread in a population of distasteful animals because they would be shared by the individual which the predator samples and those nearby which survive as a result. Examples of distasteful animals in which aggregation occurs are the red and black cotton-stainer bugs of the genus *Dysdercus*.

Most aposematic animals are mimicked by palatable, harmless species. The coloration and sounds of wasps and bees are mimicked by flies and moths (Figure 107) and the unpleasant tasting monarch butterfly *Danaus plexippus* is mimicked by the viceroy butterfly *Limenitis archippus* (Brower 1958). Some of these mimics also modify their behaviour so as to increase the resemblance to the model species (see review by Wickler 1968).

Timing activities, positioning and vigilance

Timing

Most mammals are nocturnal or crepuscular. It seems likely that the principal reason for this is that it is easier for mammals to avoid predators at night. Animals which live in the inter-tidal region may be more vulnerable to predation during the day and when the tide is out. Many littoral species are more active at night and when the tide is in, although this may also be due to the reduced risk of desiccation. Crypsis and hiding in holes or in vegetation, are often combined with the precise timing of activity periods such that the animal can feed and yet minimise the risk of predation. Prey animals may modify activity rhythms according to those of predators, for example by hiding at the time of day when an individual predator is most likely to appear. They can also minimise the chances that they will be killed, by synchronising their activity with that of other prey individuals. The mass emergence, from streams or lakes, of adult mayflies on particular days is one type of example of synchrony. The predators cannot kill them all at once. Another is the synchronisation

107 The hornet clearwing moth mimics a large wasp in morphology and behaviour. Photograph: G. C. Bellamy.

of chick departure from the nest cliffs by Brünnich's guillemots (*Uria lomvia*) (Figure 108). Daan and Tinbergen (1980) reported that most of the chicks jumped, with parental encouragement, at 1630–2130 h. Some of those which landed on the slopes below the nesting ledges were killed by predators, but all of those which jumped at other times were killed.

Positioning

The spatial positioning of individual animals is of great importance, for secondary defence mechanisms might be necessary at any time. Animals remain near refuges and avoid places where predators might conceal themselves. Some animals surround themselves with physical impediments to attack, for example those which feed in thorny bushes, or near the tips of branches, or near precipices. If a predator attacks it must risk impaling itself or, if a ground predator, falling. Another widespread mechanism is for an individual to keep alternative prey between any predator and itself. Hiding within a group of conspecifics may be advantageous because another individual will be taken first, or because the predator will avoid mechanical damage due to collision with many prey, or because of collaborative defence possibilities (see Chapter 8). When Seghers (1974) studied five populations of guppies (*Poecilia reticulata*) in Trinidad he found that some were more subject to predation than others. Guppies from populations where predation was infrequent were least likely to school.

108 A juvenile Brünnich's guillemot *Uria lomvia* gliding down towards the sea from its nest-site on a cliff in Spitzbergen. The parent follows closely, using its wings and feet to slow its fall. After Daan and Tinbergen 1980.

Vigilance

Animals with sensory systems which allow them to detect predators at a distance can maintain vigilance and thus initiate secondary defence mechanisms before the predator attacks. Animals which feed in exposed places, like grazing ungulates in grassland or monkeys in the tops of trees, spend some time in feeding and some time in sniffing, listening and looking around for terrestrial or aerial predators. This vigilance may be possible at the same time as feeding or other activity but it often necessitates cessation of activity to look around, orient the ears or flare the nostrils. A predator may be detected directly or the prey species may hear an alarm call, or see signs of escape behaviour. As described in Chapter 8, the proportion of time which must be devoted to looking out for predators is usually lower if groups of other prey animals, often conspecifics, are nearby.

Defence behaviour after predator detection

Animals which are already in refuges, or which are camouflaged, often respond, when a predator is detected, by keeping still. This has the effect that the predator is less likely to detect them by visual, auditory, lateral-line or electrical receptors. Those animals with warning coloration or other aposematic signals, however, may keep still so as to escape detection or may move so as to emphasise the warning to the predator. The location of dangerous predators is sometimes monitored and advertised by mobbing behaviour. If the predator is close to the place where the prey animals live, it may be advantageous to keep it under surveillance. Mobbing usually involves harrassment as well, so vocalisation which encourages other prey animals to join in is beneficial (see discussion by Harvey and Greenwood 1978). Other active responses to predators are listed below.

Withdrawal to a retreat

The rabbit grazing near its burrow and the human child playing outside its dwelling will run inside if they see, hear or smell something which may indicate imminent danger from a predator. The distance travelled is small and the route to safety is well known. Those animals which are in permanent contact with their retreat have an even simpler response. Caddis larvae, bagworms in the moth family Psychidae and various marine invertebrates such as snails or serpulid worms make or secrete their own case, shell or tube. When they are startled they withdraw those parts of the body which are exposed. Tortoises withdraw into shells which are a part of their body and a variety of armoured animals curl up, e.g. armadillos, pangolins, some lizards, woodlice and millipedes (Figure 109).

Flight and evasion

If a deer is grazing in an area of open grassland and a wolf approaches from the distance, the deer may fail to respond, by running away, until the wolf is within 200m. Similarly, the rabbit mentioned in the previous section may refrain from running into its burrow until a fox is within 50m. These distances have been called flight distances by Hediger (1955) and others. The flight distance may differ according to the behaviour of the predator, for the sight of a wolf which is stalking may elicit flight at a greater distance than the sight of a wolf which is resting or which is following a regularly used path. Adaptations for rapid departure by swimming, running, jumping or flying from places which may be dangerous have a considerable effect on the body form of many prey species. The behaviour shown often involves expenditure of all available energy on the escape but it is important that the direction of escape be carefully selected. Most prey species will have explored possible avenues for escape before the advent of the predator.

109 The armadillo *Tolypeutes conurus* curls up when predators approach. The horny layer on the dorsal body surface offers protection against most predators but the animals are readily caught and killed by man. Modified after Bourlière 1955.

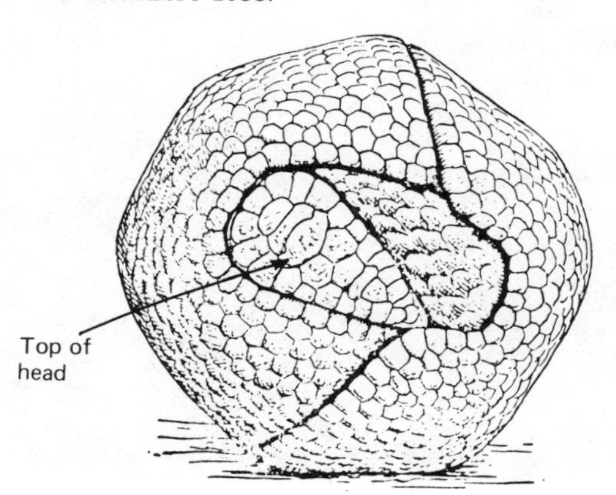

If there is a possibility that the speed of the prey is insufficient for it to escape from the predator, other evasive tactics may be used. Departure from the predator's medium is one effective tactic. Water voles escape from land predators by swimming; diving ducks or grebes go under water when threatened by an aerial predator and flying fish emerge into the air when chased by predatory fish. Small birds can escape from ground predators by flying slowly to a tree but they must fly rapidly to cover if a hawk or falcon chases them. Some largely sedentary marine animals such as the scallop *Pecten* and the sea anemone *Stomphia* can swim away from their starfish predators (Robson 1962).

Unpredictability of movement is advantageous to prey animals pursued by a predator with a speed advantage. Snipe (*Gallinago gallinago*) and ptarmigan (*Lagopus mutus*) follow a zig-zag path when they fly away from a predator and are thus difficult for a hawk or falcon to catch as well as being difficult to shoot. Pigeons can double back when escaping from birds of prey, an ability which has been exaggerated, by breeding, in the 'tumblers'. The ability of some nocturnal moths to detect the calls of an approaching bat was first described by von Uexküll (1940). Subsequent work by Roeder and Treat (1961) showed that the moths twist and fall in flight when they hear bat ultrasounds. Miller and Olesen (1979) have shown that green lacewings (*Chrysopa carnea*) 'nose dive' when they hear bats (Figure 110) but can momentarily interrupt their fall with a wing flip and thus further deceive a pursuing bat (Figure 111). Olesen and Miller (1979) found that these actions are a consequence of temporary uncoupling of motor neurons to the continuously active flight pattern generator when the bat's call is heard.

Another, quite different, example of unpredictability in prey is the use of flash coloration. When some grasshoppers and moths take off they are very conspicuous because their hind wings are brightly coloured but when they land these bright colours are concealed and the animal is inconspicuous. If a predator chases a brightly coloured prey, or a prey which makes much noise as it flies, it may be unable to locate that prey when these obvious characteristics are no longer shown. The importance to prey of unpredictability of behaviour was emphasised by Chance and Russell (1959) who referred to a variety of examples of elaborate flight behaviour and deflection of attack as protean behaviour.

Use of displays which deter attack

When aposematically coloured animals are approached by a potential predator they may behave in such a way as to accentuate their coloration. For example, the spotted skunk (*Spilogale gracilis*) will stand on its fore legs and raise its back legs into the air thus displaying its black and white back (Bourlière 1955 and Figure 112). Other animals with effective deterrents to attack may signal their presence by making noises, for example hissing or rattle-shaking by snakes and quill-rattling by porcupines. More obvious weapons, such as the sting of a scorpion, the horns of some ungulates and the teeth of many animals, are displayed as a deterrent to attack. The behaviour may be more complex in group-living animals, for example the defensive rings formed by musk oxen when wolves approach or by eland when confronted by hyaenas (see Chapter 8).

110 Successive positions (1–6) of a bat and a lacewing shown by flashes at 70msec intervals. The lacewing hears the bat and starts to nose-dive but is caught. During the 70ms interval before the first exposure the bat vocalises twice but during the last 70ms before capture it vocalises 13 times in order to pinpoint the position of its prey using sonar. Photograph: L. Miller.

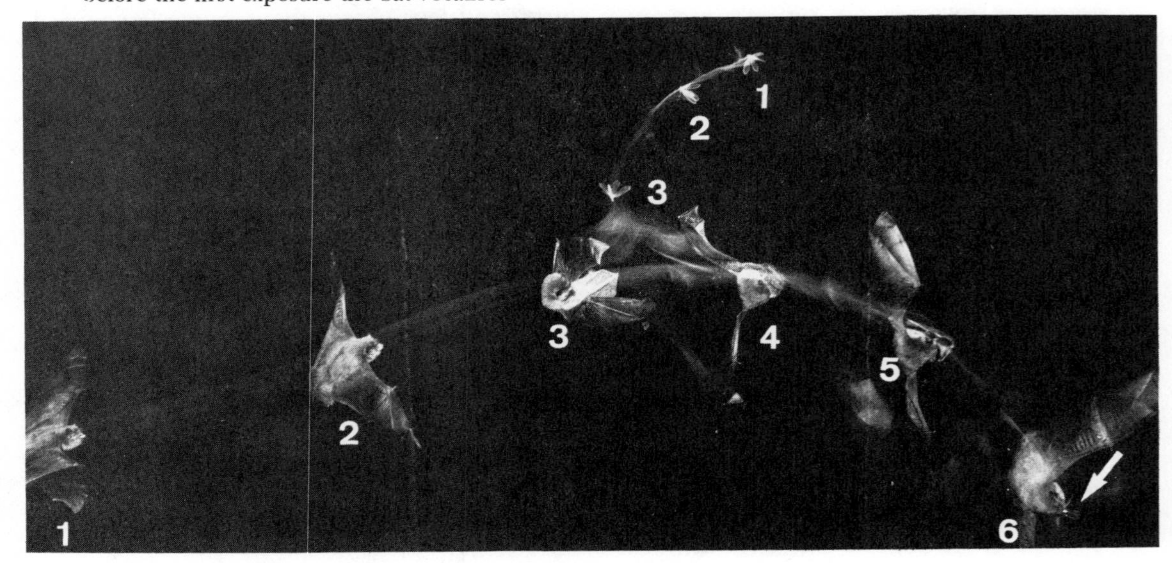

The ubiquity of frightening, or deimatic (Maldonado 1970) displays amongst species which possess weapons is not surprising. The cost of fighting is high so genes which promoted display as a preliminary would spread in most populations. The displays would be more effective if they exaggerated the size and apparent fighting power of the animal. Mammals can make themselves look bigger by erecting their fur. A domestic cat which is approached by a dog will do this and will also vocalise, spit and bare teeth and claws. Chimpanzees respond to the approach of a leopard by hair erection, loud vocalisation and arm threats (Gandini and Baldwin 1978). Defensive behaviour of this kind is also shown by many other animals which have no very powerful weapons. A toad raises itself to its maximum height when a snake, or snake-like model, approaches (Ewert and Traud 1979) and the orthoperan insect *Mygalopsis ferruginea* can deter approaching lizards by raising itself from the ground and stridulating (Sandow and Bailey 1978 and Figure 113).

As described in Chapter 2, markings which are coloured and positioned to resemble eyes may effectively prevent a predator attack, especially if they are exposed very rapidly as in the movement of a hawk-moth's wings. The eye-spots mimic the face of a vertebrate predator. Other predator mimics include caterpillars which look like snakes and birds which sound like snakes. Members of the tit family Paridae and the wryneck *Jynx torquilla*, which nest in holes in trees, will hiss and bang the walls of the nest-hole if a predator appears at the nest entrance. An example of a marine animal which mimics a predator is the fish *Calloplesiops altivelis* which is so coloured that when it assumes a certain posture near a hole in rock or coral it resembles the moray eel *Gymnothorax meleagris* (McCosker 1977).

Feigning death

One aspect of anti-predator behaviour has been observed with sufficient frequency to have resulted in a new phrase, 'playing possum', entering the English language. The American opossums *Didelphis* lie still on their sides when

111 Sequence of six positions of bat and lacewing as in Figure 110 but the lacewing interrupts its fall with a wing flip (position 5) and the bat misses it. Modified after Miller and Olesen 1979.

a predator is near them (e.g. see Hamilton 1963). Immobility in such situations, which is sometimes called thanatosis, is also shown by various beetles, bugs, grasshoppers, spiders, lizards and birds (Figure 114). Many birds will lie quietly if the head is covered by a human hand, whilst the bird is held on its side, and the hand is then withdrawn slowly. This behaviour is sometimes called, inappropriately, hypnosis. The behaviour is also shown to predators other than man and it often results in a withdrawal by the predator and hence an opportunity for the bird to escape. An individual which is apparently dead is also less likely to be bitten severely by a predator and may be ignored altogether by those animals which will eat only freshly killed prey.

Behaviour which deflects attack

Avian predators attacking large insects are most likely to peck at conspicuous eyes so butterflies with false eyes on the hind wings may escape with wing damage, when attacked, instead of being killed (Cott 1940, Carpenter 1941). A large eye-mark on the side near the tail is used by the John Dory fish (*Zeus faber*) in elaborate attack-deflection behaviour. The fish swims slowly backwards in the sea so that the tail with its eye-mark appears to be the head. When a predator comes close, the John Dory darts off in the opposite direction from that which the predator expects. Another deception is that used by cephalopod molluscs such as *Octopus* and *Sepia*. When the animal is disturbed by a predator, just before it swims rapidly away it ejects a cloud of ink-like pigment which is approximately the same size as the animal. The attack is deflected towards the ink which then disperses and obscures the flight path of the cephalopod. Tail autotomy by lizards may dis-

112 When approached, the striped skunk *Spilogale gracilis* displays its black and white coloration before releasing the malodorous secretion from its anal glands. After Bourlière 1955.

113 The grasshopper-like *Mygalopsis ferruginea* (*a*, resting, size x 0.7) responds to an approaching lizard (*b*, directly in front; *c*, from the side) by raising its legs and stridulating. Animals assuming these postures are attacked less frequently by lizards than are those which do not. Modified after Sandow and Bailey 1978.

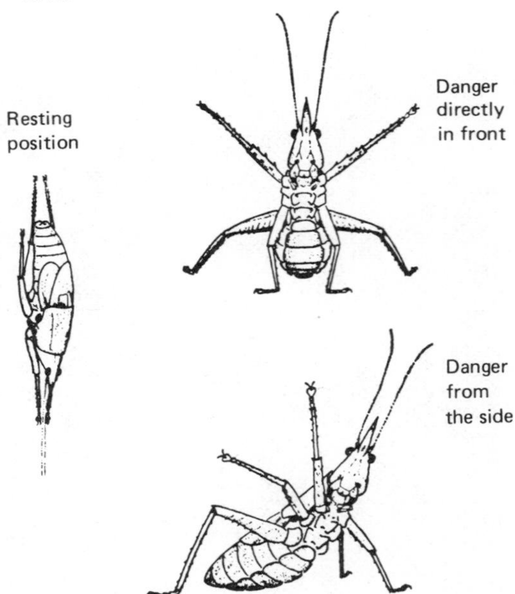

tract the predator as well as allowing the escape of an animal which has been grasped by the tail. Parents sometimes deflect attack towards themselves and away from their eggs or young by means of display. The broken wing display by many wading birds and the miming of injury by the fish *Amia* are examples.

Deflection of attack towards conspecifics may occur in group-living animals (see Chapter 8). If an animal moves in such a way as to interpose another member of the group between itself and the predator, or if an individual which is pursued runs or flies into a group and hence causes another to flee, the predator may direct its attack at the nearest or most conspicuous individual. Alarm calls may also have the effect of making other members of the group take flight at the same time as the caller, so that the approaching predator has many alternative prey to attack (Charnov and Krebs 1975).

Retaliation

Animals with weapons for mechanical attack, such as armoured jaws, antlers, or spines may use them during a predator's attack. Chemical weapons such as stings or spraying mechanisms may be used in the same way. The weapons may also be used in predatory attacks, e.g. in wasps, or in intraspecific encounters, e.g. in ungulates such as the oryx. The retaliatory behaviours of very many animals have been described by nineteenth century zoologists who were endeavouring to collect specimens. The characteristics of those animals whose behaviour is an effective deterrent to future attack on the same species by man are learned by children following unpleasant experiences or dire warnings from parents. Much of the recent information about such mechanisms has concerned those retaliatory behaviours which are directed at predators other than man. The use of chemical secretions and sprays by arthropods such as the bombadier beetle *Brachinus crepitans*, the nasute soldiers of various species of termites and the whip-scorpions are described in detail by Roth and Eisner (1962) and by Eisner and Meinwald (1966). One elaborate retaliatory behaviour sequence described by Eisner (1971) is that of the harvestman *Vonones sayi*. When grasped, for example by an ant, it secretes two separate substances, mixes them together and then paints the mixture on its assailant with its fore leg. This causes the predator to withdraw rapidly. Molluscs may also produce noxious secretions when attacked and those of the gastropods *Acteon* and *Haminoea* are toxic to small planktonic animals (Fretter and Graham 1962). Organs which have a major function which is not that of fighting may sometimes be used in retaliation during a predatory attack. Branch (1978) has described the vigorous defence of two species of limpet *Patella granatina* and *P. oculus*. They smash their shells down on starfish or predatory gastropods which are attacking them and thus damage them. Behaviour patterns with other functions may also be used in defence. Pinel and Treit (1978) describe how rats which were 'attacked' by electric shocks from a metal prod would bury it in bedding material.

114 This western fence lizard (*Sceloporus occidentalis*) feigned death after being caught.

Exploration and fear responses

Efficient exploitation of food sources, other maintenance of the body, avoidance of physical hazards, avoidance of predation and performance in social interactions require that the individual should familiarise itself with its surroundings. Activities involved in such familiarisation are referred to as *exploration* (for example see Barnett 1963). Individuals also explore their own abilities by monitoring the effects of carrying out an activity. The latter form of exploration includes some activities which may be called play as well as many different behaviours concerned with each of the functional systems. Social relationships may be very complex and therefore require much exploration. Learning occurs as a result of exploratory behaviour, for subsequent behaviour is modified consequent upon the experience obtained during exploration.

The major function of exploratory behaviour for many animals is to improve their ability to escape predation. As mentioned earlier in this chapter, animals which are cryptically coloured must ensure that they find a good resting site. Those which hide must find the hiding place, or a site and materials for its construction. Those which might flee, display, deflect attack or fight must learn the characteristics of their surroundings so that they can act appropriately and efficiently. Exploration must occur before there is imminent danger of predator attack.

Exploration may occur in a situation where predator attack is probable. In such situations, just as in those which are physically hazardous, an adrenal response often occurs so that blood levels of adrenaline and glucocorticoids increase (Chapter 5). The animals will then show more scanning behaviour and increased responsiveness to any environmental change. Observers may call such behavioural changes fear responses, or they may restrict the term to the behaviours which occur after the detection of a predator, a dangerous competitor, or a physical hazard. As Hinde (1970) points out, exploratory behaviour and fear responses are overlapping categories. Most behavioural changes which have been called fear responses are associated with adrenal activity. It is therefore most useful if the term *fear* is taken to include (1) those activities which follow adrenal activity and which are a preparation for danger, as well as (2) those which are a response to detectable danger. All usage of the word fear is vague for it refers to a set of causal factors whose levels are very difficult to estimate. It is inadequate to try to describe behaviour or motivational state solely by reference to 'fear responses' or to 'fear'. The diversity of uses of the word is discussed by Archer (1979*a*). The section below on general components of anti-predator responses includes descriptions of behaviours and physiological mechanisms which other authors call fear responses and mechanisms but which are described here in more precise terms.

Exploration in relation to defence

Evidence from studies of wild animals, of the importance of exploration as a prelude to successful anti-predator behaviour includes the high mortality of birds and mammals which move into unfamiliar areas. There are also many anecdotal observations of apparent exploration by animals when they go to a new place or encounter a new object within their home range. There is, however, a dearth of field studies of exploration and its effects on responses to predators. In the laboratory, on the other hand, many hundreds of studies of exploratory behaviour have been carried out. The majority of these experiments have involved human experimenters placing a laboratory rat into an arena approximately 1m in diameter and them recording ambulation and defaecation by the rat. The initiator of this experimental procedure (Hall 1934) called such an arena an 'open field' and the term is still used despite the fact that the arena is closed and scarcely field-like. In similar experiments where the rat was placed in a small box or pipe in an unfamiliar arena, the time taken for the animal to emerge was measured. The results of such experiments are very difficult to interpret for the behaviour shown will be a combination of the response to being handled by the experimenter and exploration. The function of this exploration may be feeding, finding a mate, finding social partners, or avoiding predators. Each of these aspects of the response will be affected differently by immediate and long-term previous experience. The interpretation of such experiments in terms of a single variable, referred to as 'emotionality', has been criticised by Archer (1973) and Daly (1973). One very clear result of these experiments is to confirm Barnett's (1963) observations that rats explore a novel arena even in the presence of food and after food deprivation. The proportion of time spent moving around the arena declines with time in the arena and with successive visits to it. The possible role of the hippocampus in this decline in exploration is shown by experiments in which rats with dorsal hippocampal lesions continued running around a strange cage and sniffing in the corners for one or more hours instead of stopping this behaviour after 10–15 min as normal rats do (Teitelbaum and Milner 1963).

Another technique for investigating exploratory behaviour in the laboratory is to introduce novel objects into the home cage. Glickman and Sroges (1966) found that the amount of investigation shown to various small, novel objects by animals of 100 different species was highest amongst primates and carnivores, less for rodents and lowest for reptiles. In this study and in a study of marsupials by Russell and Pearce (1971), investigation of objects was greatest amongst carnivores and least amongst herbivores. This suggests that the objects were treated as potential food by most animals. In contrast to these studies of zoo animals, wild animals such as rats and buntings will often completely avoid novel objects for some time (Shorten 1954, Andrew 1956*b*). Their behaviour is most likely to be anti-predator in function.

Startle responses

As mentioned in Chapter 4, any detectable, novel change in an individual's surroundings elicits an orientation reaction. The changes in receptor orientation, EEG, heart rate, muscle tonus etc., prepare that individual for anti-predator or any other kind of behaviour. If the change in surroundings is one which might be caused by a predator, vulnerable prey species must show further responses. Some of the behavioural and physiological changes are mediated via adrenal activity (Chapter 5 and Archer 1976) whilst others are brought about by direct brain control. Experimental investigations of these changes require that a repeatable stimulus be used. For example Harlow and Zimmerman (1959) presented a moving clockwork toy to young rhesus monkeys and other authors have used sounds and lights. Low intensity sounds elicit orientation reactions from various mammals, but higher intensity sounds with a rapid rise time elicit a much more obvious startle response (Soltysik, Jaworska, Kowalska and Radom 1961, Fleshler 1965).

When a small light-bulb is suddenly illuminated on the wall of the home pen of young domestic chicks they show a brief orientation reaction and then a startle response (Broom 1969*a*,*b*). The startle response is either a period of immobility, or a flight response with loud calling, or, most frequently, the former followed by the latter. These behavioural startle responses are usually associated with tachycardia, an acceleration of heart rate. Starlings presented with a distress call (Thompson, Grant, Pearson and Corner 1968), chick embryos presented with noxious odours (Tolhurst and Vince 1976), rats given electric shocks (Teyler 1971, Chalmers and Levine 1974) and people subjected to a variety of noxious stimuli (Wilson 1964) all show tachycardia. The response should be assessed by comparing the heart rate after disturbance with that predicted if no disturbance occurs (Figure 115). These heart-rate responses contrast with the bradycardia, slowing of heart rate, which is associated with orienting reactions (e.g. Lacey 1967, Horvath, Kirby and Smith 1971).

Brain activity and anti-predator behaviour

The control by the brain of startle responses and subsequent anti-predator behaviour has been studied by stimulation, ablation and recording experiments. The regions of the brain which are involved appear to be concerned also with attack behaviour, at least in the cat, which is a favourite subject for study. Since a cat which is attacked may flee and then retaliate this is not surprising. The major behavioural response which is measured in these studies is slow withdrawal or active flight but it has seldom been possible to decide whether the brain area under investigation is concerned with defence behaviour in general or with just this particular form of defence. An example of a study in which single unit recording was used is the work of Adams (1968). Units were found in the mid brain of a cat which fired only when a cat was defending itself against attack by another cat, whilst other units also fired when the cat was startled in various ways. Lesions in the mid brain disrupt defensive behaviour, whilst lesions in the hypothalamus have complex effects on attack and defence.

A summary of early brain stimulation studies in which defensive behaviour was elicited from cats is shown in Figure 116. In addition to the areas shown in the figure, stimulation of the thalamus can also lead to crouching and escape behaviour by cats (Roberts 1962). Previous work by Roberts (1958) had shown that there were sites in the hypothalamus and in the thalamus where stimulation elicited behaviour from a cat which was like that produced by detection of an approaching dog. There were sites in the posterior hypothalamus where apparent searching for an escape route was elicited. Many aspects of the flight response were shown by cats stimulated in the dorsal hypothalamus by Clemente and Chase (1973). The cats were confined in a cage during the experiment and showed rapid breathing, pupil dilation, urination or defaecation as well as attempts at flight behaviour. As shown in Figure 116, the amygdala is involved in flight behaviour. Many studies also demonstrate its involvement in attack behaviour but some sites are not concerned with attack. Hernández-Peón, O'Flaherty and Mazzuchelli-O'Flaherty (1967) applied the neurotransmitter acetylcholine to amygdalar neu-

115 A 5-day-old chick shows an increase in heart-rate, tachycardia, when disturbed by a light and a jolt (first arrow). The response is assessed by comparing the heart-rate after disturbance with that extrapolated from the previous heart-rate pattern. Modified after Forrester 1979.

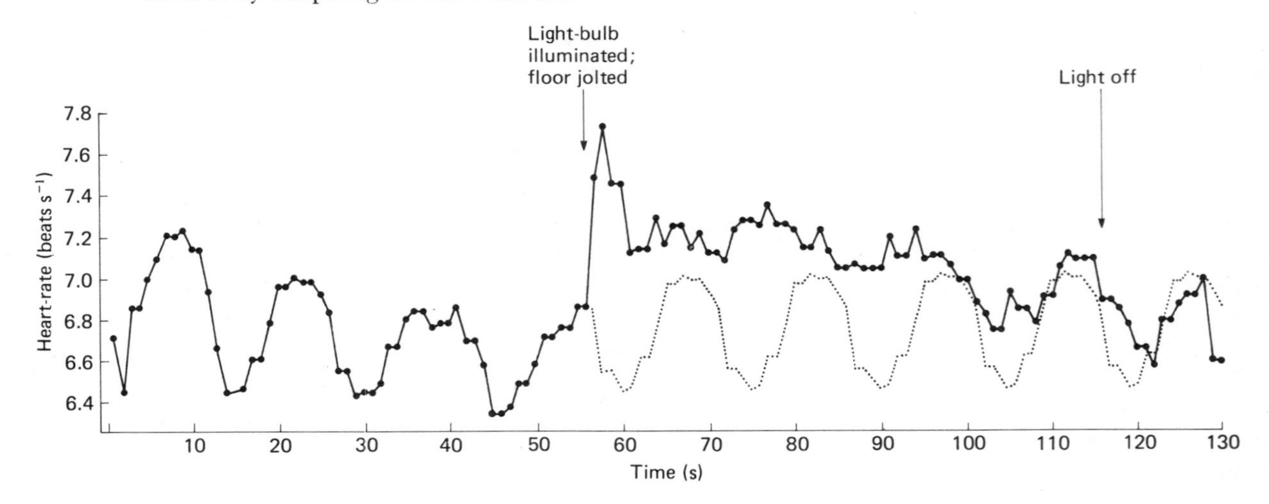

rons of cats and elicited flight but not attack.

A variety of components of anti-predator behaviour by domestic fowl were elicited in brain stimulation studies by von Holst and von St Paul (1960), who did not determine the site of stimulation but were able to demonstrate more and more extreme responses with increasing levels of electrical stimulation (Figure 117). They also described how stimulation could alter responses to a model predator, a stuffed polecat. Studies on newly hatched domestic chicks show that startle and escape responses can be elicited from various points in the septal and anterior commisure regions (Cannon and Salzen 1971). The anterior and dorsomedial hypothalamus of the young chick includes some sites where stimulation elicits a flight response and others where a freeze-and-hide response is elicited (Andrew and Oades 1973). Adult fowl and ducks, stimulated in the hypothalamus region, showed escape at some sites but attack at others. In the mid brain, however, escape but not attack was shown (Putkonen 1967, Phillips and Youngren 1971, 1973). The systems which control anti-

116 Diagram of cat brain showing regions where flight and defensive hissing are elicited by electrical stimulation. Modified after Fernandez de Molina and Hunsperger 1959.

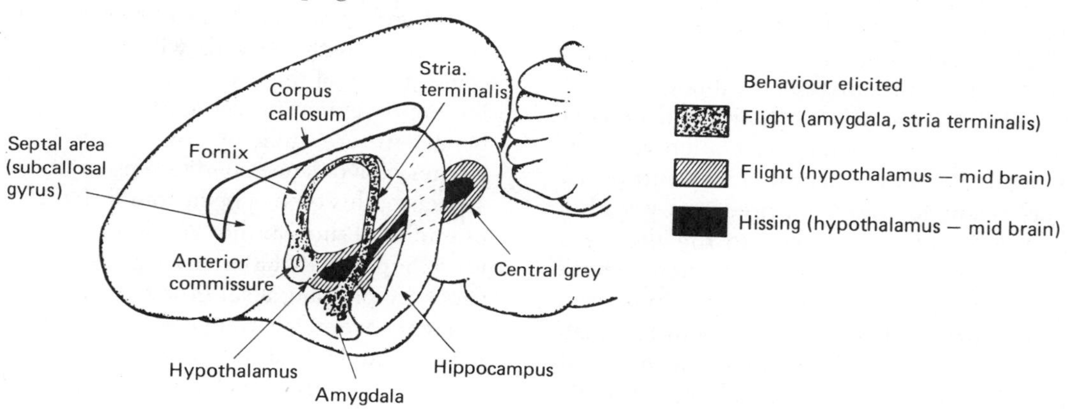

117 An adult fowl shows more and more extreme escape behaviour as the voltage applied to a single brain site is increased. Modified after von Holst and von St Paul 1960.

predator behaviour in bird and mammal seem similar in many respects but there is still much to discover about their mechanisms.

Habituation and sensitisation

The orientation reaction, which is the prelude to anti-predator behaviour, is shown when any unexpected environmental change is detected. If that change recurs repeatedly it ceases to be unexpected so the animal becomes less and less likely to show the reaction. This waning of a response which could still be shown is called *habituation.* Sokolov (1960) describes several examples of habituation of the orientation reaction in dogs when specific stimuli are presented repeatedly. Startle responses may also habituate to very specific environmental changes and reappear if a change occurs which is slightly different in quality or duration (Schleidt 1955, Broom 1968). Although it is advantageous for animals to show rapid habituation of the, very general, orientation reaction, the rate and extent of habituation of anti-predator behaviour to cues from predators would be expected to differ according to the degree of hazard. A person in a wood might show a startle response when the sound of a falling pine cone is first heard but the response would habituate if the sound recurred frequently with no adverse effects on that person. Habituation to the sight of a hunting leopard, however, is unlikely to occur. In fact, there would be the reverse effect of sensitisation to environmental changes which would not have elicited a startle response prior to the sight of the leopard.

Clark (1960) showed that the ragworm *Nereis* habituated its withdrawal response when a shadow was repeatedly cast on it. Evans (1965), however, showed that repeated electric shocks could *sensitise* the ragworm so that it increased its response to shadows. The alternative possibilities of habituation or sensitisation to repeated stimuli have been incorporated by Groves and Thompson (1970) into their 'dual process theory'. They propose that habituation and sensitisation occur simultaneously in any repeated stimulation situation. They use this theory to explain observations of initial increases in responsiveness followed by decreases after a larger number of stimulus presentations. Habituation occurs in many situations in addition to those in which anti-predator behaviour is shown and research in this area continues to show that many factors affect what used to be considered to be a simple form of learning. For further information about the mechanisms of habituation see Horn (1967), Peeke and Herz (1973) and the references mentioned above.

The development of anti-predator behaviour

A baboon which is attacked by a leopard but which survives, learns something about the speed and attack methods of leopards. Future encounters with leopards will be different as a consequence of the experience. Field studies of the improvement in techniques of anti-predator behaviour are difficult, but many laboratory studies of avoidance conditioning mimic certain aspects of this type of behavioural change. Some examples of such studies are described in Chapter 5. Suppose that a rat in a cage learns to avoid shock by pressing a set of levers in a particular sequence when a certain combination of stimuli is presented. That rat may be using abilities which would also be of use were it faced with successive exposures to a predator. The effects of experience are often easiest to study in developing animals since some aspects of previous experience can then be controlled.

In species in which parental care is prolonged, the most effective anti-predator behaviour which young animals can show is often to recognise and maintain contact with the parents and the place where they are put by the parents. Responsiveness to parents is discussed in Chapters 1 and 9. The overriding importance of finding the parent leads to some apparently anomalous responses to predators. Lorenz (1935) describes how newly hatched precocial birds, such as goslings, show no avoidance behaviour to a variety of moving objects including, for example, man, a goose predator. Schaller and Emlen (1962) pushed objects towards the young of ten species of precocial birds and found very little avoidance reaction at 10 h of age. Avoidance in their experiments, and in those of Bateson (1964) on chicks exposed to novel moving objects in a runway, increased considerably during the first week of life but changed little after this time (Figure 118). The startle responses shown by young domestic chicks in their home pen, when a light-bulb on the wall was illuminated for the first time, increased in magnitude and duration with increasing age from 1 to 10 days (Broom 1969*a*). The change with age in all of these studies was greatest during the first three days. As the animal gets older, it learns the characteristics of its immediate environment. If the degree of novelty of any environmental change depends upon a comparison with a 'model' of the familiar surroundings (Salzen 1962), novelty will increase whilst the model is established and hence responsiveness increases with age (Broom 1969*a*).

118 The avoidance responses of young, naive precocial birds, when an object is thrust towards them, increase from a low level at 10 h of age to a steady, higher level by 2 to 4 days. *a*, Chinese ring-necked pheasant; *b*, *Coturnix* quail; *c*, Khirgiz pheasant; *d*, wild turkey; *e*, green pheasant; *f*, white rock domestic chicks. Modified after Schaller and Emlen 1962.

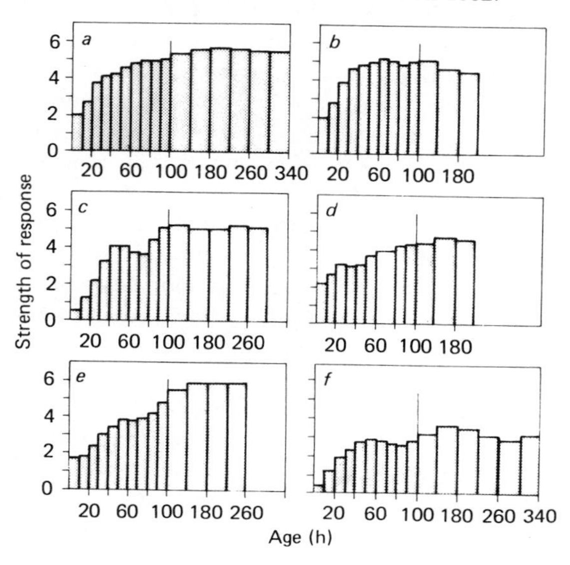

A system which results in the young animal showing anti-predator behaviour to any novel change in its surroundings will operate when a real predator appears. Responsiveness to potential predators by young animals does not depend solely upon degree of novelty. It is also affected by the nature of the sensory input which results from the presence of the predator. Bodily damage, a looming object, or a loud noise will elicit more extreme anti-predator behaviour than the sight of small objects, the sound of quieter noises, etc. (Hebb 1946). The existence of perceptual analysers in the brain which enable an individual to recognise dangerous events, for example a looming object (Ewert 1971), is discussed in Chapter 2. Some of these analysers are known to develop in a different way according to specific visual experience but others probably depend on general environmental factors such as light and metabolic state at key times during embryological development. The average individual, therefore, would have some predator detection mechanisms whether or not there had been previous visual experience of the predator. An elaborate and specific possibility is a detector for flying birds of prey. Lorenz (1939) and Goethe (1940) reported that young precocial birds did not respond to flying geese but showed a flight response when a hawk or falcon passed overhead. Experiments by Schleidt (1961) showed that this result could have been due to habituation to the geese but not to hawks. Further work on naive mallard ducklings by Green, Green and Carr (1966) and by Mueller and Parker (1980), however, indicates that hawk and goose silhouettes can be distinguished and do elicit different reactions. The reactions were not clear anti-predator reactions in either case. The linking of the perceptual recognition to the command which initiates such reactions, therefore, may require experience which the experimental subjects in these experiments did not have.

The general effects of experience on anti-predator behaviour have been examined in many experiments with young animals. If the occurrence of such behaviour depends, in part, on the gradual formation of a model of familiar surroundings, with which future events can be compared, individuals reared in different surroundings will form different models which may be formed at various rates. Experiments with rhesus monkeys, chimpanzees and domestic chicks indicate that the magnitude and duration of responses to novel objects or novel environmental changes is less for individuals with more complex rearing conditions (Harlow and Zimmermann 1959, Menzel, Davenport and Rogers 1963, Broom 1969*b*). The large literature on the effects on later behaviour in a strange pen ('open field') of handling rodent pups during infancy, leads to the same general conclusion. It seems that the major effect of handling a rodent pup is to change the amount of time that the mother spends licking and sniffing that pup (Richards 1966, Barnett and Burn 1967). Perhaps as a consequence of this, handled pups show a less marked response in novel situations (review by Denenberg 1964). The extent and type of maternal care also has effects on the responses of young primates confronted with possible danger. For example Ainsworth, Bell and Stayton (1971) found that one-year-old human infants, whose mothers were inadequate in their responses to them, explored less and fled to mother more readily in strange and hence potentially dangerous situations.

Defensive responses by farm animals to man

To the ancestors of present-day farm animals, man must have been recognised as a dangerous predator. Despite many generations in domestication, the anti-predator functional system still has considerable effects on the behaviour of farm animals. In some situations, predation by wild animals or dogs is still possible but most anti-predator behaviour now is directed against environmental changes whose origin is largely unknown to the animals, or against man. The occurrence of unexpected sounds or the arrival of olfactory or visual portents of danger may elicit adrenal and behavioural responses which are the same as those shown by their wild ancestors. The detection of a person who remains at a considerable distance from the animal may elicit no such responses but the close approach which is often necessary on farms must often elicit a defensive response. The fact that people are often treated as predators by farm animals is ignored by many of those who work on farms or who attempt to understand the behaviour of the animals but the competent stockman has learned by observation about such behaviour. Good stockmen also know how to minimise the anti-predator behaviour which is directed at them, for if they can do this they can often improve production (Seabrook 1972, 1977) and enjoy their work more.

The importance of the abilities of the cowman in obtaining the best possible production from a dairy herd was stressed by Albright (1971). Milk let-down occurred faster if the cows saw the familiar milker (Baryshnikov and Kokorina 1959). Yield was higher if the cowman was deliberate in movements, talked quietly to the cows and followed a regular routine with no panic activity (Seabrook 1977). Other effects of a 'good' cowman were that cows entered the milking parlour, and were more approachable in the field. It seems likely that a 'poor' cowman elicits a greater adrenal response and corresponding defensive behaviour so that management of the dairy herd is more difficult and milk production is impaired. Studies of the defensive responses of cows to cowmen and to disturbing aspects of yard and milking parlour design (Albright 1979), are helping to improve management efficiency and milk production.

The detection of a person in an animal-house may be a traumatic experience for comparatively small animals such as chickens or turkeys. If a person enters a house rapidly and noisily the nearest birds may show a violent escape response. Whether the birds are in separate cages, or in one large aggregation on the floor of the house, a wave of violent escape behaviour, sometimes called 'hysteria', may be transmitted down the house. Such behaviour can be minimised by a quiet knock before entry and by slow, quiet movements whilst entering and when within the house. The economic importance of quiet slow movements in poultry and other livestock houses is obvious to the turkey farmer who has found the corpses of dead birds under a heap of turkeys disturbed by the entry of a noisy person. The effects of defensive behaviour and adrenal activity on larger animals may be obvious, for example, if a fleeing calf breaks a leg, but many effects are less obvious. Intermittent high adrenal activity may affect adversely milk or egg production, growth rate and disease resistance as well as ease of handling (Kiley-Worthington 1971).

8

Functions of social behaviour and dispersal

The distribution of the individuals of an animal species, within the area where it is possible for them to live, is usually clumped or spaced out rather than random (Figure 119). Examples of animals which are spaced out include more or less regularly distributed territorial animals. The word overdispersed is sometimes used as a synonym for spaced out. Plants are often present as large, local concentrations so that animals feeding on them are aggregated in their vicinity and are absent or rare in areas between the plant concentrations. Such aggregations of herbivorous animals may last for a very short time or for many years according to the food availability. Animals which prey on the herbivores might also be aggregated. The same arguments obtain for other essential resources and, sometimes, for factors like temperature and humidity which affect animal distribution.

If individual animals can detect and respond to the presence of other members of their own species they may compete with one another for resources. Some of the most obvious activities of animals are contests between individuals clumped in a social group or spaced out in territories (see later in the chapter). The likelihood of contests must depend upon the distribution of resources and since this differs in different examples, it is not surprising that there is great variation in animal distribution patterns. The questions which have given rise to much discussion are why some animals spend some or all of their lives in very close proximity to conspecifics and why others actively move away from conspecifics or enforce separation by defending a territory.

Before considering the functions of social behaviour it is necessary to define it. If many individual animals are attracted by a chemical emanating from a food source or keep moving around an area until they reach a dark, damp place, they will aggregate. This will occur even if the individuals do not respond to one another

119 When the distribution of distances between individual animals is considered, the animals might be found to be randomly distributed. More frequently, however, the number of smaller than average distances is high, i.e. the animals are clumped, or the number of larger than average distances is high, i.e. the animals are spaced out.

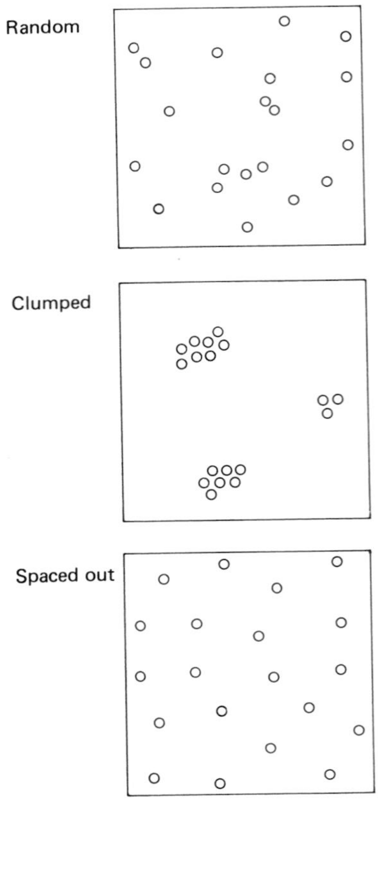

and it is not social behaviour. If, however, an individual finds conspecifics in a favourable place, has the ability to recognise them and avoid them but does not do so, then it is showing some signs of social behaviour. An animal which recognises and remains with conspecifics when it encounters them is showing social behaviour and the aggregation thus produced is a social group. Such animals often behave in a way which increases their chances of finding conspecifics. Many invertebrates occur in aggregations and it is often assumed that the individuals are not showing social behaviour. The brittle-star *Ophiothrix fragilis* (Figure 120) forms dense aggregations on the sea-bed. When individuals were removed from the aggregation to a distance of 20 cm they walked across current until they regained contact with the patch, whose longest edges were usually parallel with the current direction. Since they stopped and clung to members of their own species but did not stop when they made contact with stones, hydroids, starfish or other animals, their behaviour towards their own species can be called social (Broom 1975).

Questions about why social groups and territorial behaviour exist can be answered in two ways: firstly by considering the possible

120 An aggregation of brittle-stars *Ophiothrix fragilis* on the sea-bed. Each individual uses some of its five arms to hold on to others and extends some arms into the current of water so as to obtain food particles suspended in it. Photograph: G. F. Warner.

evolutionary origin of the behaviour, and secondly by considering the selection pressures favouring the behaviour now. The selection pressures which resulted in the initiation of the behaviour in the species may still be important but they need not be. The environment and genotype of the species may have changed so that factors which did not act when the behaviour arose may now be the most important factors maintaining it. In the first part of this chapter, some of the factors which might influence the occurrence of group-living are considered. Those associated with habitat modification, food finding and food acquisition, e.g. approaching, grasping, handling and ingesting; anti-predator behaviour, mate finding and rearing offspring, and other systems such as temperature regulation are discussed. Ideas about the possible origins of group living and some of its advantages are then reviewed. In the latter part of the chapter the occurrence of spacing out behaviour, the advantages of territorial behaviour and the functions of migration and dispersal are surveyed.

Modifying the local environment

As soon as animals aggregate they start to change the characteristics of their immediate surroundings. Animals such as woodlice which move around until they find a dark, damp area might aggregate in a crevice which becomes much more humid because of their presence. Merely by aggregating they would benefit individually because the loss of water would be less from an individual in a cluster of woodlice than from a single animal. This point was discussed at length by Allee (1938) who reviewed various experiments in which the physical or chemical surroundings of individuals were found to be different in a group situation. The small freshwater crustacean *Daphnia* dies in conditions which are too alkaline. Its respiratory product, carbon dioxide, makes its surroundings more acidic so that individuals survive when in a group but die if put individually into alkaline water.

Since animals are most active and function most efficiently within a fairly narrow temperature range, temperature regulatory mechanisms are important amongst both cold-blooded and warm-blooded animals. If animals which are warm aggregate closely they can reduce heat loss. Various group-living animals show such behaviour in cold conditions. Emperor penguins (*Aptenodytes forsteri*) huddle together during their Antarctic winter breeding season; wrens (*Troglodytes troglodytes*) and tree creepers (*Certhia familiaris*) congregate in tree holes in cold conditions; honeybees cluster together in the winter. Ribbands (1953) describes how bees form tighter clusters in colder conditions and how the outer bees respond to a drop in temperature. The change in their behaviour is detected by the inner bees which walk about and vibrate their abdomens, thus increasing the temperature. In hot conditions, bees cool the hive by fanning their wings, sometimes utilising droplets of water and thus cooling the hive by evaporation. Lindauer (1954, 1961) reported that bee clusters in winter maintained 20–30°C despite external temperatures as low as −28°C and maintained a hive temperature in summer of 35°C even when the ground surface temperature was 70°C. Termites construct nests which may have elaborate capillary-like chambers. These allow air flow to all parts of the nest and also result in precise temperature regulation. Lüscher (1961) found that the temperature within the fungus garden chambers in the nest of *Macrotermes natalensis* (Figure 121) in Africa varied by no more than one degree about a mean of 30°C whilst the carbon dioxide level remained almost constant. Subsequent work by Ruelle (1964) showed that termite building activity also played a considerable part in keeping mean summer and winter temperatures very constant inside the nest. There were, however, daily fluctuations in temperature, of 2°C and in carbon dioxide level.

A quite different way in which individuals in a group can benefit from the immediate effect of their neighbours on their surroundings is the utilisation of turbulences produced in water by swimming fish or in air by flying birds. The positions sought by fish in shoals may be determined by hydrodynamic constraints. Breder (1965) suggested that each fish may benefit from the vortices made by other school members in front of the shoal and thus require less energy to swim at a given speed within a school than in isolation. This has been questioned, however, by Partridge and Pitcher (1979). It is possible that large birds such as geese, cranes and the larger gulls reduce turbulence when they fly in V-formation.

The effects of other individuals are sometimes necessary in order that food can be acquired. Larvae of the fruit fly *Drosophila* feed easily if the substratum is broken up by the movement of other larvae but, if isolated, they find difficulty in feeding on an unbroken substratum. Hence they survive better in a group (Manning 1979). Many grazing animals such as rabbits or prairie dogs can feed only in places where the grass is continually cropped by the grazing of a group of animals.

121 The nest of the termite *Macrotermes natalensis* is constructed in such a way that the inside temperature varies little during the year. Modified after Lüscher 1961.

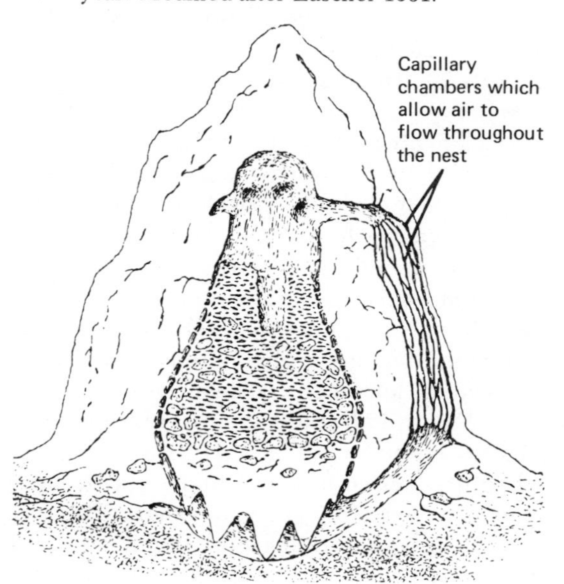

Animals which feed on particles suspended in water currents run the risk of being washed away by the current. Warner (1971) has pointed out that aggregations of brittle-stars, which feed on suspended matter, occur in places where the currents are too strong for individuals to feed and this has been supported by experiment (Broom 1975). If the animals can hold on to other individuals in a dense patch, they can extend two or three of their arms into the water current and catch suspended food particles. Isolated individuals, however, have to hold on to the substratum with all five arms and thus cannot feed. The forest of arms extending into the water above a dense patch of brittle-stars also slows down the current in the vicinity of the arms and this facilitates the trapping of suspended particles.

Many aspects of life are helped by modifying the habitat extensively in a small area so as to make a place in which to live. The construction of nests and other habitations by groups of birds, social insects and men are obvious examples of this. The massive nests of weaver birds in Africa, the nests of ants, bees, wasps and termites, and the villages and cities of man, are local modifications of habitat which can be carried out only by groups of individuals. These examples of advantageous habitat modification are exceptional amongst animal species for the mere accumulation of waste products causes disadvantageous changes in the immediate surroundings of most groups of animals.

Food finding and acquisition

Joining groups

Food sources can sometimes be detected from great distances; a mosquito in savannah may detect human odour when it is hundreds of metres from a human settlement and can readily find the settlement by flying upwind (see page 130). Some food is not found so easily but animals may be able to improve their chances of finding it by responding to the sight, sound or smell of conspecifics. If a group of conspecifics is detected, there is a higher probability that there is a food source in their vicinity than in any other area. Thus an individual may find more food by joining groups than by avoiding them (Murton 1971*b*). Such behaviour is shown most frequently by very mobile animals, for example vultures which will join other vultures on the ground and will respond to the sight of a vulture descending towards the ground. If the food of a species occurs in local concentrations, each sufficient for several individuals, the group-joining behaviour pattern will be advantageous.

Discovering food locations by watching others

Animals living in groups have the opportunity to obtain information about food localities by observing the behaviour of other members of the group. The pioneering work of von Frisch on bee communication is well known. In 1946 he published the results of experiments which indicated that worker bees returning from a food source performed a dance on the hive which included information about the direction of that food source, and its distance from the hive. Gould (1976) has reviewed the controversy surrounding these results and has described rigorously controlled experiments which demonstrated that workers in the hive can utilize this information to find the food sources (Gould 1975).

The possibility that other assemblages of animals might act as 'food information centres', which individuals could use to discover new food sources, was proposed by Ward (1965) and elaborated by Zahavi (1971*a*) and Ward and Zahavi (1973). They suggested that hungry individuals at communal roosts and breeding colonies of birds might be able to find food by following individuals flying away from the roost or colony towards feeding sites (see Hamilton and Gilbert 1969). Most birds which roost or breed communally feed in groups, or at least within sight of other members of their species. If this theory is correct, the departure of birds from roosts should sometimes involve recruitment to departing flocks and this has been observed for the seed-eating weaver-bird (*Quelea quelea*) in Nigeria and for the insectivorous pied wagtail (*Motacilla alba*) in England (Broom *et al.* 1976). The transfer from a poor feeding site to a better feeding site by following others away from a roost need occur only occasionally to be very advantageous to the bird which shows this behaviour.

If birds can benefit from each others' knowledge about food sources, it might be predicted that the members of smaller roosts will join larger roosts in adverse conditions and this has been reported for the starling (*Sturnus vulgaris*) (Wynne-Edwards 1962), for *Quelea* (Ward 1965) and for the pied wagtail (Moffat 1931, Broom *et al.* 1976). Ward (1972) has suggested that the mass drinking flights of sandgrouse (Pteroclididae) might also allow information exchange. The food information centre hypothesis is difficult to prove or disprove for any aggregation of animals, although de Groot's (1980) laboratory experiments show that naive *Quelea* can follow knowledgeable birds to food. If this occurs only once during the lifetime of the average individual, but it results in that individual surviving to breed during the next season rather than dying, then genes promoting the behaviour will spread in the population. It is possible, however, that the benefits will be smaller and the costs of fruitless following will outweigh them.

Colonially nesting birds might also be able to find food by following others. Horn (1968) suggested that Brewer's blackbirds (*Euphagus cyanocephalus*), whose food was clumped spatially and temporally when it consisted of insects hatching in local concentrations on a lake, could benefit from colonial nesting in this way. Krebs (1974) found that great blue herons (*Ardea herodias*) often left their colonial nest-sites in groups (Figure 122) and that birds from adjacent nest-sites often fed in groups. The favoured feeding sites varied from day to day and groups of herons were readily joined by others. This resulted in groups forming and building-up in numbers in places where the food availability was higher than in other possible feeding sites. A laboratory study by Krebs, MacRoberts and Cullen (1972) showed that individual great tits (*Parus major*) were more likely to find concealed food if they were in groups of four than if they were alone or in a pair. It seems likely that small birds feeding on clumped prey in woodland benefit from flocking because they can see or hear when other members of the group find a cache of food. Many other group-living animals may find food in this way. When one member of a group of chimpanzees in a large enclosure was shown a hidden food source it was readily followed to the food by the other members of the group when they were released (Menzel 1974). In this study, the behaviour of the 'leader', who was shown the food, included running towards the food and then stopping and looking back at the other members of the group, tapping them and generally soliciting them to follow. The behaviour of the 'leader' varied according to the quantity and quality of food which had been disclosed to it. This provided information used by the other chimpanzees, especially if there were two 'leaders' which had been shown good or poor food caches (Menzel and Halperin 1975).

122 Great blue herons which were going to seek food were most likely to leave the nesting colony soon after another individual had left. Intervals between successive departures of less than 5 min were much commoner than would be expected if birds left the colony independently. After Krebs 1974.

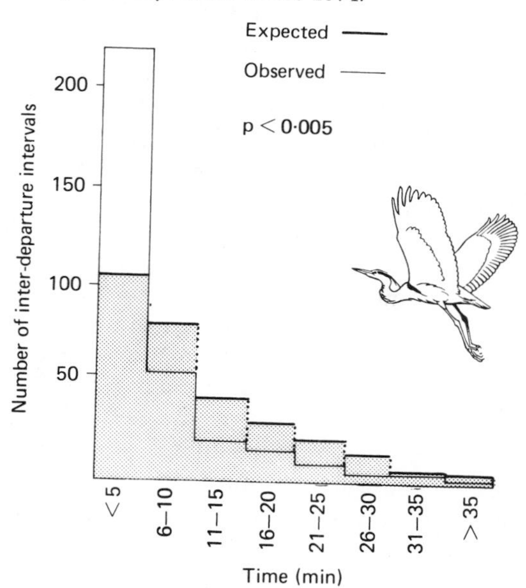

Discovering feeding methods by watching others

Individuals in groups can find out how to obtain food, as well as where the food is, by watching other individuals. Welty (1934) showed that goldfish learned faster to swim through a maze for a food reward if they were given the opportunity to watch other goldfish successfully negotiating the maze. Birds which start to utilize a new food source may do so by imitation or by 'local enhancement', the increased tendency to respond to a part of the environment as a consequence of observing the response of another individual to it (Hinde 1970). Examples of such new methods of obtaining food include the spread of the opening of milk bottle tops by tits *Parus* species (Fisher & Hinde 1949) and of the consumption of the berries of the daphne bush (*Daphne mezereum*) by greenfinches (*Chloris chloris*) (Pettersson 1956; for other examples see Chapter 6). Local enhancement or the actual copying of feeding techniques are likely to be very important in the development of young animals and group-living provides more experts to copy. Murton (1971*b*) suggested that young birds feeding at the front of flocks of wood pigeons (*Columba palumbus*) copied the older birds behind them and Thompson, Vertinsky and Krebs (1974) included social learning as a factor in their simulation model for the survival value of flocking in birds.

Collaboration in acquiring or handling food

One of the most obvious advantages of group living, which is shown by many groups of predators, is collaboration in acquiring and handling food. Some of the previous examples of group behaviour in this chapter include elements of such collaboration and the subject is discussed further in Chapter 6 and by Alcock (1980) and Curio (1976). Groups of predators may be able to herd, drive or lure prey species into an ambush, to alternate energetically expensive chasing behaviour, to co-ordinate activities which restrict prey movement before capture, and to synchronise attacks. As a result of such behaviour each individual may be able to feed on prey which it could not catch unaided (Kruuk 1972). For example, a wolf pack may catch an adult moose which weighs eight times as much as an individual wolf and which can run faster (Mech 1970). Alternatively, a predator may be able to obtain prey items more efficiently in terms of energy expenditure per unit of calorific intake, by collaborating with other individuals. If all the members of a group of white pelicans (*Pelecanus onocrotalus*) move forward and scoop the water at the same time, small fish which dart away from one pelican's bill are likely to be caught by another (Figure 123). Either of these advantages could apply to the group construction of webs by spiders (Kullmann 1972). Spiders which have built a large web can combine to attack and subdue large prey caught in the web. Smaller prey species may have less chance of avoiding capture in a large web than they would if the individual spiders had each built individual small webs.

123 White pelicans collaborate during feeding by adopting a horseshoe formation and then scooping the water synchronously so that fish escaping from one pelican will often be caught by another.

Avoiding depleted food sources

Whatever the pattern of food distribution, if an individual needs to expend energy looking for food items it is disadvantageous to have to spend time searching in areas which have already been depleted by others. One way of minimising this problem is to hunt in a group which includes most of the individuals in that area. Cody (1971) found that flocks of finches feeding during the non-breeding season on seeds in the Mojave desert avoided hunting again in areas completely denuded of seeds and delayed return to an area where a new crop of seeds would appear. When a large flock of starlings or *Quelea* is feeding in a field, there is a continual movement of birds from the rear of the flock to the front so that the flock appears to roll across the field. The same behaviour is shown by locusts in a swarm. As a result, a swathe of land depleted of food is cut across the field and the flock or swarm systematically moves over the field, exploiting the food source. The number of prey items missed by such behaviour must be much smaller than the number which would be missed if all the individuals hunted separately but direct evidence of advantage of such behaviour to individuals is lacking.

A long-term means of avoiding depleted food sources is the mechanism for population assessment and subsequent migration which was proposed by Wynne-Edwards (1962). If an individual is able to assess the population density of its species and it finds that this is high in relation to food availability it is advantageous for that individual to migrate to a place where food is more plentiful. Wynne-Edwards proposed that communal roosts and many other social gatherings of animals allowed the members of the gathering to assess population density. He coined the term 'epideictic display' for very obvious gatherings such as the aerial gyrations which occur prior to roosting in some birds and for the choruses of sound produced by aggregations of frogs, birds or insects. Whilst food depletion might be detected directly and migration might result from this, it is also possible that some animals leave crowded areas before resource depletion, as Wynne-Edwards suggests. They may be able to monitor the food situation by watching other individuals at the roost.

Collaboration in defence of a food source

If an individual finds a concentration of food, it can be driven away from that food by a group of individuals which tolerate one another but attack non-group members. A group is less likely to be displaced from a food source and hence, in some mobile species, group defence is essential for adequate feeding (Emlen 1973). Observations of feeding by cedar waxwings (*Bombycilla cedrorum*), American robins (*Turdus migratorius*) and starlings showed that individuals were more likely to gain access to defended food if they were in a flock (Moore 1977). Group defence was temporary in this study but with some sorts of food distribution group defence of a territory, which provides a food supply for the group, may occur. Such group territories are discussed later in this chapter.

Adverse effects of grouping on food acquisition

For many animals the efficiency of finding and acquiring food is less rather than greater if a group of conspecifics is present. All of the advantages listed above may be outweighed by the disadvantage of having to compete with other individuals who want to eat the same item. This difficulty is exacerbated in large groups where there are more competitors and where more elaborate social responses are needed to prevent frequent, energetically expensive encounters with other individuals. For some species of animals, for example those whose food is evenly distributed throughout the habitat, is quite difficult to find, but is easy to acquire, the disadvantages of food competition exert a strong selection pressure against the evolution of social grouping behaviour. Grouping is also disadvantageous to animals which hunt, by stealth, prey which provide a meal for only one predator. Examples include kingfishers and many other predators.

Avoiding capture by predators

Hiding within a group

Many observers have noticed that predators are more likely to take prey from the edge of a group than from its centre (see Milinski 1977) provided that individuals are not very small in relation to the predator. If they are very small the predator might take many prey at once as does a shark taking mouthfuls from the dense centre of a shoal of small fish (Bullis 1961). Where predators capture prey singly, however, an individual of the prey species can improve its chance of surviving a predator attack by moving to the centre of a group (Williams 1964, 1966). As Hamilton (1971) has pointed out, the simplest behaviour pattern which achieves this beneficial effect is to move towards the closest conspecific. Hamilton considered an example of a situation in which a predator, previously hidden from the prey, appears within the group of prey individuals or within attacking range. Frogs in a lily pond might be especially vulnerable to attack by a water snake at certain times of day. They might, therefore, sit on the bank of the pond at that time. It is assumed that they cannot venture away from the edge of the pond because of other predators. If the snake, as Hamilton puts it, 'rears its head out of the water, surveys the disconsolate line and snatches the nearest one' then the problem for a frog is to avoid being the nearest one. A frog which is some distance from its nearest neighbour, as in Figure 124*a*, is vulnerable to attack by a predator which may appear from any direction. The frog is vulnerable because it has around it a large area, its *domain of danger*, within which the snake will be nearer to it than to any other animal. If this frog jumps between its two nearest neighbours then they are at much greater risk of attack and its domain of danger is reduced. Provided that the snake takes a small proportion of the frog population on any occasion, there will be a strong selection pressure favouring the spread in that population of any genes which promote hiding between neighbours. If individuals in a group show such behaviour when predator attack is likely, the result is that the mean distance between individuals will decrease and the group becomes more compact, as in Figure 124*b*.

Hamilton's model, that explained above, is a simple description of a complex situation but it does allow some predictions about anti-predator behaviour. Lazarus (1972, 1978) showed that the amount of time that individual white-fronted geese (*Anser albifrons*) were vigilant for predators declined as flock size increased. If the domain of danger is important it would be predicted that, where it is large, more time would be spent in vigilance. For white-fronted geese and for pink-fronted geese (*Anser brachyrhynchus*) Lazarus (1978) found that the birds which were at the greatest distances from their neighbours were vigilant for longest. Jennings and Evans (1980) found that time spent in vigilance by starlings was greater if the birds were on the periphery of the flock than if they were in the centre.

124 If a snake in a pond may appear anywhere along the edge and always takes the nearest of the frogs sitting around the edge, all individuals are in high danger in (*a*), but the danger is lower for those which have jumped between others in (*b*). Example from Hamilton 1971.

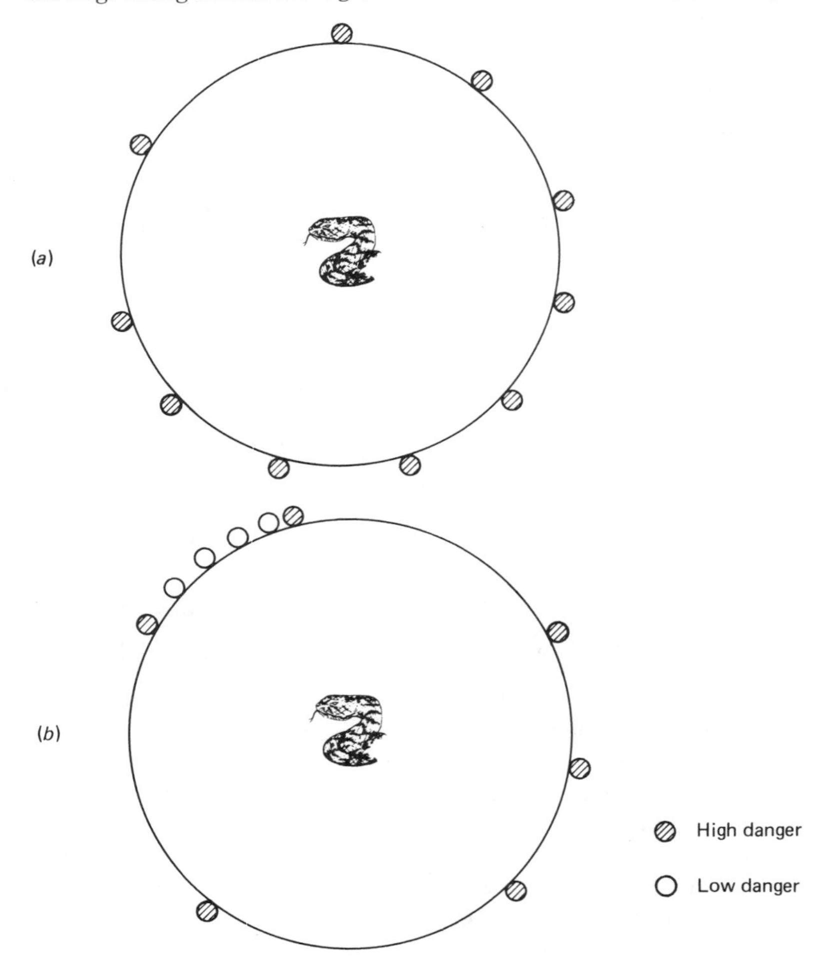

Tinbergen (1951) described how starlings form compact flocks when a peregrine falcon (*Falco peregrinus*) is about to stoop on them. Other workers have reported such behaviour in birds and many observers have described the rapid predator-induced concentration of fish in a shoal, or antelopes in a herd, which is caused by individuals moving to the centre of the group. The members of a herd of antelopes or cattle can reduce their domains of danger by walking towards or between other individuals. This will hold for a stationary herd when predators appear or for a moving herd which might encounter a hidden predator. A bird at the edge of a flock, or detached from the flock, is the most likely prey for a falcon. The predator may attack it because it is encountered first (Hamilton 1971), because it is the most obvious prey and the flock is confusing (Allee 1938), because the falcon will not risk damage by diving into the centre of the flock (Tinbergen 1951), because the birds on the edge are weaker individuals and less able to evade capture than those in the middle, or for several of these reasons. Whatever the reason for the behaviour of a predator, if individuals at the centre of a group are safer from attack, centre-seeking behaviour will become widespread in the population.

There must be a lower limit to inter-individual distance because of the likelihood of interference between animals at very close proximity (see later in this chapter). This limit will be lower in stationary animals so it can be predicted that groups might be even more compact if they rest together for a long time. Flocks of wading birds which feed on mud-flats often spend the high tide period in dense aggregations. Individuals in nocturnal roosts of birds may also be closely packed together. In the pied wagtail roost shown in Figure 125, many birds roosted on the clinker between the two or three moving sprinklers whose ledges were occupied,

125 Pied wagtails roosted close together at this sewage purification works despite the fact that the movement of the distributors bearing the water sprinkler pipes forced the individuals on the ground to fly every 7 min throughout the night. Note the regular distribution of the birds on the girders.

despite the fact that there were 45 other, almost identical sprinklers which could have been used for roosting. The advantages of keeping in a group were emphasised when man acted as a predator. The individuals at the edge of the group were reached first and a man trying to catch the roosting birds by hand was sometimes successful if the birds were roosting in isolation, but he was usually thwarted in any attempt to catch the birds in the centre of the roost. This was because the alarm calls of those individuals which were awake, aroused the others (Zahavi 1971*b*, Broom *et al.* 1975).

Nesting within a group

Colonial breeding also results in predators encountering edge individuals before centre individuals. The advantage of nesting in a colony may be further enhanced by responding to alarm calls and by collaboration in defence, as described later in this chapter. Nests in the centre of colonies were more successful than those on the edges in Adélie penguins (*Pygoscelis adeliae*) (Taylor 1962), black-headed gulls (*Larus ridibundus*) (Patterson 1965), kittiwakes (*Rissa tridactyla*) (Coulson 1968), and sand martins or bank swallows (*Riparia riparia*) (Hoogland and Sherman 1976). The work of Horn (1968) on Brewer's blackbirds (*Euphagus cyanocephalus*), which nested in colonies in sagebrush vegetation, showed that nests which were close together were subject to more than average predation if they were in long narrow colonies, but to less than average predation if they were in more circular colonies. The long narrow colonies obviously have more edge nests than do circular colonies but the advantage obtained from being far from the edge may be due in part to group defence of a colony as well as to the fact that edge nests are encountered first.

If a predator can take the contents of several nests, the high risk zone around the edge of the colony is wider. As Tinbergen, Impekoven and Franck (1967) showed experimentally, small prey items which are only a few steps apart for the predator are more at risk than those which are many steps apart. Some predators might, therefore, take more nests from the edge of a colony than they would from an equal number of scattered nests. Birds which nest in the centre of a large colony, however, will have an advantage over those whose nests are scattered if the major predators approach from the side of the colony rather than drop in out of the sky. Their advantage is increased if they synchronise their egg laying so that there are too many eggs and young at one time for the local predator population to take all of them (Fraser Darling 1938, Patterson 1965).

Responding to alarm signals

Individuals which detect and recognise alarm signals, or reactions to a predator, by members of their own or other species, are better able to modify their behaviour to minimise the chances that they will be caught. Whether they respond by freezing in a place where they are concealed, or fleeing towards cover, or reparing for defence against attack, the forewarning can be very valuable. This advantage can be gained only if other individuals which can give an alarm signal are nearby. Galton (1871) described the possession of many near neighbours, who would give warning of any impending attack, as a major advantage of gregariousness and exemplified this with descriptions of the behaviour of African cattle. Studies on flocks of birds have shown that doves (*Streptopelia senegalensis*) responded faster to a falcon-shaped model if they were in groups (Siegfried and Underhill 1975). Goshawks (*Accipiter gentilis*) were seldom successful when attacking flocks of 11 or more wood-pigeons but often caught single pigeons or individuals from groups of 2 to 10 (Kenward 1978 and Figure 126). A problem of interpretation with such studies, however, is that the pigeons which are single or in small groups may be of inferior ability.

When a predator is detected by one of a flock of pigeons, the other flock members respond to the take-off movements of that individual. They distinguish between a sudden take-off and similar movements which are preceded by pre-flight movements in a bird which is not alarmed (Davis 1975). The greater safety of birds in larger flocks is emphasised by the observations by Lazarus (1979) on the weaverbird (*Quelea*). Captive birds in small flocks take flight when a hawk is seen but those in larger flocks show the preliminary movements of take-off or merely show an orienting reaction (see Chapter 7). These reduced responses are a way of economising on energy. Surveillance behaviour is also reduced in larger groups, for example, starlings in groups spent 12–15% of their time on surveillance activities but isolated birds spent 47% (Powell 1974). The extra time is available for feeding and other activities. Many of the baboons in a troop, or the ants in a colony, may devote little time to looking out for predators because other members of the group are doing this. Baboons (*Papio cynocephalus*) may respond to alarm signals of impala and vice versa. Members of mixed groups of baboons and impala may benefit from the presence of the other species because the baboons have good colour vision whilst the impala have especially good olfactory and hearing abilities (Altmann and Altmann 1970). When an alarm signal is detected, it often provides information about the nature of the danger and allows appropriate response, e.g. the different hawk and ground predator alarm calls of many birds (Marler 1957); or the snake, ground predator and avian predator alarm calls of vervet monkeys (Struhsaker 1967).

126 A goshawk was most likely to catch a lone pigeon and least likely to catch one in a large flock. After Kenward 1978.

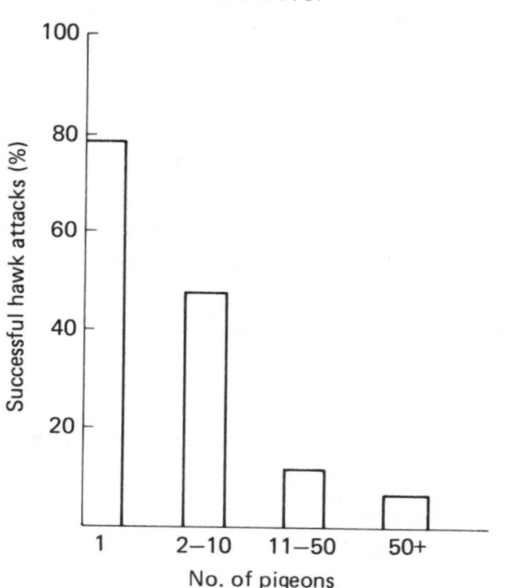

Collaboration in defence

An attack by one wood ant (*Formica rufa*) is a small deterrent to a clumsy human intruder but an attack by 200 ants is likely to put that intruder to flight. The collaboration in this case takes the form of synchronisation of attack and concentration in one area of the intruder's person. Ants and bees also benefit from the fact that whilst a large intruder may kill one or more of their number initially, he will probably learn the characteristics of his small adversaries during the attack and be less likely to intrude upon that nest, or that type of animal, again. As well as synchronising defence, social animals can collaborate by organising themselves into a predator-repellent unit. They can warn the predator of their capabilities by displaying weapons, as do the larvae of the owlfly (*Ascoloptynx furciger*) (Henry 1972, Figure 127); or by combining to make a loud noise, such as the rhythmic drumming produced by the concerted efforts of several hundred wasps in their nest (Evans and West Eberhard 1970). A predator which would attack a single individual might be deterred by the sight or sound of a group. Alternatively, or additionally, they can adopt a formation which makes it difficult for predators to attack, for example the defensive ring of horned ungulates such as buffalo or musk oxen. The social insects carry group defence to an extreme by having specialised soldiers which adopt defensive positions around nests or groups of foraging workers (see review by Wilson 1971). Even amongst aphids there is evidence for the existence of a soldier caste which is present as first instar larvae only and which attacks intruders which might harm its parthenogenetically produced sisters Aoki, S. (1978). *New Ent.*, *27*, 7–12).

The advantages of participation in group defence are sometimes obvious; for example Kruuk (1972) describes how buffalo and eland show group defence which results in few of them being killed by hunting dogs or hyaenas. In other African herbivores, group defence is sometimes successful but the defenders may be more at risk than those who run so the defence occurs most frequently when offspring, or other close kin, are attacked. Individuals in breeding colonies often collaborate in attempting to drive off predators. Kruuk (1964) showed that the more gulls in a mobbing group, the less successful was the predator. An experimental study of egg predation by Andersson and Wiklund (1978) showed that eggs left in artificial nests within, or close to, colonies of fieldfares (*Turdus pilaris*) were much less likely to be taken by predators than were those which were not near a fieldfare colony since the birds vigorously defended the nest colony area.

127 The display of weapons by grouped owlfly larvae may deter predator attack. After Henry 1972.

Adverse effects of grouping on predator avoidance

Herds of wildebeeste, troops of gibbons, swarms of bees, or starling roosts are very much more obvious to us than are scattered individuals of the same species. Any of the previously listed advantages of group living, with regard to predator avoidance, may be counterbalanced by the disadvantages which result from the greater ease with which predators can find individuals of group-living prey species. As Brock and Riffenburgh (1960) have pointed out, predators which move about randomly will encounter the prey species less frequently if the prey are aggregated than if the prey are scattered. This confers little benefit on an individual of the prey species, however, unless the predator is unable to consume many of the prey when it does find them. The argument was put forward to account for fish schooling and seems less likely to be valid for most aggregations on land where many predators can see for large distances and can return readily to an area by recognising landmarks. Many large aggregations of prey species are attended regularly by predators. Many species are easily found when they are in a large group because they live in open areas with little cover, or because they are noisy, or visually conspicuous, e.g. the pre-roost gatherings and aerial gyrations of starlings. This behaviour does not seem to be adapted for predator avoidance, although it is possible that individuals in groups may sometimes derive benefit from advertising to the predator that they are in such a well organised group.

Improving mate finding and breeding success

Mate finding

Animals which spend the period prior to mate selection in a mixed-sex social group will encounter potential mates in that group, whereas those which are solitary may seldom meet other conspecifics. Social grouping will clearly be advantageous to those individuals which can establish a pair-bond with another member of that group. There may, however, be much more competition for mates so the weaker animals may be less successful in a group than they would be if encounters were rare, but usually followed by mating. In a group, members of the sex which is most active in mate selection, usually the females, have many opportunities for testing potential mates. They will not mate unless the partner is proved to be above average with respect to benefits to quantity and quality of offspring. There are other ways of testing a potential mate, for example by assessing the quality of its territory. Some individuals must benefit from the higher quality mate they can obtain if selecting in a social situation.

A special case of mate testing occurs in lek species which include members of ten families of birds and certain antelopes, bats, and fruit flies (Spieth 1968). The males may compete with one another for very small territories, as in the black grouse (*Lyrurus tetrix*) (Kruijt and Hogan 1967), spruce grouse (*Centrocercus urophasianus*) (Wiley 1973) and Uganda kob (*Kobus kob*) (Buechner and Roth 1974). In other leks, males compete for song perches in a singing assembly, as in humming birds such as the little hermit (*Phaethornis longuemareus*) (Snow 1968) and the hammer-headed bat (*Hypsignathus monstrosus*) (Bradbury 1977). The female selects a male, mates with him and then departs to prepare for offspring production. The males generally take no part in parental care. In some leks formed by male wild-turkeys (*Meleagris gallopavo*) Watts and Stokes (1977) found that males from sibling groups display in unison and larger groups are usually able to drive off smaller groups. The females mate with the dominant member of the most successful group so that the subordinate members of that group have no offspring themselves. They are, however, 0.25 related (see Chapter 1) to the offspring which are produced, as compared with a 0.5 relationship to their own offspring. Selection will favour the spread of genes which promote the brother-helping behaviour provided that the result of the behaviour is an increase in the production of the brother's offspring which is at least twice as great as the number of offspring which could be fathered if brother-helping at the lek did not occur. Watts and Stokes did once observe a male other than the dominant one from a sibling group mating with a female whilst his dominant brother was mating elsewhere in the display group but this was exceptional. It is possible, however, that brother-helping improves future performance in contests and that this advantage, together with any sneaked matings, results in greater reproductive output than that of an individual which is driven from the lek and can never get back.

Whilst the altruistic behaviour of the non-reproducing turkeys is not reciprocated by the dominant brother, there is an example of males collaborating in achieving mating (Packer 1977) and other examples of reciprocal altruism (Trivers 1971) in finding and mating with a partner may exist, especially amongst primates. Packer watched subordinate male olive baboons (*Papio anubis*) which would not normally be able to gain access to oestrus females because the dominant male prevents them from doing so. On many occasions, one subordinate enlisted the help of another which engaged the dominant male in fighting and chasing whilst the first subordinate copulated with the female. On other occasions, the help of the individual which had copulated would be enlisted by the other subordinate so that he gained access to the female. Such benefits, in terms of offspring production, could be achieved only in a species which lives in long-lasting social groups. As Bertram (1975, 1976) has pointed out, a form of co-operation which is harder to recognise, but which is probably much commoner, is refraining from competition with another individual in return for later reciprocation of such restraint.

Rearing young

Group-living may afford opportunities for collaboration in rearing young but it certainly makes behaviour which harms the young of other individuals easier than it would be in non-social species. Tolerance towards unrelated young by adults which are well able to kill them is one of the most remarkable characteristics of animals which breed in groups. To kill such young might benefit the adults' own young by reducing competition but, even in the absence of advantages to the young of being in a group of age-mates, it seems likely that the disruptive effect on the social group caused by such behaviour might result in its breakdown with consequent loss of all advantages. In other words, killing young is not an ESS.

Adult lionesses will allow any cubs in their pride, which are of a similar age to their own cubs or younger, to suckle them (Schaller 1972, Bertram 1975). This occurs despite the fact that the average degree of relatedness between a lioness and the cubs of another lioness is 0.075 as compared with 0.5 with her own cubs. Domestic sows kept in groups will sometimes allow cross-suckling, as will African hunting dogs (*Lycaon pictus*) (Kühme 1965), but mechanisms for preventing its occurrence are widespread in mammals. Other forms of help towards the rearing of young, by individuals other than the parents, have been recorded for several species of birds and mammals. Mammalian examples include provision of food for young and defence of them by hunting dogs, lions, wolves (Mech 1970) rhesus monkeys and other primates (Spencer-Booth 1970, Hrdy 1976; Clutton-Brock and Harvey 1976).

Attention was drawn to avian examples of helpers at the nest many years ago by Skutch (1935, 1961). The Mexican jay (*Aphelocoma ultramarina*) is found during the breeding season, in flocks of 8 to 20, which usually include one or two breeding pairs together with the young from previous seasons. These young carried out about half of the feeding visits to nests watched by Brown (1970, 1972). Amongst Florida scrub jays (*A. coerulescens*) male helpers contribute more food to the nest than do female helpers and pairs with helpers live longer and produce more young (Stallcup and Woolfenden 1978). Pairs of colonial red throated bee-eaters (*Merops bulocki*) reared 16% more young per nest if they had helpers (Fry 1972) and in the Tasmanian native hen (*Tribonyx mortierii*) trios composed of a female and two brothers produced larger clutches and reared more young than did pairs in similar conditions (see Table in review by Emlen 1978). The role of older young which are present during the rearing of subsequent broods in the Arabian babbler (*Turdoides squamiceps*) seemed often to be a hindrance rather than a help to rearing the young (Zahavi 1974) but there is no doubt that many young birds and mammals do survive better because of the efforts of members of their social group other than their parents.

In addition to the immediate advantages of being fed and protected, young animals may derive other benefits from the presence of members of a social group. As mentioned earlier in this chapter, individuals can learn from others in their group about how to find food and how to avoid predators. Those which do not acquire such skills are less likely to survive and reproduce than those which do. Selection, therefore, will favour genes which promote both the ability to learn from other members of the group and staying with the group rather than leaving as soon as some degree of independence is achieved. As explained in detail by Hinde (1974), the presence of adults and peers during the period when most social learning occurs is important for survival and later reproductive success in most primate species including man.

The most extreme examples of young individuals which depend upon adults other than their mother come from the social insects. There is no doubt that in most nests of social insects, however, the workers are closely related to the young reproductives. In fact, as explained by Hamilton (1964*a*,*b*), due to the fact that the males in social ants, bees and wasps are haploid, i.e. they have only half of the normal complement of chromosomes, the workers and female reproductives are related to their mothers by the usual factor of 0.5 but are related to one another by a factor of 0.75 (see Chapter 1). Hence it is not surprising that behaviour patterns amongst workers which help those female reproductives which are their sisters to survive are many and complex (Wilson 1971).

The evolution of social behaviour

Origins of group-living

Selection pressures change during the course of evolution. Genes which promote a particular biological characteristic with a certain function may have spread in the population in the past because of the action of selection pressures related to that function. That characteristic might now have a secondary function so another selection pressure is acting. The fact that we can now suggest, and find evidence for, a variety of functions of living in groups does not mean that each of these was concerned with the evolutionary origins of group living in any species.

Group living must have evolved many times in species subject to many different selection pressures but the first step in its evolution must have been the appearance of genes which promoted the initiation of aggregation. Later stages must relate to the continuation of an aggregation once it was formed. Since individuals reproduce, the first question about animal distribution is not 'why aggregate?' but 'why separate?' It is generally assumed that competition for resources between parent and offspring usually resulted in their separation. In some animals, advantages similar to those listed for group living, such as environment modification, food acquisition, predator avoidance and helping offspring, were sufficient to counterbalance disadvantages due to competition and colonial animals, which are permanently attached to one another, such as hydroids evolved. The parallels between colonial animals and animals which are separate individuals living in groups are instructive but there are also considerable differences. One difference concerns the degree of relatedness of the members of the colony or group. Since members of insect societies are very closely related they are most frequently compared with colonial animals. The idea of an insect society as a 'super organism', more like an individual multicellular animal than like a group of animals, has been criticised by Wilson (1971). He does admit, however, that the idea has played a part in the development of recent ideas about factors affecting social organisation.

When might individuals aggregate? The most likely occasion for most mobile animals is when there is a local abundance of food. Once some individuals have found food, others may be best able to find it by recognising the presence of those individuals. Thus genes which promote the recognition of and approach to conspecifics will be likely to spread in the population. A less frequent occasion which requires more complex behaviour from the individuals concerned, is when there is a limited number of hiding places. It has been suggested that the avoidance of detection by predators or the use of early warning of predator attack might have been functions which led to the initiation of aggregation behaviour (e.g., Treisman 1975*a,b* Lazarus 1979). Aggregation at a food source or hiding place seems a more likely first step but it is not possible to discover the temporal ordering of events in the evolution of group living. The initial function of group-living may well have been different in different species and several selection pressures may have acted at once.

Once aggregation behaviour occurs other selection pressures, such as those associated with predator avoidance or utilisation of information about food sources, may result in the maintenance and modification of such behaviour. If, as Hamilton (1971) proposes, animals behaved in such a way that they minimised their domain of danger, the result would be to concentrate an aggregation which might otherwise be diffuse and temporary. The other anti-predator behaviour mentioned earlier in this chapter would also reduce the advantages of leaving the group. It could be best for animals at a long-lasting food source to remain there, rather than separating, either because the risk of not finding another food source was great or because the risk of predation whilst searching was greater than if they remained aggregated. Most frequently, the food source would not last for long so the individuals would need to move away from it to find another. Individuals which left the food source separately might be very vulnerable to predation so they might do best to remain in the midst of the group as all searched for food together. The group might remain together irrespective of predation pressure. As Ward and Zahavi (1973) have explained, an individual which knew of no other food source would benefit by following any other in the group which departed towards another food source when the first became depleted. The animal which knew of another food source would do best if it did not advertise it, unless reciprocal altruism is invoked, but it would be difficult for it to avoid being followed when it left the group. These points and those in the following sections are summarized in Figure 128.

Altruism to relatives

Altruistic behaviour towards relatives is much more likely to occur if the relatives are nearby. Most of this altruistic behaviour involves improving the chances that juveniles will find food and avoid predation. Its evolution is

128 Possible origins of social behaviour and steps in its evolution.

Path A A1 Approach and join groups because they have probably found food

A2 (i) Predator approaches, get in the middle of the group

and/or
A2 (ii) Stay in group because early warnings by others can be used

and/or
A2 (iii) Food depleted, one individual leaves, follow it for it may know where to find more food

Path B B1 Parents remain with offspring / Offspring remain with parents } Parental care increases survival chances of offspring

Offspring which stay with parents after the stage at which independence is possible can learn about social techniques as well as feeding and predator avoidance methods

Then A2 as above

? B2 Animal stays in group because close relatives survive better if it stays much easier to understand if we compare survival rates, in the animal population, of genes which have different effects. Hamilton (1964*a,b*) discusses whether a gene which promotes some aspect of parental care will survive through successive generations. He argues that it is necessary to think in terms of 'inclusive fitness', the contribution of a genotype to the gene pool, when trying to explain why a particular characteristic is shown by an individual. Dawkins (1978) argues even more explicitly that selection must be considered to be acting on the replicators themselves. The implications of Hamilton's approach are that a characteristic may be shown by an individual, not only if it benefits that individual directly but also if it benefits close relatives. Thus an animal may risk its life if in doing so it saves three offspring, each related to it by a factor of 0.5, which can then go on to breed. Similarly, there are circumstances where behaviour will occur if it benefits siblings, also related by 0.5, nephews, nieces or grandchildren, related by 0.25, or even first cousins, related by 0.125.

Certain social groups may therefore exist because they provide the opportunity for individuals to help their relatives. As always, any such benefits must counterbalance the disadvantages due to increased competition for resources or, if those relatives are not offspring, reduced chances of producing offspring. Many examples have been quoted of species in which altruistic behaviour towards siblings or the offspring of siblings is shown. The individuals which are most likely to show this behaviour are those which would be unlikely to produce any offspring themselves in that breeding season, for example young adult birds and worker ants or bees. They then show altruistic behaviour if it improves the survival and future reproductive chances of the young.

Social insects

The initial stage in the evolution of social behaviour in social insects must have occurred a very long time ago and has probably occurred several times amongst Hymenoptera (ants, bees and wasps) as well as among Isoptera (termites). Individual insects often show extreme forms of altruistic behaviour, and Wilson (1971) lists the following groups.

1. Soldier termites which are the first to defend winged reproductives from attack by ants.
2. Injured workers of the ant *Solenopsis* which attack more readily than do uninjured workers and leave the nest before dying.
3. Workers of many species, which approach danger when there is a disturbance.
4. Honeybees, some polybiine wasps and the ant *Pogonomyrmex*, which die when they sting a mammalian predator.
5. Worker bees, fed on sugar water, which donate their own tissue proteins to the larvae which they feed.
6. Workers of many species whose egg laying is inhibited by the presence of a laying queen.
7. Honey ants which spend much of their lives acting as a communal stomach for others in the nest.

As previously mentioned, the social Hymenoptera are haplodiploid, i.e. males are haploid but females diploid, a characteristic shared with certain Rotifera, mites, white flies, coccids, thrips and beetles (Hamilton 1972). This makes altruistic behaviour by workers even more likely than in a diploid species and leads to a situation where the workers can be said to manipulate the queen when there is any conflict of interest (see explanation by Trivers and Hare 1976). Although haplodiploidy is an important factor in maintaining the societies of social insects today, it was only one of several factors influencing the evolution of such societies. The females have stings and group living, as mentioned earlier, maximises their effectiveness in defence. It seems likely, therefore, that the ancestral females laid eggs, provisioned the larvae and then remained with them until they were fully grown. The young were then able to help to defend her as she laid further eggs. Once this situation existed, if haplodiploidy appeared by mutation it would be likely to persist in the population and favour the evolution of further altruistic traits.

Reciprocal altruism

Altruism need not always be directed at relatives. D. L. Krebs (1970) has surveyed examples of altruistic behaviour in man and examples of apparently altruistic behaviour towards non-relatives have been reported for many other species. Trivers (1971) has elaborated the concept of reciprocal altruism, epitomised by the example: I'll save you from drowning today if you might save me tomorrow. Provided that the cost of the altruistic act is quite small, and the benefit great, the probability of reciprocation need only be small for selection to favour genes which promote the behaviour. Trivers points out that reciprocal altruism will be most likely in long-lived species in which social contacts are prolonged so that there will be many opportunities for reciprocation. If reciprocal altruism is widespread it increases the likelihood that social groups will persist and improves social cohesiveness. Altruistic behaviour of any sort, therefore, may be directed towards relatives or non-relatives in the expectation of future reciprocation. One example is allogrooming in primates. Another is provided by flocks of birds, looking for food in woodland, which may call when food is found and thus share it with others in the group. Individuals of various species may help one another by giving alarm calls or defending against predators. The example of baboons collaborating to mate with a female (Packer 1977) and lions tolerating mating by others in the pride (Bertram 1975) have already been mentioned. Many examples of reciprocal altruism are vulnerable to cheats, i.e. individuals which receive a benefit without reciprocating. Punishment of cheats may be costly and this cost will reduce the likelihood that some possible forms of reciprocal altruism will occur, (Zahavi personal communication). Mutually advantageous reciprocation does seem to occur, however, for some behaviour, e.g. allogrooming in primates.

It is likely that reciprocal altruism is especially important and extensive in man because of our abilities to communicate and to recognise cheats. As a consequence of these abilities, a gene which promoted some form of reciprocal altruism would have been especially likely to spread in a human population. If someone helps me I can ask others to help that person and can even write a letter saying that the bearer has been shown to be a useful member of my community. Such acts are more likely to occur if some future reciprocation is possible and the strong are more likely to be helped than the weak. Many codes of conduct based on reciprocal altruism have been devised in human societies and these have been effective in promoting cohesion within the society which utilizes them. As communication improves, reciprocation is possible from more and more people so mankind becomes more like a single society with weaker and weaker barriers between previously discrete sections.

Alarm signals

In practice it is often difficult to determine whether or not a behaviour can reasonably be called reciprocal altruism. As mentioned earlier, the possibility of detecting alarm signals given by another member of the group may be very advantageous to the hearer. The problem, which has been discussed for many years, is to determine the advantage to the giver of the signal, for that action could increase the chances that the giver will be attacked. Maynard Smith (1965) suggested that alarm calls 'probably originated as a signal from parents to their offspring during the breeding season'. The fact that females in a group of Belding's ground squirrels (*Spermophilus beldingi*) gave alarm calls quicker if some of their young were present supports this idea (Sherman 1977). Since alarm calls are still given in winter flocks, Maynard Smith postulated that a gene promoting this behaviour would spread if winter flocks included a lot of close relatives but he felt that this was not sufficient reason for the high frequency of occurrence of alarm calls. A similar conclusion was reached by Hamilton (1964*a*) who also considered that the risk to the caller was not great. Most birds give alarm calls, in response to a ground predator, after leaving the ground. For calls given at the approach of a flying predator, however, the risk to the caller seems to have been sufficient to have resulted in selection for calls which are difficult to locate (Marler 1955). Trivers (1971) suggested that reciprocal altruism could account for alarm calling and also that a predator which heard an alarm call might be discouraged from continuing to hunt in that area. Alcock (1980) and Dawkins (1976) considered that an alarm call might result in none of the members of the group being detected, whereas if some had continued with their previous activity the predator might have seen the group and perhaps taken the caller. It may be that the caller can derive direct benefit from calling, in a species which flees when startled, for if the caller saw the predator and then fled without calling it might be at greater risk from the predator than if it called and fled with the rest (Charnov and Krebs 1975). The functions of alarm calls would seem to differ according to the nature of the predator and the response. It is very difficult to determine their evolutionary origin. Some apparent alarm signals, like the stotting jumps of gazelles, may even serve to draw attention to that individual (Smythe 1970). This may emphasise its physical prowess and deter the predator from chasing it (Zahavi 1977, Woodland, Jaafar and Knight 1980).

Territories and home ranges

Individual distance

Most animals refrain from making direct contact with conspecifics and actively preserve a minimum distance around them whether moving or resting. This minimum distance within which approach elicits attack or avoidance, was called the *individual distance* by Hediger (1941, 1955). He described black-headed gulls (*Larus ridibundus*) sitting in a row on a parapet. None was closer than 30 cm from its neighbour and most were exactly that distance. Flamingos (*Phoenicopterus ruber*) in a zoo are usually at least 60 cm apart and swallows (*Hirundo rustica*) sitting on a wire or thin branch are seldom less than 15 cm apart (see Figure 125). Some monkeys maintain an individual distance in some circumstances but huddle together in others. Individuals forced to rest closer together than their individual distance spend much time hiding, averting their gaze from other individuals and failing to carry out normal behaviour.

Possible advantages of maintaining an individual distance include reductions in 1) damage to the body due to contact, 2) interference and competition whilst feeding, 3) impedence when starting to flee, 4) disease or parasite transmission, and 5) the chance of rape. Hall (1966) emphasised the differences between people of different nationalities in their tolerance of close proximity by others. Northern Europeans have the largest individual distance, southern Europeans a shorter distance and Japanese, especially in commuter trains, tolerate most contact. Walther (1977) and others have described differences in individual distance according to activity. Male Thomson's gazelles (*Gazella thomsoni*) walk 2 m apart, rest 3 m apart but graze 9 m apart. Walther attributes this last high figure to the similarity of the grazing posture to the threat display.

In some circumstances close aggregations do occur, e.g. the mats of feeding brittle-stars, clumps of bees in cold conditions, heaps of juvenile King crabs (*Paralithodes camtschatica*) (Powell and Nickerson 1965) or pods of striped mullet (*Mugil cephalus*) (Wilson 1975).

Individual distance might be regarded as a form of territory, in that it is a defended area, but it is mobile, being linked to the position of the individual rather than to a topographically definable location. The territories held in leks (see p. 189) may persist for less time than the area around itself defended by a roosting bird. Likewise the territory around the nest of a colonial sea-bird may be no larger than that maintained by individual distance outside the territory. Both of these territories, however, are areas of ground which do not move around with the individual.

Territory and its demarcation

The functions mentioned above for the maintenance of individual distance are sometimes better served by excluding other individuals from a larger area. As appreciated by Howard (1920) and later explained in more detail by Nice (1941), Hinde (1956) and others, if competitors can be kept out of an area which includes food sources, the owner can monopolise that food and become familiar with the area so that there is no need to travel around unknown terrain. Hence the food sources can be exploited efficiently and suitable cover can readily be found if a predator attacks. In a familiar area the best resting places can be utilised, thus reducing the risk of predation and environmental hazards such as temperature extremes. The extent of these advantages will vary according to food distribution patterns and the probabilities of predator attack and other hazards. An area which is defended by fighting, or by demarcation which other individuals detect and which acts as a deterrent to entry, is referred to as a *territory* (Figure 129).

The demarcation of a territory may be visual as in cleaner fish (Limbaugh 1961, Potts 1973) and other reef fish (Reese 1975), auditory as in many birds (Howard 1920), olfactory as in scent marking by many mammals (Ralls 1971,

129 The territories held by eleven roebucks (*Capreolus capreolus*) during one season. The territories, some of which overlap, were demarcated and defended by fraying trees with the antlers, probably by deposition of pheromone, by barking, by display and by fighting. Modified after Bramley, 1970.

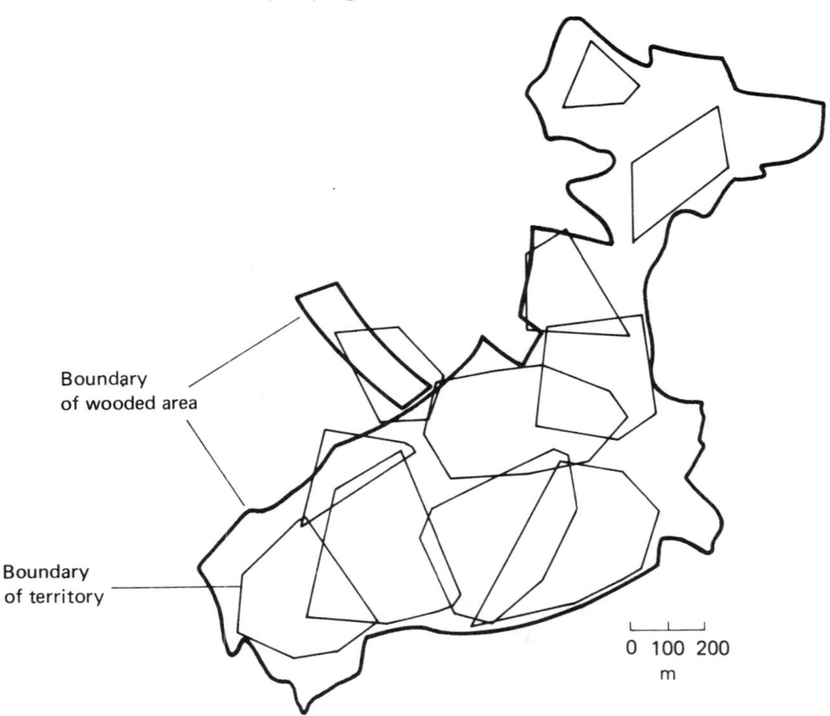

Johnson 1973, Stoddart 1980), or electrical as in electric fish (Westby 1975). All these forms of demarcation are energetically costly to the territory holder. Fighting may result in physical damage and it requires energy. Singing, patrolling to deposit scent marks, and maintaining an electrical field use energy and take up time which might otherwise be spent feeding or carrying out other essential activities. These costs may be greater than any benefit which accrues from territorial behaviour so some animals are never territorial and others defend territories in certain circumstances only.

Group living and territorial behaviour are not always mutually exclusive for territories are sometimes defended by groups of animals. Ellefson (1968) studied white-handed gibbons (*Hylobates lar*) in Malaysia and showed that family groups of parents and their young defended territories of about 100 ha. Such territories included fruit trees and were defended by calling and by chasing intruders. Hall (1968) reports patas monkey (*Erythrocebus patas*) groups chasing away other groups. However he also reviews inter-group behaviour in old-world primates and comments on the frequent tolerance of the presence of other groups by primate species. The home ranges of many primate groups have core areas which are used intensively but which are rarely entered by other individuals.

Home range

Those animals which restrict their movements to a discrete area are said to have a *home range*. This may or may not be a territory but it is often difficult for an observer to be sure whether or not there is defence, or has been defence, of an area. The Soay sheep, which are feral on the small island of St Kilda, were observed to remain in several ewe-groups within well defined areas without any contest between individuals in the different groups (Grubb and Jewell 1966). It is possible, however, that they did fight or display when the distribution pattern was initiated.

Some species which move around within a home range do not defend an area but threaten or attack conspecifics which approach them. Warner (1970) found that the mangrove crabs *Aratus pisoni* and *Goniopsis cruentata* chose paths around the mangrove roots so that they did not come close to any larger individual and attacked any smaller individual which came close to them.

Territory and foraging

Bird territories which include feeding sites range from 0.07 ha in the least flycatcher (*Empidonax minimus*) to 9300 ha among golden eagles (*Aquila chrysaetos*) in the mountains of California. There is an approximately linear relationship between the logarithm of body weight and the logarithm of territory area (Schoener 1968) but many other factors also affect territory size. Herbivores usually have smaller territories than carnivores and the area defended must be much larger if food is scarce. Holmes (1970) found that territories of the dunlin (*Calidris alpina*), a small wading bird, were five times larger at an arctic locality in Alaska with few ponds but an unpredictable food supply (Figure 130*b*) than at a subarctic locality with many ponds and abundant food (Figure, 130*a*). Another example of the relationship between food supply and density of animals is shown by the work of Kitchen (1974) who found that those male pronghorn antelopes whose territories provided the most forage attracted the largest number of females (Figure 131). The territories of golden-winged sunbirds (*Nectarinia reichenowi*) vary greatly in size but each contains approximately the same number of the *Leonotis* flowers on whose nectar the sunbird feeds (Gill and Wolf 1975). The nectar available is just sufficient for the birds' daily energy requirements (Wolf *et al.* 1975).

The relationships between the costs and benefits of territorial behaviour are discussed in detail by Davies (1978*b*). He points out that where there are benefits in the form of resources within the territory, the optimum territory size will be that at which the net benefit is at its highest. There is a maximum level for benefit, measured immediately as food supply but ultimately as fitness, but the cost of territorial defence continues to increase as the area increases (Figure 132). There is no evidence that animals hold extra territories to spite other individuals, i.e. to reduce the amount of territory held by others. They do sometimes defend a territory which is larger than present requirements but which will be needed when a mate and offspring are present or when resources are scarcer.

Food availability and distribution often vary from day to day so animals sometimes alternate between territorial and non-territorial behaviour. Zahavi (1971*b*) watched white wagtails (*Motacilla alba*) wintering in Israel and found that they flocked when food was scattered but defended piles of food when these were provided. This work was extended by Davies (1976) who showed that pied wagtails, a different subspecies, would leave their territories when food there was scarce and join flocks (Figure 133). Even when feeding in the flock they continued to visit their territories intermittently and to drive off intruders.

130 Territories of dunlin, a small wading bird, were much smaller (*a*) at the Kolomak River, Alaska where food was abundant, than (*b*) at Barrow, Alaska where the supply was unpredictable. Modified after Holmes 1970.

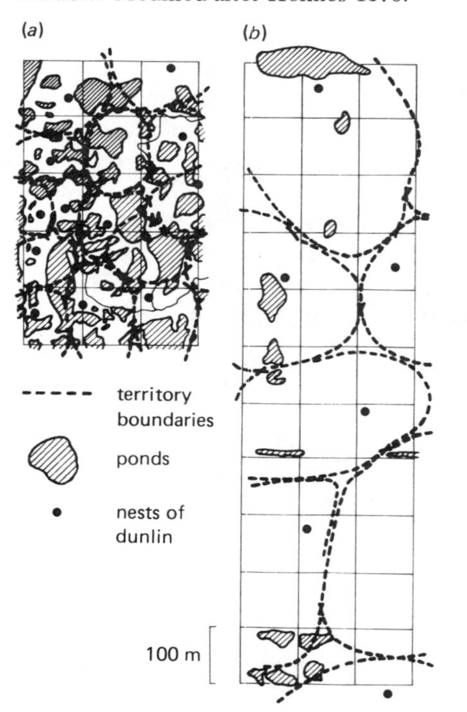

Territory and breeding

Very small territories with no resources include those in leks (see p. 189) and some territories held by butterflies. Male speckled-wood butterflies (*Pararge aegeria*) defend sunspots in woodland against males and pair with visiting females (Davies 1978*a*). Other small territories include those of colonial nesters such as sea-birds. Sea-birds suffer restrictions in the number of nesting places which are not vulnerable to ground predators. Their nests are often crowded together but there is always a minimum inter-nest distance. In the gannet (*Sula bassana*) it is 90 cm, this being determined by the distance which can be reached by two adjacent incubating birds stretching out their necks (Nelson 1965, 1978).

Many animals cannot breed unless they hold a territory. Male sticklebacks, for example cannot attract a mate unless they have a territory and have constructed an adequate nest (ter Pelwijk & Tinbergen 1937). Females may select mates on the basis of the size of territory which the male can defend, e.g. more gravid female

131 Pronghorn antelope bucks whose territory during the rut included the most food per unit area attracted the most does. Modified after Kitchen 1974.

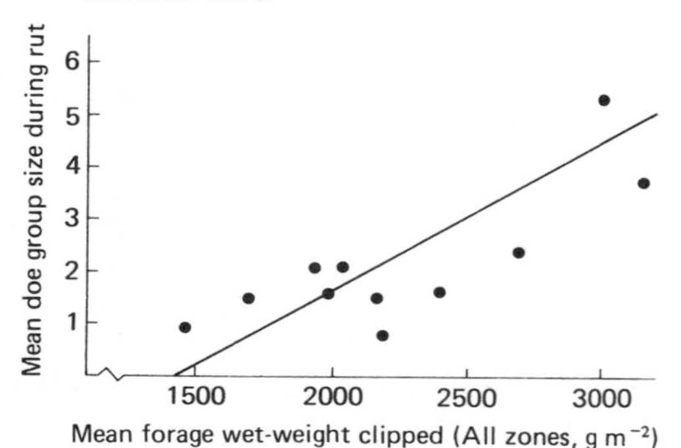

sticklebacks lay eggs in the nests of males with large territories than in those of males with small territories (van den Assem 1967). A gene which promotes such behaviour in female sticklebacks will spread in the population provided that males with large territories do help these females to produce fitter offspring than those females which mate with a male whose territory is small.

In certain circumstances, e.g. the red grouse (*Lagopus scoticus*) observed by Jenkins, Watson and Miller (1963), individuals which have no territory not only fail to mate but die much sooner than territory holders. Population size of grouse on an area of moorland and of great tits (*Parus major*) in a woodland was found to depend upon the number of territories which the area could support. When Watson (1967) removed territory-holding red grouse, adjacent territory holders enlarged their territories initially but new pairs soon arrived and took over the empty territory. Most of the newcomers had not previously held a territory so were now more likely to survive. Watson therefore proposed that territorial behaviour in red grouse sets an upper limit to the population size. Subsequent studies (Watson and Moss 1979, 1980) have shown that the grouse's spacing behaviour changes as population density changes and may be a major cause of population cycles. Krebs (1971) removed pairs of great tits holding territories in a wood and found that they were replaced by pairs which had previously held hedgerow territories (Figure 134). Krebs found

132 The benefits to an individual obtained from a territory must have a maximum level but the costs of territory defence, in terms of energy, risk, and hence reduced fitness increase with territory size. Thus there is an optimum territory size (*X*) where net benefit, i.e. benefit minus cost, is at a maximum. After Davies 1978*b*.

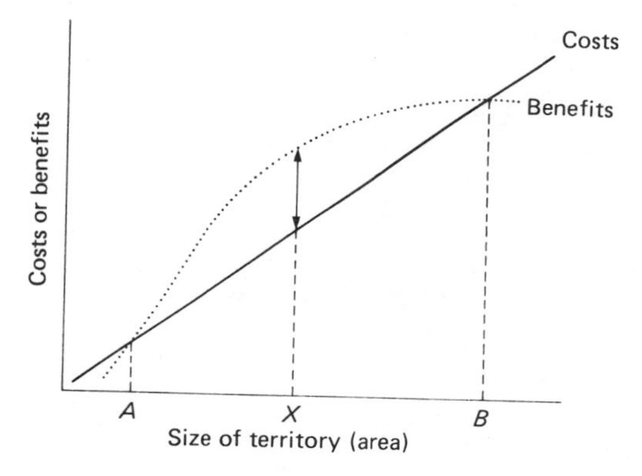

133 Pied wagtails were least likely to leave their winter territories and join feeding flocks when their feeding rate in the territory was high. After Davies 1976.

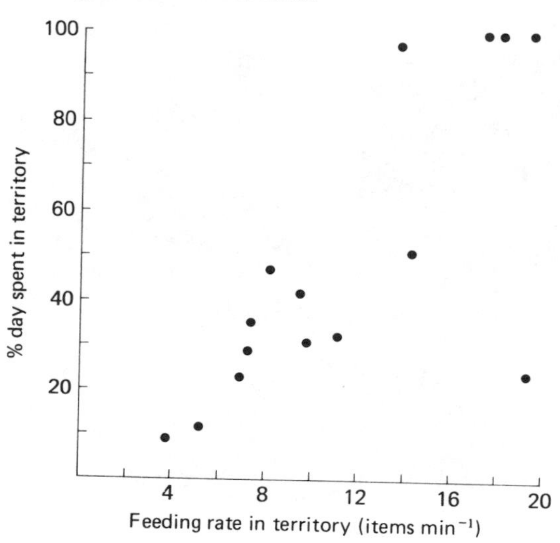

134 When the pairs of great tits holding the territories marked X on map (*a*) were removed, new pairs formed territories (stippled on map *b*) in vacated areas within 3 days. Modified after Krebs 1971.

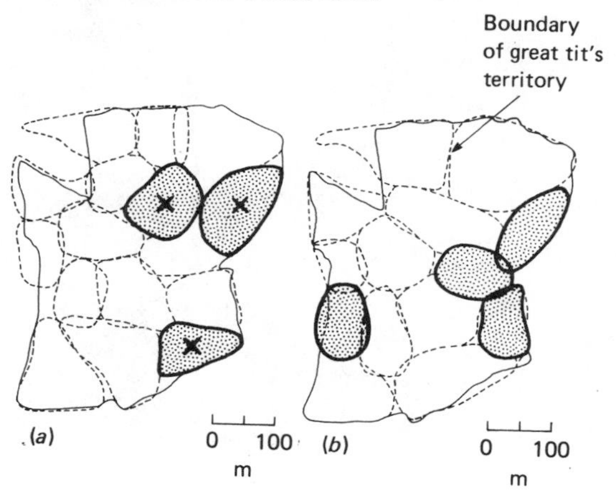

that only 22% of pairs breeding in hedgerows were successful whereas the success rate of those breeding in woodland was 92%. Territorial behaviour limited the number of birds that could breed in the most favoured habitat and forced some to breed in the less favoured hedgerows.

Many birds keep others out of their territories by singing. Krebs, Ashcroft and Webber (1978) found that, after the removal of great-tit territory owners, invasion of the territory was delayed by about 12 h if a single great-tit song was played on a loudspeaker within the territory but by about 24 h if a repertoire of great-tit songs was played.

Interspecific territoriality

Territory holders usually defend strongly against conspecifics but sometimes they also defend against other species, especially those with similar diets. Reese (1975) describes several butterfly fish on the Great Barrier Reef which will defend territories against other species. Figure 135 shows a *Chaetodon trifasciatus* entering the territory of another coral-feeding butterfly fish, *Megaprotodon strigangulus*, and then being chased away. Examples of birds which show interspecific territorial defence include the displacement of redwinged blackbird (*Agelaius phoeniceus*) by yellow-headed blackbird (*Xanthocephalus xanthocephalus*) in northwestern America (Orians and Willson 1964) and the responses of various *Acrocephalus* warblers to other species. Catchpole (1977, 1978) found that sedge-warblers (*A. schoenobaenus*) responded by approach and singing when reed warbler (*A. scirpaceus*) song was played to them and that reed warblers responded to marsh war-

135 When one of a pair (A1 and A2) of a species of butterfly-fish approached the territory of an individual of another species (B) it was driven off. Both of these species feed on coral. Modified after Reese (1975) who based the drawing on a film sequence.

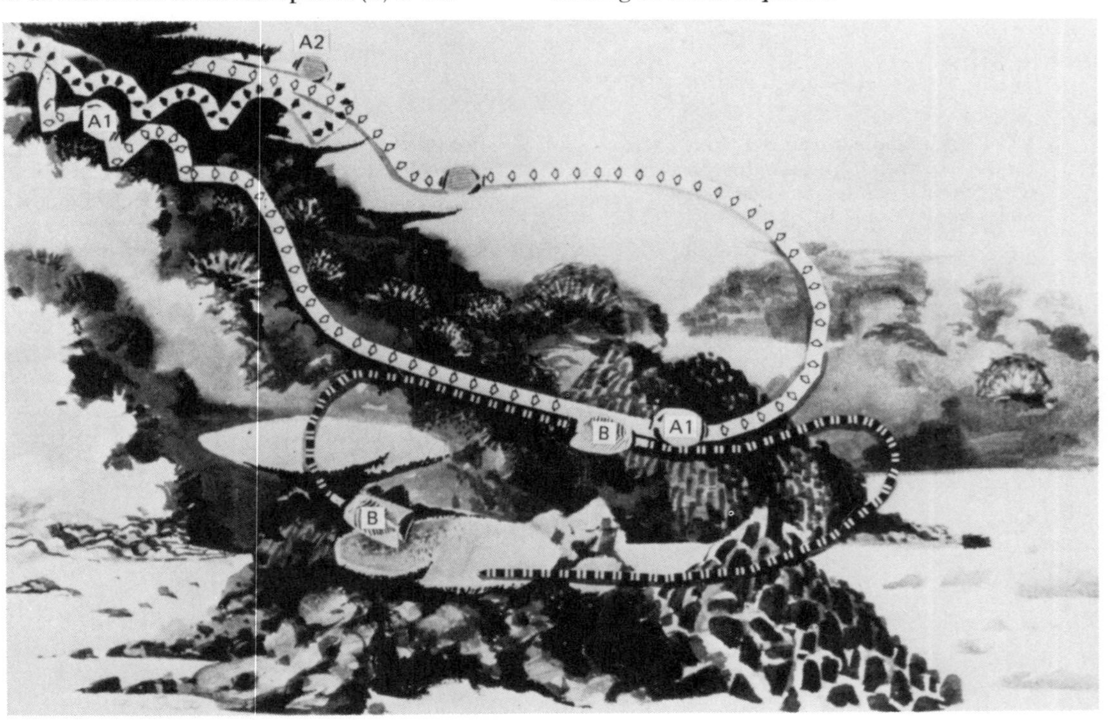

bler (*A. palustris*) song but only in areas where both species occurred. Various primates also show some territorial exclusion of other species. For example, ring-tailed lemurs (*Lemur catta*) were observed by Jolly (1966) to bar the way of Verraux's sifaka (*Propithecus verrauxi*) in Madagascar.

Migration and other types of dispersal

Individuals in a species compete with one another for resources so it is not surprising that mechanisms promoting spatial separation of offspring from their parents and from one another have evolved. Such dispersal, which is further discussed in Chapter 9, is usually a necessary prelude to territorial defence by single or paired animals and may involve travelling a small or a large distance. In troops of baboons and prides of lions the young males disperse much further than the young females, which may in fact remain in their natal group. In many other species all individuals disperse from their rearing site but various factors affect the extent of that dispersal. Once an individual has moved to a point where it is no longer influenced by its parents or siblings, its further dispersal is likely to be governed mostly by food availability and distribution (Chapter 6). Dispersal or emigration from an established range will occur, amongst species which are sufficiently mobile, when the risk of staying is larger than the risk of migrating. An alternative to this method of minimising the effects of unfavourable conditions is to reduce the metabolic rate and to become dormant. Emigration may occur if a risk of starvation is predicted by difficulties in food finding, if predation risks become apparent because of frequent near-misses by predators, if physical conditions deteriorate, or if a breeding site cannot be obtained. Lack of food seems most likely to be the cue for migration but the likelihood of future lack of food may sometimes be predicted from physical conditions, e.g. adverse weather, or by detecting an excessively large population size (Wynne-Edwards 1962). For theoretical discussions of strategies and methods of dispersal see MacArthur and Wilson (1967), Gadgil (1971) and Baker (1978).

Migration over long distances is an impressive and well documented aspect of behaviour amongst birds, insects, fish and some other very mobile animals. It often necessitates an ability to orient in places which have not been visited previously and such migration methods have been the subject of much research (Matthews 1968, Schmidt-Koenig 1975). The pattern of breeding in the temperate zone and wintering in the tropics is unlikely to have started in a temperate area with an extremely unfavourable season. The first stages of migration probably evolved in tropical species which, due to competition for resources such as food, moved into cooler areas during their more favourable season and then returned when conditions started to worsen. Later, cues from changes in day-length or temperature could have been used to predict the onset of poor conditions in the temperate area. The summer in the temperate zone is good for breeding since day-length is greater than in the tropics so there is more time for food gathering. Some species, therefore, could have migrated greater and greater distances to breed until, as we find today, in summer most temperate parts of the world have become populated partly by migrants.

9

Reproductive behaviour, including parent–offspring interactions

Throughout this book, reproductive success has been emphasised as the yard-stick against which the effectiveness of biological mechanisms is measured. The initial result of the efficient functioning of most of the systems described in Chapters 2 to 7 is the survival and perhaps growth of the individual. This efficiency will result in the survival of genes into the next generation provided that the reproductive functional system is also efficient. This chapter deals with reproductive behaviour, its evolution, control, development and economic importance. The major parts of the functional system considered are behaviour leading to mating, mating and interactions between parents and offspring. These topics have been grouped together in one chapter in order to emphasise their inter-relationships. For example, some aspects of courtship and mating depend upon the extent of subsequent parental care and some parental behaviour depends upon preparations made before mating which were also involved in courtship. Before considering behaviour leading to mating, there are general discussions of the consequences of sexual reproduction and the various mating systems.

Sexual reproduction and mating systems

In order to discuss strategies of reproductive behaviour it is necessary to attempt to quantify the effort expended on reproduction. *Reproductive effort* in a given season can be estimated by measuring all the resources expended by an individual on reproduction in that season. If reproductive effort is very high in one season, a result may be a considerable diminution in the chances of reproductive success in the following season (Gadgil and Bossert 1970, Trivers 1972). A part of reproductive effort is parental investment. Trivers (1972) defined *parental investment* as 'any investment by the parent in an individual offspring that increases the offspring's chance of surviving (and hence reproductive success) at the cost of the parent's ability to invest in other offspring'. Parental investment thus includes the resources required for egg or sperm production, preparing to care for the young and doing so.

Some consequences of sexual reproduction

Most multicellular animals grow to a particular size and then start to reproduce sexually. The advantages of sexual reproduction were enumerated by Fisher (1930) and by Muller (1932). As explained in detail by Maynard Smith (1971, 1978*a*,*b*), there are situations where genes which suppress the individual's sexual reproduction will survive for many generations but, ultimately, that path is a dead end in a changing world. Evolutionary consequences of sexual reproduction include competition between individuals for mates, differences among mates in the optimal extent of collaboration and differences between parents and offspring in the optimal extent of parental investment (Wilson 1975).

Mate finding, mate selection and competition for mates, which are discussed later in this chapter, would all occur if the two gametes, with any associated structures, were equal in size. The fact that male and female gametes are unequal in size has additional consequences. If the two sexes do not contribute resources equally towards the production of offspring, the difference in quality between a better and a poorer mate is of greater importance to the sex which contributes more, because a mistake which resulted in the production of poor quality offspring would cost an individual of this sex a higher proportion of its lifetime reproductive effort. The members of this sex, therefore, may show more elaborate mate selection methods. It is more expensive, in terms of resources, to produce an egg with its yolk than to produce a sperm, so this aspect of parental investment, which may be the whole investment in some species, is larger in females than in males. Bateman (1948) argued that, as a consequence of this difference, females should be more discriminating than males when selecting a mate and males should be eager to mate with more than one female. He supported his argument with experiments on *Drosophila* in which five chromosomally marked males and females were put together and the markers were located in each offspring so that the parenthood was established. He found that only 4% of females failed to produce offspring but 21% of males failed and some males produced three times as many offspring as the most successful female. All males made attempts to copulate and the results were not explicable by some males restricting the access of others to females following inter-male contests, so the females must have accepted some in preference to others.

Trivers (1972) quotes examples from field studies of greater variation in the reproductive success of males than in that of females. Since females have to invest more in egg production than males invest in sperm production, all animals should be, fundamentally, *polygynous*, i.e. the males should try to mate with more than one female. Polygyny will be more likely as the difference in parental investment between male and female increases. Fertilised females derive little advantage from further mating if embryo development occurs within the female or within her care but males can produce large numbers of sperm and can continue to fertilise females. A variety of ecological factors influence the likelihood of *monogamy*, i.e. where each individual mates with one member of the opposite sex, and *polygamy*. The latter term includes polygyny and *polyandry*, where each female mates with several males.

There are complex inter-relations between the type of mating system which is used, the mechanisms of mate finding and selection, the extent of competition for mates, the type of courtship which occurs, the amount of parental investment, and the way in which the young develop. All of these are also related to the questions of how often the individual will attempt to breed during its life, how many offspring will be produced at one time (Lack 1968) and the sex ratio of those offspring (Wilson 1975, Trivers and Hare 1976, Williams 1979). The evolutionary approach to studies of interactions leading up to mating and of interactions between parents and offspring has made it easier to link together the various components of the reproductive functional system. It has helped to explain how the system works in individuals now, as well as why it has evolved to its present state in any species.

Mating systems

Despite the argument, based upon relative parental investment by the sexes, that polygyny should be the normal mating system in animals, monogamy is common in some groups, for example, in 90% of birds. Parental investment in many birds, however, is almost as great in the males as it is in the female. Young altricial birds in nests are very vulnerable to predation and survive best if they can develop rapidly. If the female alone can seldom rear two young, it is advantageous for males to help with parental care. Armstrong (1955) suggested that wrens (*Troglodytes troglodytes*) are monogamous, when both parents are needed to collect enough food for the young to develop. In long-lived birds, long-term mate fidelity seems to result in great efficiency of reproduction due to rapid mating and effective co-operation between mates (Coulson 1966, Davis 1976). Crook (1964), who compared 90 species of weaver bird, Ploceinae, found that the primarily insectivorous species which inhabit forests are monogamous whereas savannah-dwelling seed-eaters are colonial and polygynous. The monogamous species defend and forage in territories whilst the seed-eaters forage in flocks and probably obtain food more readily; hence the males do not have to feed the young.

Habitat quality may affect mating systems. Verner (1964) found that female long-billed marsh wrens (*Telmatodytes palustris*) would mate with a male which was already mated, rather than with an unmated male, if the quality of the territory of the first male was much better than that of the second. Verner and Willson (1966) therefore proposed that there is a polygyny threshold in habitat quality (Figure 136). This idea was developed by Orians (1969) who quotes supporting evidence from his own studies of red-winged *Agelaius phoeniceus* and yellow headed *Xanthocephalus xanthocephalus* blackbirds. Orians proposes that polygyny should be commoner in habitats where food is locally abundant, especially if the number of nest-sites is restricted and that it should be prevalent in precocial birds. In such situations,

136 The contribution of a male to the female's offspring production is greater in a poorer quality environment so monogamy is favoured there. If it is possible for a female to increase her fitness by bigamous mating with a male holding a high quality territory she should do so. Modified after Orians 1969.

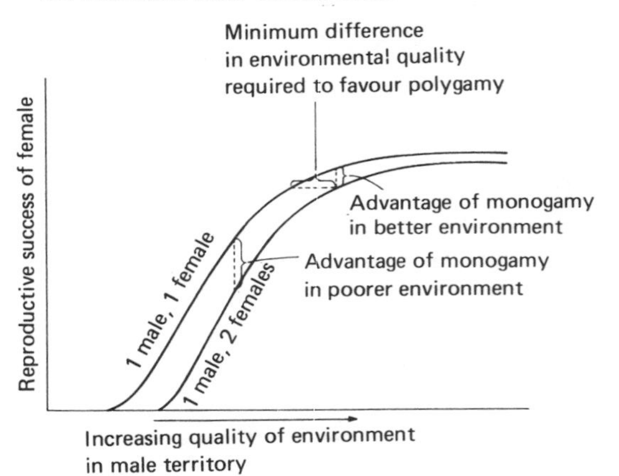

genes which have the effect of favouring polygyny if the male himself, or the territory which he controls, is of high enough quality, should spread in the population. It seems that this has happened in habitats where there is a period of rapid resource increase. In addition to marsh and savannah dwellers, such as those mentioned, polygyny occurs in many upland game birds (Wiley 1974), in the ruff *Philomachus pugnax* and in a few sandpipers (Lack 1968).

Emlen and Oring (1977) summarise the conditions necessary for the occurrence of polygamy in terms of the relative economics of defending one or several mates. This depends upon the spatial distribution of resources, the sex ratio and the time at which mates become available. If females show synchrony in their sexual receptivity, polygyny is more difficult. Emlen and Oring describe three major types of polygyny, involving resource defence by males, harem defence, or selection by females from male assemblies such as leks (see Chapter 8). It is not surprising that polygyny is the normal mating system in mammals for the female's parental investment is always large. Eisenberg (1966) lists a few cases of monogamy, for example marmosets (Callitrichidae), gibbons (Hylobatinae), beavers (Castoridae), one seal and a few terrestrial carnivores.

An intermediate between polygyny and polyandry, in which females lay a second batch of eggs which the male incubates or guards but where the male may also mate with other females was called *rapid multiple clutch polygamy* by Emlen and Oring. This is common in fish and occurs occasionally in birds. Polyandry has not been reported for mammals (Eisenberg 1966) but Jenni (1974) describes it in several species of birds. Most are in the order Charadriiformes and all have precocial young (see Chapter 1). The females of the American jacana (*Jacana spinosa*) defend large territories in marsh areas which include the smaller territories of several males. The males perform all incubation and parental care but they defend their territories against other males. In addition to such resource defence polyandry, Emlen and Oring describe harem polyandry and female access polyandry. For a review of mating systems in relation to resource availability see Bradbury (1980).

One consequence of recent work on mating systems is that it has become apparent that they can vary within species and within individuals according to the conditions which obtain at any time. Emlen and Oring list examples of variation between polygyny and monogamy; from monogamy to rapid multiple clutch polygamy to polyandry, and from lekking to mate or resource defence polygyny. Detailed study of individuals emphasises the plasticity of mating systems and raises questions about the factors which affect reproductive behaviour, both during development and at the time that decisions about the behaviour are taken.

Behaviour leading to mating

Many animals show no reproductive behaviour for large parts of the year because reproduction is energetically impossible or any offspring produced could not survive at that time. This discontinuity of reproduction is usually a consequence of temporal fluctuations in the availability of resources. Seasonality of breeding will not be discussed in detail in this chapter but it is an important example of the wide-reaching effects of environmental variables on reproductive behaviour.

Finding and competing for a mate

Once an individual has reached the appropriate physiological state for reproductive behaviour to occur, the first changes in behaviour are those which increase the chances that a mate will be encountered. Mate-finding behaviour involves active searching, self advertisement or both. An element of competition is a part of such behaviour, for others of the same sex will be showing mate-finding behaviour and the more efficient searchers and advertisers will produce more offspring per unit of energy expended. Since two or more rivals will often find the same potential mate, competitive encounters will occur. These involve the major energetic cost of mate-finding in some species, whilst in other species the major cost is in searching. The type of mate-finding and the extent of competition which occur in a species depends upon a variety of factors, especially the way in which individuals are distributed. Within a species alternative competition methods may be used, for example, some male bluegill sunfish (*Lepomis macrochirus*) compete aggressively for nest sites whilst others mimic females, sneak into territories and fertilise eggs (Dominey 1980).

Finding a mate at a resource

Searching behaviour has been discussed extensively in Chapter 6 and many of the mechanisms used for food-finding can also be used for mate-finding. Indeed, in many species mates are usually found by searching for a resource such as food and then waiting for a potential mate to find the same resource site. The tsetse fly *Glossina palpalis* in Africa rests in a tree until a large moving object passes near. Such objects are usually the ungulates upon whose blood they feed. Males and females meet by finding the same antelope, or other mammal and males often follow antelopes whilst waiting for a female to appear. Vehicles such as Land-Rovers are also large moving objects and are often followed by groups of male tsetse flies.

Another fly which finds its mate indirectly by searching for a resource is the dung-fly (*Scathophaga stercoraria*) which has been studied by Parker (1970, 1974, 1978). These flies feed on other insects but the eggs are laid in dung-pats. The males and females meet at the dung pats but the sex ratio is 4 or 5 males to each female so that females usually encounter many males when they reach a dung-pat. The females search for a dung-pat by olfaction, as soon as their eggs are mature. They mate once or twice and lay a batch of eggs at one visit. Dung is detected by smell more readily if it is fresh and the optimum time for egg-laying appears to be about one hour after the dung has been deposited. The male dung-fly, therefore, needs to find the dung as early as possible and to position itself so that any approaching female can be detected and approached before its rivals do so (Figures 137 and 138). A similar situation obtains for the European toad *Bufo bufo* in which males and females meet at the breeding pond (Eibl-Eibesfeldt 1950). The smaller males often wait for females on the land adjacent to a pond (Davies and Halliday 1979). Competition between males in both dung-flies and toads is discussed later in this section.

As discussed in Chapter 8, territorial behaviour often involves resource defence. Hence the first part of mate-finding in a territorial breeder is to find an area which includes adequate food and a nest-site. Similar habitat selection by members of the other sex will result in encounters between potential mates. The likelihood of encounter in territorial breeders is often increased by self-advertisement. Mate-finding may also be facilitated by returning to a known area to breed. Young animals may return to the breeding colony where they were reared, for prospective mates are likely to be found there. Older animals return for the same reason. Some birds return to the same nests where they encounter their mates of the previous season.

Detecting a mate at a distance

Where potential mates are sufficiently conspicious, searching behaviour can be directed towards them. Mates search for and find the singing males of many song birds, whales and crickets (Chapter 3), the odour-producing females of moths such as silk-moths *Bombyx mori* (Chapter 2) and mammals such as dogs, and the visual display producers such as the male grouse in leks (Chapter 8) or the male synchronously flashing fire-flies *Pteroptyx* (Buck and Buck 1976). The searcher merely has to move around in the appropriate area until the signal is detected. The advertiser has to find a suitable site for efficient advertisement. Any mechanism which renders the individual conspicuous to potential mates also increases vulnerability to predation so the advertiser has to make decisions in which the benefits of advertisement are balanced against the risks of pre-

137 Male dung-fly *Scathophaga stercoraria* waiting on grass near dung-pat.

dation. The functioning of communicative mechanisms has been extensively studied and discussed (Thorpe 1961, Marler and Hamilton 1966, Catchpole 1979, Lewis and Gower 1980). Animals which live in social groups may be surrounded by potential mates which are not usually in the appropriate physiological state for breeding. Mate-finding, in this situation, involves the recognition of sexual receptivity. Competition usually occurs before mating is possible.

Competing for a mate: dung-flies

Whenever there is competition between individuals for a resource there will be alternative behavioural strategies. The best strategy for any one individual in a given situation will depend upon the strategies employed by opponents. As explained in Chapter 1, for any problem involving competition it is useful to look for an evolutionarily stable strategy (ESS) (Maynard Smith 1976*b*). This approach has been used in each of the three examples of competition for mates quoted here.

A male dung-fly (Parker 1978) can mate by finding a female as she arrives at a dung-pat or by taking over a female from another male during mating or oviposition. Take-overs involve some cost, in that vigorous fighting is necessary but the last male to mate with a female displaces some of the sperm deposited during previous matings and fertilises 80% of the eggs. Mating takes about 35 min and the best oviposition time is about one hour after deposition of the dung. Hence during the first 20 min the male's behaviour is devoted almost entirely to maximising his chance of encountering a newly arrived female. Some females alight near the dung-pat and walk to it whilst others fly directly to it. Parker found that the proportion of females captured in various zones around the dung-pats was equal to the proportion of males searching there at this time. After the first 20 min, takeovers gradually become more profitable so the searching behaviour of males should be related to the distribution of copulating pairs as well as to that of female arrivals. Parker calculated the egg-fertilisation benefits from takeovers and from awaiting the arrival of new females, at intervals after dung deposition. He found that the frequency of occurrence of takeovers and of the searching behaviour necessary for takeovers, corresponded with the benefits. The males move their females away from the dung-pat during copulation if there are many other males nearby, thus reducing the likelihood of being dispossessed. The same effect results from their remaining passively on the females during oviposition.

138 Dung-flies mating on cow dung-pat.

Parker has applied Charnov's (1976) marginal value theorem (Chapter 6) to male behaviour. The questions at issue here are how long should a male continue to copulate and how long should he remain in the vicinity of a dung-pat before leaving to search elsewhere? The behaviour shown appears to be the ESS calculated for copulation duration, on the basis of the number of eggs fertilised. The calculated ESS for dung-pat leaving behaviour is (1) for all competitors to remain at the dung-pat initially but (2) for each individual to monitor availability of resources at the dung-pat in relation to those available elsewhere and to leave or stay according to the number of rivals remaining. It seems that rules of this kind do govern the dung-flies' behaviour. The results of this work vindicate the ESS approach but it must be remembered that in this, as in other studies, the logical calculation of the ESS depends upon the precision of the observations and measurements. It is necessary to know much about the limitations of the animals in order to calculate realistic strategies.

Competing for a mate: toads and deer

Mate-finding behaviour is of less importance than competitive ability for the male toad and is of very little importance to the red deer stag. The male toad searches for females on land and in water (Figure 139). When he finds a female he clasps her with his fore arms (amplexus) and is carried around for several days before the egg-strings are laid and fertilised externally by the male. Other males try to take over the female (Figure 140) and may be repulsed or may squeeze themselves between the male and female, thus dispossessing him. Davies and Halliday (1979) found that whilst all of 62 marked females spawned, only 20% of males did so. Small toads were able to pair with females arriving at the pond but large toads were more successful at takeovers within the pond. A male toad within the pond has the option of moving around searching for females or concentrating his efforts near the spawning grounds where females go when spawning is imminent. The ESS calculated for the average-sized toad depends upon the possibility of displacing a paired male, the numbers of paired and unpaired females in the pond and the number of pairs and males at the spawn site. Since the distribution of male toads fits the model proposed it seems that they do use an ESS which involves frequent sampling of the spawning area and of the remainder of the pond.

In competitions between male toads they appear to use the pitch of an adversary's croak when assessing body size, and hence fighting ability (Davies and Halliday 1978). Red deer stags also use sound in this way for they respond to the pitch of roar produced by a rival when deciding whether or not to engage him in battle (Clutton-Brock and Albon 1979). They also use body weight and the number of points on the

139 Male toads searching for females. After Davies and Halliday 1979.

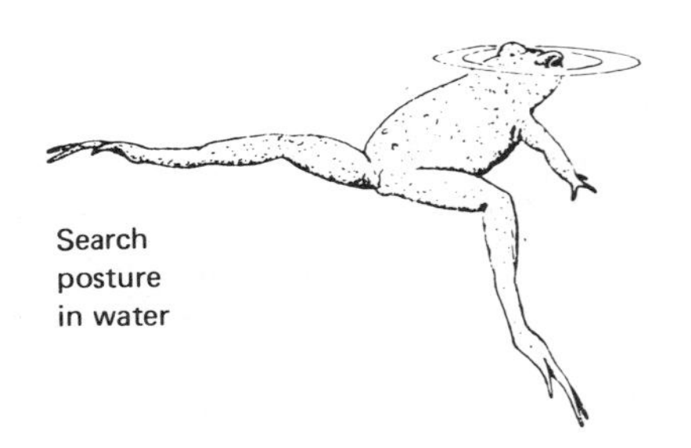

140 Male toads hold on to a female and defend against attacking males (darker) but they may be supplanted. Modified after Davies and Halliday 1979.

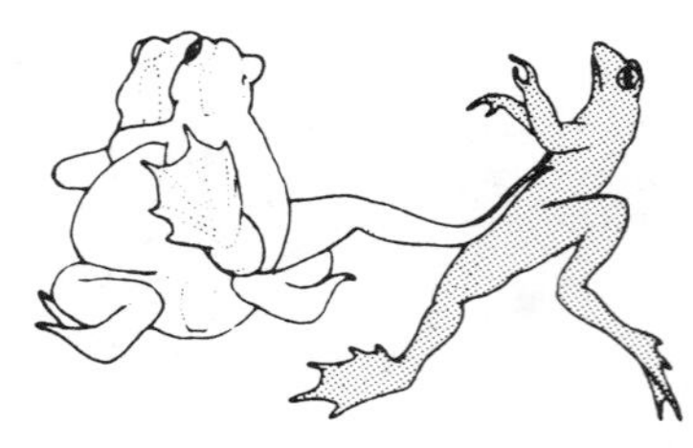

antlers when assessing a rival but the antler characteristics are not as important in this species as they are in the mule deer (*Odocoileus hemionus hemionus*) (Wachtel, Bekoff and Fuenzalida 1978). Those red deer stags which successfully defend or take over harems of hinds during the period of rut, sire offspring during that year, whereas few of the other stags mate at all. Clutton-Brock, Albon, Gibson and Guinness (1979) have estimated the costs and benefits of fighting. Stags avoid fighting if display suffices to drive off a rival and also if the rival seems very likely to win. The assessment of the likelihood of winning or losing, however, sometimes necessitates initiation of combat with antlers and some escalation of the fight. Fights of long duration, with consequent high risk of injury, occur only when many hinds are at stake. The ESS involves opponent assessment and decisions about fight escalation based upon this and upon the potential reproductive benefits.

Mate selection

If males expend time and energy actively fighting and competing for a female, then that female obtains a mate who is, on average, more able in certain respects than are the unsuccessful males. This is a form of passive selection by females. In many cases, a female is able to influence the likelihood that any particular male will mate with her by fleeing, preventing copulation or by attacking him. Males sometimes appear haphazard in their attempts to secure mates but they also show selection by refraining completely from approaching or courting some individuals and by preferring one although several are available. Mating selection mechanisms affect the chances that resources will be wasted by addressing sexual behaviour to the wrong species or the wrong sex. Mate selection also alters the likelihood of inbreeding, as against outbreeding, and provides an opportunity for obtaining a high quality mate.

Reproductive isolation, the avoidance of inter-specific hybridisation, may occur because of spatial separation from potential hybridisers, or mate-selection mechanisms, or by physiological mechanisms after copulation (Mayr 1963). An example of two very similar species whose males are readily distinguished during display are the colonial, polygynous weaver finches *Ploceus cucullatus* and *Ploceus nigerrimus* (Crook 1963). During pair formation the males show an advertisement display which emphasises those aspects of coloration which differ in the two species. The male *P. cucullatus* postures before the female with wings rigid and elevated, dashes to the nest and back repeatedly, then sings. In contrast, the male *P. nigerrimus* sings at the female with wings quivering and then dashes to the nest and back. Hybridisation between the species is very rare.

In many birds and in some insects, amphibians and mammals, females recognise males of their own species by specific characteristics of their vocalisations (Marler 1957, Konishi 1970). Preferences for mates which are of the same subspecies, breed or colour form have also been demonstrated. Lill and Wood-Gush (1975) found preferences amongst domestic fowl for mates of their own, rather than other, strains, whilst Barlow and Rogers (1978) showed that normal and gold morphs of the midas cichlid fish (*Cichlasoma citrinellum*), whose parents were of the same colour as themselves, chose mates of that colour. Similar results were obtained by Cooke, Finney and Rockwell (1976) for the blue and white morphs of the lesser snow goose (*Anser caerulescens*). Even in the absence of obvious anatomical differences between forms, the freshwater form of the three-spined stickleback (*Gasterosteus aculeatus*) preferred to mate with the same form, rather than with a form which came into the same stream to breed but which spent the rest of its life in the sea, i.e. an anadromous form (Hay and McPhail 1975). The advantage to an individual stickleback of showing such *assortative mating* is that its offspring will survive better in the habitat for which they are best adapted if they are not hybrids. The nature of the advantage of assortative mating is not always clear, however.

Inbreeding or outbreeding

Many animals show preferences for mates of the same species, breed or morph but few choose mates which are close relatives. Mating with parents, offspring or siblings is called incest. The disadvantage is usually considered to be the increased homozygosity which will occur in the offspring, since deleterious recessive genes are then likely to be expressed. Evidence for such effects in man and other species is reviewed by Crow and Kimura (1970) and Bodmer and Cavalli-Sforza (1976). As explained in Chapter 1, all individuals of a species share some genes. The proportion of genes shared will be greater in small than in larger populations, i.e. the degree of *inbreeding* is higher.

One argument in favour of some inbreeding is that useful combinations of genes are not broken up by inbreeding as many are when very dissimilar individuals mate. This has already been implied as a disadvantage of inter-specific breeding. This advantage may account for some of the mate choice by colour morphs etc. within a species. Smith (1979) has pointed out that this advantage could lead to the normal occurrence of incest in populations where the effects of homozygous deleterious genes are small. He quotes the example of the fallow deer *Dama dama* in Britain where small park herds have been maintained for centuries so that deleterious genes have disappeared from the population as their homozygous bearers have died. Since males return to a rutting stand and daughters seem to accompany their mothers to rutting stands it is likely that mating between fathers and daughters is frequent. Other examples of behavioural mechanisms which increase the likelihood of breeding with close or distant relatives may have been overlooked because it has been assumed that all inbreeding is disadvantageous. There is no doubt that many long-lived species avoid incest when they are selecting a mate and young animals of one or both sexes may disperse away from the parents and siblings at or before sexual maturity (Itani 1972, Wilson 1975). However, as Bateson (1979) points out, dispersal and mate-selection mechanisms may result in the choice of a mate which is distantly related rather than one which is closely related or unrelated. The optimum degree of relatedness at any time would vary from species to species according to the recent history of the population and the degree of divergence in specialised characteristics which can be tolerated before there is a reduction in the fitness of the offspring.

Mate quality

Selection of mates of high quality should result in better quality offspring which will themselves be likely to produce more offspring than if no mate selection had occurred. In species in which the mate shows parental care, it will be advantageous to select one which will be good at care-giving behaviour, for this will increase the chances of offspring survival and may also increase the likelihood that the selector will survive to produce more offspring at a later date. The selection may be carried out, as mentioned previously, by awaiting the result of a contest and mating with the victor, or individuals may be selected on the basis of some anatomical or behavioural characteristic. Where males defend territories, females select territory holders rather than males with no territory. In species like the red deer, females are most likely to mate with the harem holder. It is sometimes assumed that females are largely passive in a harem situation but even in a species like the elephant seal (*Mirounga angustirostris*), in which 4% of males achieve 85% of matings (Le Boeuf 1974), females show clear selection. As shown in Figure 141, females are much less likely to protest if the male which is at the head of the competitive order mounts her than if another male does so. Since protests increase the likelihood that the mount will be interrupted before copulation is completed (Figure 142) the female's behaviour reduces the chance that subordinate males will father her offspring (Cox and Le Boeuf 1977). In order to be able to discriminate against subordinate males, the females must be able to recognise them. The males of the cockroach *Nauphoeta cinerea* establish competitive hierarchies at high population densities (Ewing 1972) and females select males which produce a different odour from subordinate males (Breed, Smith and Ball 1980). In primates and most other group-living animals, the behaviour of the victors after competitive encounters between males is different from that of the vanquished. If females can detect this difference and can subsequently recognise individuals the information can be used in future mate choice. For male primates there are sometimes positive correlations between rank in a competitive order and the proportion of copulations achieved, e.g. in baboons (*Papio cynocephalus*) (DeVore, 1965*a*) and langurs (*Presbytis* sp.) (Jay 1965). In other studies however, e.g., on bonnet macaques (*Macaca radiata*) (Simonds 1965) and rhesus monkeys (*M. mulatta*) (Loy 1971), there was no such correlation. Struhsaker (1967) found that the presence or absence of this correlation varied from troop to troop of vervet monkeys (*Cercopithecus aethiops*) and Wallis (1979) has shown that social status and matings achieved with females varies during the course of a year in mangabeys (*Cercocebus aterrimus*).

The direct selection of mates on the basis of their displays has been described by Tinbergen (1951) and many others. If the vigour of a male's display is not sufficient to indicate his quality as a potential mate he will not be chosen by a female which has a chance of finding an attractive mate. The likely parental ability of a male must often be judged from the quality of his courtship and Erickson and Martinez-Vargas (1975) argue that, in doves at least, such judgement by the female is efficient since the parental and courtship behaviour depend on the same hormones. In some animals, however, the male's reproductive effort is more tangible for it involves a resource-containing territory, which the male has defended, or courtship feeding. The males of some spiders and insect predators bring a prey item to the female. Acceptance of the male by the female hang-fly (*Bittacus apicalis*) depends upon the size of the prey (Thornhill 1976). If it is large the female takes longer to eat it, so copulation lasts longer and egg production is accelerated. The female is sustained for longer by a larger meal and the quality of the male which provides it must be higher. Females

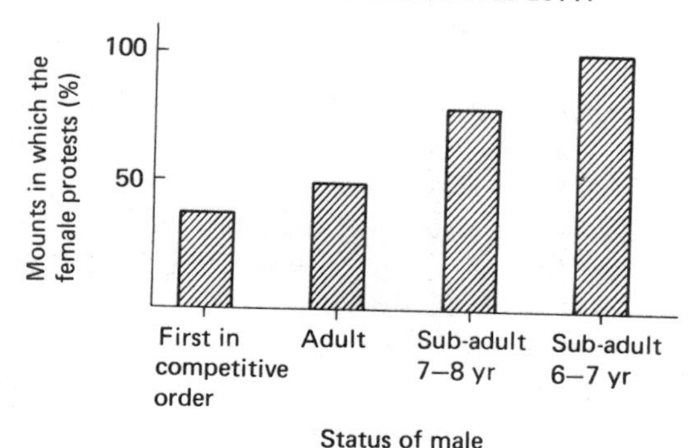

141 The protests of female elephant seals (*Mirounga angustirostris*) when mounted, on the breeding grounds, are inversely related to the status of the male. Modified after Halliday 1978; data from Cox and Le Boeuf 1977.

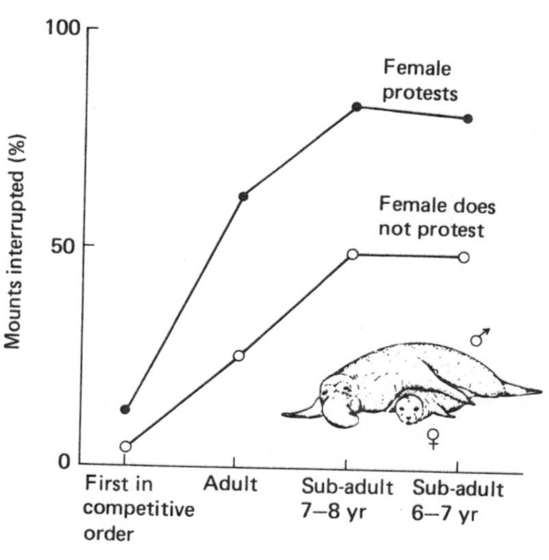

142 Mounts by elephant-seal males are more likely to be interrupted if the male is subordinate and if the female protests. Modified after Halliday 1978; data from Cox and Le Boeuf 1977.

of the common tern (*Sterna hirundo*) also use courtship feeding by males when selecting a mate and Nisbet (1973) has found that there is a positive correlation between the amount of food fed to the female during courtship and the amount of food fed to the young by the male.

Effects of experience on mate selection

The mate-selection mechanisms for most of the species mentioned above involve sophisticated discriminations between individuals. It is not surprising that experience plays a considerable part in the development of the requisite discriminatory ability. In order to avoid mating with very close relatives, individuals need to be able to learn the characteristics of their relatives. If very different individuals are to be avoided, a level of familiarity must be established so that the unfamiliar are recognised. When assessing the quality of a mate, for example by vigour of display or by gifts provided, previous observations of display and knowledge of the difficulty of acquiring the gifts could be used. If a mate proves inadequate, knowledge of the characteristics of that mate may be of value when selecting another mate.

The effects of early rearing experience on mate selection have long been known to animal breeders and were described by Whitman (1919). When Whitman allowed a male passenger pigeon (*Ectopistes migratorius;* now extinct) to be reared by ring doves (*Streptopelia risoria*) he found that it would mate only with ring doves. In observations on a variety of waterfowl, Lorenz (1935), Fabricius (1951) and Ramsay (1951) described similar occurrences. An extensive study of ducks was carried out by Schutz (1965). Male mallard (*Anas platyrhynchos*) and other sexually dimorphic, dabbling ducks which were reared with their own species until up to 21 days and then with another species until 49 or 70 days, showed most sexual behaviour, when mature, to the other species. Females of these species, however, whatever their rearing conditions, reacted to male courtship and usually accepted their own species. Male courtship patterns appeared to be the same irrespective of the species to which they were directed (de Lannoy, 1967). Males which had been completely isolated for their first nine weeks and males reared in a group of their own and other species, addressed most of their courtship behaviour to females of their own species. In the Chilean teal, (*Anas flavirostris*) whose sexes are alike and conspicuous, both male and female modify sexual preferences as do male mallards.

Even in experiments with birds reared by hand or with moving models, the rearing conditions modify sexual preferences. Guiton (1966) reared cockerels in groups or by hand in isolation. After 47 days, testosterone-injected birds would show courtship behaviour which was directed preferentially towards a gloved hand in the isolation-reared birds. After three days grouping with pullets, however, these cockerels were equally likely to court a hand or a pullet and after six months social rearing they mated only with pullets. Schein (1963) and Schleidt (1970) reported similar displays by turkeys and Kruijt (1971) described copulation by male jungle fowl (*Gallus gallus spadiceus*) with the artificial models with which they had been reared. Klinghammer (1967) found that hand-reared male and female ring doves would select a human hand as sexual partner (Figures 143, 144). He also showed that male mourning doves (*Zenaidura macroura*) separated from their parents before day 8 and then hand-reared would court a human hand when sexually mature unless kept with female mourning doves from day 52 onwards.

It is apparent from a variety of studies of birds that mate preferences are affected by exposure more at certain ages than at others and more in males than in females. In studies of zebra finches (*Poephila castanotis*) Immelmann (1965) reared males with society finches, a subspecies of the Bengalese finch (*Lonchura striata*), until 33–66 days. He then isolated these males until they were mature at 120 days when he put them with a female of each species. They displayed to and attempted to mate with a female society finch rather than a female of their own species.

Later work by Immelmann (1972, 1977) showed that the effects of early rearing experience on mating preferences could be reversed. If male zebra finches were kept with Bengalese finch foster parents for less than 40 days and were then put with female zebra finches for three or seven days, some of them showed later preferences for zebra finches. A brief exposure like this did not reverse the preference if the time spent with the foster parents was more than 40 days. A longer exposure to female zebra finches, lasting 30 or 60 days, did reverse the

143 Hand-reared male ring-doves (*Streptopelia risoria*) show bow-coo displays towards a human hand and will copulate with it. Modified after Klinghammer 1967.

144 Female ring doves invite courtship and copulation from a hand if they have been hand-reared. Modified after Klinghammer 1967.

preference provided that the time with foster parents was not more than 71 days (see Immelmann and Suomi 1981).

Female zebra finches reared with society finches in the same way as the males, by Immelmann (1965), would accept male zebra finches which courted them. In a detailed study, Sonnemann and Sjölander (1977) showed that mate selection by females, however, was affected by early experience. Some female zebra finches reared with Bengalese finches for 40 days responded to male zebra finches. Others responded to male Bengalese finches which had been reared by zebra finches and hence courted them, whilst others would mate with neither. In multiple-choice tests these zebra finch females spent most time in front of male Bengalese finches.

All of these results show that there is considerable plasticity in the system but that such a mechanism does make it likely that individuals of one or both sexes will try to mate with an animal like their parents or siblings. The system must also account for the observation that actual parents and siblings are not chosen as mates. Bateson (1978) reared male and female Japanese quail in groups of ten for 35 days and then isolated them. When mature the males were tested for their preferences among brown females. Some of these were the ones with which they had been reared, some were unfamilar brown females and others were unfamilar white females. Using measures of approach and of copulation, males were shown to prefer unfamiliar brown females and to avoid white females. Gallagher (1976, 1977) has also shown that males reared with white females prefer them as mates. Bateson proposed that there might be an optimal level of discrepancy from the familiar when selecting a mate (Figure 145). If the rearing group was large, however, unfamiliar individuals were not preferred, possibly because they resembled one or more members of the rearing group (Bateson, 1981).

145 The familiarity of individuals (or objects) which is established during rearing affects mate selection when sexually mature. It is postulated that there is an optimal discrepancy, in terms of difference from the familiar, such that mating does not normally occur with very familiar individuals and the least familiar individuals are avoided. A flatter top to the curve seems more likely. Modified after Bateson 1979.

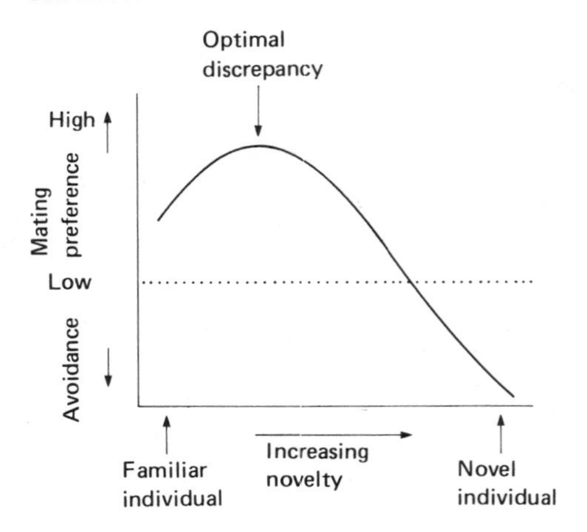

The age at which the characteristics of the parents might be learned could be as early as the sensory and brain mechanism development would allow. If siblings are to be recognised, however, it may be necessary to see them after they develop some of their adult characteristics. Bateson (1978) has drawn attention to the early stage at which such recognition appears to occur in the fast developing Japanese quail (Gallagher 1977) in contrast to the later stages in cockerels (Vidal 1976) and ducks (Schutz 1965) at which exposure affects later mating preferences. Effects of early experience on later mate selection have been described in fish (e.g. Kop and Heuts 1973, Barlow and Rogers 1978), and various mammals including man (Mainardi, Marsan and Pasquali 1965, Gilder and Slater 1978, Shepher 1971) but there is still much scope for detailed tests of preferences in relation to previous social experience.

Functions of courtship

Courtship is behaviour which indicates willingness to pair, with subsequent mating, and which is sufficiently different from other behaviour to allow a potential mate to recognise it. The extensive studies of courtship by ethologists (see Bastock 1967) have led to many ideas about the evolution of behaviour and the control of behaviour in individuals. The major functions of courtship are to allow mate recognition, to encourage acceptance and to initiate and synchronise reproductive mechanisms. The duration and complexity of courtship varies greatly from species to species and the components of courtship displays may have several functions and may be used in several different situations (Beer 1975). One frequent area of overlap is where the same displays are directed at rivals as well as at potential mates. Communication between rivals, such as that between male Siamese fighting fish (*Betta splendens*) (Simpson, 1968), is similar in many respects to that between males and females.

Attracting and holding attention

The first stage in courtship is to attract attention. Next, species, sex and reproductive state must be made clear to the potential mate. The necessity for such advertisement and responses to it during mate selection have been described earlier in this chapter. Even after a mate has been found, very conspicuous displays are used to attract the attention of the mate. A male peacock may spread his tail or a female old-world monkey may present her bottom to a male. These displays also serve another important function, emphasised by McKinney (1975), which is to indicate who is being courted. The monkey's presentation is directional. The peacock may orient himself towards a female and drakes in the genus *Anas* emphasise this in their grunt–whistle display by splashing water as they orient (Figure 146). In man, binocular contact is used to inform an individual that he or she is being courted.

Attention-holding displays occur in very many species and are usually continuous chains of sounds or movements like the zig-zag dance of the male stickleback. At this stage in display some assessment of mate quality can occur so elaborate movements which emphasise vigour, speed, etc. may be shown. There is scope for more individual variation in attention-holding displays than there is in those displays in which the displayer must be recognised as being of the right species and perhaps as not being a close relative. The communication of such information in displays is discussed by Hinde (1974) and the control of movements during displays is described in Chapter 3.

Close approach and leading

The minimal responses of a female to a male's display are to refrain from attacking him and to remain in his vicinity. The next objective of a male which has succeeded in reaching this stage is to lead the female away from his competitors to a place where courtship can continue. In sticklebacks the male leads the female to a nest and ducks may use short flights (McKinney, 1975) or what Johnsgard (1960) calls 'turn-back-of-head'. In human courtship, either sex may initiate such leading and, as in many other species, assessment of mate quality continues throughout this phase. The males of some species, such as dung-flies (Parker, 1978) and hamadryas baboons (*Papio hamadryas*) (Kummer, 1968) forcibly remove females from areas

146 The grunt-whistle display of the yellow-billed pintail (*Anas georgica*). Redrawn after Johnsgard 1965.

where disturbance by other males is more likely and legends of cave-men would have us believe that the same behaviour was frequent in man. Even where the male uses force, the initiation of contact between male and female first requires that the female allows the male to come within a distance which will enable him to catch her. In this situation, and in those where each sex allows the close approach of the other without force being used, the normal barriers to encroachment upon individual distance (Chapter 8) must be overcome. This is another important function of courtship. During the display period, each individual must reassure the other that no injury will result from close approach. Sometimes this reassurance is facilitated by the presentation of gifts, either by the stronger to the weaker as in gulls, or by the weaker to the stronger as in male spiders or mantids to their larger females. In the latter cases, males can mate whilst the female devours the gift but may be eaten before mating, if there is no gift, or during mating whilst insemination continues if the gift is not large.

Effects on hormone levels

Courtship has a direct effect on the behaviour of the mate but it may also have indirect effects, via hormonal changes. The acts of courtship may also be necessary to bring the displaying individual into full reproductive state and it may affect neighbours in a colony so as to increase the likelihood that breeding is synchronised (Beer 1975). The fact that male courtship could initiate ovulation in female doves was discovered by Craig (1911). Extensive experimentation by Lehrman and his collaborators (Lehrman, Brody and Wortis 1961, Lehrman 1965) have shown how a courting male and presence of nesting material induce ovulation and incubation behaviour (Figure 147, 148). The amount of courtship by a male has direct effects on ovarian development and on the number of ovulations (Erickson and Lehrman 1964, Erickson 1970) and one hour of courtship by two males has a greater effect than one hour by one male (Barfield 1971*a*). The bow-coo (Figure 143) and other displays are described for different stages of the ring dove reproductive cycle by Lovari and Hutchison (1975) and their precise effects on female hormones and behaviour are reviewed by Cheng (1979). Female reproductive develop-

147 When male ring doves (*Streptopelia risoria*) were put with a female and started to show courtship (day 0) ovulation was induced. The presence of nesting material accelerated the onset of ovulation and no ovulation occurred in the absence of a male. Modified after Lehrman *et al.* 1961.

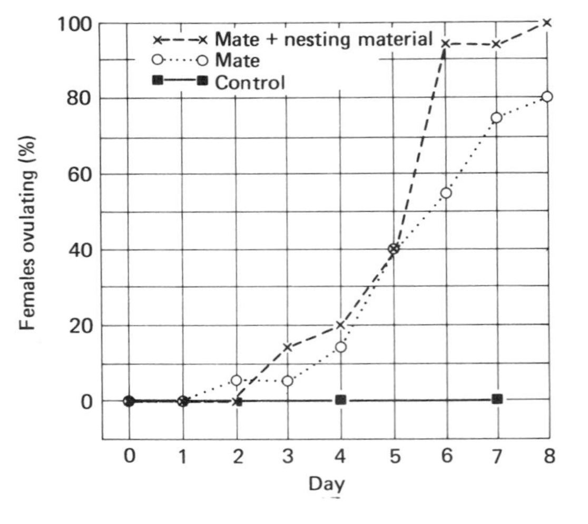

148 In the same experiment as that in Figure 147 the time course of the development of incubation behaviour was very similar to that for ovulation. Modified after Lehrman *et al.* 1961.

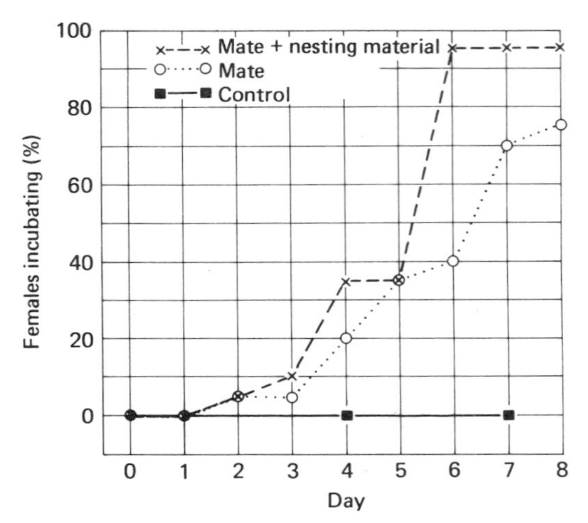

ment is accelerated by male display in other birds such as canaries (Hinde and Steel 1966, 1978) and budgerigars (Brockway 1969). Male sexual development in budgerigars is also accelerated by hearing the warbling songs of other males. In mammals such as rats and man, ovulation occurs in the absence of male courtship but in rabbits and many marsupials, insectivores and carnivores, copulation is a necessary prelude to ovulation.

Development of courtship behaviour

Sexual maturity is a prerequisite for most of the functions of courtship discussed above. Except for the possibilities that practice might be important in the development of courtship, or that pairing might occur before sexual maturity, there is no necessity for most components of courtship to be shown before mating is imminent. Developmental studies of many species show, however, that courtship movements start to appear in the behavioural repertoire at an early age. Kruijt (1964) describes in detail how sexual behaviour develops in jungle fowl and also reports that males reared in isolation for ten months are unable to perform normal copulatory behaviour. Similar inadequacies have been described for guinea-pigs (Valenstein, Riss and Young 1955), rats (Gruendal and Arnold 1969), cats (Rosenblatt 1965) and rhesus monkeys (Mason 1960, Harlow 1965), reared with varying degrees of social deprivation. The attempts at courtship and mating shown after social deprivation earlier in life include incorrect orientation, abnormal sequences of movement and some atypical movements. It is difficult, however, for an observer to differentiate between modification of mating preference, with consequent incomplete courtship, and real inability to show courtship and mating behaviour. The effects of experience, during development and when adult, on mating behaviour in mammals is reviewed by Diakow (1974). Experiments on the effects of experience on developing motor patterns in courtship, such as bird song, are discussed in Chapter 3 and by Marler and Hamilton (1966).

The control of courtship and mating

Sexual behaviour occupies a small proportion of the lives of most individuals and as mentioned at the beginning of this chapter, in some species it may not occur at all during seasons which are unfavourable for offspring production. Nevertheless, some of the components of the physiological control systems are present all the time. The likelihood of sexual behaviour by intact animals can be increased by the injection of hormones or by brain stimulation provided that an appropriate sexual partner is present. The sexual behaviour control system depends for its function on complex sensory input and on hormones as well as on brain mechanisms. Many of the components of the system must be inhibitory. Some of the motor patterns can be very similar on successive occasions (see Chapter 3) but complex decisions must be taken about the combination of different motor patterns and about the timing of the behaviour so various aspects of previous experience must influence these decisions (Rosenblatt 1965).

The location of control centres

Complete mating sequences can be carried out by decapitated male insects and this allows copulation to be completed in those species, such as mantids, in which the female may eat the male (Roeder 1963). Even in mammals the centres controlling the patterning of some male and female mating movements are located in the spinal cord (see review by Beach 1967). Hart (1967) severed the spinal cord of male dogs below the brain and subsequently observed penile erection, pelvic thrusting, back arching and ejaculation. Not all species can carry out copulatory movements without brain control for when Hutchison and Poynton (1963) transected the brain of the clawed toad (*Xenopus laevis*) behind the medulla, no clasping and mating occurred.

In mammals, copulatory movements are inhibited at most times by the cortex and the hypothalamus. Sexual behaviour by male rats is abolished by lesions in the preoptic region (Heimer and Larsson 1967) but is facilitated by lesions caudal to the hypothalamus (Heimer and Larsson 1964, Goodman, Jansen and Dewsbury 1971). In female rats, preoptic lesions increase sexual behaviour (Powers and Valenstein 1972) so the control system is not located in the same place in males and females. Lesions in the temporal lobe of male cats lead to attempts at copulation with various objects or animals as well as with female cats (Green, Clemente and de Groot 1957) so mating behaviour must depend upon the normal functioning of this region of the cortex. Brain stimulation experiments emphasise the importance of the hypothalamus in sexual behaviour but, as in the lesion experiments, the role of the various parts of the hypothalamus is not clear. Caggiula and Hoebel (1966) elicited sexual behaviour by male rats towards females when the posterior hypothalamus was stimulated electrically and Roberts, Steinberg and Mears (1967) obtained exaggerated sexual behaviour towards females and objects when they stimulated the anteromedial hypothalamus of a male opossum. Recording studies show that electrical activity in the lateral hypothalamus coincides with sexual behaviour or stimulation (Porter *et al.* 1957, Barraclough and Cross 1963).

It seems that the patterning, co-ordinating and directing of mating behaviour in mammals is controlled by a system with components in the spinal cord, hypothalamus and temporal cortex. Courtship behaviour, however, is controlled by a variety of other parts of the brain whose locations are little known. The research which has been carried out yields complex results and with the possible exception of the control of song and other vocalisations in birds (Phillips and Peck 1975, Nottebohm *et al.* 1976), is difficult to interpret unless longer term effects of hormones are also studied.

The role of hormones

How to study hormones

Ways of investigating how hormones are involved in the regulation of behaviour are listed here and are exemplified later in this chapter.

1. If there are large differences in the secretory activities of glands or neurosecretory centres, it is usually possible to distinguish fluctuations in the overall size or in the histology of those secretory areas and to relate them to the behaviour at the time.
2. In recent years it has become possible to measure hormone levels in the blood, e.g. by radioimmunoassay (Follett, Scanes and Cunningham 1972) and to relate those levels to behaviour and physical changes.
3. If the area which secretes a particular hormone is removed, the effects of the lack of further secretion of the hormone can be assessed.
4. Hormone inhibitors, such as the anti-androgen cyproterone acetate, can be injected into an individual and the effects of a temporary absence of the hormone studied (Zucker 1966).
5. Synthesised hormones, or those obtained from other individuals, can be injected into the body of an individual.
6. Hormones can be implanted in known areas of the brain so that they diffuse slowly away from the implant (Harris, Michael and Scott 1958).
7. The hormones from two individuals can each be supplied to the other by simultaneous cross-transfusion of blood so that the animals are in parabiosis (Terkel and Rosenblatt 1972).
8. Radioactively labelled hormones can be injected into an animal and their target area in the brain discovered by autoradiography (Michael 1961).

More than one of these techniques, together with detailed studies of behaviour and brain stimulation, ablation or recording, are often used in conjunction. For example, many studies in which sex hormones are injected or implanted are accompanied by gonadectomy so that the effects on behaviour of the injected or implanted hormone can be assessed.

Some ways in which hormones affect behaviour

Hormones affect growth and metabolism but there are various ways in which they have more direct effects on behaviour (Lashley 1938, Hinde 1970). Recent work suggests that effects of hormones on specific mechanisms within the brain may be the most important way in which hormones affect behaviour (Hutchison 1976, 1978) and such work will be discussed in detail later in this section. During development, hormones exert an influence in a different way by modifying the growth of the nervous system. This area of research is reviewed by Joffe (1969) and Beach (1975).

Another way in which the brain might be influenced by hormones is by affecting organs so that there is a change in sensory input. An increase in the density of cornified papillae on the glans penis of male rats occurs when androgen levels rise and coincides with an increase in copulatory behaviour (Beach and Levinson 1950). Following castration, with associated reduction in androgen levels, the papillae disappear (Phoenix, Copenhaver and Brenner 1976). The sensory input has not been measured, in this case, but Komisaruk, Adler and Hutchison (1972) and Kow and Pfaff (1973) have demonstrated that sensory input from the pudendal nerve of female rats is increased by oestradiol treatment. Other hormone-induced changes in anatomy, however, may not result in changes in sensory input. Hinde, Bell and Steel (1963) and Steel and Hinde (1964) showed that increases in the sensitivity of the brood patch in female canaries coincide with increased oestrogen levels but Hutchison and Konishi (personal communication) have found that oestrogen does not alter the sensory output of tactile receptors so the change in sensitivity must be due to a central effect.

Other ways in which hormones affect behaviour are the general effects on the individual's ability to engage in vigorous activity which are caused by adrenal hormones and the effects on anatomy which alter the response of other animals to that individual. An example of the latter type of effect is the oestrogen-induced production of odoriferous substances by female rhesus monkeys. Michael and Keverne (1968) showed that males would press a lever 250 times in order to gain access to an unfamiliar female which he could smell. He would not do so if she produced no odour or if he was deprived of his sense of smell.

Oestrus cycles and behaviour

In studies of mammalian reproductive behaviour, the oestrus cycle of the female is a major variable. In most domestic dogs, the period when bitches are receptive to males occurs once every six months and lasts for one to two weeks. Female rats are receptive during one day in a cycle lasting four to five days and their activity is greatest on that day (Figure 149). The menstrual cycle is 28 days in women and in rhesus monkeys and 34–35 days in chimpanzees. During this time there are fluctuations in the likelihood of copulation but no precise days of oestrus. Diamond, Diamond and Mast (1972) showed that visual sensitivity in women is greatest at the time of ovulation but studies of human sexual behaviour show that there is much variation in the time of the menstrual cycle at which copulation is most likely. Female rhesus monkeys, which could obtain access to their male partner by pressing a lever, did so earlier at the time of ovulation. Males mated with them most at this time (Figure 150, Keverne 1976). Females of some primate species show considerable swelling of perineal sexual skin at one time during their cycle whilst others, e.g. women and baboons, show visible menstrual bleeding (see review by Chalmers 1979).

The three characteristics of female mammals which affect the likelihood of successful mating behaviour were summarised by Beach (1976) under the headings attractivity, proceptivity and receptivity. Attractivity, or *attractiveness*, is measured by assessing the extent to which she evokes sexual responses from a male. *Proceptivity* refers to invitation behaviours by

149 Five days during the oestrus cycle of a female rat showing that activity counts reach a peak at the time of receptivity to males and the occurrence of cornified epithelial cells in vaginal smears. Modified after Munn 1950.

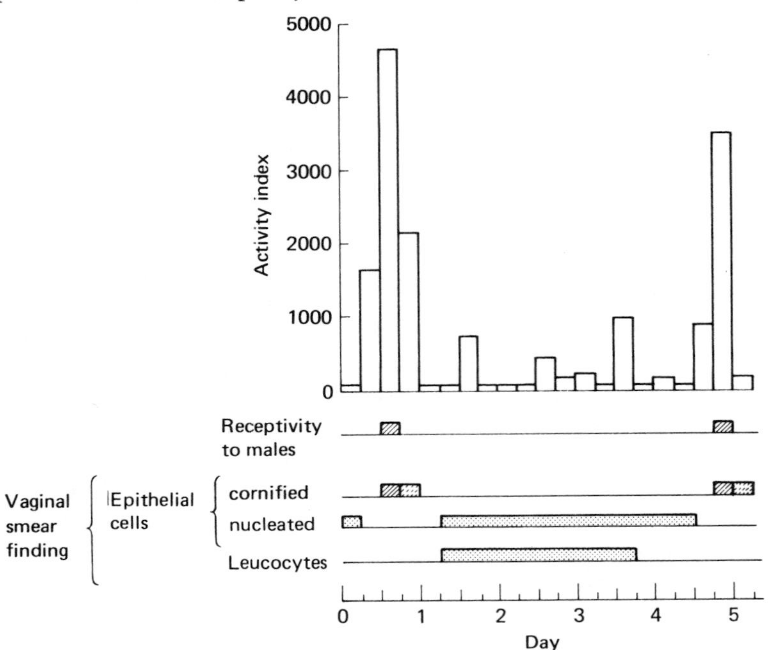

150 The upper line shows that female rhesus monkeys which could gain access to a male by pressing a lever, did so earlier at the time of ovulation. The histogram shows that male ejaculations were more frequent during the middle part of the cycle. Modified after Keverne 1976.

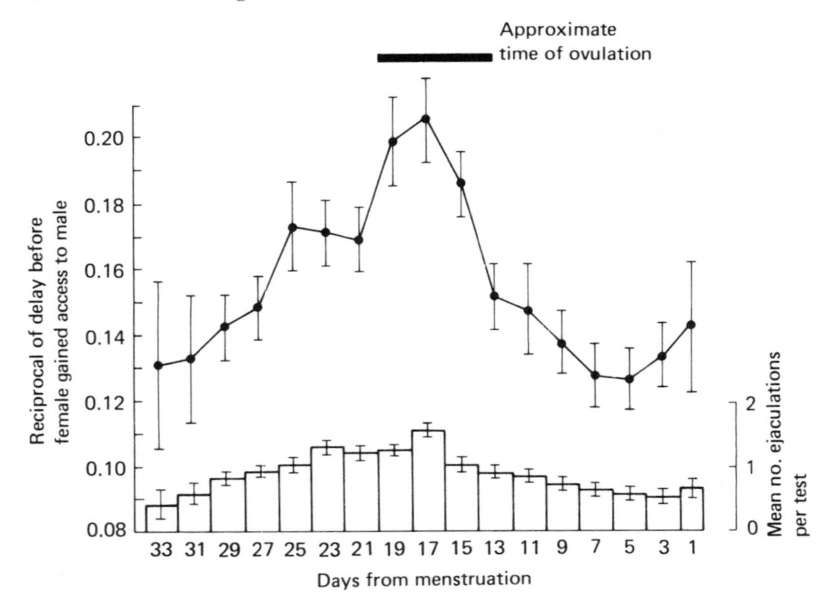

means of which females solicit copulation by a male. The responses of a male will be affected by such soliciting behaviour but they also depend upon the odours which she produces and her other physical characteristics. *Receptivity* concerns the willingness of the female to accept male copulatory attempts. These characteristics of female mammals are differentially affected by oestrogen, progesterone, luteinising hormone releasing factor and adrenal androgen, according to species (Johnson and Phoenix 1976, Slater 1978).

Effects of hormones on brain function

Hormones which are released into the blood can alter behaviour by effects on specific areas of the nervous system. The effects of oestrogen, progesterone and testosterone on neural activity in the brain have been reviewed by Komisaruk (1971) and by Pfaff, Lewis, Diakow and Keiner (1973) whilst the work of Hart (1973) suggests strongly that testosterone acts directly on mechanisms in the spinal cord which control some aspects of mating in male rats. Since many effects of sex hormones on behaviour continue for several days, the best way to investigate their effect on the brain is to implant pellets containing hormone into a part of the brain so that the hormone will gradually diffuse out of the pellet (see Figure 151). This technique was pioneered by Harris *et al.* (1958) who found that stilboestrol, a synthetic oestrogen, induced copulatory behaviour if implanted in the anterobasal hypothalamus of female cats. Subsequent work by Lisk (1962) showed that the anterior hypothalamus was similarly sensitive in ovariectomised rats, but in rabbits the oestrogen-sensitive area is in the posterior

151 Three sections through part of the brain of a male dove showing the positions of testosterone propionate implants. In the most anterior section (on the left) complete courtship occurred if the implant was in the pre-optic region. The anterior hypothalamus is also involved in the control of courtship but the posterior hypothalamus (right hand section) plays little part. Modified after Hutchison 1976.

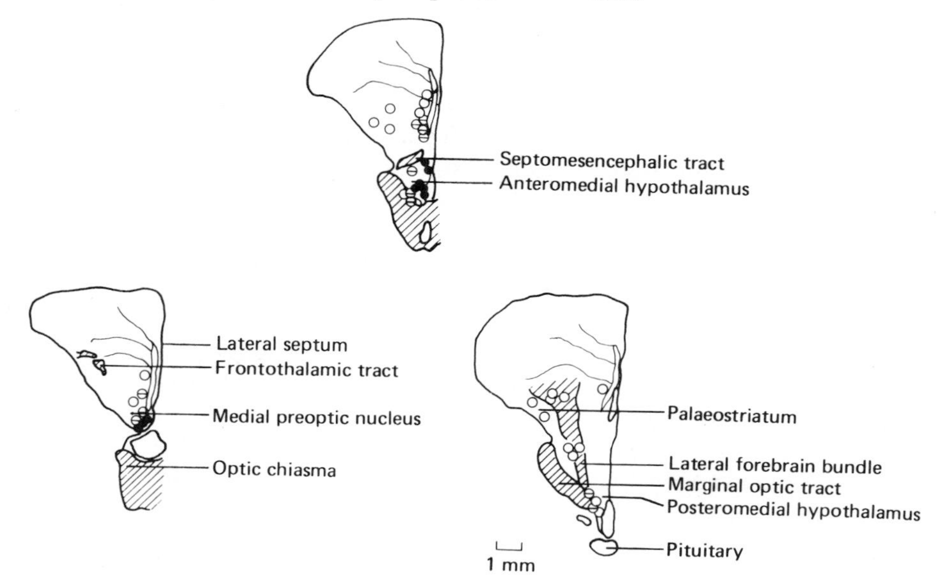

hypothalamus (Palka and Sawyer 1966). Male copulatory behaviour could be induced in rats by implants of testosterone, in the form of testosterone propionate, in an area between the medial preoptic nucleus and the posterior hypothalamus (Davidson 1966, Lisk 1967).

In the male ring dove, androgen sensitivity associated with copulatory behaviour is located in the preoptic and anterior hypothalamic areas (Barfield 1971*b*). Hutchison (1967, 1969, 1970, 1974, 1976) has described the effects of testosterone-propionate implants in the preoptic and hypothalamic region on the courtship behaviour of male doves. Castrated doves show a little chasing, bowing and nest soliciting immediately after castration but these displays disappear by 20 days later. Preoptic and anterior hypothalamic implants of testosterone, 15 or 30 days after castration, restored courtship behaviour but later implants had less effect, i.e. the sensitivity to androgen declines after castration. The birds were also less sensitive to androgen implants in winter than in summer. The regions of the brain in which implants affected courtship are shown in Figure 151. Hutchison was able to demonstrate that aggressive components of courtship, such as chasing, required higher hypothalamic concentrations of androgen before they occurred than did nest-oriented components. The hypothalamus must contain a population of steroid-sensitive cells but Hutchison (1976) proposes that they act by collectively affecting systems outside the hypothalamus which are concerned with the integration of sexual behaviour.

Parent–offspring interactions

Fish like the cod which release their eggs and sperm in the sea, are not likely to recognise or interact with their offspring. Their parental investment (see earlier in this chapter) is limited to the energy required to produce the eggs or sperm and they show no parental care. The amount of parental care shown by the members of a species varies from nil to the prolonged and energetically expensive care of many birds and mammals (see Chapter 1). As emphasised in the section on mating systems, the relative magnitude of parental investment by male and female has effects on the likelihood of monogamy or polygamy. The relationship between parental investment and the probability that an individual will show parental care or desert its mate and offspring, is discussed further in the next section. The subsequent section summarises the different aims of parents and offspring and the extent to which they might conflict. Some detailed examples of interactions between offspring and their parents are then discussed. Finally, before considering economic aspects of the study of reproductive behaviour, the effects of experience on parental behaviour and the evidence for hormonal mechanisms are reviewed.

To care for or to desert?

In extreme forms of polygamy, one member of a pair makes no parental investment after gamete release. The competition between the sexes is clearly won by one of them, in the sense that it is free to attempt to procreate elsewhere whilst its mate shows parental care. In many species of birds and mammals, however, both members of a pair make some investment in their offspring after gamete release. There is variation amongst species in this, as already discussed, and there is also individual variation in the relative contribution of the two members of a pair. Trivers (1972) proposed that decisions, at any time, about the amount of parental care to show, should depend upon the amount already invested. Dawkins and Carlisle (1976) pointed out that this is a fallacious argument since future expectations of reproductive costs, rather than past investments, should affect behaviour.

The decision about whether to desert the mate and offspring should depend upon the likelihood of the survival of the offspring if deserted and the chances of producing more offspring as a consequence of the desertion. This idea has been explored by Maynard Smith (1977, 1978*a*,*b*) and by Grafen and Sibly (1978). Since the optimal behaviour for one parent depends upon what the other parent is doing it is useful to look for an ESS. It is necessary to be aware, however, that this ESS may be so complex, due to dependence on many variables, that it may result in much individual variation.

Maynard Smith (1977) lists four main factors which affect whether or not there will be parental care and which parent will provide it. These are (1) the effectiveness of parental care by one or two parents, (2) the chance that a male deserting a female after mating will be able to mate again, (3) the fact that a female which exhausts her food reserves in laying eggs is less able to guard them, and (4) whether a male can be confident that a particular batch of eggs was fertilised by him. Maynard Smith considers three types of situation. In two of these, breeding is seasonal and reproductive success is limited by parental care or by both the amount invested in the egg and by parental care. In the third, breeding is continuous rather than seasonal. Logical analysis of these situations shows that a possible ESS for each is either the 'duck' ESS in which the male deserts and the female cares for the young, or the 'stickleback' ESS in which only the male cares for the young.

When offspring survival is much better if two parents are present than if only one shows parental care, desertion is unlikely except at extreme sex ratios (Grafen and Sibly 1978). There may, however, be differential investment and attempts by males to achieve additional matings without any necessity for additional parental care. The sex which is worse at parental care will be likely to desert first but only if there will be time to find a mate and to produce further offspring before any unfavourable season. Grafen and Sibly's model explaining the occurrence of mate desertion is simple in that breeding success is measured in terms of the number of offspring surviving to first breeding. They point out that it is not possible to consider distant generations, or to assume what the breeding strategy for the next generation will be, when designing such a model. In some studies of parent–offspring interactions it is possible to attempt to assess the quality of the offspring produced and hence to predict what their future breeding success might be.

Dawkins (1976) considered the ways in which parental care by the female in ducks and by the male in fish such as sticklebacks might have evolved. He pointed out that after copulation the female duck is left with the zygotes but after a male stickleback has enticed a female into his nest and then fertilised the eggs whilst she departs, he is left with the eggs.

The aims of parents and of offspring

Genes which affect behaviour towards offspring will spread in the population if their effects promote the survival of the maximum number of future descendants which bear such genes. In any particular breeding period when offspring have been produced, the behaviour of parents would be expected, therefore, to maximise the fitness of these offspring, provided that this effect on parental inclusive fitness is not counterbalanced by adverse effects on the parent's prospect for offspring production in future breeding periods. That is to say that parental altruism is not unlimited if that parent could live to breed another day (Hamilton 1964*a,b*, Trivers 1974). Nevertheless, in some species parents do devote a high proportion of their resources for long periods to behaviour, and physiological functions like milk production, which benefit the present offspring. Before considering these in detail it is necessary to understand the requirements of the young.

The requirements of the young

The parental care needed by offspring depends upon their stage of development at birth or at hatching. For example, help with thermoregulation is much more important if the young are altricial than if they are precocial (see page 20, Figure 9). The problems which must be solved by a developing cod and a developing chick are discussed in Chapter 1. An important problem which was mentioned briefly for the chick, is to recognise, find and stay with the mother. As explained in Chapter 2 the young chick has perceptual mechanisms when it hatches which result in a greater likelihood of approach to objects with certain combinations of visual and auditory characteristics. Young precocial birds do interact with parents and siblings before hatching (Drent 1970, Dimond and Adam 1972, Impekoven 1976). Irrespective of such interaction, the mother is an object which has several of the preferred characteristics so she is approached by most chicks.

The mother hen helps her chicks with thermoregulation, food finding and protection against predators (see Chapters 5, 6 and 7 for further details). In order to obtain these benefits it is advantageous for the chick to be able to distinguish its mother from other objects, including other hens. The learning process which results in the establishment of this social preference is usually called imprinting (Lorenz 1935, Bateson 1966), although the use of this term should not be taken to imply that the mechanisms of this learning process are distinguishable in some way from that of any other. The timing of the process of familiarisation with the mother and with other aspects of the young bird's surroundings has been the subject of much research (reviewed by Bateson 1979).

The problems of a young rodent or a young monkey or human child have also been discussed in previous chapters but the predominating importance of establishing and maintaining appropriate contact with the mother or with both parents needs further discussion. If the young are in a nest, as are those of a rat or gerbil, then from birth onwards they need to have behavioural mechanisms for finding the mothers' teats and for signalling to the mother or father if they are cold, separated from the nest, or hungry. The young monkey must also hold on to the mother when she moves. Young mammals and birds must manipulate the behaviour of the parent so as to maximise their chances of surviving and breeding successfully. In order to do this each must encourage the parents to feed it, to forage efficiently, to protect it and to help it to learn how to behave efficiently. Siblings compete with one another for some resources provided by parents, e.g. food and priority of protection, but they do not compete when learning by observing parental behaviour and they may help one another during trial-and-error learning. Young animals benefit from prolonging the period when parents or other relatives feed and protect them.

The requirements of a parent

In order that a parent can care for its offspring it is necessary that the offspring be recognisable, first as young conspecifics (see Chapter 2) and second as its own young. Few male animals can be certain of their paternity but those which show parental care have usually minimised the chance of cuckoldry by preventing mating between their mate and other males prior to birth or egg laying. Male rodents, for example, may have to live with the pregnant female before they are able to show parental care (Elwood 1977, Labov 1979). Females are also vulnerable to deception about the identity of their own young whenever they lose contact with them temporarily. Recognition of individual characteristics is the surest way of avoiding the risk of expending resources rearing another's young but this takes some time. Animals which make a nest in a discrete site can assume that the young in that nest are their own, rather than identifying them individually, but there will always be the possibility of a cuckoo in the nest.

Once accepted, the young can be fed, protected from predators and hazards, kept warm and helped to learn new skills. The latter aspect of parental care often consists of allowing the young to remain close to the parent whilst it is gathering food or carrying out some activity which the young animal could observe and copy. The parent is faced with a series of decisions regarding the nature and extent of its parental care. The first, which is often decided before the birth or hatching, is how many young to rear. This question, which is discussed by

Smith and Fretwell (1974) and by Brockelman (1975), is of importance to all parents for the optimum number will differ according to their ability to acquire resources. They will often have to share food amongst young which are competing with one another, otherwise they may rear only one large offspring. A further aspect of this problem is behaviour towards weaklings. A weak individual might endanger the rest of the offspring or the parent and might not be worth the energy needed to rear it so it might be left behind, starved or even killed (O'Connor 1978).

A second question for a parent concerns the sex ratio of the offspring. Factors during embryological development in the female may promote survival of one sex and it would be possible for a parent to favour offspring of one sex during rearing (Maynard Smith 1978*a*, Williams 1979).

A third problem for parents is how long to continue parental care. The costs of most aspects of parental care increase as the offspring get larger but the ability of the offspring to fend for themselves also increases as they approach adulthood. If the parents may be able to breed again a point will be reached, during offspring development, when the parents benefit most by ceasing parental care and commencing any recuperation which is necessary as a prelude to further breeding. This problem is related to another which concerns the possibility that after they have reached adulthood, the offspring may compete directly with the parents for resources or may increase the risk, to the parent, of predation. If the parents can breed again they should drive the young away unless the young will help with the next brood (see pages 190–1). If no further breeding is possible, the parents should die so that they do not compete with their offspring. Breeding should be continued so long as it can be carried out at least half as efficiently as that of the offspring. Dawkins (1976) discusses these topics and also suggests that the menopause in women may have evolved because older women can contribute more to the survival of the relevant genes by helping grandchildren and other relatives than by continuing to try to breed.

Parent–offspring conflict

Each offspring solicits parental investment at a level which is usually higher than that which the parent will make so there is some parent–offspring conflict (Trivers 1974). Parents will ration investment in any individual so that other offspring present at that time receive a share and they will also conserve their resources in a way which is related to their potential for future offspring production. Put another way, this argument states that it is not worthwhile collecting food for an offspring, or defending that offspring against a predator, to such an extent that the parent dies, if that parent could produce more than one offspring by refraining from sacrificing itself.

Much of the parent–offspring interaction involves soliciting, e.g. begging for food, by the offspring and a complete satisfaction of the offspring's need by the parent. This may involve the parent giving a little more than it needs to give in order to ensure offspring survival. The time at which parent–offspring conflict is most obvious is at the termination of parental care. Soon after a young rhesus monkey is born, the benefit of maternal care to that young monkey, in terms of its survival, growth quality and hence future reproductive success, is large and the cost to the mother of feeding and protecting the baby, in terms of the reduction of her future reproductive success, is quite small. Hence the ratio of that benefit to that cost is large. The decline in benefit to cost ratio as the baby matures is shown in Figure 152. Older babies might survive without milk and if she were to continue to feed them, they would cost the mother much more because they would consume a great deal of milk. The point at which maternal care should end, which is optimal as far as offspring are concerned, is as shown in Figure 152 if the future offspring of the mother are full siblings of those offspring. If they are half-siblings it would benefit the offspring if maternal care continued until the ratio of benefit to cost dropped to half of that value.

The idea of parent–offspring conflict was criticised by Alexander (1974) on the basis that parents would always win such a conflict because a conflictor gene which reduces the fitness of all siblings could not spread. This argument was criticised on logical grounds by Dawkins and has been thoroughly explored and disproved by Parker and Macnair (1978). Theoretical possibilities for conflict between siblings in one or more broods have been studied by Macnair and Parker (1978, 1979) and Parker and Macnair (1978) for situations where breeding is monogamous or where females are promiscuous. The situations where evolutionary retaliation by parents can occur are considered in a fourth paper (Parker and Macnair 1979). An ESS for each situation has been calculated. Important factors affecting the ESS are the costs to the offspring of soliciting, the costs to the parent of ignoring soliciting, and any initial allocation of parental investment by the parent. Whilst 'offspring wins' and 'parent wins' ESSs are possible, the commonest ESS is predicted to be of the type where the parent gives more than its optimum but less than the offspring demands. Such compromises do seem to be shown by studies of parent–offspring interactions in most species. Many young animals attempt to extend the duration of parental care by deception, for example, by acting as if they are at an earlier developmental stage than they really are, when begging for food. The manipulation of parental behaviour by offspring is probably widespread and the idea that parents alone determine the course of parent–offspring interactions must seldom be correct (see review by Stamps and Metcalf 1980).

152 Parents will try to terminate parental care before offspring want them to do so. The ratio of the benefit to the offspring, in terms of its future reproductive success, to the cost to the parent, in terms of reduction in future reproductive success, is high at first. The actual level varies from species to species. Parental care should cease at levels between 1 and 0.5 for monogamous species. Modified after Trivers 1974.

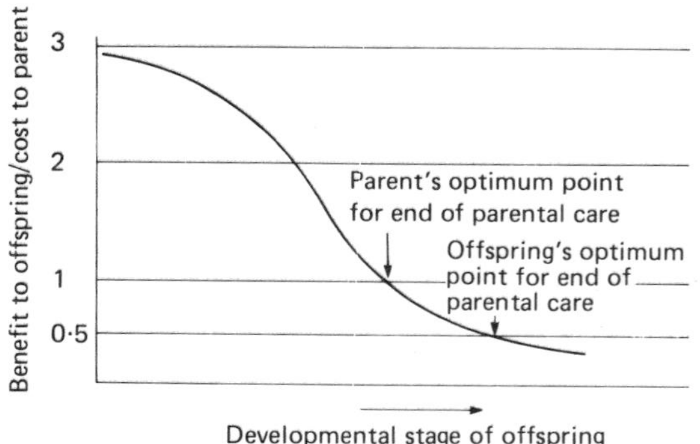

Examples of parent–offspring interactions in mammals

The examples which will be described in this section are: firstly, the comparatively simple interactions, during the first two days after parturition, between a mother ungulate and her precocial offspring; secondly, the interactions between a mother or father and altricial young, such as those of rodents, which are initially in a nest; and thirdly, the interactions between parents and young primates in a complex social situation. Parent–offspring interactions have been widely reviewed, those in mammals in general by Rheingold (1963) and Walser (1977), in rodents by Moltz (1975), Rosenblatt, Siegel and Mayer (1979) and Elwood (1981), in cats by Rosenblatt (1976) and in rhesus monkeys by Hinde (1974).

Cows and other ungulates

When a bovine ungulate, such as the domestic cow, is about to give birth she usually moves away from the herd and if possible finds a sheltered place. Eventually, a typical cow will lie down, give birth and after recovering from her exertions, stand again. She then proceeds to lick the calf and to lick up birth fluids. The calf is licked all over its body with the result that birth fluids, which matt its hair, are removed. This results in an improvement in the insulating qualities of the hair as well as the removal of odoriferous substances which might attract predators. Whilst licking, the cow learns to recognise the olfactory and other characteristics of the calf. The licking is often vigorous enough to lift part of the calf's body from the ground and this may encourage the calf to stand. In a study of several hundred calvings by dairy cows (Edwards and Broom, in press) first standing by most calves occurred 30–90 min after birth. Licking continues after the calf stands but declines to a low level by the sixth hour after birth (Figure 153). At some time, usually between the second and the sixth hour, the afterbirth is produced. As in most other mammals the mother cow eats the afterbirth. This process may take about 20 min and has obvious advantages as an anti-predator behaviour, although it is apparent that the cow's dentition is poorly adapted for consuming animal tissue. In order to avoid the possibility that the cow might choke, the afterbirth may be removed by a conscientious stockman.

Whether birth occurs in a field or in a building, most cows spend 90% of their time within 2m of their calves during the first two to three hours but may move some distance away on the second day (Broom and Leaver 1977). If the calf moves away from the cow during the first few hours the cow often vocalises and approaches it so contact is maintained more by cow behaviour than by calf behaviour. The red deer (*Cervus elephas*) hind remains within 50m of the calf for the first few hours and remains about 100m away for several days after this (Clutton-Brock and Guinness 1975). In red deer and to a lesser extent in cattle, the young are 'hiders' whereas in sheep and many antelopes they are 'followers' which move with the mother away from the birth site after a few hours. Soon after the calf stands it approaches the mother and starts nuzzling around her ventral regions. The mother usually stands still whilst this occurs and, especially in multiparous animals, i.e. those which have reared young previously, may turn so as to increase the chance that the calf will encounter the udder. Calves will nuzzle between the fore legs and suck any protuberance on the mother's body. They may not find a teat during their first exploratory investigation of the mother. The likelihood of success depends on the size and height above ground of the udder, the size and orientation of the teats and the behaviour of the mother. After the first few hours the mother cow starts grazing again and returns to the calf at intervals, standing still if it suckles (Edwards and Broom in prep). Older calves may spend much of their time with other calves and take the initiative when resuming contact with the mother in order to suckle (Kiley-Worthington, personal communication). Grass consumption by calves commences on the first or second day of life so both milk and grass are consumed for some time. Ungulate mothers do, however, actively wean their young by refusing to stand and by pushing them away whilst allowing them to remain in the same social group.

153 The mother cow spends much time licking her calf during the first hour after birth but this behaviour declines with time. S. A. Edwards and D. M. Broom, in press.

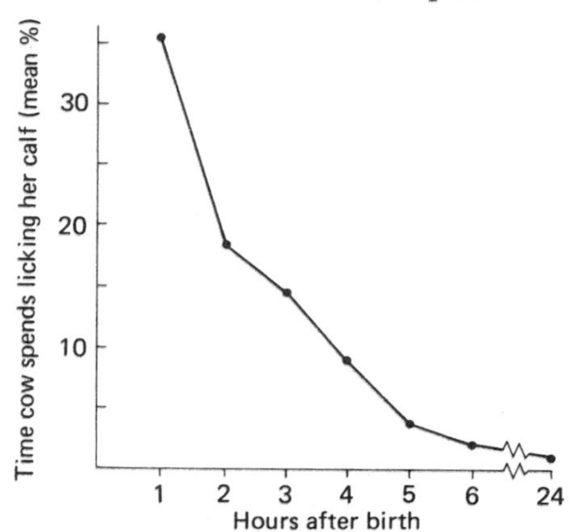

Gerbils and others with altricial young

The construction by the mother of a nest, in an appropriately secluded place, precedes parturition in rodents. During birth, rats, gerbils and other small rodents adopt a sitting posture with the head lowered so that vigorous licking of the vaginal area and the emerging pups is possible. The thorough licking of pups and consumption of foetal membranes and afterbirth are accomplished rapidly. The pups of the Mongolian gerbil weigh about 3g at birth; they are naked for the first seven to ten days, their ears open after five days and their eyes after 15 (McManus 1971, Kaplan and Hyland 1972, Norris and Adams 1972).

Soon after the end of parturition the mother rodent, cat or dog lies on her side with legs outstretched around the litter. The young have just sufficient motor ability to move to the nipples. They follow paths of thermal, tactile and olfactory stimuli provided by the skin temperature, hair patterns and odoriferous substances on the belly of the mother (Rosenblatt 1976, Blass, Hall and Teicher 1979). They may be steered by the mother licking this region. Milk ejection by the mother usually necessitates contact and suckling by the young and may not occur unless several nipples are sucked at once (see study by Drewett and Trew 1978 on anaesthetised females). Such a physiological mechanism prevents the mother from wasting time rearing a litter of one or two pups when she could initiate the production of another larger litter. Litter size also affects other aspects of parental behaviour (Elwood and Broom 1978). Since the behaviour of very young rodent pups is restricted to nipple finding, and little else, it is essential to them that they have some means of regaining the nest if separated from it. As described on page 112, they respond to separation from their warm nest and siblings by producing ultrasonic calls to which the parents respond by retrieving them.

Most female rodents keep the father away from the nest on the day of parturition or for a few days longer, but in some species he may then enter the nest and show parental care (Wiesner and Sheard 1933, Elwood 1975). In the Mongolian gerbil, interactions between mother and offspring and the rate of development of the pups are affected by the presence of the father, the presence of other adults or juveniles; the size of the litter, the amount of space available in the cage, the presence of a nesting compartment, the pregnancy of the mother, and the age, parity and rearing conditions of the parents (Elwood and Broom 1978, McLoughlin 1980).

Parental behaviour in altricial animals changes as the pups develop and Schneirla, Rosenblatt and Tobach (1963), in their study of cats, divided it into three intervals. In the first of these, the mother initiates nursing and she seldom resists any approach to her nipples by the kittens. In the second interval either mother or kittens may initiate contact and suckling, whilst in the third interval all approaches are made by the kittens and they are seldom allowed to suckle. Weaning is not the end of parental care, especially in carnivores whose prey-catching ability is poor when young and the termination of parental investment by the parent by driving the young away may not occur until a new litter is due to be born.

Primates

New-born primates are intermediate in developmental state between the ungulates and the rodents described above. A neonatal rhesus monkey, chimpanzee or gorilla can cling to the mother when she moves, although she usually gives some support with her arms and thighs. Human babies have some grasping and clinging ability but few can support their own weight. They also show the arm movements necessary for grasping the mother's body if they feel themselves falling, a response which is called the Moro reflex (Prechtl 1965). In addition to these mechanisms for retaining contact with the mother, neonatal primates encourage the return of mother by crying when separation occurs. The initial food-finding behaviour of a human or monkey baby consists of upward climbing and 'rooting' movements of the head. The side-to-side movements of the head increase the chance that the mouth or cheeks will make contact with one of the mother's nipples. If the area around the mouth is touched, the baby turns towards the touch and when the nipple touches the soft palate, sucking is initiated and most other movements are inhibited (Prechtl 1965, Hinde 1974). The mother interacts with the baby during all of these movements by responding to its cries and movements and holding it in a position which allows suckling (Figure 154). Other full-grown individuals in a group of monkeys are unlikely to have much contact with a very young infant but may do so later.

Father rhesus monkeys tolerate the activities of infants and protect them against predators but, unlike marmosets and some other species (Box 1977), male rhesus monkeys seldom carry infants or play with them. Sub-adult females, however, may interact with infants and their behaviour may alter mother-infant relationships (Rowell, Hinde and Spencer-Booth 1964, Spencer-Booth, Hinde and Bruce 1965). The care and exploitation of primate infants by males and by females other then mother is reviewed by Hrdy (1976). A major effect of such aunting' behaviour is that an infant may survive if the mother is absent or if she fails to respond to it.

During the first few days of life, primate babies show visual preferences (Chapter 2 e.g. Fantz 1965), prefer to hold furry surfaces (Harlow and Zimmerman 1959), respond differentially to sounds (e.g. in human babies, heartbeat Salk 1973 and voices Hutt *et al.* 1968), and can learn to change their response to odours (Lipsitt 1967). The process of attachment between human mother and baby involves: increasingly specific responses to smiling and other signals, recognition of the baby by the mother and, later, of the mother by the baby, and changes in the behaviour of both mother and baby as development proceeds. It has been studied by Bowlby (1969, 1973) and by Ainsworth (1979).

Much detailed experimental work has been carried out on parent–offspring interactions in rhesus monkeys. Hinde and his collaborators have studied captive rhesus monkeys in groups whose composition was similar to that of small troops in the wild. The male, two to four females and their young were housed in a pen which was 6m long. After describing birth, suckling, clinging and carrying behaviour, Hinde, Rowell and Spencer-Booth (1964) described the first occasions when the infant left the mother at a mean age of eight days. The increases in the time spent off the mother and in the proportion of that time when the infant remained more than 60cm from her are shown in Figure 155. The mothers sometimes prevented such departures and followed the infant during the first ten weeks but after this time, mainte-

154 Mother vervet monkey (*Cercopithecus aethiops*) holding her baby in a position which allows it to suckle.

155 Baby rhesus monkeys start to leave their mother when they are about one week old but they are seldom seen more than 60cm away from the mother during the first five weeks. After this time the duration and distance of separation increases. Modified after Hinde *et al.* 1964.

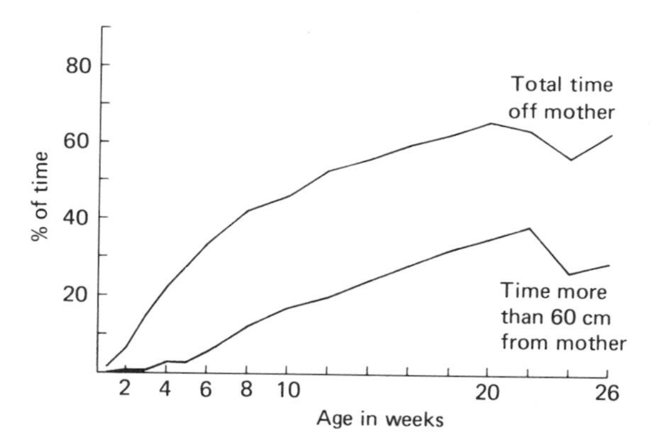

nance of proximity to the mother was largely due to the infant (Hinde and Atkinson 1970). A similar change is apparent when the making and breaking of ventro-ventral contact between mother and baby is analysed. Figure 156 shows that the net initiative for maintaining contact is the mother's during the first 12 weeks but is the baby's thereafter (Hinde and White 1974). The mother sometimes rejects attempts at contact by the infant. When Hinde (1974) compared the roles of mother and infant in regulation of contact and proximity it was apparent that the increasing independence of the infant was due more to the mother's behaviour than to the infant's. These results, however, were obtained in situations where there was little anti-predator behaviour and no likelihood that the mother and infant would be separated.

If the mother was removed from the group when the infant was about 24 weeks old, the infant became inactive and adopted a huddled posture. When the mother was returned after a few days, the infant made many 'whoo' calls and clung to the mother. She allowed this clinging initially but started to reject infant approaches after some time (Spencer-Booth and Hinde 1971, Hinde and Spencer-Booth 1971). The changes in mother–infant interactions depended in part upon their previous relationship; thus the infants which had been rejected most previously were the most disturbed at separation and reunion. If the infant, rather then the mother, was removed from the social group, the infant seemed to show less distress at reunion than did one whose mother had been removed (Hinde and Davies 1972*a*,*b*). If the infant knows that it has been removed, it needs merely to restore contact with the mother on its return. If mother leaves, however, on her return the infant behaves in a way which minimises the chance that she will do so again for the mother must be left in no doubt as to the distress of the infant. The results of these studies are generally consistent with Trivers' (1974) suggestions about the attempts by offspring to manipulate parental behaviour but they are also of particular interest in relation to the work of Bowlby and others on child development.

156 The net initiation of ventro-ventral contact between mother and infant rhesus monkey is by the mother during the first 12 weeks of the infant's life but by the infant after that. Modified after Hinde and White 1974.

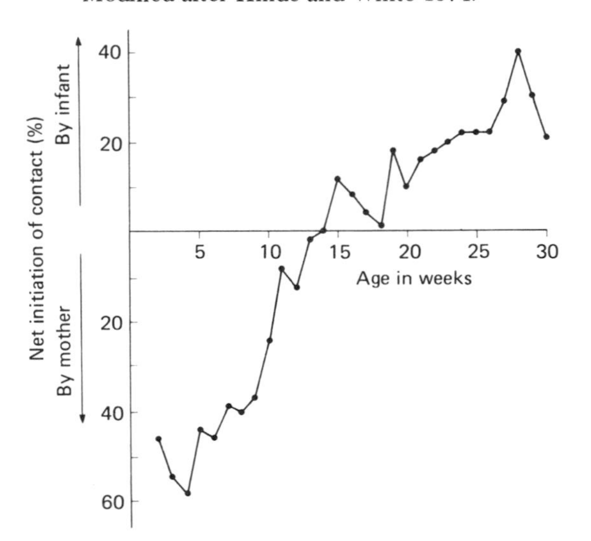

Effects of experience on parental behaviour

The way in which a rhesus monkey or human mother treats her first baby changes rapidly in the first few days. Mothers may hold their babies ineptly at first (Bertrand 1969), for example Hinde *et al.* (1964) described the behaviour of a rhesus mother which held her baby tightly in a position which did not allow suckling. Human mothers learn how to minimise crying and later maximise smiling. As development proceeds, both parents and offspring modify their behaviour according to their previous experience in comparable situations. These are short-term effects of experience which raise the question of the effects of such experiences on the behaviour of the parents towards subsequent offspring. When Seay (1966) compared the behaviour towards young of *primiparous*, i.e. rearing first young and *multiparous*, i.e. previous experience of rearing young, rhesus-monkey mothers, he found no large differences between them. There were, however, some differences in facial expressions, gestures and maternal rejections of infant approaches. White and Hinde (1975) found little effect of parity on rhesus monkey behaviour but Hinde (1974) points out that for group-living monkeys, subadult females have often had considerable experience of interaction with infants before they themselves give birth.

In a study of Mongolian gerbil families, McLoughlin (1980) found that younger, primiparous females were more active, burrowed more, sniffed pups more, wrestled with pups more and investigated the genital areas of male pups more than did older, multiparous females. Some of these differences were still apparent when primiparous and multiparous animals of similar age were compared but many aspects of parental behaviour are not affected obviously by previous experience of rearing young (Grota 1973). Primiparous females show maternal behaviour when the normal combinations of hormones are present but when their hormonal state was drastically altered shortly before parturition, many failed to show maternal behaviour to foster litters which were offered to them. Almost all multiparous females, however, showed maternal behaviour to such litters after ovariectomy and progesterone injection (Moltz and Wiener 1966, Moltz, Levin and Leon 1969, Moltz 1975).

The maternal behaviour of rhesus monkeys is drastically impaired by severe *social deprivation* during early development (Harlow and Harlow 1965). Infant monkeys reared without mother or peers show abnormal sexual behaviour and if ever successfully mated, show little response to their babies. Animals reared with peers but without a mother show some inadequacies as mothers (Chamove, Rosenblum and Harlow 1973). The parental behaviour of other mammals is affected less by lack of early social experience. Stern and Hoffman (1970) studied guinea-pigs, which are precocial and can be reared in isolation, but were unable to detect any differences between isolation-reared and social-reared animals, in maternal behaviour. Sharpe (1975), working with rats, found that the sex of litter-mates affected nest-building behaviour after parturition, those reared in a bisexual litter producing larger nests. In cattle, maternal behaviour seemed not to be different according to whether or not the animal had interacted with its own mother after birth (Albright, Brown, Traylor and Wilson 1975). There were, however, signs that inadequacies of social behaviour after prolonged isolation-rearing as calves were also reflected in behaviour towards a heifer's own calf (Broom and Leaver 1977 and unpublished).

The control of parental behaviour

The links between different aspects of reproductive behaviour are emphasised by the fact that some brain regions and hormones are concerned with the control of courtship, mating and parental behaviour. This is true for neonatal effects of hormones on sexual responsiveness, ovulation and nest-building in mice (Lisk, Russel, Kohler and Hanks 1973) as well as for the control of reproduction in adult birds and mammals. As has been mentioned above, implants of testosterone and oestrogen in the hypothalamus of male or female doves initiates courtship behaviour. Testosterone implants into the male doves' anterior hypothalamus can also lead to the collection of nest material (Erickson and Hutchison 1977) and progesterone implants in the hypothalamus result in incubation behaviour by male doves (Komisaruk 1967). Brain mechanisms of parental behaviour have been investigated less than have those of courtship and mating but there has been extensive work on the role of hormones in reproduction in doves and in canaries. The changes in levels of the gonadotrophin luteinising hormone (LH), of progesterone and of oestrogen, in the blood of female doves are shown in Figure 157. Events during reproduction in doves, starting with courtship and proceeding to nest-building, mating, egg laying, incubation and chick feeding are summarised in Figure 158. The extensive series of papers from Lehrman's laboratory providing the evidence for the interactions between stimuli, brain mechanisms and hormones portrayed here is reviewed by Cheng (1979). A similar, but more detailed, summary of Hinde and Steel's work is shown in Figure 159.

Parental behaviour in rodents can occur in the absence of any of the steroid hormones which are usually present at parturition. Rosenblatt (1967) demonstrated that intact, ovariectomised or hypophysectomised females and intact or castrated males show parental behaviour after 10–15 days of exposure to rat pups. The transfer of blood to virgin females from female rats which had just given birth, resulted in parental responses to pups within 48 h (Terkel and Rosenblatt 1968) but blood transferred from individuals which had initiated parental behaviour as a result of exposure to pups did not have this effect (Terkel and Rosenblatt 1971). Parental behaviour can be initiated and organised in the absence of hormones but it seems likely that its timing is hormonally controlled (Dewsbury 1978). The rapidity of its initiation in virgin female rats was greatest if the blood from a female which had given birth was supplied continuously by parabiosis (Terkel and Rosenblatt 1972).

Parental behaviour in rats has been successfully elicited in ovariectomised females by the daily injection of oestradiol, progesterone and prolactin (Moltz, Lubin, Leon and Numan 1970, Zarrow, Gandelman and Denenberg 1971). The delay before parental behaviour was shown was 35–40 h if all three hormones were present but was much longer, or no such behaviour was shown, if one of the hormones was missing. The sites in the brain at which this combination of hormones act are not known with certainty but experiments by Numan (1974) show that female rats lesioned in the medial pre-optic area ignore their offspring and show aberrant nest-building behaviour. Other lesion studies implicate the stria terminalis, septal area, olfactory lobes and cingulate cortex (Slotnick 1967, Carlson and Thomas 1968,

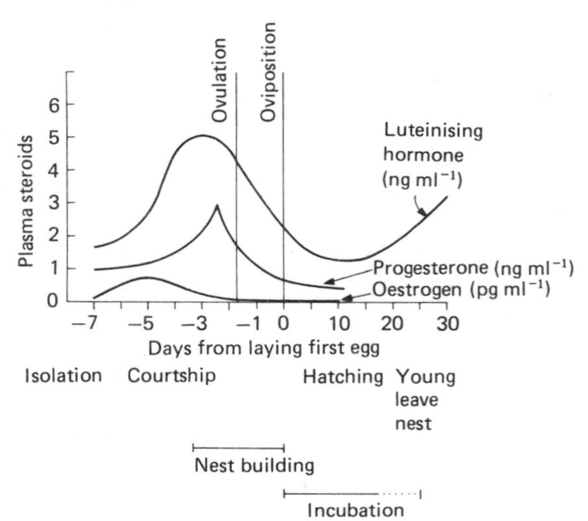

157 The hormone levels in a female ring dove from the time that it is put with a male to the latter stages of parental care. Modified after Cheng 1977.

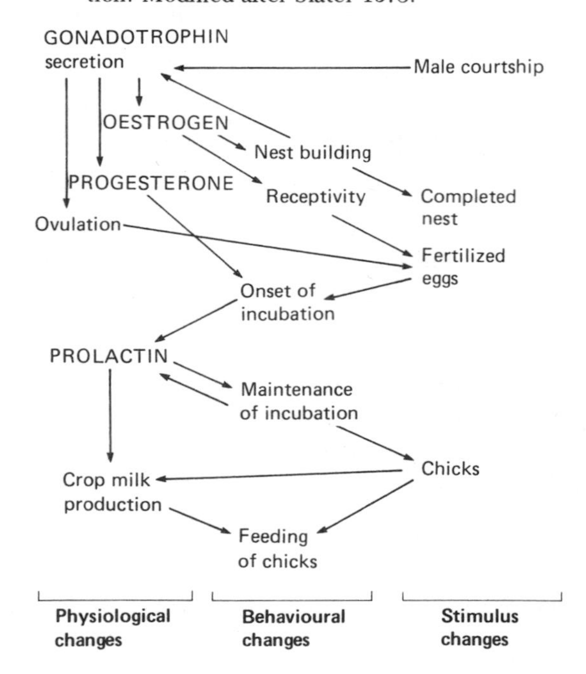

158 Scheme showing the interactions between external stimuli, physiological changes and behavioural changes during ring-dove reproduction. Modified after Slater 1978.

Fleming and Rosenbatt 1974, Carlson 1977) in parental behaviour. As with all lesion studies, it is difficult to know which effects are specific to parental behaviour without an extensive study of the behaviour of the lesioned animals.

Promoting reproductive success in farm animals

One of the major areas where behaviour studies contribute to animal husbandry concerns attempts to maximise the overall reproductive rate. Every time that an animal which could have become pregnant fails to do so, or a young animal is weak or dies, the farmer loses money. Mating and parental behaviour have been observed by farmers and veterinarians for many years but more recent, systematic observation has improved reproductive efficiency.

159 Scheme showing the interactions between external stimuli, internal hormone conditions and reproductive development and behaviour in female canaries. The dotted lines indicate effects not demonstrated conclusively. After Hinde 1970.

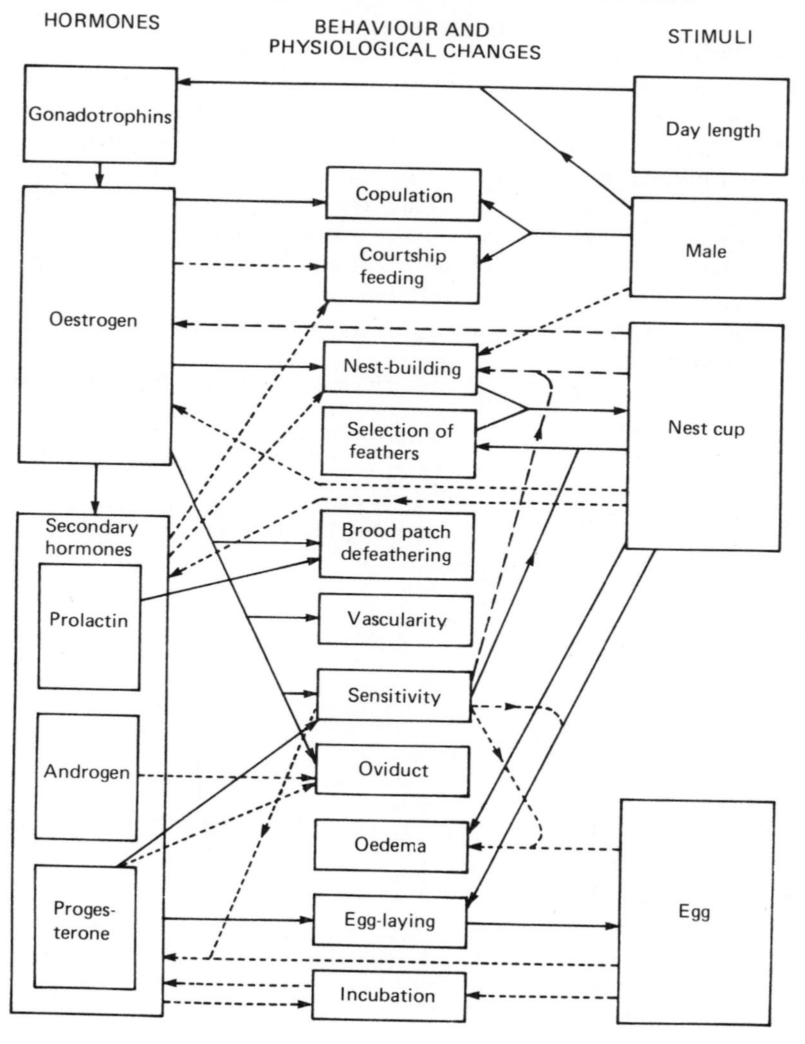

Mating

The males of ungulates kept on farms are more expensive to keep, in relation to their productivity and are considerably more dangerous than are females. As a consequence, farmers wish to ensure that, whilst all the females must be mated, the number of males kept for the purpose should be as small as possible. A further advantage of this practice is to improve the average quality of the stock by using only the best males for mating or the production of sperm with which females will be artificially inseminated.

Oestrus detection

If the male is to be put with the female for the smallest possible time, or if artificial insemination is to be used, the stockman needs to be able to recognise oestrus in the females. This is valuable even if males are present because animals which never come into oestrus must be removed from the herd and if the time of oestrus is known, parturition dates can be calculated. Behavioural studies of cattle and pigs have helped in this recognition. Observations of mounting behaviour have helped in the development of mounting markers. These are put on the males in such a way that they mark the females which have been mounted. In this case, the males detect oestrus.

In the absence of males, oestrus is sometimes detectable because it is accompanied by some vulval swelling and vulval secretions but behavioural changes are easier to recognise. During oestrus in cattle there are increases in general activity level, vocalisations, licking other animals, mounting other cows, and standing to be mounted (Fraser 1968, 1974). Standing for mounting was considered to be the best indicator of oestrus by Mylrea and Beilharz (1964) and this finding has been confirmed in a detailed study of heifers, i.e. first calvers, by Esslemont, Glencross, Bryant and Pope (1980). They found that mounting behaviour, resting the chin on another heifer, sniffing and licking other heifers all increased over a period lasting about two days. Standing to be mounted, however, showed a peak incidence lasting for less than one day (Figure 160). Although the incidence of this behaviour was variable, a cowman who watched a herd for two one-hour periods during each day would be likely to detect oestrus efficiently for most animals. A less efficient but still useful method of detecting which cows stand for mounting is to place on the rump of the animals a paint-filled device which releases its contents when mounting by a member of the herd occurs.

When a sow or a gilt, i.e. a mature female which has not yet reared a litter, is in oestrus and is mounted by a boar, she stands immobile as soon as pressure is applied to her back. Some gilts in oestrus (48%) respond to pressure applied to the back, by becoming immobile in the absence of a boar. More gilts (70%) show the response if the vocalisations of a courting boar, the 'chant du coeur', are played to them or (80%) if the odour of a boar is smelled (Signoret and du Mesnil du Buisson 1961). The odoriferous substance of the boar is produced by the submaxillary gland. It affects the female only if

160 The frequency of standing to be mounted shown by heifers increases rapidly to a peak at a time when there are other behavioural and physiological signs of oestrus. It then declines rapidly so this behaviour is an indicator of the times of oestrus which is useful to stockmen. Modified after Esslemont *et al.* 1980.

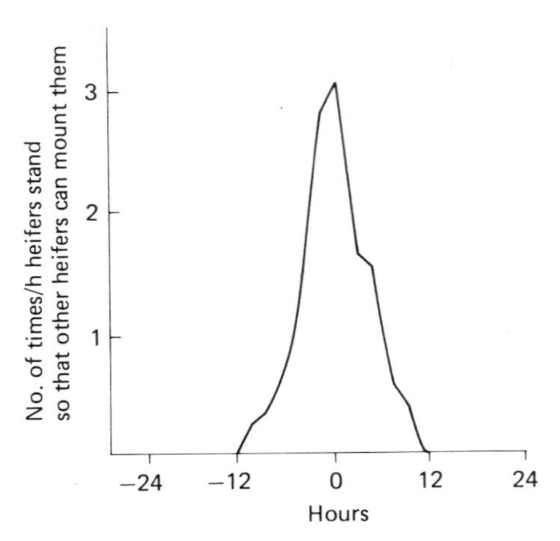

it is warmed above 20°C (Signoret 1965) and has been identified as 5α-androst 16 ene-3-one (Patterson 1968). Farmers can now obtain commercially an aerosol spray of this substance which can be used, in conjunction with the application of pressure on the back of the sow or gilt, to test for oestrus. The substance, which resembles perspiration in odour, also has interesting effects on the preferences of women selecting chairs in a doctor's waiting room (T. O. Clark, personal communication).

Oestrus detection in mares is comparatively easy because they show a characteristic urinating stance, they urinate in bursts so that their heels are splashed, they elevate their tails and they show clitoris flashing movements (Fraser, 1968, 1974). The oestrus period lasts for four to eight days as compared with one day or less in cows and pigs. In goats, oestrus is indicated by bursts of lateral flagging of the tail, increased bleating and roaming behaviour. In sheep, however, ewes may consort with a ram but few other signs of oestrus are shown (Fraser 1968, 1974). The occurrence of oestrus is changed by the presence of a male in domestic ungulates. Just as Whitten (1956) found that the odours of male mice induced and synchronised oestrus in female mice, so sheep, goats, pigs and cows have been found to come into oestrus earlier if a vasectomised male is kept in the herd (review by Kiley-Worthington 1977).

Male mating behaviour

If bulls or boars are to supply semen for artificial insemination their behaviour towards model females, or real females fitted with semen collection devices, must be investigated. The stimuli which maximise sexual arousal and the optimum characteristics of a 'teaser' object with which the male will mate, have been the subject of much study (Sambraus 1971, Signoret, Baldwin, Fraser and Hafez 1975).

Where males are put with females for mating, it is economically important to discover both the minimum number which is needed to ensure that the females are mated and the best social grouping of males to put together. Although oestrus in ewes is very difficult for man to detect, rams readily detect it by olfaction (Lindsay 1965). Tomkins and Bryant (1972) found that if one ram was put with six ewes, whose oestrus had been synchronised hormonally, the ewes were mounted a mean of 8.5 times during their 48 h oestrus period and this resulted in 75% of them conceiving. If a ram was put with 30 oestrus ewes, most ewes were mounted but the mean frequency was 3.7 times and only 56% conceived. In a subsequent study using ram to ewe ratios of 1:6, 1:12, 1:18, and 1:30, Bryant and Tomkins (1975) confirmed this result, finding that the percentages of ewes lambing were 87%, 65%, 63%, and 56% respectively. They concluded that the number of mounts was an important factor influencing the likelihood of conception and that ratios of more than six ewes per ram are economically undesirable. Arnold and Dudziński (1978) emphasise that the ratio of rams to ewes required is larger in range conditions, where the flock splits into sub-flocks and spreads over a large area, than in small fields. In some circumstances it seems that ewes compete with one another for the attentions of a ram (Hulet, Ercanbrack, Price and Wilson 1962) and there are individual differences in the abilities of rams to make a ewe stand for mating (Inkster 1957). Rams seem to respond to any ewe which stands for mating, irrespective of her other behaviour (Signoret 1975) and rams certainly compete with one another. If there are three rams present, one may mate with twice as many ewes as do either of the others (Bourke 1967). This competition occurs in cattle also and may interfere with mating. Blockey (1979) found that fewer cows became pregnant when they were with a group of bulls of mixed ages, including one older individual which dominated the others, than if they were with bulls of similar age and size.

Effects of experience

The effects of previous experience on mating behaviour has been the subject of a few research studies, especially in attempts to train males to mate with artificial females for semen collection. Banks (1965) described improvement with practice in ram-mating behaviour and Hemsworth, Beilharz and Brown (1978) found that sow-rearing conditions affected mating success. Differences amongst individuals in sexual behaviour and success in reproduction are sometimes a consequence of diet (e.g. Wierzbowski 1978) but there is scope for further work on the effects of rearing conditions and husbandry methods on mating behaviour.

Parental behaviour and the requirements of young

Parturition

Since domestic ungulates, especially cows and horses, may need assistance at parturition and since it is sometimes desirable to separate the mother from the herd and put her in a sheltered place at this time, the advent of parturition must be recognised. As mentioned earlier in this chapter, females about to give birth often separate themselves from the herd, seek shelter and vocalise. Anatomical changes are the most useful guide to the precise time of parturition. The time of day of parturition in cattle is not predictable. Edwards (1979) found that 522 Friesian cows showed no bias towards day or night in the time of calving. Amongst older cows, however, there was a lower incidence of calving during the period at which they would have been milked, a time when milking machines were audible. The cows must have some control over the timing of calving. In a study of Hereford cows in Manitoba, Canada, T. A. Yarney, W. M. Palmer and others have found that cows calved at night more often if they were fed regularly in mid morning and late afternoon rather than in late morning and late evening.

Young obtaining colostrum

The most important function of the mother, as far as neonatal offspring are concerned, is to supply milk. The first milk or colostrum is vital for the survival of most young mammals because it includes immunoglobulins. These proteins, which confer protection against disease and cannot cross the placenta in ungulates, must be absorbed from ingested milk. Such absorption is most efficient during the first 12 h of life so the investigation of factors which affect the likelihood of occurrence of suckling at this time is of importance to the farming industry. The terminology used here is that proposed by Cowie *et al.* (1951) which is accepted generally in scientific publications: the behaviour of the young animal when extracting milk from the mother is called suckling whilst the behaviour of the mother at that time is called nursing. A correlation between the time of first suckling and the level of immunoglobulins in the blood of 48 h was reported by Selman, McEwan and Fisher (1970*a*) and Edwards (1980).

Many calves, especially those of older cows, failed to suckle during six hours of observation after parturition (Edwards and Broom 1979). Difficulties in finding the udder and grasping the teat were a major cause of this failure. Large teats on large pendulous udders were the hardest for calves to find (Selman, McEwan and Fisher 1970*b*) but sometimes mothers failed to stand still or actively rejected the calf (Edwards, unpublished). It is advantageous for the cowman to put the calf to the mother's teat during the first six hours in order to minimise the chances that the calf will contract diseases. Arnold and Morgan (1975) found that 14–20% of ewes fail to stand for suckling by their lambs with the result that some lambs must be hand-reared if they are to survive. Hand-rearing of calves is possible but colostrum absorption is facilitated if the cow is with the calf during the first 24 h (Selman, McEwan and Fisher 1971).

Responses to young by females

Another problem for a young ungulate is that too many individuals in the herd will take an interest in it so that it is trampled (see reviews by Kiley-Worthington 1977, Arnold and Dudziński 1978). Mothers which are about to give birth sometimes steal the offspring of another in the herd. Welch and Kilgour (1971) and Arnold and Morgan (1975) reported that 10% or more of ewes did this also and then neglected their own lamb when it was born so that it died. This problem, together with the desire of farmers sometimes to encourage fostering of young, has lead to extensive study of the methods of recognition of offspring by mothers and vice versa. Collias (1956) reported maternal rejection of lambs separated from their mothers for more than a few hours after birth. Le Neindre and Garel (1976) found that any calf was rejected by cows unless presented during the three hours after parturition whilst Hudson and Mullard (1977) found that after only five minutes contact with her calf, a cow would accept it again up to 12–24 h later.

Ewes and sows seem to use olfaction in their initial recognition of their offspring (Morgan, Boundy, Arnold and Lindsay, 1975, Meese and Baldwin 1975). After a few days, sheep and goats use the auditory (Poindron and Carrick 1976, Lenhardt 1977) and visual (Alexander and Shillito 1977) characteristics of their young in order to recognise them. Lambs recognise their mothers during the first day of life (Shillito and Alexander 1975) and seem to use auditory cues initially but visual cues when a few weeks old (Alexander 1977). The best method of persuading a ewe or a cow to accept an alien calf is to put the amniotic fluids from the mother's own young on to the alien (Collias 1956, Hudson 1977). If this is not possible, the ablation of the olfactory bulb in the brain or, more practically, the masking of odours will facilitate acceptance of alien young (Baldwin and Shillito 1974, Crowley and Darby 1970).

Restricting reproduction in pests

A knowledge of the reproductive biology of a pest species often helps in the design of control measures because the behaviour may be especially predictable at this time. Once it was known that the female gypsy moth (*Porthetria dispar*) crawled up the trunks of apple trees in order to lay eggs it was possible to trap the moths with tar-rings. Once Knipling (1955) had discovered that sterile males of the screw-worm fly (*Cochliomyia hominivorax*), a serious pest of livestock in the USA and elsewhere, would be accepted as sole mate by females, vast numbers of sterile males could be released and the pest controlled (see review by Coppel and Mertins 1977). The first step in such studies is to investigate the sequence of events, physiological and behavioural, during reproduction of the pest. Any study of this kind on a pest species, such as the courtship of the greenhouse whitefly (*Trialeurodes vaporariorum*) (Las 1980), might provide information which would lead to improved control methods. One area of research which has yielded particularly valuable results is the study of sex attractant pheromones in insects.

The major use of chemical sex attractants has been to lure insects of one sex into a trap. The simplest way of doing this is to use a live insect as bait on a sticky board. Alternatively a chemical, collected from insects or synthesised, is used. The trap may contain an insecticide, or other means of killing the insect. One of the best examples of this is the use of a pheromone extracted from the male boll weevil (*Anthonomus grandis*) or of the synthetic pheromone 'Grandlure' (Tumlinson *et al.* 1970, Birch *et al.* 1974). The population of boll weevils, a major pest of cotton, can be reduced considerably by the use of such traps. They are especially useful for catching weevils as they emerge from hibernation. The substance 'Grandlure' has also been used to attract weevils, early in the season, to a trap area which has been treated with insecticide. Other examples of the use of natural or synthetic pheromones in traps are the use of Trimedlure and methyl eugenol against the fruit flies *Ceratitis capitata* and *Dacus dorsalis* (Beroza 1970*a*); the use of pheromone traps against the tortricid redbanded apple leafroller moth

(*Argotaenia velutinana*) (Trammel, Roelofs and Glass 1974); and the use of (Z,E)-9, 12-tetradecadienyl acetate in traps for the flour moth (*Ephestia cautella*) in grain stores (Read and Haines 1976).

A variation of the pheromonal trapping method is to lure insects into traps which contain a chemosterilant so that when the insects escape again, they mate but produce no offspring. This technique has been used successfully for houseflies (Meifert, LaBreque, Smith and Morgan 1967) and for the Mexican fruit fly (Sanchez Riviello and Shaw 1966), but Finch and Skinner (1973) were not able to use it for the cabbage-root fly (*Delia brassicae*). The use of large quantities of chemical attractants to confuse insects so that their mating is disrupted was first suggested by Beroza (1960). The synthetic pheromone 'Disparlure' has been used to reduce mating in gypsy moths and the method may prove to be economically viable (Beroza *et al.* 1973, 1974).

The reproductive behaviour of most pest species has not been studied adequately. There is, therefore, much scope for the study of behaviour to be used to improve methods of controlling insects and other animals which are injurious to man and our crops.

10

The organisation of social groups

This chapter concerns the description of how social groups are organised, the behaviour of individuals which leads to this organisation and the effects of the social organisation of others on the various individuals in the group. The term *organisation* includes the size and composition of the group, the spatial distribution of the individuals and the relationships among them. These relationships are a consequence of the physical and behavioural characteristics of each individual and of the ways in which others respond to them. They also depend on the results of previous interactions within the group.

The term *structure* is often used with reference to social groups. In this book the *physical structure* of a social group refers to its size and composition. *Social structure* refers to all of the relationships among individuals in the group.

The first step in trying to understand the organisation of a social group is to describe the group, so this topic is considered in the next section. The following three sections deal with physical structure, some of the ways of describing the role of individuals in the group and the occurrence of competition in groups. Since social behaviour is much affected by experience during development, this is considered next. The meaning and significance of crowding and the implications of the study of social behaviour for animal husbandry are the subjects of the last two sections.

In some studies it is necessary to use spatial definitions of social groups although this may be complicated by the presence of individuals on the periphery of the group and by the intermittent merging and separation of subgroups. In long-term studies it is often best to refer to individuals with established relationships as a group or subgroup.

The description of social groups

Spatial description

Most studies of social groups start with some description of the spatial distribution of the individuals in relation to the habitat available to them. As discussed in Chapter 8, individuals which remain closer together than they are forced to do by habitat constraints, despite having the ability to recognise conspecifics and separate from them, are showing the simplest form of social behaviour. In addition to helping observers to make decisions about what constitutes a social group, measurements of the spacing of individuals provide information about relationships within the group. Mates and potential mates, or parents and offspring, may be close together and individuals of similar age or sex may associate.

If useful information is to be obtained from a description of the distances between individuals and their pattern of distribution, it is essential to record as much detail as possible about the circumstances when the description is made. The same group of animals may be distributed in different ways at different times according to the availability of food, the danger of predation, the hormonal states of the group members, etc. Distances between individual rooks (*Corvus frugilegus*) were greater if food items were distributed sparsely (Patterson, Dunnet and Fordham 1971) and the adult males in baboon troops may be positioned on one side of a troop if danger threatens from that direction (Hall and DeVore 1965, see also Chapter 8). The spatial distribution of animals may also be altered by a variety of other factors such as temperature (see Chapter 8) and the presence of biting flies. A high incidence of fly attacks sometimes results in horses approaching one another and standing parallel, head to tail, so that each disturbs the flies from the other. Any conclusions from observations of spatial distribution within a group must take account of such factors.

Interactions and relationships

The responses of the members of a group to the behaviour of others may occur irrespective of which individual is showing the behaviour or they may be very different according to the individual. An alarm call or a sudden movement may elicit a similar response from all those in the group. Likewise, the movements of members of fish shoals whilst swimming or of flocks of flying birds are likely to be responses to other individuals which occur irrespective of their identity. When a baboon walks through a troop, however, young males may move away from certain individual males and approach or ignore other troop members. Similarly, cows in a field may lick some animals which approach them, but ignore or turn away from others. Individual recognition and appropriate modification of behaviour occurs in many social groups, probably in many more taxa of animals than we know about at present.

When describing responsiveness it is important to record which individuals are involved and the context in which the interaction occurred. When this is done, the description of the group often becomes a collection of descriptions of *dyadic* relationships, i.e. those which involve two individuals. Sometimes interactions amongst more than two individuals are described but most research workers try to split these so that their data can be summarised in terms of dyads. It is desirable, however, to preserve more information than just that concerning dyads if this is possible. An interaction between animal A and animal B in a monkey troop may be different because of the presence of C and D. If A grooms B, it is possible that it would not normally do so but that it approached C to groom it, was then deterred by the presence of D and groomed B instead. The incidence of fighting amongst adjacent males may be greater if one has been bitten by another individual so the whole social context of an interaction must be described, rather than just the dyadic interactions. A third example of the importance of describing more than just dyadic interactions is that of the captive rhesus monkey mothers which considerably restricted movements of their infants in response to the attempts of an adolescent female to hold and groom these infants (Hinde *et al.* 1964).

The simplest form of *interaction* is where one individual detects the presence of the behaviour of another and modifies its own behaviour. This need not involve proximity or any display. The behaviour of individuals in many groups is affected by what they can detect of their fellow members' positions and activities. In many interactions, however, two or more individuals detect and respond to one another.

One form of interaction which is widespread in social groups but which is not easy for a human observer to record is eye-to-eye contact. Studies of many species show that some individuals can deter others from approaching merely by looking at them. The fact that members of primate groups spend much time looking at other members led Chance (1967) to propose that the 'attention structure' of a group is an indicator of its social organisation. This idea, which was elaborated by Chance and Jolly (1970), is important in that it emphasises that the frequency with which group members look at the different members of a troop is a useful measure when assessing social relationships. This measure is not, however, as Chance and Jolly imply, sufficient in itself to explain social structure. Whilst it helps in description, other behavioural measures are required in order to understand why some individuals receive more attention and others receive little (see Hinde 1974 and Chalmers 1979 for further discussion).

The result of the description of interactions among individuals is some understanding of the relationships amongst them. That understanding will be better if the quality of the interactions and their patterning in time is described. When monkeys groom one another, it is informative to record whether the grooming is hasty and rough or careful and gentle. The frequency, timing and duration of grooming episodes of various kinds amongst the group members should also be recorded. If, as is likely, particular types of interaction recur in the same dyad or larger subgroup, the individuals will learn what sequence of events to expect. As soon as there is an expected sequence in an interaction, variations in the sequence will assume a meaning to the participants. Hinde and Stevenson-Hinde (1976) stress the importance of such learning (see later in this chapter) and point out that the course of an interaction sequence depends upon each individual's view of itself and predictions about the likely behaviour and expectations of the other participants. It is important for observers of social behaviour to appreciate that relationships persist, so most interactions between individuals which do recognise one another will depend in part upon their previous relationship.

The complexity of relationships and the fundamental importance of analysing them scientifically is discussed further by Hinde (1979). Writing of social species, man especially, he states that 'Each individual's life is in fact a continuing dialectic between the self he is or believes himself to be and the relationships he forms, with each affecting the other'. He defines a relationship as 'a potential for patterns of interactions which may be of a certain general type but whose precise form will depend on events in the future'. For simplicity, Hinde considers only dyadic relationships but he does state that 'dyadic relationships can never be fully understood in isolation from their social context'. Studies of behaviour in dyads are very useful for improving understanding of some relationships, for example those between some mothers and offspring or some mated pairs. Even in these examples, however, the relationships may be different in a social group situation. As mentioned earlier, in most social groups it is desirable, where possible, to consider interactions and relationships amongst all the members of the group, not just those between pairs of individuals.

Social orders and castes

If each member of a social group possesses some anatomical or behavioural characteristic to a different degree, it is possible to rank the individuals. Such a ranking is a simple description of the group which may be helpful when trying to understand the various aspects of social organisation in the group. Behavioural characteristics which have been used include the order in a group progression from place to place, the frequency with which each individual grooms the others, or the frequency with which each displaces the others in competitive encounters for food. A common result of such analysis of behaviour is that a linear order is produced.

The very fact that an individual may behave towards another in a different way according to which other individuals are near shows that a linear order of any kind cannot be more than a crude description of social structure. The linear order is often used as a simple way to summarise a set of observations initially but the limitations of such a description should be considered when interpreting the results. Another type of description of social structure is to classify the individuals into categories of individuals which are considered to have certain roles in the group. An extreme form of this, which may be determined by anatomical considerations, is to divide the group into different castes. This has been done when discussing social insects and there is no doubt that the behaviours of ants which are called small workers, large workers, or soldiers are different in several respects. Predetermined decisions about castes, or about age and sex categorisation, may, however, make it less likely that an observer will be impartial when trying to assess relationships amongst individuals in a social group.

The size and composition of social groups

Group size

Social aggregations range in numbers up to well over a million, for example some shoals of fish, flocks of starlings (*Sturnus vulgaris*) going to roost in central London, flocks of sooty shearwaters (*Puffinus griseus*) gathering on the Californian coast prior to migration, colonies of the African driver ant (*Dorylus wilverthi*), or brittle-stars (*Ophiothrix fragilis*) holding on to one another whilst suspension feeding in currents (see Chapter 8). The advantages and disadvantages of living socially are discussed in Chapter 8 but since most social groups are much smaller than those listed above, it is useful to consider whether large social aggregations are best considered as sets of smaller groups and what limits the size of a social group. The fact that group-living offers a different balance of the various advantages and disadvantages in different species and in the same species at different times means that care must be taken in any generalisation about social organisation.

The upper limits to group size may include factors related to any of the disadvantages of group living. The brittle-stars are limited in number by the presence of a suitable substratum and food-laden current (Warner 1971). Some ant colonies are limited in size by the dimensions of the cavity in the tree where they site their nest (Wilson 1971). Herds of ungulates (Geist 1974, Jarman 1974) and troops of primates (Altmann 1974) cannot grow too large because there are limits to the amounts of renewable food resources. In some circumstances, large groups may attract many predators or become especially susceptible to parasites or disease. The distribution of group sizes might reflect resource availability, the likelihood of death due to predation or some other hazard, or a combination of these factors. Since so many variables affect group size it is not surprising that groups of different sizes might be found in different conditions.

Since the mechanisms for change in group size are birth, death, immigration and emigra-

161 The observed distribution of sizes of baboon groups is similar to the truncated negative binomial distribution predicted by Cohen's model. Data from Cohen 1969.

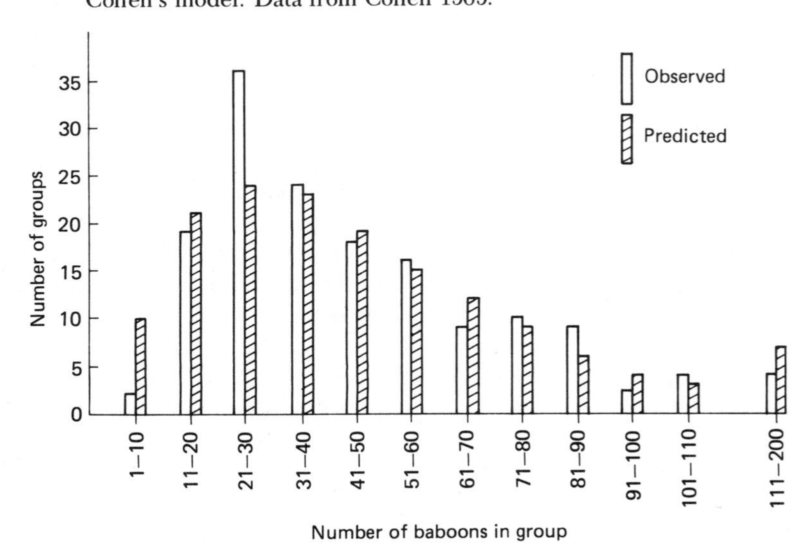

tion it is useful to construct models which predict the distributions of group sizes for particular species. Cohen (1969) devised models of this kind and compared their predictions to group sizes in primates. One situation which he considered was that typified by baboons, e.g. the *Papio cynocephalus* described by Altmann and Altmann (1970). In these groups there are several females which give birth and members of the group are born, die and migrate from group to group at rates which are not dependent on the size of the group to which they belong. Cohen's model predicts that the distribution of group sizes should be negatively binomial but truncated because there would be no groups of zero or one. As shown in Figure 161, observed baboon group sizes fit the predicted distribution quite well. In gibbon groups where only one infant is born at a time, regardless of group size, it is predicted that group sizes will follow a truncated Poisson distribution. The data from Carpenter (1940) and Ellefson (1974) fit this well (Figure 162), as do Carpenter's (1962) data on howler monkeys after an epidemic of disease had killed most infants and juveniles.

Group composition

The composition of social groups is usually described in terms of the age, sex and degrees of relatedness of the group members. In some fish shoals or bird flocks the individuals may group together irrespective of these factors so a description of group composition would merely mirror the population structure. Other shoals and flocks are composed of individuals of one age group or one sex. Some bird flocks and herds of mammals are aggregations of subgroups whose members have been together for some time and have complex relationships. The composition of the more permanent animal groups has been extensively reviewed elsewhere (Box 1973, Wilson 1975) but some of the types of groups will be described here.

162 The observed distribution of sizes of gibbon groups follows the truncated Poisson distribution predicted by Cohen's model. Data from Cohen 1969.

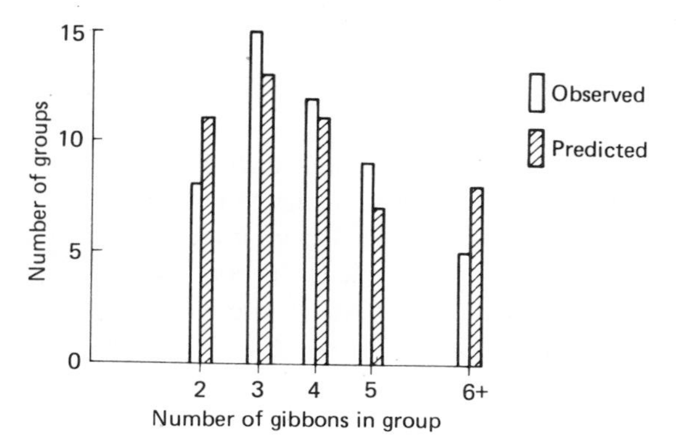

(*a*) *Monogamous family groups* (see page 203 for discussion of monogamy). The minimum number is a pair which remain together outside the breeding season and the maximum number is a pair with all of their non-breeding offspring. Examples include many monogamous birds whose young remain with them temporarily and some primates such as the gibbons (Hylobatidae) (Carpenter 1940) and marmosets (Callitrichidae) (Epple 1975).

(*b*) *Multiple monogamous family groups*. The basic unit is as in (*a*) but brief or prolonged associations of family groups occur. Many geese and swans move around in family groups but spend much of the winter months in flocks, within which families come and go e.g. Bewick's swan (*Cygnus columbianus*) (Evans 1979, Bateson, Lotwick and Scott 1980). Wolf packs are usually family based in this way (Mech 1970). Breeding colonies of sea-birds such as the arctic tern (*Sterna paradisaea*) include adult pairs which collaborate in colony defence and multi-family flocks may be formed after the young are full-grown.

(*c*) *Matriarchal groups of females with young*. Throughout most of the year red deer (*Cervus elephas*) hinds, their daughters and their sons of less than three-years-old remain in a group. Adult males often group together as well and males try to form harems during the rut (Fraser Darling 1937).

(*d*) *Single males with females and young*. In the South American vicuna (*Vicugna vicugna*) the male maintains his harem throughout the year

(Koford 1957). Similar groups are formed by several primate species, for example the blue monkey (*Cercopithecus mitis*) (Aldrich-Blake 1970). These groups sometimes congregate at good feeding sites but they separate again later. Some males of certain species, such as the gorilla (Schaller 1963, 1965), may tolerate the presence of their sons until they are fully grown.
(*e*) *Bands of single males with their females and young.* Hamadryas baboons (*Papio hamadryas*) often spend much of their time in bands which are aggregates of the groups described in (*d*). Even larger herds may form at sleeping places (Kummer 1968). Gelada 'baboons' (*Theropithecus gelada*) also form such semi-permanent bands but the larger herds are found at feeding sites (Altmann 1974).
(*f*) *Multi-male, multi-female groups.* Examples of species of animals in which individuals spend most of their lives in social groups containing adult males and females include certain spiders (Burgess 1978), insects (Wilson 1971), fish (Barlow 1974), birds and mammals. The most social of birds are the babblers *Turdoides* (Zahavi 1974) and anis *Crotophaga* (Vehrencamp 1977), in which groups remain together during breeding and non-breeding periods. Examples of mammals which form groups of this kind include kangaroos and wallabies such as the whiptail wallaby (*Macropus parryi*) (Kaufman 1974), prairie dogs (*Cynomys ludovicianus*) (King 1959), dolphins (Norris 1966), peccaries (*Peccari angulatus*) (Eisenberg 1966), and a variety of primate species such as the Japanese macaque (*Macaca fuscata*) (Sugiyama 1976).
(*g*) *Multi-male and multi-female groups where most females are sterile workers or soldiers.* The organisation of the social wasps, bees, ants and termites is sufficiently different from (*f*) to warrant a separate category. Most males leave the colony before they reproduce and there is usually one reproductive queen at any time. Insect societies are discussed at length by Wilson (1971) and summarised by Wilson (1975).

Not all social species fit neatly into the categories of social group composition listed above. Some intermediate and mixed categories exist and some species live in different sorts of social group according to the conditions or the time of year. The effects of ecological factors on the composition of a social group and on the evolution of mechanisms which regulate social composition, are discussed for a variety of species by Wilson (1975). Some of the most detailed of such analyses have been carried on primates (Altmann 1974, Clutton-Brock and Harvey 1976, 1977*a* Struhsaker and Leland 1979).

Group cohesion and fission

There are many behaviour patterns which maximise the chance that an individual group-member will be able to stay in the group. There are also behaviour patterns which result in the break-up of a group or, more often, the departure of certain of its members.

Cohesion in social groups is more likely to be maintained if individual members can detect the immediate benefits of group membership. An individual would be less likely to leave a group with which it had collaborated recently with success. Collaboration in food acquisition, anti-predator behaviour, etc. is discussed in Chapters 6, 7 and 8.

Cohesion in fish shoals

The term 'shoal' refers to any social group of fish whilst a co-ordinated, synchronised, polarised group is referred to as a school (Pitcher 1979). When a school is moving, each member must monitor continually the positions of the others and adjust its course and speed so that it does not become separated from them (Figure 163). Fish of a schooling species which are not in a school, will approach a school when they see one. The more individuals in the school, the more attractive it is to a solitary fish (Keenleyside 1955). The mechanisms which ensure that fish do not bang into one another and do not become separated from the school were said to be primarily visual by Shaw (1962) but blind fish can school (Pitcher *et al.* 1976). Experiments with mirrors, temporary blind-folding and lateral-line-nerve section have now demonstrated the relative roles of vision and lateral-line organs in schooling (Pitcher 1979, Partridge and Pitcher 1980). Schools of fish of different species have been described as having a characteristic spacing, velocity and three-dimensional shape (Pitcher 1973, Radakov 1974) but many species show much variability in school structure (Breder 1967). Detailed studies of schools of at least six fish show that positive correlations for the behaviour of individual members are high and the three-dimensional structure is similar for several species (Partridge 1980).

Cohesion in bird flocks

Another example of considerable group cohesion is that in flocks of birds. Individuals react to one another using visual cues where this is possible and using auditory signals when it is not. Flocks of small birds in woodland often call and the calls seem to be answered. Even if a bird is hidden from other flock members by intervening foliage it can retain auditory contact. At night, members of migrating flocks also call at intervals and presumably use the calls of others to avoid separation from the flock. Ornithologists in Northern Europe often detect the arrival of the first flocks of redwings (*Turdus musicus*) of the winter by hearing their 'seep' calls overhead at night. If individuals do become separated from their social group they must use their navigation, searching and recognition abilities in order to regain contact with that group or to find another into which they will be accepted.

Cohesion in primate groups

In addition to having the ability to achieve or maintain contact with a social group it is often necessary for an individual to be able to modify or preserve relationships with other group members in such a way that it is not driven out of the group and the other members are encouraged to stay in the group. If an individual returns to a group after foraging elsewhere, or if it

163 These saithe (*Pollachius virens*) are schooling in a large water tank. They are marked so that the schooling behaviour of each individual can be studied. Photograph: T. J. Pitcher.

comes into close proximity with other individuals after being separated from them, it may show behaviour which appears to be analogous with human greetings. For example a chimpanzee may hold out her hand as in Figure 164, present her genital area (Figure 165) or kiss (Figure 166). The use of greeting and reassurance behaviour by wild chimpanzees is described in detail by van Lawick-Goodall (1968). Social ants, bees and wasps use complex antennal signals when they encounter one another, and many social birds and mammals use some sort of greeting display.

An activity of group-living primates which may be important in maintaining group cohesion is allogrooming, i.e. where one animal grooms another. It is likely that individuals which are groomed by others benefit in that their coat is kept free from parasites and dirt. Studies of captive ring-tailed lemurs (*Lemur catta*), lion-tailed macaques (*Macaca silenus*) and Celebes black 'apes' (*Cynopithecus niger*) show that parts of the body which are inaccessible during autogrooming are groomed preferentially by others (Hutchins and Barash 1976). The removal of dirt from wounds by bonnet macaques (*Macaca radiata*) has been reported by Simonds (1965) and S. A. Altmann (personal communication) has observed that baboons (*Papio cynocophalus*) which are not allogroomed become infested badly with ticks and other ectoparasites. The high frequency of allogrooming in some primate groups, however, suggests that grooming has secondary functions (see later in this chapter). Adult males of various species of monkeys are groomed more often than they themselves groom and the reverse is true for adult females (Chalmers 1973, Struhsaker and Leland 1979). The observation that male mangabeys (*Cercocebus albigena*) groom females more in groups where social organisation is more variable (Wallis 1979, Struhsaker and Leland 1979) might be accounted for if individuals groom more in order to gain or maintain acceptance in a group. Allogrooming in marmosets (*Callithrix jacchus*), which live in monogamous family groups, is less frequent than in monkeys living in larger groups. It is not mutual, i.e., one individual is groomer and the other groomee at any one time. Allogrooming is most frequently

164 An adult female chimpanzee with a newborn baby approaches and greets a male in a group of three. She holds out her hand and he responds by holding out his. Drawing of a photograph in Van Lawick-Goodall 1968.

165 The adolescent female chimpanzee has approached the seated, mature female and greets her by presenting her genital area, which the adult is just about to raise her hand and touch. Drawing of a photograph in Van Lawick-Goodall 1968.

166 A juvenile female chimpanzee greets a juvenile male with a kiss. Drawing of a photograph in Van Lawick-Goodall 1968.

shown by the parental pair, especially during the period just before birth and post-partum oestrus and Box (1978) considers that it accentuates the reproductive bond between them. Box and Morris (1980), in a study of pairs of captive tamarins (*Saguinus mystax*), found considerable individual variation in the extent of allogrooming.

In order that collaborative, mutually beneficial activities can be carried out, individuals need to know what the others are doing and what they are likely to do next. Communicative mechanisms in a variety of species are reviewed by Hinde (1972) and Sebeok (1977). The use of facial expression and of vocalisations during social interactions in primates is also discussed by Andrew (1963), Redican (1975) and Marler (1976, 1978).

Group fission

Social groups may split up for a variety of reasons. Some flocks of birds, shoals of fish and herds of ungulates may decrease in size, or split completely, at the end of the non-breeding season because pairs leave to breed, either briefly or for a substantial period when they hold breeding territories. Large groups may break up because of a change in food distribution or a reduction in predation pressure. Sometimes individuals may become separated accidentally from a group. In species with permanent social groups, individuals which reach sexual maturity may leave or be expelled from the group. Dispersal behaviour is discussed from an ecological and evolutionary viewpoint in Chapter 8.

The reproductives which leave a group may be of both sexes, as in social insects and in animals living in social groups composed of a monogamous pair plus young. More frequently, however, one sex is more likely to leave than is the other. In the Arabian babbler, in which it is usual for one female only in the social group to breed, the young females are most likely to leave when they reach breeding age (Zahavi 1974). Females move between groups more often than do males in the chimpanzee (Nishida and Kawanaka 1972, Pusey 1980, Wrangham 1980). The female with her offspring seems to be the only stable group in chimpanzees but related males occupy and may defend large ranges so that it may be very difficult for any male which does move to become established in a new group.

In most mammals which live in large groups it is the young male which leaves the social group. The most detailed studies have been carried out on primates and the benefits to the young male of moving are discussed by Clutton-Brock and Harvey (1976) and Chalmers (1979). The marked difference between male and female Japanese macaques in the likelihood of their remaining in their natal group is shown in Figure 167. Sugiyama (1976) describes how juvenile males move to the peripheral part of a group at two or three years of age. They visit the central area, occupied by their mothers and other females, less and less during the following year and they venture away from the group more and more frequently. In some groups, all males left before the age of six years. These males remain outside the group for much of the next two to five years but they gradually spend more time on the periphery of a group, often not their natal group and may gain acceptance into it. Sometimes, groups of Japanese macaques leave a group together and form a multi-

167 Most young male Japanese macaques *Macaca fuscata* leave their natal group before reaching sexual maturity. From Sugiyama 1976, quoting data from Chsawa.

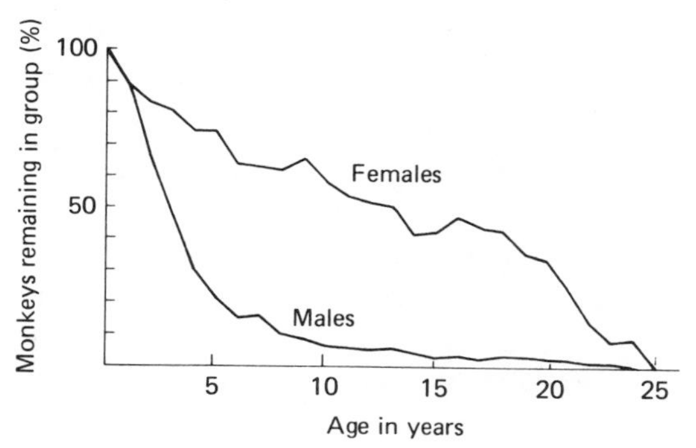

male group. Occasionally large groups split into smaller groups (Furuya 1969). The departure of young males from natal groups has been described for many other monkey species, for example, Packer (1975) found that every male in a group of baboons which he studied emigrated from its natal group during adolescence. Fission of large groups also seems to be widespread and Eisenberg, Muckenhirn and Rudran (1972) considered that most arboreal monkey groups return periodically, during many years, to a one-male state.

Leaders, initiators and controllers

The term *leader* is sometimes used to refer to the individual which is in front during an orderly group progression. This is the clearest descriptive meaning of the word and it would be simpler if its use was confined to this usage. Another way in which 'leader' has been used has been to refer to a temporal *initiator*, i.e. the individual which is the first to react in a way which elicits similar or different responses by others. The individual which flees first when a predator is detected may or may not be the leader of the exodus so it is better to call it the initiator of the movement. A third way in which 'leader' is used concerns the control of group behaviour. The majority of members of a group may refrain from initiating an activity, or may refuse to follow when one individual starts to move away, unless a certain individual initiates or joins in the movement. It may require much observation of group behaviour before it is possible to identify this *controller* and it is likely that some group activities will be controlled whilst others are not.

There may not be any single controller in a group but if there is one, that controller need not be the leader during movement or the initiator of activities. In a group of ungulates or primates, the initiator of departure from a place where flies are biting may be the individual which is bitten most, whilst the initiator of flight from a predator will often be the individual which detects it first. Movement to a feeding place or source of water may be led regularly by the same individual but the timing of such movements may depend upon the behaviour of a controller. The controller may be the strongest individual but it is more likely that other group members would respond to control by the individual who is best able to take optimal decisions. The controller of a group is therefore likely to be older than most and experienced in optimising behaviour. Examples of social groups where control is exerted by force and the controller is at the head of a competitive order are quoted later in this section but the majority of activities of many social groups are not controlled by such individuals. In human terms, the controller is equivalent to the captain of a football team who may also be the leader when his team comes on to the pitch. Leaders of armies and nations were, at one time, the individuals at the forefront in battle but the way in which the word leader is now used is confusing and imprecise so its definition must be narrowed for the description of social behaviour.

Leaders

Most fish schools and bird flocks seem to have no clear leader, for changes in direction are accompanied by changes in leadership (Shaw 1962, Allee, 1938). The leader of a wolf pack during a chase is often the largest male (Mech 1970) and large males may lead groups of macaques and langurs (Hinde 1974). In studies of most social mammals, as a result of group shape the leader is not obvious and the description of the order of progression is neglected. Amongst herds of ungulates, however, a leader is often apparent when the whole herd moves from place to place. Fraser Darling (1937) described how a mature red-deer hind consistently led a herd of hinds and calves and another hind was often in the rear. Some of the followers were certainly the daughters of the leader and Scott (1945) found that young sheep followed their mothers or other older animals.

Some consistency of leadership has been observed in many studies, especially those on domestic animals. Squires (1974) observed the same individuals leading during many of the movements of 1000 Merino ewes in a sparsely vegetated area of Australia. The order of entrance of cows into milking parlours was consistent in some studies (Dietrich, Snyder, Meadows and Albright 1965, Soffié, Thinès and Marneffe 1976), sometimes for several lactations (Ferguson *et al.* 1967, Wisniewski and Albright 1978). More variation in entry order was found in other studies (e.g. Gadbury 1975).

Leaders are often the older, more experienced members of a group. The leader in groups of New Forest ponies were usually mares which Tyler (1972) considered to be familiar with the terrain. In some studies the leaders are also the individuals which are highest in an order based on competitive interactions, for example in sheep (Squires and Daws 1975, see review by Syme and Syme 1979). The sheep which were observed most often to be leaders by Arnold (1977) were those which usually separated from others whilst resting or feeding.

The movements of groups of domestic animals sometimes refer to free movement and sometimes to driven herds. The order of progression is different in these two circumstances and this may account for some variability in results, e.g. in order of entrance to milking parlours. In driven herds, dairy cows which are high in a competitive order are often in the centre of the herd. Those which are low in competitive order bring up the rear and one or other of the individuals in the middle of the order leads the herd (Kilgour and Scott 1959, Beilharz and Mylrea 1963).

Initiators and controllers

The most obvious examples of animals in groups acting as initiators of group activity are those occasions where a predator is detected. Other examples are of the beginning of feeding after resting, drinking after feeding, or resting after feeding. When one of a herd of cows lying in a pasture stands and starts to graze, many of the others do so as well within a few minutes. The likelihood that the followers are in exactly the same physiological state as the initiator is small so, in the absence of events which might trigger the change in behaviour in all of the animals, it must be assumed that social facilitation is occurring. Studies of several groups of pigs by Meese and Ewbank (1973*a*) showed that the initiator of movements towards food, novel objects or sleeping huts could be any of several individuals.

In hamadryas baboons Kummer (1968) found that in groups of an older and a younger adult male with their females, movements were usually initiated by the younger male. If the older male followed, the females also did so but if not, the females stayed where they were and the younger male returned after a short excursion. The same distinction between initiator and controller of group movement is apparent in Schaller's (1963) description of gorilla behaviour. It is the mature, silverback male which determines, ultimately, whether or not an initiator of movement is followed by the rest of the group. Control of movement in baboons (*Papio anubis*) is exercised largely by the older females who are the most stable elements in the group (Rowell 1966*b*). Herds of female New Forest ponies sometimes move as a result of control by the oldest mare and sometimes as a result of driving by the stallion (Tyler 1972). Movements by other ungulates are often difficult to attribute to control by a single individual. The red-deer herds seem to be controlled by one hind and it is likely that older individuals are influential in determining the timing and direction of movements in most ungulate herds.

Competition within social groups

This discussion of competition in social groups begins with sections on when, why and how animals compete. Although it is difficult to describe the relationships in some social groups in terms of social orders, this approach is widespread and, in some species at least, it is useful. The sections on the constancy of social orders and the variety of social orders introduce some of the techniques and terms used in such work. The inadequacy of simple explanations for complex social behaviour and the loose usage of terms such as dominance are emphasised. The advantages and disadvantages to individuals of competitive orders and the uses and limitations of the competitive order concept are then considered.

Why and when to compete

Membership of a social group may involve collaboration with other group members with regard to some functional systems for part or all of the time. Where any resource is not effectively limitless for all group members or where any spatial position is preferred by more than one individual, however, they will compete for it. The resource may be food, water or a mate and the position may be one which is desirable for predator avoidance, reduction of physical hazard, or improvement of energetic efficiency.

Behavioural competition need not involve any physical confrontation between rivals. For example, as Syme (1974) points out, in many species competition for a food item is often resolved by the fastest mover acquiring it. Nevertheless, the occurrence of threat displays and fights is frequently reported by those who describe social behaviour. The literature is misleading in this respect for the incidence of such competitive orders and the rank which is ascribed to individuals on the basis of the order are overemphasised in descriptions of the behaviour of most social species. Social animals allocate a very small proportion of their available energy to competitive confrontations throughout most of the year and evolutionarily stable strategies (ESSs) include provision for losers as well as for winners (see next section). A reason for particular interest in competitive encounters, even if they are rare, is that the outcome of a single fight will sometimes have a lasting effect on the relationships of the participants with all group members. These relationships may affect behaviour frequently in some species but only rarely, e.g. during extreme food shortage, in other species.

The time of occurrence of competitive encounters may coincide with that at which there is rivalry for a resource or position but it need not do so. A fight might put both individuals at risk at a time when both need to eat rapidly in order to survive or when both are trying to find the best position for predator avoidance. As a consequence, competitive encounters sometimes occur when there is no obvious resource or position at stake. In some species, individuals threaten or fight as soon as they meet whilst in others there are long delays before any encounter. Domestic animals are often put together in groups and it was this situation which Schjelderup-Ebbe (1922) studied in domestic fowl. He found that they pecked one another often initially but the incidence of pecking declined with time. Since the individuals which were pecked were likely, subsequently, to avoid those who had pecked them, he concluded that they recognised one another individually and remembered the results of encounters. Guhl (1942, 1953) confirmed that these initial interactions affected later relationships in chickens and various authors have described similar high frequencies of encounters when other farm and laboratory animals are first put together.

In groups of primates and ungulates, competitive encounters and amicable associations may occur during the same day. Harcourt (1977) described how adult male mountain gorillas competed for food or copulation with females during one part of the day whereas they rested side-by-side and groomed one another at another time. As in man, the nature of the interaction depended upon the context. Cows will also butt and lick another individual at different times on the same day. The requisite motivational state for competitive interactions by cows is more likely to occur at the time when a limited quantity of food is provided but as Bouissou (1976) points out it is also more likely after any disturbance by people.

How to compete

A gene which made it likely that the bearer would fight to the death against any rival would be unlikely to spread in the population unless the increased fitness conferred by winning the fight was very large. The importance of the cost of competing is stressed by Maynard Smith and Price (1973) in their theoretical discussion of animal conflicts. As explained on pages 11–12, they calculate the ESS for contests which are symmetrical, i.e. neither contestant is at an advantage. Most competitive encounters, however, are asymmetrical and in this situation the ESS is to decide whether you are likely to win or not and then escalate the fight if you are but retreat rapidly if you are not (Maynard Smith 1976*b*). This suggests an explanation as to why display, threat and ritualised fights are much commoner than real fights, for the protagonists are trying to assess one another's abilities without risking bodily damage or energetically expensive fighting.

The minimal asymmetry between rivals is priority of place at the resource. This is exemplified by Davies' (1978*a*) study of the territorial defence in the speckled wood butterfly (page 198) and also by scent-marking. Scent marks, which are used to demarcate territories by mammals such as the fox (*Vulpes vulpes*) (MacDonald 1979), may convey the simple message that this area is already occupied by a fox which has visited this spot recently. Such a message may suffice to deter entry by an intruder fox (see review by MacDonald 1980). Another example of a situation where possession is sufficient asymmetry to deter any attack is Kummer's (1968) finding that male hamadryas baboons will not attempt to wrest a female from a male with whom she is already consorting.

In some situations, where there is no large initial asymmetry in the qualities of two contestants, chance may determine the result of the first encounter. One individual may be lucky in landing a blow or in obtaining an item of food for which they are competing. Once this has happened an asymmetry does exist in any further contests between these two individuals and, as Rowell (1974) explains, the victor may be more forceful and the loser more hesitant when they compete again. This difference may persist or increase during successive encounters.

In many competitive situations, the result is determined by the relative size of the bodies or the weapons of the contestants. Riechert (1981) found that the funnel web spider *Agelenopsis aperta* will take over the web of another individual if it can. The owner of a web assesses the size of an opponent who arrives on the web by monitoring the vibrations which it causes and fleeing if the newcomer is obviously larger. When Riechert put lead weights on a newcomer spider so that it appeared to be heavier than it really was, web owners were deceived and departed without a fight. Assessment at a distance, using vocalisations, is described for toads and red deer in Chapter 9. It is beneficial for individuals which are at an advantage to advertise their physical superiority. They may show off their full body size, their large chela in the case of a fiddler crab *Uca* (Crane 1975), their large comb in the case of a cockerel (Guhl 1953) or their large horns in the case of the bighorn sheep (*Ovis canadensis*) (Geist 1971). Sometimes an individual may be able to deceive its rival by a display which makes it appear larger than it really is. Confident displays (Figure 168) may be adopted by both contestants initially but

168 A confident wolf (*Canis lupus*) may be recognised by the position of its tail and ears.

if the contest escalates to the point of fighting the potential loser should concede defeat rather than risk injury. Even where contests are largely ritualised, injury may occur, for example in fiddler crabs (Jones 1980). In species which do fight, rather than just display, a result may be possible in a contest where fighting cannot occur. Lindsay, Dunsmore, Williams and Syme (1976) found that rams separated by a fence displayed at one another and the mating behaviour of the losers in this contest was inhibited. Real fights are sometimes decided by a brief bout of pushing or by one bite, especially in those animals which possess weapons which can kill a conspecific. Less well-armed animals may fight for longer since neither animal risks so much by escalating the contest.

The constancy of social orders

Persistence of orders

When groups of adult chickens are put together, the relationships in competitive situations which they establish during the first day may persist for long periods. Guhl (1968) reports that a competitive order established amongst ten hens usually persists for many months and he suggests that 'social inertia' develops in that it is much more difficult for an individual to change its relationships in an established group than it is in the early period after group formation. This is difficult to prove but in red deer, when antlers were removed from stags and replaced by larger or smaller antlers, the changes in competitive order were less than would have been predicted from consideration of antler size (Lincoln, Youngson and Short 1970). In a long-term study of a herd of cattle, Sambraus and Osterkorn (1974) found considerable stability in competitive order over a period of eight years. The mean annual reversal of rank was 14%. In most studies, however, if an individual causes another to withdraw in the majority of contests, it is considered to be higher in a competitive order. For some pairs of animals the individual classified as loser might have won four out of ten contests. The use of the words *dominant* and *subordinate* seem inappropriate to describe such a relationship because they imply uniformity in the result of contests.

A variety of factors may affect the result of any particular contest. In pigs, contests in which the initial loser wins may increase in frequency as the individuals become more familiar with one another (Bryant and Ewbank 1972). There may also be changes in results of contests as some individuals become older and stronger whilst others become ill or senile. A hungry individual, fighting to obtain food, may reverse the results of previous contests if the opponent is well fed and hence has less to lose. For example, a well-fed cow will allow itself to be displaced at a food trough by a cow which it would displace on most other occasions. King (1965) found that cockerels whose access to food was suddenly and severely restricted no longer showed a clear competitive order but pecked at one another in a much more random way. Hormonal fluctuations may also affect contest results. Testosterone treatment has some effect on the results of contests in red deer (Lincoln *et al*. 1970) and much effect on some individuals in domestic fowl (Allee, Foreman and Banks 1955, Guhl 1964). Competitive orders can also be reversed by using shocks to make feeding by the hen at the top aversive in the presence of a hen at the bottom (Smith and Hale 1959).

Types of order

When the occurrence of certain competitive interactions amongst individuals in a social group is recorded, it is possible to describe some sort of social structure on the basis of the results of the interactions. The simplest structure (Carpenter 1971) is that where a *despot* wins against the others who do not compete with one another, for example in some groups of male lizards (Figure 169). The structure which Schjelderup-Ebbe (1922) described as being frequent in chickens was a *linear peck-order* in which each individual pecks those below it but not those above it. Schein and Fohrman (1955) described similar orders based on butting behaviour in cattle. Most authors have also found that non-linear orders can occur, for example *triangular relationships* in which A beats B, B beats C and C beats A or more complex relationships (Bouissou 1965).

More-or-less linear competitive orders have been described for some primates, for example vervet monkeys (*Cercopithecus aethiops*) (Struhsaker 1967), yellow baboons (*Papio cynocephalus*) (Hausfater 1975), and Barbary macaques (*Macaca sylvanus* (Deag 1977). Linear orders, once established, may result in few interactions between individuals far apart in the order but continuing encounters between those of similar rank, e.g. in cattle (Schein and Fohrman 1955) and in pigs (Meese and Ewbank 1973*b*). One example of a visible linear order which is the result of competition is the teat order in piglets. During the first few days of life, piglets compete for positions on the udder of the mother. The anterior teats, which give more milk, are preferred and the order is stable by the end of week two in most litters (McBride 1963, Hemsworth, Winfield and Mullaney 1976).

A variety of social orders

Many behavioural measures have been used in studies of competitive encounters.

169 The lizard in the foreground is a despot in that he wins all contests with the other members of his social group. In this species *Leiocephalus carinatus* the contests consist of fights or tail curling and other threat displays. After Evans 1953.

These include fighting until one individual flees, displaying until one flees, chasing, one avoiding the path of another, one showing a submissive display which reduces the likelihood of attack by the other, one showing a confident posture or display whilst the other does not, one scent-marking whilst the other does not, one mounting the other, one grooming the other, displacement at a food or water source, displacement whilst mating; displacement from a preferred position, and giving precedence at a resource or preferred position. These interactions refer to dyads but, as mentioned earlier in this chapter, events of this kind should be described by referring to the social context as completely as possible. In many studies, an order based on one measure has been found to be different from an order based on another measure (Syme 1974). Kummer (1957, 1968) found that rank based on agonistic encounters between male hamadryas baboons was not related to rank based on access to oestrus females. A poor negative correlation between wins in fights and submission or avoidance behaviour was described by Rowell (1966*a*) for baboons and Barton, Donaldson, Ross and Albright (1974) for cattle. Bernstein (1970) was unable to find close correlations between rank orders based on agonistic encounters, mounting sequences and grooming interactions. Consequent upon such results, the validity of any real rank order was questioned by Gartlan (1968), Bernstein (1970) and Rowell (1974) who also expressed doubts about whether groups of animals in the wild would show social structures like those found in captive groups.

Other studies, however, had shown that correlations between measures of behaviour in competitive situations can be found. Simpson (1973) found correlations between displays, suppression of displays, displacement and grooming in chimpanzees. S. M. Richards (1974) calculated that competitive orders based on fights, displays, avoidance, submissive displays, and four measures of priority of access to food were similar in captive rhesus monkeys. Deag (1977) obtained similar results, for measures similar to those of Richards, in a study of wild Barbary macaques. It is apparent from these studies and from work on domestic animals, that it is useful when describing relationships in groups to record the results of competitive interactions and to rank animals on the basis of such interactions. Such a ranking is, however, only one part of the description of relationships and it may be a small part in some groups. When a competitive order has been determined it is often called a dominance order. This term may be misleading for whilst 'dominance' is usually assessed in terms of competitive interactions, when used it is often equated with control. As mentioned earlier in this chapter, it may be that the animal which is highest in a competitive order controls the activities of the group but this need not be the case. Important activities such as movement of the group to a place where food is better or predator avoidance is easier may be controlled by an old female, even though a male is the clear head of the competitive order.

Advantages of competitive orders

The following is a list, modified and extended from that of Deag (1977) of the possible advantages which may accrue to the individual which is at the top of a competitive order.

(1) *Priority of access to a potential mate.* The winner of a competition for a mate (see Chapter 9) has an obvious advantage over the losers. Guhl and Warren (1946) found that the cockerels which won most fights produced more progeny than did other cockerels in the flock.

(2) *Opportunity to reduce the reproductive success of others.* As a result of harassment during oestrus, the female gelada 'baboons' at the top of the competitive order produced more offspring than did others (Figure 170, Dunbar and Dunbar 1977).

(3) *Priority of access to food or water, immediately or in the event of a shortage, and greater feeding efficiency due to less disturbance by other group members.* Murton *et al.* (1966) showed that woodpigeons (*Columba palumbus*) which were high in the competitive order positioned themselves at the centre of a feeding flock and ate more food than did those at the flock periphery. Bouissou (1970, 1971) and Leaver and Broom (unpublished) found that the cows which won most competitive encounters obtained most food in a competitive feeding situation and put on most weight. Priority of access by high-ranking individuals to food has been described in many studies of social animals although there are also studies in which no such advantage has been detected, e.g. Friend and Polan (1978) in a study of cattle.

(4) *Priority at preferred resting places where avoidance of predators and other hazards is easiest.* The woodpigeons in the centre of the flock, mentioned above, are less vulnerable to predator attack. In bird roosts, the individuals which occupy the sites which are safest from predation and which provide most shelter from adverse weather conditions are those which have been able to compete successfully for such sites (Wynne-Edwards 1962, Broom *et al.* 1976).

(5) *Greater freedom to move around the area occupied by the group and to interact with other individuals.* Restricted activity by individuals low in a competitive order has been described for pigs (Fraser 1974), cattle (Syme, Syme,

170 By showing harassing behaviour which inhibits ovulation or leads to abortion, female gelada 'baboons' *Theropithecus gelada* at the top of the competitive order were able to produce the most offspring. Modified after Dunbar and Dunbar 1977.

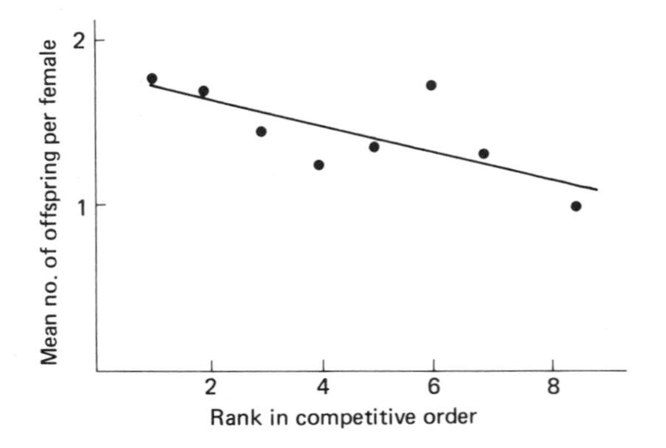

Waite and Pearson 1975) and various other animals including primates.
(6) *The possibility of intervening in fights involving kin* (see Deag 1977).
(7) *Reduced likelihood of involvement in repeated contests which may be physically or physiologically damaging.* The idea that a rapidly established competitive order would reduce the overall frequency of damaging, energetically costly contests in social groups was proposed by Guhl (e.g., 1964) and gained wide acceptance. Once the rank of an individual is known to group members, priority at resources etc, is decided without the necessity for further fighting.

A high rank in a competitive order may be disadvantageous if repeated fighting is necessary to maintain that position. The competitive order may result in other disadvantages to high ranking individuals. Firstly, the efforts of an individual in competing for food or water may be more than are necessary to allow adequate access for itself and may harm offspring or other kin. Secondly, too vigorous reinforcement of rank by attacks may increase the distance between individuals, so making offspring or kin more vulnerable to predation (Deag 1977). The disadvantages to low ranking individuals are lack of advantages (1) to (6) above and the possibility of injury or other damage.

Involvement in vigorous contests has been shown to have effects on adrenal functioning and might therefore be called 'stressful', in a variety of animals (see Chapter 5 for explanation of the measures described below). For example, adrenal weight was higher in cockerels which did not initiate many competitive encounters (McBride, Foenander and Slee 1970) and plasma corticosterone levels were higher, on average, during days 2–7 after cockerels were grouped than they were 28 days days later (Williams, Siegel and Cross 1977). Studies of rhesus monkeys have shown that adrenocorticotrophic hormone (ACTH) levels in the blood were higher amongst individuals which received more attacks and threats than amongst high ranking individuals. If the lowest ranking animal was removed, ACTH levels increased in some of the remaining low ranking animals (Sassenrath 1970). These results with captive animals, however, do not necessarily mean that such adrenal responses occur in the wild and, as discussed in Chapter 7, the adrenal response indicates physiological state but need not be damaging to the individual.

Losers in contests can minimise their loss of fitness by retreat before injury, by avoidance of fighting or by leaving the group. In some groups where much fighting occurs, individuals which lose if they fight behave so as to minimise encounters and then have more energy available for feeding. Such individuals may produce more young ultimately than those high in the competitive order. Practice in fighting when younger, even though it involves losing, may result in more efficient fighting when older. Guhl's view that competitive orders reduce the overall incidence of damaging contests may well be correct for many social groups. Low ranking animals might be less subject to attack by other group members if such an order exists. Studies of many animals in which some competitive order has been described show that all animals in the group are able to feed adequately and to breed. Perhaps, in these cases, collaborative behaviour is of great importance to all group members so that it is not in the best interests of the high ranking individuals to injure or drive away lower ranking members.

The competitive order in relation to other measures of describing social structure

Whilst competitive orders are very clear and have a considerable effect on the fitness of individuals in some groups, they are of minor importance or are absent in other groups. No study of social structure can be complete unless competitive interactions are studied as part of the description of social relationships within the group but it is essential to study all types of interactions in order to understand the relationships. The rank of an individual in a competitive order need not be related to the extent to which that individual controls the behaviour of others and neither the highest ranking individual nor the controller need be the leader of a progression during locomotion or the initiator of a group activity. The term 'dominant' has been used most frequently to refer to rank in a competitive order. Other meanings have often been implied as well and there may be several different competitive orders so it is a confusing term. If it is to be used, dominance should refer only to rank in a specified competitive order. With increased sophistication of description of social groups it has become apparent that studies in which only one parameter of social behaviour is measured are of limited use. A much better understanding of relationships within the group can be obtained by the use of a wide range of measures.

The development of social skills

During development and to a lesser extent throughout much of adult life, experience leads to improvements in sensory functioning, motor control, the ability to respond to novel stimuli and the size of the repertoire of abilities learned. Social behaviour depends on each of these, hence the sections on behaviour development in previous chapters are relevant to considerations of how social behaviour develops. For example, a gerbil pup whose eye-opening is accelerated by the presence of the father or as a result of the litter size being three rather than five, will be able to engage in some social behaviour earlier as a consequence (Elwood and Broom 1978). Any individual which, as a result of its rearing conditions during development, is better able to cope with rapidly changing, recurring variables will be more likely to succeed in acquiring social skills. Social behaviour necessitates the recognition of specific cues, the execution of appropriate responses and the prediction of complex sequences of events. It is, therefore, the most difficult type of behaviour for most animals. As a consequence there is intraspecific variability in social behaviour.

The gradual appearance of components of social behaviour has been described in detail for the Burmese red jungle-fowl (*Gallus gallus spadiceus*) by Kruijt (1964, 1971). The development of sexual behaviour is discussed in Chapter 9 and the ontogeny of fighting is detailed here. Adult cocks show leaping with wing-flapping, pecking and kicking during a fight. The precursor of leaping and wing-flapping is a small jump, called hop by Kruijt, which is performed by the young chicks. Since jumping by one chick often induces other chicks to do the same, apparently accidental clashes of chicks which are jumping may occur during the latter part of the first week of life. After this age, jumps are sometimes directed at other chicks and the first fights occur. After two or three weeks of age, pecking movements, confined previously to feeding behaviour, are included in the fights. Kicks are incorporated into fights at a latter stage and the precision of all the movements increases. Experienced fighters appear to be more proficient at kicking than are those with less experience, hence the careful training which used to occur for cock-fighting. The other aspect of fighting behaviour investigated by Kruijt was the choice of opponent. Cockerels reared in groups fought other cockerels but those reared with moveable cylinders or other objects in their cages fought these.

Effects of isolation-rearing

The simplest experiments on the development of social behaviour are those in which an individual is reared in spatial, visual or complete isolation and any inadequacies of social behaviour are assessed when it is put in a group. Male jungle-fowl reared in visual isolation until they were six to nine months old were able to show most, but not all, displays although the timing and duration were not always appropriate (Kruijt 1964). Isolation rearing increased inter-individual distance when domestic chicks were grouped (Pattie 1936; Baron and Kish 1960; Shea 1976). The avoidance of conspecifics after isolation was also shown by rhesus monkeys (Mason 1960) and the social responses of rhesus monkeys after six months spatial isolation were drastically altered (Figure 171, Harlow 1969). When competitive interactions between spatial-isolation-reared and socially reared rhesus monkeys (Mason 1961), deer mice *Peromyscus* (Rosen and Hart 1963) and cattle (Donaldson, Black and Albright 1966, Broom and Leaver 1978) were studied, the isolation-reared individuals nearly always lost the competitions. Mason (1961) concluded that 'one of the consequences of social restriction is failure to acquire effective elementary communicative skills which seem to coordinate and control the form and direction of social interactions'. Detailed observations of interactions in cattle showed that, during an encounter, previously group-reared animals held their ears forward more and looked at their rival more than did previously isolation-reared animals. The lack of adequate response when approached resulted in the isolation-reared cattle losing most of their initial encounters and avoiding such encounters later (Broom unpublished). If calves reared in spatial isolation have experience of a moving model which they can push and butt, their competitive ability when confronted with other calves is much improved (Waterhouse 1978, Waterhouse and Broom unpublished).

171 During the two months after grouping, rhesus monkeys which had previously been spatial-isolation-reared for their first six months showed less social behaviour of various kinds than did previously group-reared monkeys. Modified after Harlow 1969.

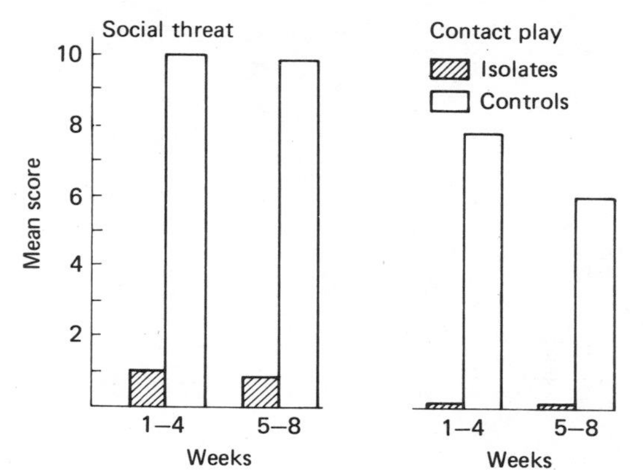

Learning from parents and peers

Since practice in interactive behaviour can have a considerable effect on later competitive behaviour, it is likely that many other social skills have identifiable precursors during development. Some influences of parental behaviour on development of young are described in Chapter 9. Hinde (1979) discusses the development of relationships in man and stresses that the social behaviour of an individual is a consequence of its physiological state and its concept of its role in the social group. A young baby soon learns about what others around expect of it and adults teach it how to behave so a role concept is acquired (see Hinde 1979). Whilst Bowlby (1969) considers that experience gained in the social world of early childhood is used widely in adulthood, Harré (1974) rejects this idea by saying that social skills acquired early in life are of no use later. Hinde emphasises the differences between the social behaviour necessary for child and adult but he considers, as would most biologists, that some influence is carried over from childhood to adulthood.

The effects of parents on the development of an individual's concept of its social role are often difficult to assess by behavioural observation because the parents have a direct effect on the way in which other social group members behave towards their offspring. Kawai (1958) reported that the rank in a competitive order of young Japanese macaques depended upon the rank of the mother. However, Cheney (1977) found that high ranking baboon mothers aided their offspring in contests more often than did low ranking mothers. In such circumstances it is difficult to determine how much the behaviour of the young animal is a consequence of watching parental and other interactions and assuming a role accordingly.

Young social animals must also learn skills and formulate their concept of their social role as a result of interactions with their peers. Almost every interactive behaviour among young animals has been called 'play', a term which is very difficult to define. The vigorous activities of young animals include chasing, fighting without causing injury, advancing towards another individual and retreating without contact, acrobatics of various kinds and many vocalisations. The idea that these might have a function in improving physical fitness and social skills was proposed by Brownlee (1954), as a result of work with cattle and reiterated by Dolphinow and Bishop (1970) and by Fagen (1976). Although such functions seem likely and there is some supporting evidence from work on children (Sylva, Bruner and Genova 1974 and other papers in Bruner, Jolly and Sylva 1974), it is difficult to demonstrate them experimentally (Chalmers 1979). One type of social behaviour which is likely to be improved by juvenile chasing, mock fighting etc. is collaborative hunting. Some of the techniques used by lions, hyaenas, hunting dogs or other group-hunting predators seem to form part of juvenile behaviour although others must be perfected as a result of practice whilst actually hunting. Animals which live in groups may be able to learn by imitation, skills used for feeding, predator avoidance etc. as well as the social skills mentioned above. The opportunity to benefit from the wisdom and innovations of group members is one of the advantages of group-living and some examples of known occurrences of this kind are described in Chapter 8 (for a review see Galef 1976).

The effects of crowding

The frequency of encounters between individuals which moved around randomly would be related directly to their population density. Greater density would mean more likelihood of approaching to within the minimal *individual distance* (page 195) which another animal will tolerate. Such close approaches often elicit attacks. The movements of animals are not random, however, and an increase in density may be followed by behavioural modification which prevents an increase in density in close encounters. Groups of individuals whose activities are restricted by the physical presence of others are said to be *crowded* but crowding is not necessarily associated with increased competition. Hence it is not easy to predict how changes in density will affect social behaviour.

There will be variation amongst species and amongst individuals in individual distance and in the extent to which behaviour is modified at high densities. As Stokols (1972) points out, high social density does not necessarily result in *over-crowding*, i.e. crowding with adverse effects on the fitness of individuals. The measurement of overall social density will also give a misleading impression of the likelihood of adverse effects if the distribution of resources is not considered. If there is only one food source or sheltered area in otherwise spacious surroundings the individuals may be temporarily over-crowded whilst the resource is used. It is essential to consider the quality of living space (Box 1973) as well as social density.

Group-size effects

For many species of animals there may be a social density threshold above which the incidence of competitive behaviour depends solely upon the numbers of individuals present. This effect is exemplified by the work of Al-Rawi and Craig (1975) on domestic chickens. When birds were housed such that the amount of floor-space

172 Different numbers of domestic hens were put in cages with a constant floor-space allocation of $0.4m^2$ per bird and the frequency of pecks and threats resulting in avoidance was greater the larger the number of birds. Modified after Al-Rawi and Craig 1975.

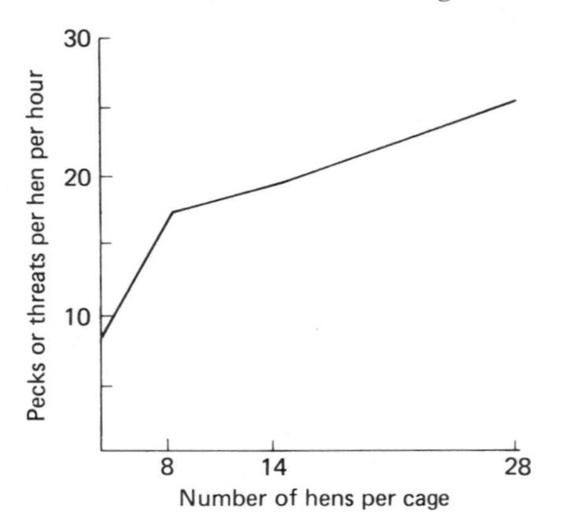

per bird was 0.4m^2, the frequency of interactions in a flock increased as flock size increased (Figure 172). Despite the fact that each bird had the same amount of space, at this social density the social organisation requires that each individual interacts with most of the others present. If the group-size reaches a high enough level it becomes impossible for an individual to interact with all the others. An increase in the size of a flock of chickens from 100 to 400, with a similar increase in total space, did not affect the frequency of competitive interactions (Craig and Guhl 1969). The results of such experiments on the effects of group-size depend upon the density and upon the social organisation of the species so the literature includes a variety of different results.

Social density effects

Studies of the effects of social density, like those above on group-size, will also depend upon the social organisation and on the modifiability of social behaviour. As a consequence, comparisons among species are difficult. The observation of adverse effects of crowding in one species does not allow the prediction that similar crowding will have adverse effects in another species. Research in which the possible adverse effects of crowding were investigated was stimulated by the work of Christian (e.g. 1955, 1961) and Calhoun (1962*a*,*b*) on fighting, adrenal function and reproduction in rodents. Christian found that in wild-caught and domesticated rats the incidence of fighting, the size of adrenal glands, the amounts of adrenocorticotrophic hormone (ACTH) and corticosterone secreted and the extent of reproductive suppression was proportional to the degree of crowding. Calhoun reported similar results and also described inadequate parental behaviour, e.g. no nest building, at high population density. Pathological consequences of these changes include prolonged high blood-pressure, kidney failure and inadequacies of inflammatory and immune responses which lead to increased susceptibility to parasitism and disease (Selye 1950, Archer 1979*b*).

In the wild, the most likely response of an individual which is frequently attacked is to leave the group. Clough (1965) describes how large female lemmings (*Lemmus lemmus*) defeated and chased away other individuals. Many individuals left the home area when the population increased. The consequence, in this species, is the initiation of large-scale migration after a period of successful breeding. In a study of wild voles (*Microtus*) Frank (1957) described behaviour modification when the population density increased. First territory size was reduced and then communal nests where several females raised young were established. Laboratory and farm animals which are attacked by another group member have no opportunity to leave the area occupied by their group. They can, however, modify their behaviour so as to minimise the likelihood of fighting. Lloyd and Christian (1967) described how house mice (*Mus musculus*) formed aggregations and remained inactive at high densities. Inactivity was also observed when cattle were very crowded (Donaldson, Albright and Ross 1972). Such adaptations ameliorate over-crowding effects but over-crowding is still a serious problem for those who keep animals in captivity. The behavioural adaptations of man to crowded conditions may be inadequate for some individuals in some circumstances.

The significance of social organisation studies for animal husbandry

A general trend in the husbandry of farm animals is to increase the size and density of animal groups. As a consequence, questions about how well the animals can adapt their social behaviour, so that normal growth and reproduction can occur in these circumstances, are now of considerable importance. Management methods which result in social behaviour which is so disrupted that animals injure one another or impair one another's growth, reproductive ability or life expectancy are undesirable on economic and humanitarian grounds. It is important, therefore, to carry out experimental and observational studies which lead to improvements in rearing methods, provision of appropriate facilities for normal behaviour and useful decisions about inter-group transfers and group composition.

When trying to understand why social behaviour problems arise in groups of farm animals it is instructive to compare the group composition with those which have been observed in *feral* populations, i.e. those of domestic animals now living in the wild state. Many of the problems with domestic fowl arise from individuals pecking one another. McBride, Parer and Foenander (1969) studied a population of 600–1500 feral fowl on an island off the coast of Queensland, Australia. The social groups in the non-breeding period consisted of harems of about 12 hens with one cock. Cocks without harems were driven away with some vigour and spent their time in isolation or at the periphery of groups. Hens were social all the time except when laying and the cocks accompanied them when they visited their nests. In this study and in that of Wood-Gush, Duncan and Savory (1978) on another feral population, competitive interactions were very infrequent except when cocks intruded into the harem or territory of another. A very low incidence of fighting was also found among the feral Soay sheep studied by Grubb and Jewell (1966) on the island of St Kilda off Scotland. Here, ewes and rams remained separate except during the breeding period. More prolonged association of males and females in spring breeding was observed in the feral cattle of the Camargue in France (Schloeth 1961). The densities of herds on farms must be higher than in these feral groups but perhaps there is something to be learned about adaptations within the species for particular group compositions at different times of year and about the desirability of constant composition in some social groups.

The consequences of social orders

Many studies have established that there can be a positive correlation between the position of an individual in a competitive order and its production. For example egg production in hens (Guhl and Allee 1944), growth rate in goats (Pretorius 1970), growth rate in pigs (James 1967), avoidance of weight loss in bulls (Blockey and Lade 1974), and milk production by cows in some studies (Barton *et al.* 1974) but not in others (Beilharz, Butcher and Freeman 1966, Dickson, Barr and Wieckert 1967). In some studies the overall correlation may not be clear but some individuals which are low in the competitive order may die or grow much less well than the average animal. The loss, or weakness and poor growth, of some animals may be expensive for the farmer and should be avoided on moral grounds whenever possible. This situation may

173 The paths of two cows, after food is provided in a wagon, are shown. Animal (a) is high in a competitive order and she obtains sufficient food in 65 min. Animal (b), which is low in a competitive order, walks much further because of displacement at the food wagon and takes 120 min to feed. Modified after Albright 1969.

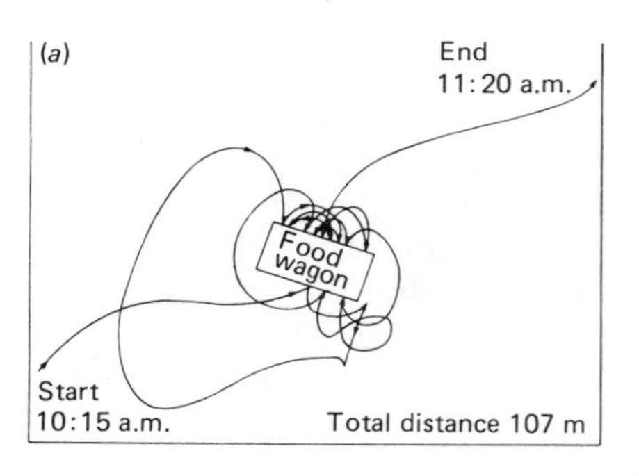

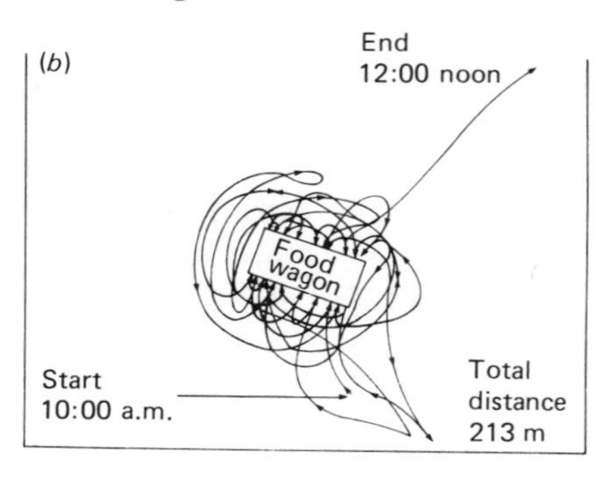

arise because the low ranking individuals are not able to gain access to resources or they may be the subject of frequent attacks. This problem is discussed by Albright (1969) and it can sometimes be solved by reducing the likelihood of competition at the source of food or water. If the food trough does not provide sufficient room for all of the animals to feed, low ranking individuals may be at a considerable disadvantage. They may have to walk much further and be disturbed much more often during feeding (Figure 173, Larsen 1963) or they may fail to get any of the highest quality food (Leaver and Broom unpublished). The situation is improved for these animals, by the use of a food trough which is much longer so that there is a place for each individual. If the long trough is placed next to a fence it can be replenished easily by farm staff. Hence there are two advantages for 'fence-line feeding'.

Where individuals are subject to frequent debilitating attacks, it may be possible to reduce the incidence of these by increasing the amounts of space available or by including refuges in the pen. Calves which are subject to attack may have to keep running around a bare pen but they may avoid most attacks if they remain in cubicles at the back of a pen (Waterhouse and Broom unpublished). Another way of reducing the severity of attacks is to reduce the efficiency of the attacker's weapons. Cattle which are dehorned and chickens whose sharp beaks are blunted by cutting off the tips are less likely to damage one another. A further method of reducing damage to low ranking individuals is to remove the most troublesome attacker. This may be an effective cure for the problem but another individual may take its place.

Effects of rearing conditions and inter-group transfer

Social behaviour develops in different ways according to experience during development and, as mentioned earlier in this chapter, it is drastically affected by isolation-rearing. The performance of isolation-reared heifers, in a situation where they have to compete with group-reared animals, is poor (Donaldson *et al.* 1966, Broom and Leaver 1978). The important general point here, however, is that competitive feeding situations should be avoided. Competition during rearing may result in one individual obtaining little food and Stephens (1974) found that calves which had competed vigorously and successfully for access to a teat supplying milk suffered a considerable growth check at weaning.

The complex structure of established social groups takes some time to develop so if animals are frequently moved from one group to another they may devote much energy to social interactions and they may suffer reduced metabolic efficiency due to excessive adrenal activity. When McBride, Arnold, Alexander and Lynch (1967) mixed two flocks of sheep they showed little sign of integration during the first week after mixing and some signs of the original grouping were still apparent after 17 days (Figure 174). The establishment of a competitive or-

174 When two groups of 100 Merino ewes were put together at a density of 15ha.^{1} they rested (camped) as shown on the first, seventh and seventeenth days. Complete flock integration took 20 days. Modified after McBride *et al.* 1967.

Day 1

Day 7

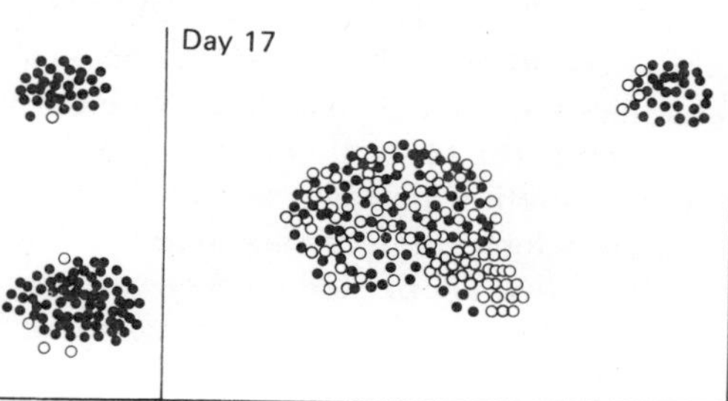

der and the consequent decline in competitive interactions when cows are first put together in a herd takes about a week (Schein and Fohrman 1955). During the period when this competitive order is being established Schein, Hyde and Fohrman (1955) reported a 5% decline in milk yield. In subsequent studies, Arave and Albright (1975) and Brakel and Leis (1976) have found similar declines. Kay, Collis, Anderson and Grant (1977) found no such decline but did detect increases in the white-cell count in the milk, a stress indicator, during the week of increased fighting. It seems that cows adapt behaviourally to inter-group movement but this adaptation is probably much less traumatic if the individuals have met one another before. Reintroductions are less likely to cause production losses than complete changes in herd composition.

Group size and density

The work by Craig and collaborators, mentioned earlier in this chapter, has shown that a complex interaction between group size, density and food container size affects the incidence of competitive behaviour and production in the domestic fowl (Al-Rawi, Craig and Adams 1976, Craig and Polley 1977). Larger group size led to increased competition, increased mortality and reduced egg production. Higher density leads to reduced weight gain in cockerels. Other studies of this kind are reviewed by Syme and Syme (1979). The optimum group size and density for growth and egg production will, by definition, result in the average individual being healthy and not subject to serious attack by others in the group. Some individuals, however, might be badly pecked or unable to feed even in these conditions. The welfare of the weakest individuals is often not considered when overall production is the factor determining husbandry methods. Another problem is that the housing conditions which are necessary to reduce the incidence of severely pecked chickens are expensive. If public morality requires that more humane husbandry methods be used for chickens then the cost of chickens and eggs will increase. For a review of the effects of crowding and other aspects of rearing conditions, on the physiology and behaviour of chickens and other farm animals see Bareham (1975).

Many studies have demonstrated that the growth rate of pigs can be adversely affected by crowding. Crowding increases fighting (Ewbank and Bryant 1972), lowers food consumption (Bryant and Ewbank 1974) and reduces the efficiency with which food is used for weight gain (Heitman, Hahn, Kelly and Bond 1961). As with chickens, both group size and stocking density are important but density seems to be the more important factor. Bryant (1972) summarises the consensus of opinion that with *ad libitum* feeding of pigs of over 50kg, weight gains are impaired if the area available per pig is less than $0.7m^2$.

Crowding is less of a problem with grazing animals like cattle but the question of the optimum size for a group of cows is important for farmers. Darwin (1839) noticed that 10000–15000 cattle on a ranch in South America subdivided themselves into groups of 40–100 which kept together. This observation and many studies of groups on modern farms, have been considered by Albright (1978) in his recommendation that herds of more than 100 should be split into smaller groups. He points out that it is difficult for animals to recognise individually all the members of large groups and he quotes examples of severe competition for resources in very large herds. This subject requires further thought and study.

It should be apparent from this section on the application of social behaviour studies to animal husbandry that precise observation of behaviour can be of great value. It is also clear, however, that there is much still to be discovered about farm animal behaviour. Farm animals are convenient subjects for behaviour studies and if modern analytical techniques are used the results will help farmers to manage their animals efficiently and humanely.

References

Abu Gideiri, Y. B. (1971) The development of locomotory mechanisms in *Bufo regularis*. *Behaviour 38*, 121–31.

Adams, D. B. (1968) The activity of single cells in the midbrain and hypothalamus of the cat during affective defense behaviour. *Archs ital. Biol. 106*, 243–69.

Adrian, E. D. and Zotterman, Y. (1926) The impulses produced by sensory nerve endings. Part 3. Impulses set up by touch and pressure. *J. Physiol., Lond, 61*, 465–83.

Aidley, D. J. (1978) *The Physiology of Excitable Cells*, 2nd edn. Cambridge: Cambridge University Press.

Ainley, D. G. (1974) The comfort behaviour of Adélie and other penguins. *Behaviour 50*, 16–51.

Ainsworth, M. D., Bell, S. M. V. and Stayton, D. J. (1971) Individual differences in strange-situation behaviour of one-year olds. In *The origins of human social relations* (H. R. Schaffer, ed.). London: Academic Press.

Ainsworth, M. D. S. (1979) Attachment as related to mother–infant interaction. *Adv. Study Behav. 9*, 1–51.

Akindele, M. O., Evans, J. I. and Oswald, I. (1970) Monoamine oxidase inhibitors, sleep and mood. *Electroenceph. clin. Neurophysiol. 29*, 47–56.

Akinlosotu, T. A. (1973) The role of Diaretiella rapae (Mackintosh) in the control of the cabbage aphid. Ph.D. thesis, University of London.

Albright, J. L. (1969) Social environment and growth. In *Animal growth and nutrition* (E. S. E. Hafez and I. A. Dyer, eds.). Philadelphia: Lea and Febiger.

Albright, J. L. (1971) Effects of varying the environment on the behaviour and performance of dairy cattle. *Proc. N.Z. Soc. Anim. Prod, 31*, 10–21.

Albright, J. L. (1978) Optimum group size for high-producing cows appears to be near 100. *Hoard's Dairyman* 25/4/78, 534–5.

Albright, J. L. (1979) Catering to the cow. In *Proceedings of Tri-State Dairy Management Workshop Program*. Cincinnati: Ohio.

Albright, J. L., Brown, C. M., Traylor, D. L. and Wilson, J. C. (1975) Effects of early experience upon later maternal behavior and temperament in cows. *J. Dairy Sci. 58*, 749.

Alcock, J. (1980) *Animal Behavior: an evolutionary approach*, 2nd edn. Sunderland, Mass.: Sinauer.

Aldrich-Blake, F. P. G. (1970) Problems of social structure in forest monkeys. In *Social Behaviour in Birds and Mammals* (J. H. Crook, ed.). London: Academic Press.

Alexander, G. (1977) Role of auditory and visual cues in mutual recognition between ewes and lambs in merino sheep. *Appl. Anim. Ethol. 3*, 65–81.

Alexander, G. and Shillito, E. E. (1977) Importance of visual clues from the various body regions in maternal recognition of the young in merino sheep (*Ovis aries*). *Appl. Anim. Ethol. 3*, 137–43.

Alexander, R. D. (1974) The evolution of social behavior. *Ann. Rev. Ecol. Syst. 5*, 325–83.

Allee, W. C. (1938) *The Social Life of Animals*. New York: W. W. Norton.

Allee, W. C., Foreman, D. and Banks, E. M. (1955) Effects of an androgen on dominance and subordinance in six common breeds of *Gallus gallus*. *Physiol, Zool. 28*, 89–115.

Allin, J. T. and Banks, E. M. (1972) Functional aspects of ultrasound production by infant albino rats. *Anim. Behav. 20*, 175–85.

Allison, T. and Van Twyver, H. (1970) The evolution of sleep. *Nat. Hist., N.Y. 79*, 56–65.

Al-Rawi, B. and Craig, J. V. (1975) Agonistic behavior of caged chickens related to group size and area per bird. *Appl. Anim. Ethol. 2*, 69–80.

Al-Rawi, B., Craig, J. V. and Adams, A. W. (1976) Agonistic behavior and egg production of caged layers: genetic strain and group-size effects. *Poult. Sci. 55*, 796–807.

Altmann, S. A. (1965) Sociobiology of rhesus monkeys. II. Stochastics of social communication. *J. theoret. Biol. 8*, 490–522.

Altmann, S. A. Ed., (1967) *Social Communication Among Primates*. Chicago: University of Chicago Press.

Altmann, S. A. (1974) Baboons, space, time, and energy. *Am. Zool. 14*, 221–48.

Altmann, S. A. and Altmann, J. (1970) *Baboon Ecology*. Chicago: University of Chicago Press.

Altmann, S. A. and Wagner, S. S. (1978) A general model of optimal diet. *Rec. Adv. Primatol. 4*, 407–14.

Anand, B. K. and Brobeck, J. R. (1951) Hypothalamic control of food intake in rats and cats. *Yale J. Biol. Med. 24*, 123–40.

Anderson, L. T. (1973) An analysis of habitat preference in

mice as a function of prior experience. *Behaviour 47*, 302–39.
Andersson, B. (1953) The effect of injections of hypertonic NaCl solutions into different parts of the hypothalamus of goats. *Acta physiol. scand. 28*, 188–201.
Andersson, B. (1978) Regulation of water intake. *Physiol. Rev. 58*, 582–603.
Andersson, B., Grant, R. and Larsson, S. (1956) Central control of heat loss mechanisms in the goat. *Acta physiol. scand. 37*, 261–80.
Andersson, B. and McCann, S. M. (1955) Drinking, antidiuresis and milk ejection from electrical stimulation within the hypothalamus of the goat. *Acta physiol. scand. 35*, 191–201.
Andersson, B. and Wyrwicka, W. (1957) The elicitation of a drinking motor conditioned reaction by electrical stimulation of the hypothalamic 'drinking area' in the goat. *Acta physiol. scand. 41*, 194–8.
Andersson, M. and Wiklund, C. C. (1978) Clumping versus spacing out: experiments on nest predation in fieldfares (*Turdus pilaris*). *Anim. Behav. 26*, 1207–12.
Andrew, R. J. (1956*a*) Normal and irrelevant toilet behaviour in *Emberiza* spp. *Br. J. Anim. Behav. 4*, 85–91.
Andrew, R. J. (1956*b*) Fear responses in *Emberiza* spp. *Br. J. Anim. Behav. 4*, 125–32.
Andrew, R. J. (1963) The origin and evolution of the calls and facial expressions of the primates. *Behaviour 20*, 1–109.
Andrew, R. J. (1976) Attentional processes and animal behaviour. In *Growing points in ethology* (P. P. G. Bateson and R. A. Hinde, eds.). Cambridge: Cambridge University Press.
Andrew, R. J. and Oades, R. D. (1973) Escape, hiding, and freezing behaviour elicited by electrical stimulation of the chick diencephalon. *Brain Behav. Evol. 8*, 191–210.
Antelman, S. M., Rowland, N. E. and Fisher, A. E. (1976) Stress related recovery from lateral hypothalamic aphagia. *Brain Res. 100*, 346–50.
Appley, M. H. and Trumbull, R. Eds., (1967) *Psychological Stress: Issues in Research*. New York: Appleton Century Crofts.
Arave, C. W. and Albright, J. L. (1975) Dominance rank and physiological traits as affected by shifting cows from one group to another. *Proc. Indiana Acad. Sci, 84*, 475.
A.R.C. (1965) *The Nutrient Requirements of Farm Livestock*. No. 2. *Ruminants*. London: Agricultural Research Council.
Archer, J. (1973) Tests for emotionality in rats and mice: a review. *Anim. Behav. 21*, 205–35.
Archer, J. (1974) The effects of testosterone on the distractability of chicks by irrelevant and relevant novel stimuli. *Anim. Behav. 22*, 397–404.
Archer, J. (1976) The organization of aggression and fear in vertebrates. In *Perspectives in Ethology*, Vol. 2 (P. P. G. Bateson and P. Klopfer, eds.). New York: Plenum Press.
Archer, J. (1979*a*) Behavioural aspects of fear in animals and man. In *Fear in Animals and Man* (W. Sluckin, ed.). Princeton, N.J.: Van Nostrand Rheinhold.
Archer, J. (1979*b*) *Animals Under Stress*. London: Arnold.
Ariëns Kappers, C. U., Huber, G. C. and Crosby, E. C. (1936) *The Comparative Anatomy of the Nervous System of Vertebrates, Including Man*, Vol. II. New York: MacMillan.
Armsby, H. P. (1928) *The Nutrition of Farm Animals*. New York: Macmillan.
Armstrong, E. A. (1955) *The Wren*. London: Collins.
Arnold, G. W. (1964) Factors within plant associations affecting the behaviour and performance of grazing animals. In *Grazing in Terrestrial and Marine Environments* (D. J. Crisp, ed.). Oxford: Blackwell.
Arnold, G. W. (1977) An analysis of spatial leadership in a small field in a small flock of sheep. *Appl. Anim. Ethol. 3*, 263–70.
Arnold, G. W. and Dudziński, M. L. (1978) *Ethology of Free Ranging Domestic Animals*. Amsterdam: Elsevier.
Arnold, G. W. and Maller, R. A. (1977) Effects of nutritional experience in early and adult life on the performance and dietary habits of sheep. *Appl. Anim. Ethol. 3*, 5–26.
Arnold, G. W. and Morgan, P. D. (1975) Behaviour of the ewe and lamb at lambing and its relationship to lamb mortality. *Appl. Anim. Ethol. 2*, 25–46.
Aschoff, J. (1960) Exogenous and endogenous components in circadian rhythms. *Cold Spring Harb. Symp. Quant. Biol. 25*, 11–26.
Aslin, R. N., Alberts, J. R. and Peterson, M. R. (1981) *Sensory and Perceptual Development: Influences of Genetic and Experiential Factors*. New York: Academic Press.
Assem, J. van den (1967) Territory in the three-spined stickleback *Gasterosteus aculeatus* L. *Behaviour Suppl. 16*, 1–164.
Attneave, F. (1959) *Applications of Information Theory to Psychology*. New York: Holt, Rinehart and Winston.
Augenstein, L. and Quastler, H. (1967) Information processing and decision making in man. *Brain Res. 6*, 587–605.
Baerends, G. P. (1976) The functional organization of behaviour. *Anim. Behav. 24*, 726–38.
Baerends, G. P., Beer, C. G. and Manning, A. Eds., (1975) *Function and Evolution in Behaviour*. Oxford: Oxford University Press.
Baeumer, E. (1955) Lebensart des Hausuhns. *Z. Tierpsychol. 12*, 387–401.

Baile, C. A. and Forbes, J. M. (1974) Control of feed intake and regulation of energy balance in ruminants. *Physiol. Rev. 54,* 160–214.

Baker, R. D. and Barker, J. M. (1978) Milk-fed calves. 4. The effect of herbage allowance and milk intake upon herbage intake and performance of grazing calves. *J. agric. Sci., Camb. 90,* 31–8.

Baker, R. R. (1978) *The evolutionary ecology of animal migration.* New York: Holmes and Meier.

Baldwin, B. A. (1972) Operant conditioning techniques for the study of thermoregulatory behaviour in sheep. *J. Physiol., Lond. 226,* 41–2P.

Baldwin, B. A. and Lipton, J. M. (1973) Central and peripheral temperatures and EEG changes during behavioural thermoregulation in pigs. *Acta neurobiol. exp. 33,* 433–47.

Baldwin, B. A. and Shillito, E. E. (1974) The effects of ablation of the olfactory bulbs on parturition and maternal behaviour in Soay sheep. *Anim. Behav. 22,* 220–3.

Baldwin, B. A. and Stephens, D. B. (1973) The effects of conditioned behaviour and environmental factors on plasma corticosteroid levels in pigs. *Physiol. Behav. 10,* 267–74.

Baldwin, B. A. and Yates, J. O. (1977) The effects of hypothalamic temperature variation and intracarotid cooling on behavioural thermoregulation in sheep. *J. Physiol., Lond. 265,* 705–20.

Banks, E. M. (1965) Some aspects of sexual behaviour in domestic sheep. *Ovis aries. Behaviour 23,* 249–79.

Baranyovits, F. (1951) Fly reactions to insecticidal deposits: a new test technique. *Nature, Lond. 168,* 960.

Bareham, J. R. (1975) Research in farm animal behaviour. *Br. vet. J. 131,* 272–83.

Barfield, R. J. (1971*a*) Gonadotrophic hormone secretion in the female ring dove in response to visual and auditory stimulation by the male. *J. Endocrinol. 49,* 305–10.

Barfield, R. J. (1971*b*) Activation of sexual and aggressive behaviour by androgen implanted into the male ring dove brain. *Endocrinology 89,* 1470–6.

Barlow, G. W. (1968) Ethological units of behavior. In *The Central Nervous System and Fish Behavior* (D. Ingle, ed.). Chicago: University of Chicago Press.

Barlow, G. W. (1974) Contrasts in social behavior between Central American cichlid fishes and coral-reef surgeon fishes. *Am. Zool. 14,* 9–34.

Barlow, G. W. (1977) Modal action patterns. In *How Animals Communicate* (T. A. Sebeok, ed.). Indianapolis: University of Indiana Press.

Barlow, G. W. Ed., (1980) *Sociobiology: beyond nature/nurture.* A.A.A.S. Selected Symposium 35.

Barlow, G. W. and Rogers, W. (1978) Female midas cichlid's choice of mate in relation to parents' and to own color. *Biol. Behav. 3,* 137–45.

Barlow, H. B. (1975) Visual experience and cortical development. *Nature, Lond. 258,* 199–203.

Barlow, H. B., Blakemore, C. and Pettigrew, J. D. (1967) The neural mechanism of binocular depth discrimination. *J. Physiol., Lond. 193,* 327–42.

Barlow, H. B. and Hill, R. M. (1963) Selective sensitivity to direction of movement in ganglion cells of the rabbit retina. *Science, N.Y. 139,* 412–14.

Barlow, H. B. and Levick, W. R. (1965) The mechanism of directionally selective units in rabbit's retina. *J. Physiol., Lond. 178,* 477–504.

Barnett, S. A. (1963) *A Study in Behaviour.* London: Methuen.

Barnett, S. A. and Burn, J. (1967) Early stimulation and maternal behaviour. *Nature, Lond. 213,* 150–2.

Baron, A. and Kish, C. B. (1960) Early social isolation as a determinant of aggregative behavior in the domestic chicken. *J. comp. physiol. Psychol. 53,* 459–63.

Barraclough, C. A. and Cross, B. A. (1963) Unit activity in the hypothalamus of the cyclic female rat: Effect of genital stimuli and progesterone. *J. Endocrinol. 26,* 339–59.

Bartholomew, G. A. (1964) The roles of physiology and behaviour in the maintenance of homeostasis in the desert environment. *Symp. Soc. exp. Biol. 18,* 7–29.

Barton, E. P., Donaldson, S. L., Ross, M. A. and Albright, J. L. (1974) Social rank and social index as related to age, body weight and milk production in dairy cows. *Proc. Indiana Acad. Sci. 83,* 473–7.

Baryshnikov, I. A. and Kokorina, E. P. (1959) Higher nervous activity and lactation. *Int. Dairy Cong. 15,* 46–53.

Bastock, M. (1967) *Courtship: A Zoological Study.* London: Heinemann.

Bastock, M., Morris, D. and Moynihan, M. (1953) Some comments on conflict and thwarting in animals. *Behaviour 6,* 66–84.

Bateman, J. J. (1948) Intra-sexual selection in *Drosophila. Heredity 2,* 349–68.

Bateson, P. P. G. (1964) Changes in chicks' responses to novel moving objects over the sensitive period for imprinting. *Anim. Behav. 12,* 479–89.

Bateson, P. P. G. (1966) The characteristics and context of imprinting. *Biol. Rev. 41,* 177–220.

Bateson, P. P. G. (1978) Sexual behaviour and optimal outbreeding. *Nature, Lond. 273,* 659–660.

Bateson, P. P. G. (1979) How do sensitive periods arise and what are they for? *Anim. Behav. 27,* 470–86.

Bateson, P. P. G. (1981) Optimal outbreeding and the development of sexual preferences in Japanese quail. *Z. Tierpsychol.* in press.

Bateson, P. P. G. and Hinde, R. A. Eds., (1976) *Growing Points in Ethology.* Cambridge: Cambridge University Press.

Bateson, P. P. G., Lotwick, W. and Scott, D. K. (1980)

Similarities between the faces of parents and offspring in Bewick's swan and the differences between mates. *J. Zool., Lond. 191*, 61—74.

Beach, F. A. Ed., (1965) *Sexual Behavior*. New York: Wiley.

Beach, F. A. (1967) Cerebral and hormonal control of reflexive mechanisms involved in copulatory behavior. *Physiol. Rev. 47*, 289–316.

Beach, F. A. (1975) Behavioural endocrinology: an emerging discipline. *Am. Scient. 63*, 178–87.

Beach, F. A. (1976) Sexual attractivity, proceptivity, and receptivity in female mammals. *Horm. Behav. 7*, 105–38.

Beach, F. A. and Levinson, G. (1950) Effects of androgen on the glans penis and mating behavior of castrated male rats. *J. exp. Zool. 114*, 159–68.

Beer, C. G. (1975) Multiple functions and gull displays. In *Function and Evolution in Behaviour* G. P. Baerends, C. G. Beer and A. Manning, eds.). Oxford: Oxford University Press.

Beilharz, R. G., Butcher, D. F. and Freeman, A. E. (1966) Social dominance and milk production in Holsteins. *J. Dairy Sci. 49*, 887–92.

Beilharz, R. G. and Mylrea, P. J. (1963) Social position and behaviour of dairy heifers in yards. *Anim. Behav. 11*, 522–8.

Bekoff, M., Ainley, D. G. and Bekoff, A. (1979) The ontogeny and organization of comfort behavior in Adélie penguins. *Wilson Bull. 91*, 255–70.

Bellairs, A. d'A. (1957) *Reptiles*. London: Hutchinson.

Bellows, R. T. (1939) Time factors in water drinking in dogs. *Am. J. Physiol. 125*, 87–97.

Belovsky, G. E. (1978) Diet optimization in a generalist herbivore: the moose. *Theoret. Pop. Biol. 14*, 105–34.

Belovsky, G. E. and Jordan, P. A. (1978) The time-energy budget of a moose. *Theoret. Pop. Biol. 14*, 76–104.

Benedictis, P. A. de, Gill, F. B., Hainsworth, F. R., Pyke, G. H. and Wolf, L. L. (1978) Optimal meal size in hummingbirds. *Am. Nat. 112*, 301–16.

Bennett, M. V. L. (1968) Similarities between chemically and electrically mediated transmission. In *Physiological and Biochemical Aspects of Nervous Integration* (F. D. Carlson, ed.). Englewood Cliffs, N.J.: Prentice Hall.

Bentley, D. R. (1969) Intracellular activity in cricket neurons during generation of song patterns. *Z. vergl. Physiol*, *62*, 267–83.

Bentley, D. R. (1971) Genetic control of an insect neuronal network. *Science, N.Y. 174*, 1139–41.

Bentley, D. R. (1975) Single gene mutations: effects on behavior, sensilla, sensory neurons, and identified interneurons. *Science, N.Y. 187*, 760–4.

Bentley, D. R. (1976) Genetic analysis of the nervous system. In *Simpler Networks and Behavior* J. C. Fentess, ed.). Sunderland, Mass.: Sinauer.

Bentley, D. R. and Hoy, R. R. (1970) Postembryonic development of adult motor patterns in crickets: a neural analysis. *Science, N.Y. 170*, 1409–11.

Bentley, D. R. and Hoy, R. R. (1972) Genetic control of the neuronal network generating cricket (*Teleogryllus, Gryllus*) song patterns. *Anim. Behav. 20*, 478–92.

Bentley, D. R. and Hoy, R. R. (1974) The neurobiology of cricket song. *Scient. Am. 231*, 34–44.

Bentley, P. J. (1966) Adaptations of amphibia to arid environments. *Science, N.Y 152*, 619–23.

Bentley, P. J. and Yorio, T. (1979) Do frogs drink? *J. exp. Biol.* 79, 41–6.

Benzer, S. (1973) Genetic dissection of behavior. *Scient. Am. 229*, (6), 24–37.

Berkson, G., Mason, W. A. and Saxon, S. V. (1963) Situation and stimulus effects on stereotyped behaviors of chimpanzees *J. comp. physiol. Psychol. 56*, 786–92.

Berlyne, D. E. (1967) Arousal and reinforcement. In *Nebraska Symposium on Motivation* (M. R. Jones, ed.). Lincoln: University of Nebraska Press.

Bernstein, I. S. (1970) Primate status hierarchies. In *Primate Behavior*, Vol. 1 (L. A. Rosenblum, ed.). New York: Academic Press.

Beroza, M. (1960) Insect attractants are taking hold. *Agr. Chem. 15*, 37–40.

Beroza, M. (1970*a*) Current usage and some recent developments with insect attractants and repellants in the U.S.D.A. In *Chemicals Controlling Insect Behavior* (M. Beroza, ed.). New York: Academic Press.

Beroza, M. (1970*b*) *Chemicals Controlling Insect Behavior*. New York: Academic Press.

Beroza, M., Hood, C. S., Trefney, D., Leonard, D. E., Knipling, E. F., Klassen, W. and Stevens, L. J. (1974) Large field trial with microencapsulated sex pheromone to prevent mating of the gypsy moth. *J. econ. Ent. 67*, 659–64.

Beroza, M., Stevens, L. J., Bierl, B. A., Phillips, F. M. and Tardif, J. G. R. (1973) Pre- and post-season field tests with disparlure the sex pheromone of the gypsy moth, to prevent mating. *Environ. Entomol.* 2, 1051–7.

Bertram, B. C. R. (1975) Social factors influencing reproduction in wild lions. *J. Zool., Lond. 177*, 463–82.

Bertram, B. C. R. (1976) Kin selection in lions and in evolution. In *Growing Points in Ethology* (P. P. G. Bateson and R. A. Hinde, eds.). Cambridge: Cambridge University Press.

Bertrand, M. (1969) The behavioural repertoire of the stump-tail macaque. *Bibl. primatol. 11*, 1–273.

Bethel, W. M. and Holmes, J. C. (1973) Altered evasive behaviour and responses to light in amphipods harbouring acanthocephalan cystocanths. *J. Parasit.* *59*, 945–56.

Beukema, J. J. (1968) Predation by the three-spined stickleback (*Gasterosteus aculeatus L.*): the influence of hunger and experience. *Behaviour 31*, 1–126.

Bindra, D. (1959) *Motivation: A Systematic Reinterpretation*. New York: Ronald Press.

Binkley, S., Kluth, E. and Menaker, M. (1971) Pineal function in sparrows: circadian rhythms and body temperature. *Science, N.Y. 174*, 311–14.

Birch, M. C. (1974) *Pheromones*. Amsterdam: North Holland.

Birch, M. C., Trammel, K., Shorey, H. H., Gaston, L. K., Hardee, D. D., Cameron, E. A., Sanders, C. J., Bedard, W. D., Wood, D. L., Burkholder, W. E. and Müller-Schwarze, D. (1974) Programs utilising pheromones in survey or control. In *Pheromones* (M. C. Birch, ed.). Amsterdam: North Holland.

Birke, L. I. A., Andrew, R. J. and Best, S. M. (1979) Distractibility changes during the oestrus cycle of the rat. *Anim. Behav. 27*, 597–601.

Blakemore, C. (1974) Developmental factors in the formation of feature extracting neurons. In *The Neurosciences Third Study Program* (F. O. Schmitt and F. G. Warden, eds.). Cambridge, Mass.: M. I. T. Press.

Blakemore, C. and Cooper, G. F. (1970) Development of the brain depends on the visual environment. *Nature, Lond. 228*, 477–8.

Blakemore, C. and Sluyters, R. C. van (1974) Reversal of the physiological effects of monocular deprivation in kittens: further evidence for a sensitive period. *J. Physiol., Lond. 237*, 195–216.

Blakemore, C., Sluyters, R. C. van, Peck, C. K. and Hein, A. (1975) Development of cat visual cortex following rotation of one eye. *Nature, Lond. 257*, 584–6.

Blass, E. M., Hall, J. G. and Teicher, M. H. (1979) The ontogeny of suckling and ingestive behaviours. *Prog. Psychobiol. physiol. Psychol. 8*, 243–99.

Blaxter, K. L. (1962) *The Energy Metabolism of Ruminants*. London: Hutchinson.

Blaxter, K. L. Ed., (1965) *Energy Metabolism*. London: Academic Press.

Blaxter, K. L. and Joyce, J. P. (1963) The accuracy and ease with which measurement of respiratory metabolism can be made with sheep. *Br. J. Nutr. 17*, 523–37.

Blest, A. D. (1957) The evolution of protective displays in the Saturnoidea and Sphingidae (Lepidoptera). *Behaviour 11*, 257–309.

Bliss, E. L. Ed., (1962) *Roots of Behavior*. New York: Harper and Row.

Blockey, M. A. de B. (1979) Observations on group mating of bulls at pasture. *Appl. Anim. Ethol. 5*, 15–34.

Blockey, M. A. de B. and Lade, A. D. (1974) Social dominance relationships among young bulls in a test of rate of gain after weaning. *Aust. vet. J. 59*, 435–7.

Bodmer, W. F. and Cavalli-Sforza, L. L. (1976) *Genetics, Evolution and Man*. San Francisco: Freeman.

Bolles, R. C. (1958) The usefulness of the drive concept. In *Nebraska Symposium on Motivation* (M. R. Jones, ed.). Lincoln: University of Nebraska Press.

Bolles, R. C. (1961) The interaction of hunger and thirst in the rat. *J. comp. physiol. Psychol. 54*, 580–4.

Bolles, R. C. (1975) *Theory of Motivation*. New York: Harper and Row.

Bolles, R. C. and Woods, P. J. (1964) The ontogeny of behaviour in the albino rat. *Anim. Behav. 12*, 427–41.

Booth, D. A. (1978*a*) Prediction of feeding behaviour from energy flows in the rat. In *Hunger Models: Computable Theory of Feeding Control* (D. A. Booth, ed.). London: Academic Press.

Booth, D. A. Ed., (1978*b*) *Hunger Models: Computable Theory of Feeding Control*. London: Academic Press.

Borchelt, P. L., Griswold, J. G. and Branchek, R. S. (1976) An analysis of sandbathing and grooming in the kangaroo rat (*Dipodomys merriami*). *Anim. Behav. 24*, 347–53.

Bouissou, M-F. (1965) Observations sur la hiérarchie sociale chez les bovins domestiques. *Annals. Biol. anim. Biochim. Biophys. 5*, 327–39.

Bouissou, M-F. (1970) Technique de mise en evidence des relations hiérarchiques dans un group de bovins domestiques. *Rev. comport. anim. 4*, 66–9.

Bouissou, M-F. (1971) Effets de l'absence d'informations optiques et de contact physique sur la manifestation des relations hiérarchiques chez les bovins domestiques. *Annls. Biol. anim. Biochim. Biophys. 11*, 191–8.

Bouissou, M-F. (1976) Effet de différentes perturbations sur le nombre d'interactions sociales échargées au sein de groupes de bovins. *Biol. Behav. 1*, 193–8.

Bourke, M. E. (1967) A study of mating behaviour of Merino rams. *Aust. J. exp. Agric. Anim. Husb. 7*, 203–5.

Bourlière, F. (1955) *The Natural History of Mammals*. London: Harrap.

Bowlby, J. (1969) *Attachment and Loss*, Vol. 1. *Attachment*. London: Hogarth Press.

Bowlby, J. (1973) *Attachment and Loss*, Vol. 2. *Separation*. London: Hogarth Press.

Box, H. O. (1973) *Organisation in Animal Communities*. London: Butterworths.

Box, H. O. (1977) Quantitative data on the carrying of young captive monkeys (*Callithrix jacchus*) by other members of their family groups. *Primates 18*, 475–84.

Box, H. O. (1978) Social behaviour in the common marmoset monkey (*Callithrix jacchus*). *Biol. Hum. Affrs. 43*, 51–64.

Box, H. O. and Morris, J. M. (1980) Behavioural observations on captive pairs of wild caught tamarins (*Saguinus mystax*). *Primates 21*, 53–65.

Bradbury, J. W. (1977) Lek mating behavior in the hammer-headed bat. *Z. Tierpsychol. 45*, 225–55.

Bradbury, J. W. (1980) Foraging, social dispersion and mating systems. In *Sociobiology: beyond nature/nuture* (G. W. Barlow, ed.). A.A.A.S. Selected Symposium 35.

Brakel, W. J. and Leis, R. A. (1976) Impact of social disorganisation on behavior, milk yield and bodyweight of dairy cows. *J. Dairy Sci. 59*, 716–21.

Bramley, P. S. (1970) Territoriality and reproductive behaviour of roe deer. *J. Reprod. Fert., Suppl. 11*, 43–70.

Branch, G. M. (1978) The responses of South African patellid limpets to invertebrate predators. *Zool. Afr. 13*, 221–32.

Brattgård, S-O. (1951) The importance of adequate stimulation for the chemical composition of ganglion cells. *Expl Cell Res. 2*, 693–5.

Brattgård, S.-O. (1952) The importance of adequate stimulation for the chemical composition of retinal ganglion cells during early post-natal development. *Acta radiobiol. Suppl. 96*, 1–80.

Brecher, G. and Waxler, S. (1949) Obesity in albino mice due to single injections of goldthio-glucose. *Proc. Soc. exp. Biol. Med. 70*, 498.

Breder, C. M. (1949) On the behaviour of young *Lobotes surinamensis*. *Copeia*, pp. 237–42.

Breder, C. M. (1965) Vortices and fish schools. *Zoologica, N.Y. 50*, 97–114.

Breder, C. M. (1967) On the survival value of fish schools. *Zoologica, N.Y. 52*, 25–40.

Breed, M. D., Smith, S. K. and Ball, B. G. (1980) Systems of mate selection in a cockroach species with male dominance hierarchies. *Anim. Behav. 28*, 130–4.

Brenner, S. (1974) The genetics of *Caenorhabditis elegans*. *Genetics 77*, 71–104.

Broadbent, D. E. (1958) *Perception and Communication*. London: Pergamon Press.

Broadbent, D. E. (1971) *Decision and Stress*. London: Academic Press.

Brock, V. E. and Riffenburgh, R. H. (1960) Fish schooling: a possible factor in reducing predation. *J. Cons. perm. int. Explor. Mer. 25*, 307–17.

Brockelman, W. Y. (1975) Competition, the fitness of offspring, and optimal clutch size. *Am. Nat. 109*, 677–99.

Brockway, B. F. (1969) Roles of budgerigar vocalisation in the integration of breeding behaviour. In *Bird Vocalisations* (R. A. Hinde, ed.). Cambridge: Cambridge University Press.

Broom, D. M. (1968) Specific habituation by chicks. *Nature, Lond. 217*, 880–1.

Broom, D. M. (1969*a*) Reactions of chicks to visual changes during the first ten days after hatching. *Anim. Behav. 17*, 307–15.

Broom, D. M. (1969*b*) Effects of visual complexity during rearing on chicks' reactions to environmental change. *Anim. Behav. 17*, 773–80.

Broom, D. M. (1975) Aggregation behaviour of the brittle-star *Ophiothrix fragilis*. *J. mar. biol. Ass. U.K. 55*, 191–7.

Broom, D. M. (1979) Methods of detecting and analysing activity rhythms. *Biol. Behav. 4*, 3–18.

Broom, D. M. (1980) Activity rhythms and position preferences of domestic chicks which can see a moving object. *Anim. Behav. 28*, 201–11.

Broom, D. M. (1981) Behavioural plasticity in developing animals. In *Development in the Nervous System* (D. R. Garrod and J. D. Feldman, eds.). Cambridge: Cambridge University Press.

Broom, D. M., Dick, W. J. A., Johnson, C. E., Sales, D. I. and Zahavi, A. (1976) Pied wagtail roosting and feeding behaviour. *Bird Study. 23*, 267–79.

Broom, D. M., Elwood, R. W., Lakin, J., Willy, S. J. and Pretlove, A. J. (1977) Developmental changes in several parameters of ultrasonic calling by young Mongolian gerbils (*Meriones unguiculatus*). *J. Zool., Lond. 183*, 281–90.

Broom, D. M. and Leaver, J. D. (1977) Mother–young interactions in dairy cattle. *Br. vet. J. 133*, 192.

Broom, D. M. and Leaver, J. D. (1978) Effects of group-rearing or partial isolation on later social behaviour of calves. *Anim. Behav. 26*, 1255–63.

Broom, D. M., Pain, B. F. and Leaver, J. D. (1975) The effects of slurry on the acceptability of swards to grazing cattle. *J. agric. Sci., Camb., 85*, 331–6.

Broughton, W. B. Ed., (1974) *Biology of Brains*. London: Institute of Biology.

Brower, J. V. Z. (1958) Experimental studies of mimicry in some North American butterflies. Part 1. The monarch, *Danaus plexippus*, and viceroy, *Limenitis archippus archippus*. *Evolution 12*, 32–47.

Brown, J. H., Calder, W. A. and Kodric-Brown, A. (1978) Correlates and consequences of body size in nectar-feeding birds. *Am. Zool. 18*, 687–700.

Brown, J. L. (1970) Cooperative breeding and altruistic behaviour in the Mexican jay, *Aphelocoma ultramarina*. *Anim. Behav. 18*, 366–78.

Brown, J. L. (1972) Communal feeding of nestlings in the Mexican jay, (*Aphelocoma ultramarina*): interflock comparisons. *Anim. Behav. 20*, 395–403.

Brown, J. L. (1975) *The Evolution of Behavior*. New York: W. W. Norton & Co.

Brown, J. S. (1961) *The Motivation of Behavior*. New York: McGraw Hill.

Brownlee, A. (1950) Studies in the behaviour of domestic cattle in Britain. *Bull. Anim. Behav. 1*, (8), 11–20.

Brownlee, A. (1954) Play in domestic cattle. An analysis of its nature. *Br. vet. J. 110*, 48–68.

Bruner, J. S., Jolly, A. and Sylva, K. Eds., (1974) *Play: its role in development and evolution.* Harmondsworth: Penguin.

Brush, A. Ed., (1978) *Chemical Zoology*, Vol. 10. New York: Academic Press.

Bryant, M. J. (1972) The social environment: behaviour and stress in housed livestock. *Vet. Rec. 90*, 351–8.

Bryant, M. J. and Ewbank, R. (1972) Some effects of stocking rate and group size upon agonistic behaviour in groups of growing pigs. *Br. vet. J. 128*, 64–70.

Bryant, M. J. and Ewbank, R. (1974) Effects of stocking rate upon the performance, general activity and ingestive behaviour of groups of growing pigs. *Br. vet. J. 130*, 139–49.

Bryant, M. J. and Tomkins, T. (1975) The flock-mating of progestogen-synchronized ewes. I. The influence of ram to ewe ratio on mating behaviour and lambing performance. *Anim. Prod. 20*, 381–90.

Buck, J. and Buck, E. (1976) Synchronous fireflies. *Scient. Am. 234*, (5), 74–85.

Buechner, H. K. and Roth, H. D. (1974) The lek system in Uganda kob antelope. *Am. Zool. 14*, 145–62.

Buggy, J., Fisher, A. E., Hoffman, W. E., Johnson, A. K. and Phillips, M. I. (1975) Ventricular obstruction: effect on drinking induced by intracranial injection of angiotensin. *Science, N.Y. 190*, 72–4.

Bullis, H. R. (1961) Observations on the feeding behaviour of white-tip sharks on schooling fishes. *Ecology 42*, 194–5.

Bullock, T. H. (1961) The origins of patterned nervous discharge. *Behaviour 17*, 48–59.

Bullock, T. H. (1975) Are we learning what actually goes on when the brain recognizes and controls? *J. exp. Zool. 194*, 13–34.

Bullock, T. H. and Diecke, F. P. J. (1956) Properties of an infrared receptor. *Science, N.Y. 115*, 541–3.

Burgess, J. W. (1978) Social behaviour in group-living spider species. *Symp. zool. Soc. Lond. 42*, 69–78.

Burrows, M. and Horridge, G. A. (1974) The organisation of inputs to motoneurons of the locust metathoracic leg. *Phil. Trans. R. Soc., B, 269*, 49–94.

Butenandt, A., Behmann, R., Stamm, D. and Hecker, E. (1959) Über den Sexualstoff des Seidenspinners *Bombyx mori* Reindarstellung und Konstitution. *Z. Naturf. 146*, 283–4.

Caffyn, Z. E. Y. (1972) Early vascular changes in the brain of the mouse after injections of goldthioglucose and biperidyl mustard. *J. Path. 106*, 49–56.

Caggiula, A. R. and Hoebel, B. G. (1966) 'Copulation reward site' in the posterior hypothalamus. *Science, N.Y. 153*, 1284–5.

Calhoun, J. B. (1962*a*) *The Ecology and Sociology of the Norway Rat.* U.S. Dep. Helth, Educ. Welf. P.H.S. Doc. 1008. Washington, D.C.: U.S. Govt Printing Office.

Calhoun, J. B. (1962*b*) Population density and social pathology. *Scient. Am. 206*, 139–48.

Campbell, B. G. Ed. (1972) *Sexual Selection and the Descent of Man 1871–1971.* Chicago: Aldine Publishing Company.

Campling, R. C. and Balch, C. C. (1961) Factors affecting the voluntary intake of food by cows. 1. Preliminary observations on the effect, on the voluntary intake of hay, of changes in the amount of the reticulo-ruminal contents. *Br. J. Nutr. 15*, 523–30.

Cane, V. R. (1978) On fitting low-order Markov chains to behaviour sequences. *Anim. Behav. 26*, 332–8.

Canning, E. V. and Wright, C. A. Eds., (1972) *Behavioural Aspects of Parasite Transmission.* London: Academic Press.

Cannon, R. E. and Salzen, E. A. (1971) Brain stimulation in newly-hatched chicks. *Anim. Behav. 19*, 375–85.

Cannon, W. B. (1915) *Bodily Changes in Pain, Hunger, Fear and Rage.* New York: Appleton.

Cannon, W. B. (1929) Organization for physiological homeostasis. *Physiol. Rev. 9*, 399–431.

Carlson, N. R. (1977) *Physiology of Behavior.* Boston: Allyn and Bacon.

Carlson, N. R. and Thomas, G. J. (1968) Maternal behaviour of mice with limbic lesions. *J. comp. physiol. Psychol. 66*, 731–7.

Carmichael, L. (1926) The development of behavior in vertebrates experimentally removed from the influence of external stimulation. *Psychol. Rev. 33*, 51–8.

Carpenter, C. C. (1971) Discussion of Session 1: Territoriality and dominance. In *Behavior and Environment: the use of Space by Animals and Men* (A. H. Esser, ed.). New York: Plenum.

Carpenter, C. R. (1940) A field study in Siam of the behavior and social relations of the gibbon (*Hylobates lar*). *Comp. psychol. Monogr. 16*, 1–212.

Carpenter, C. R. (1962) Field studies of a primate population. In *Roots of Behavior* (E. L. Bliss, ed.). New York: Harper and Row.

Carpenter, F. L. (1978) A spectrum of nectar-eater communities. *Am. Zool. 18*, 809–19.

Carpenter, G. D. H. (1941) The relative frequency of beak-marks on butterflies of different edibility to birds. *Proc. zool. Soc. Lond. 111* A, 223–31.

Carr, H. A. Ed. (1919) *Posthumous Works of Charles Otis Whitman, III.* Washington: Carnegie Institute.

Carr, W. J. and McGuigan, D. I. (1965) The stimulus basis and modification of visual cliff performance in the rat. *Anim. Behav. 13*, 25–9.

Carruthers, M. (1978) Antidotes to stress. In *Human Behaviour and Adaptation* (V. Reynolds and N. Blurton-Jones, eds.). *Symp. Soc. Study hum. Biol. 18*. London: Taylor and Francis.

Cassidy, M. D. (1978) Development of an induced food plant preference in the Indian stick insect, *Carausius morosus*. *Entomologia exp. appl. 24*, 87–93.

Castilla, J. C. (1972) Responses of *Asterias rubens* to bivalve prey in a Y-maze. *Mar. Biol. 12*, 222–8.

Castro, J. M. de and Balagura, S. (1975) Ontogeny of meal patterning in rats and its recapitulation during recovery from lateral hypothalamic lesions. *J. comp. physiol. Psychol. 89*, 791–802.

Catchpole, C. K. (1977) Aggressive responses of male sedge warblers (*Acrocephalus schoenobaenus*) to playback of species song and sympatric species song, before and after pairing. *Anim. Behav. 25*, 489–96.

Catchpole, C. K. (1978) Interspecific territorialism and competition in *Acrocephalus* warblers as revealed by playback experiments in areas of sympatry and allapatry. *Anim. Behav. 26*, 1072–80.

Catchpole, C. K. (1979) *Vocal Communication in Birds*. London: Arnold.

Chacon, E. and Stobbs, T. H. (1976) Influence of progressive defoliation of a grass sward on the eating behaviour of cattle. *Aust. J. agric. Res. 27*, 709–27.

Chacon, E. and Stobbs, T. H. (1977) The effects of fasting prior to sampling and diurnal variation on certain aspects of grazing behaviour in cattle. *Appl. Anim. Ethol. 3*, 163–71.

Chacon, E., Stobbs, T. H. and Sandland, R. L. (1976) Estimation of herbage consumption by grazing cattle using measurements of eating behaviour. *J. Br. Grassld Soc. 31*, 81–7.

Chalmers, D. V. and Levine, S. (1974) The development of heart rate responses to weak and strong shock in the preweaning rat. *Devl. Psychobiol. 7*, 519–27.

Chalmers, N. R. (1973) Differences in behaviour between some arboreal and terrestrial species of African monkeys. In *Comparative Ecology and Behaviour of Primates* (R. P. Michael and J. H. Crook, eds.). London: Academic Press.

Chalmers, N. R. (1979) *Social Behaviour in Primates*. London: Arnold.

Chambers, D. T. (1959) Grazing behaviour of calves reared at pastures. *J. agric. Sci., Camb. 53*, 417–24.

Chamove, A. S., Rosenblum, L. A. and Harlow, H. F. (1973) Monkeys (*Macaca mulatta*) raised only with peers. A pilot study. *Anim. Behav. 21*, 316–25.

Champ, B. R. and Dyte, C. E. (1976) *Report of the FAO Global Survey of Pesticide Susceptibility of Stored Grain Pests*. Rome: FAO.

Chance, M. R. A. (1967) Attention structures as the basis of primate rank order. *Man 2*, 503–18.

Chance, M. R. A. and Jolly, C. J. (1970) *Social Groups of Monkeys, Apes and Men*. London: Cape.

Chance, M. R. A. and Russell, W. M. S. (1959) 'Protean displays': a form of allaesthetic behaviour. *Proc. zool. Soc. Lond. 132*, 65–70.

Chang, K-J. and Cuatrecasas, P. (1979) Multiple opiate receptors. *J. biol. Chem. 254*, 2610–18.

Charnov, E. L. (1976) Optimal foraging: the marginal value theorem. *Theoret. Pop. Biol. 9*, 129–36.

Charnov, E. L. and Krebs, J. R. (1975) The evolution of alarm cells: altruism or manipulation? *Am. Nat. 109*, 107–12.

Chatfield, C. and Lemon, R. E. (1970) Analysing sequences of behavioural events. *J. theoret. Biol. 29*, 427–45.

Chelidze, L. R. (1975) Neuronal reactions of the rabbits' visual cortex as a function of the interval between flashes of light. In *Neuronal Models of the Orienting Reflex* (E. N. Sokolov and O. S. Vinogradova, eds.). Hillsdale, N.J.: Lawrence Erlbaum Ass.

Cheney, D. L. (1977) The acquisition of rank and the development of reciprocal alliances among free-ranging immature baboons. *Behav. Ecol. Sociobiol. 2*, 303–18.

Cheng, M-F. (1977) Role of gonadotrophin releasing hormones in the reproduction behaviour of female ring doves. *J. Endocrinol. 74*, 37–45.

Cheng, M-F. (1979) Progress and prospects in ring dove research: a personal view. *Adv. Study Behav. 9*, 97–129.

Cherrett, J. M., Ford, J. B., Herbert, I. V. and Probert, A. J. (1971) *The Control of Injurious Animals*. London: English Universities Press.

Chitty, D. Ed. (1954) *Control of Rats and Mice*, Vol. 2. *Rats*. Oxford: Oxford University Press.

Chivers, D. J. (1977) The feeding behaviour of siamang (*Symphalangus syndactylus*). In *Primate Ecology* (T. H. Clutton-Brock, ed.). London: Academic Press.

Chkikvadze, I. I. (1975) Single unit responses to flashes of light of different intensities in the visual cortex of the rabbit. In *Neuronal Models of the Orienting Reflex* (E. N. Sokolov and O. S. Vinogradova, eds.). Hillsdale, N.J.: Lawrence Erlbaum Ass.

Chow, K. L., Riesen, A. H. and Newell, F. W. (1957) Degeneration of retinal ganglion cells in infant chimpanzees reared in darkness. *J. comp. Neurol. 107*, 27–40.

Christian, J. J. (1955) Effects of population size on the adrenal glands and reproductive organs of male mice. *Am. J. Psychol. 182*, 292–300.

Christian, J. J. (1961) Phenomena associated with population density. *Proc. natn. Acad. Sci. U.S.A. 47*, 428–49.

Claridge, M. F. and Wilson, M. R. (1978) Oviposition behaviour as an ecological factor in woodland canopy leafhoppers. *Entomologia exp. appl. 24*, 87–93.

Clark, R. B. (1960) Habituation of the polychaete *Nereis* to sudden stimuli. 1. General properties of the habituation process. *Anim. Behav. 8*, 82–91.

Clemente, C. D. and Chase, M. H. (1973) Neurological substrates and aggressive behaviour. *Ann. Rev. Physiol. 35*, 329–56.

Clough, G. C. (1965) Lemmings and population problems. *Am. Scient. 53*, 199–213.

Clutton-Brock, T. H. Ed. (1977) *Primate Ecology*. London: Academic Press.

Clutton-Brock, T. H. and Albon, S. D. (1979) The roaring of red deer and the evolution of honest advertisement. *Behaviour 69*, 145–70.

Clutton-Brock, T. H., Albon, S. D., Gibson, R. M. and Guinness, F. E. (1979) The logical stag: adaptive aspects of fighting in red deer (*Cervus elephas* L.). *Anim. Behav. 27*, 211–25.

Clutton-Brock, T. H. & Guinness, F. E. (1975) Behaviour of red deer (*Cervus elephas* L.) at calving time. *Behaviour 55*, 287–300.

Clutton-Brock, T. H. and Harvey, P. H. (1976) Evolutionary rules and primate societies. In *Growing Points in Ethology* (P. P. G. Bateson and R. A. Hinde, eds.). Cambridge: Cambridge University Press.

Clutton-Brock, T. H. and Harvey, P. H. (1977*a*) Primate ecology and social organization. *J. Zool., Lond. 183*, 1–39.

Clutton-Brock, T. H. and Harvey, P. H. (1977*b*) Species differences in feeding and ranging behaviour in primates. In *Primate Ecology* (T. H. Clutton-Brock, ed.). London: Academic Press.

Cody, M. L. (1971) Finch flocks in the Mojave desert. *Theoret. Pop. Biol. 2*, 142–58.

Coghill, G. E. (1929) Anatomy and the Problem of Behaviour. Cambridge: Cambridge University Press.

Cohen, J. E. (1969) Natural primate troops and a stochastic population model. *Am. Nat. 103*, 455–78.

Cohen, S. and McFarland, D. (1979) Time-sharing as a mechanism for the control of behaviour sequences during the courtship of the three-spined stickleback (*Gasterosteus aculeatus*). *Anim. Behav. 27*, 270–83.

Collett, T. (1974) The efferent control of sensory pathways. In *Biology of Brains* (W. B. Broughton, ed.). London: Institute of Biology.

Collias, N. E. (1956) The analysis of socialization in sheep and goats. *Ecology 37*, 228–39.

Colovos, N. R., Holter, J. B., Clark, R. M., Urban, W. E. and Hayes, H. H. (1970) Energy expenditure in physical activity of cattle. In *Energy Metabolism of Farm Animals* (A. Schürch and C. Wenk, eds.). Zürich: Juris Verlag.

Cooke, F., Finney, G. H. and Rockwell, R. F. (1976) Assortative mating in lesser snow geese (*Anser caerulescens*). *Behav. Genet. 6*, 127–40.

Coppel, H. C. and Mertins, J. W. (1977) *Biological Insect Pest Suppression*. Berlin: Springer Verlag.

Cordier, R. (1964) Sensory cells. In *The Cell*, 6 (J. Brachet and A. Mirsky, eds.). New York: Academic Press.

Coss, R. G. (1972) Eye-like Schemata: their Effect on Behaviour. Ph.D. thesis, University of Reading.

Coss, R. G. (1978*a*) Perceptual determination of gaze aversion by the lesser mouse lemur (*Microcebus murinus*) the role of two facing eyes. *Behaviour 64*, 48–70.

Coss, R. G. (1978*b*) Development of face aversion by the jewel fish (*Hemichromis bimaculatus*, Gill 1862). *Z. Tierpsychol. 48*, 28–46.

Cott, H. B. (1940) *Adaptive Coloration in Animals*. London: Methuen.

Coulson, J. C. (1966) The influence of the pair-bond and age on the breeding biology of the kittiwake gull *Rissa tridactyla*, *J. Anim. Ecol. 35*, 269–79.

Coulson, J. C. (1968) Differences in the quality of birds nesting in the centre and on the edges of a colony. *Nature, Lond. 217*, 478–9.

Cowie, A. T., Folley, S. J., Cross, B. A., Harris, G. W., Jacobson, D. and Richardson, K. C. (1951) Terminology for use in lactational physiology. *Nature, Lond. 168*, 421.

Cox, C. R. and Le Boeuf, B. J. (1977) Female incitation of male competition: a mechanism of mate selection. *Am. Nat. 3*, 317–35.

Cragg, B. G. (1967) Changes in visual cortex on first exposure of rats to light. *Nature, Lond. 215*, 251–3.

Cragg, B. G. (1975) The development of synapses in the visual system of the cat. *J. comp. Neurol. 160*, 147–66.

Craig, J. V. and Guhl, A. M. (1969) Territorial behavior and social interactions of pullets kept in large flocks. *Poult. Sci. 48*, 1622–28.

Craig, J. V. and Polley, C. R. (1977) Crowding cockerels in cages: effects on weight gain, mortality and subsequent fertility. *Poult. Sci 56*, 117–20.

Craig, W. (1911) Oviposition induced by the male in pigeons. *J. Morph. 22*, 299–305.

Crain, S. M., Bornstein, M. B. and Peterson, E. R. (1968) Development of functional organization in cultured embryonic CNS tissues during chronic exposure to agents which prevent bioelectric activity. In *Ontogenesis of the Brain* (L. Jílek and S. Trojan, eds.). Prague: Charles University Press.

Crane, J. (1975) *Fiddler Crabs of the World*. Princeton, N.J.: Princeton University Press.

Crisp, D. J. Ed. (1964) *Grazing in Terrestrial and Marine Environments*. Oxford: Blackwell.

Crook, J. H. (1963) Comparative studies of the reproductive behaviour of two closely related weaver birds species (*Ploceus cucullatus* and *Ploceus nigerrimus*) and their races. *Behaviour 21*, 177–232.

Crook, J. H. (1964) The evolution of social organisation and visual communication in the weaver birds (Ploceinae). *Behaviour Suppl. 10,* 1–178.

Crook, J. H. (1965) The adaptive significance of avian social organizations. *Symp. zool. Soc. Land. 14,* 181–218.

Crook, J. H. Ed., (1970) *Social Behaviour in Birds and Mammals.* London: Academic Press.

Crook, J. H. and Michael, R. P. (1973) *The Comparative Ecology and Behaviour of Primates.* London: Academic Press.

Crow, J. F. and Kimura, M. (1970) *An Introduction to Population Genetics Theory.* New York: Harper and Row.

Crowden, A. E. and Broom, D. M. (1980) Effects of the eyefluke, *Diplostomum spathaceum,* on the behaviour of dace (*Leuciscus leuciscus*). *Anim. Behav. 28,* 287–94.

Crowley, J. P. and Darby, T. E. (1970 Observations on the fostering of calves for multiple suckling systems. *Br. vet. J. 126,* 658.

Cruze, W. W. (1935) Maturation and learning in chicks. *J. comp. Psychol 19,* 371–409.

Culshaw, A. D. and Broom, D. M. (1980) The imminence of behavioural change and startle responses of chicks. *Behaviour 3,* 64–76.

Curio, E. (1966) Die Schutzanpassungen dreier Raupen eines Schwärmers (Lepidopt., Sphingidae) auf Galapagos. *Zool. Jb. Syst. 91,* 1–29.

Curio, E. (1976) *The Ethology of Predation.* Berlin: Springer.

Daan, S. and Tinbergen, J. (1981) Young guillemots (*Uria lomvia*) leaving their arctic breeding cliffs: a daily rhythm in numbers and risk. *Ardea,*

Daly, M. (1973) Early stimulation of rodents: a critical review of present interpretations. *Br. J. Psychol. 64,* 435–60.

Darwin, C. (1839) *Journal of researches into the natural history and geology of the countries visited during the voyage of H. M. S. Beagle around the world.* London: Henry Colburn.

Daumer, K. (1958) Blumenfarben, wie sie die Bienen sehen. *Z. vergl. Physiol. 41,* 49–110.

Davidson, J. M. (1966) Activation of the male rats' sexual behaviour by intracerebral implantation of androgen. *Endocrinology 79,* 783–94.

Davies, N. B. (1976) Food, flocking and territorial behaviour of the pied wagtail (*Motacilla alba yarrellii* Gould) in winter. *J. Anim. Ecol. 45,* 235–53.

Davies, N. B. (1978*a*) Territorial defence in the speckled wood butterfly (*Pararge aegeria*): the resident always wins. *Anim. Behav. 26,* 138–47.

Davies, N. B. (1978*b*) Ecological questions about territorial behaviour. In *Behavioural Ecology: an Evolutionary Approach* J. R. Krebs and N. B. Davies, eds.). Oxford: Blackwell.

Davies, N. B. and Halliday, T. R. (1978) Deep croaks and fighting assessment in toads *Bufo bufo. Nature, Lond. 274,* 683–5.

Davies, N. B. and Halliday, T. R. (1979) Competitive mate searching in male common toads, *Bufo bufo. Anim. Behav. 27,* 1253–67.

Davis, H. (1961) Some principles of sensory receptor action. *Physiol. Rev. 41,* 391–416.

Davis, H. (1965) A model for transducer action in the cochlea. *Cold Spring Harb. Symp. Quant. Biol. 30,* 181–90.

Davis, J. M. (1975) Socially induced flight reactions in pigeons. *Anim. Behav. 23,* 597–601.

Davis, J. W. F. (1976) Breeding success and experience in the arctic skua *Stercorarius parasiticus* (L). *J. Anim. Ecol. 45,* 531–5.

Dawkins, M. (1971) Perceptual changes in chicks: another look at the 'search image' concept. *Anim. Behav. 19,* 566–74.

Dawkins, M. and Dawkins, R. (1974) Some descriptive and explanatory stochastic models of decision making. In *Motivational Control Systems Analysis* (D. J. McFarland, ed). London: Academic Press.

Dawkins, R. (1968) The ontogeny of a pecking preference in domestic chicks. *Z. Tierpsychol. 25,* 470–4.

Dawkins, R. (1969*a*) A threshold model of choice behaviour. *Anim. Behav. 17,* 120–133.

Dawkins, R. (1969*b*) The attention threshold model. *Anim. Behav. 17,* 134–41.

Dawkins, R. (1976) *The Selfish Gene.* Oxford: Oxford University Press.

Dawkins, R. (1978) Replicator selection and the extended phenotype. *Z. Tierpsychol. 47,* 61–76.

Dawkins, R. and Carlisle, T. R. (1976) Parental investment: a fallacy. *Nature, Lond. 262,* 131–3.

Dawkins, R. and Dawkins, M. (1973) Decisions and the uncertainty of behaviour. *Behaviour 45,* 83–103.

Dawkins, R. and Dawkins, M. (1976) Hierarchical organization and postural facilitation: rules for grooming in flies. *Anim. Behav. 24,* 739–55.

Dawkins, R. and Krebs, J. R. (1979) Arms races between and within species. *Proc. R. Soc., B., 205,* 489–512.

Deag, J. M. (1977) Aggression and submission in monkey societies. *Anim. Behav. 25,* 465–74.

Dejours, P. (1962) Chemoreflexes in breathing. *Physiol. Rev. 42,* 335–58.

Delius, J. D. (1969) A stochastic analysis of the maintenance behaviour of skylarks. *Anim. Behav. 33,* 137–78.

Delius, J. D. (1970) Irrelevant behaviour, information processing and arousal homeostasis. *Psychol. Forsch. 33,* 165–88.

Delius, J. D., Perchard, R. J. and Emmerton, J. (1976) Polarised light discrimination by pigeons and an

electroretinographic correlate. *J. comp. physiol. Psychol.* *90*, 560–71.
Denenberg, V. H. (1964) Critical periods, stimulus impact and emotional reactivity: a theory of infantile stimulation. *Psychol. Rev.* *71*, 335–51.
Desiraju, T., Banerjee, M. G. and Anand, B. K. (1968) Activity of single neurons in the hypothalamic feeding centers: effect of Z-deoxy-D. glucose. *Physiol. Behav.* *3*, 757–60.
Dethier, V. G. (1963) *The Physiology of Insect Senses*, London: Methuen.
Dethier, V. G. (1969) Feeding behaviour of the blowfly. *Adv. Study Behav.* *2*, 111–266.
DeVore, I. (1965*a*) Male dominance and mating behaviour in baboons. In *Sex and Behavior* (F. A. Beach, ed.). New York: Wiley.
DeVore, I. Ed., (1965*b*) *Primate Behavior*. New York: Holt, Rinehart and Winston.
Dewsbury, D. A. (1978) *Comparative Animal Behavior*. New York: McGraw Hill.
Diakow, C. (1974) Male-female interactions and the organization of mammalian mating patterns. *Adv. Study Behav.* *5*, 227–68.
Diamond, M., Diamond, L. and Mast, M. (1972) Visual sensitivity and sexual arousal levels during the menstrual cycle. *J. nerv. ment. Dis.* *155*, 170–6.
Dickson, D. P., Barr, G. R. and Wieckert, D. A. (1967) Social relationship of dairy cows in a feed lot. *Behaviour* *29*, 195–203.
Dietrich, J. P., Snyder, W. W., Meadows, C. E. and Albright, J. L. (1965) Rank order in cows. *Am. Zool.* *5*, 713.
Dijkgraaf, S. and Kalmijn, A. J. (1966) Versuche zur biologischen Bedeutung der Lorenzinischen Ampullen bei der Elasmobranchieren. *Z. vergl. Physiol.* *53*, 187–94.
Dimond, S. J. and Adam, J. H. (1972) Approach behaviour and embryonic visual experience in chicks: studies on the effect of rate of visual flicker. *Anim. Behav.* *20*, 413–20.
Dobson, C. W. and Lemon, R. E. (1979) Markov sequences in songs of American thrushes. *Anim. Behav.* *68*, 86–105.
Dolphinow, P. J. and Bishop, N. (1979) The development of motor skills and social relationships through play. *Minn. Symp. Child Psychol.* *4*, 141–98.
Dominey, W. J. (1980) Female mimicry in male bluegill sunfish – a genetic polymorphism? *Nature, Lond.* *284*, 546–8.
Donaldson, S. L., Albright, J. L. and Ross, M. A. (1972) Space and conflict in cattle. *Proc. Indiana Acad. Sci.* *81*, 352–4.
Donaldson, S. L., Black, W. C. and Albright, J. L. (1966) The effects of early feeding and rearing experiences on dominance, aggressive, and submissive behaviour in young heifer calves. *Am. Zool.* *6*, 247.
Donelly, E. D. (1954) Some factors that affect palatability in *Sericea lespedeza*. *Agron. J.* *46*, 96–7.
Dorsett, D. A., Willows, A. O. D. and Hoyle, G. (1973) The neuronal basis of behavior in *Tritonia*. IV. The central origin of a fixed action pattern demonstrated in the isolated brain. *J. Neurobiol.* *4*, 287–300.
Doty, R. W. (1968) Neural organization of deglutition. In *Handbook of Physiology*. Vol. IV, Sect. 6, *Alimentary Canal* (C. F. Code and C. L. Prosser, eds.). Washington, D.C.: American Physiological Society.
Doty, R. W. (1976) The concept of neural centers. In *Simpler Networks and Behavior* (J. C. Fentness, ed.). Sunderland, Mass.: Sinauer.
Dowling, J. E. and Boycott, B. B. (1966) Organisation of the primate retina: electron microscopy. *Proc. R. Soc.*, B *166*, 80–111.
Drent, R. H. (1970) Functional aspects of incubation in the herring gull (*Larus argentatus*). *Behaviour Suppl.* *17*, 1–132.
Drewett, R. F. and Trew, A. M. (1978) The milk ejection of the rat, as a stimulus and a response to the litter. *Anim. Behav.* *26*, 982–7.
Duffy, E. (1941) The conceptual categories of psychology: a suggestion for revision. *Psychol. Rev.* *48*, 177–203.
Duffy, E. (1962) *Activation and Behavior*. New York: Wiley.
Dunbar, R. I. M. and Dunbar, E. P. (1977) Dominance and reproductive success among female gelada baboons. *Nature, Lond.* *266*, 351–2.
Duncan, I. J. H., Horne, A. R., Hughes, B. O., and Wood-Gush, D. G. M. (1970) The pattern of food intake in female Brown Leghorn fowls as recorded in a Skinner Box. *Anim. Behav.* *18*, 245–55.
Eccles, J. C. (1973) *The Understanding of the Brain*. New York: McGraw Hill.
Edgerton, V. R., Grillner, S., Sjöstrom, A. and Zangger, P. (1976) Central control of locomotion in vertebrates. In *Neural Control of Locomotion, Adv. Behav. Biol.* *18*, 439–64. New York: Plenum.
Edmunds, M. (1974) *Defence in Animals. A Survey of Anti-Predator Defences*. Harlow: Longman.
Edney, E. B. (1954) Woodlice and the land habitat. *Biol. Rev.* *29*, 185–219.
Edwards, S. A. (1979) The timing of parturition in dairy cattle. *J. agric. Sci., Camb.* *93*, 359–63.
Edwards, S. A. (1980) Behavioural interactions between dairy cows and their newborn calves with relation to the calf serum immunoglobulin levels. Ph.D. thesis, University of Reading.
Edwards, S. A. and Broom, D. M. (1979) The period between birth and first suckling in dairy calves. *Res. vet. Sci.* *26*, 255–6.
Edwards, W. (1954) The theory of decision making. *Psychol. Bull.* *51*, 380–417.
Ehrenpreis, S. and Solnitzky, O. C. Eds., (1971) *Neurosciences Research*, Vol. 4. New York: Academic Press.

Eibl-Eibesfeldt, I. (1950) Ein Beitrag zur Paarungsbiologie der Erdkröte (*Bufo bufo* L.). *Behaviour 2*, 217–36.

Eibl-Eibesfeldt, I. (1951) Nahrungserweb und Beuteschema der Erdkröte (*Bufo bufo* L.). *Naturwissenschaften 53*, 589.

Eibl-Eibesfeldt, I. (1967) *Grundriss der vergleichenden Verhaltensforschung-Ethologie.* Munich: Piper Verlag.

Eisenberg, J. F. (1966) The social organizations of mammals. *Handb. Zool. 10*, 1–92.

Eisenberg, J. F., Muckenhirn, N. A. and Rudran, R. (1972) The relation between ecology and social structure in primates. *Science, N.Y. 176*, 863–74.

Eisner, T. (1971) Defence of phalangid: repellant administered by leg dabbing. *Science, N.Y. 173*, 650–2.

Eisner, T. and Meinwald, J. (1966) Defensive secretions of arthropods. *Science, N.Y. 153*, 1341–50.

Ellefson, J. O. (1968) Territorial behaviour in the common white-handed gibbon, *Hylobates lar* Linn. In *Primates* (P. C. Jay, ed.). New York: Holt Rinehart and Winston.

Ellefson, J. O. (1974) A natural history of white-handed gibbons in the Malayan peninsula. In *Gibbon and Siamang*, Vol. 3. (D. M. Rumbaugh, ed.). Basel: Karger.

Elner, R. W. and Hughes, R. N. (1978) Energy maximisation in the diet of the shore crab, *Carcinus maenas* (L). *J. Anim. Ecol. 47*, 103–16.

Elwood, R. W. (1975) Paternal and maternal behaviour in the mongolian gerbil. *Anim. Behav. 23*, 766–72.

Elwood, R. W. (1977) Changes in responses of male and female gerbils (*Meriones unguiculatus*) towards test pups during the pregnancy of the female. *Anim. Behav. 25*, 46–51.

Elwood, R. W. Ed. (1981) *Parental Behaviour of Rodents.* New York: Wiley.

Elwood, R. W. and Broom, D. M. (1978) The influence of litter size and parental behaviour on the development of Mongolian gerbil pups. *Anim Behav. 26*, 438–54.

Emlen, J. M. (1966) The role of time and energy in food preference. *Am. Nat. 100*, 611–17.

Emlen, J. M. (1973) *Ecology: an Evolutionary Approach.* Reading, Mass.: Addison Wesley.

Emlen, S. T. (1978) The evolution of cooperative breeding in birds. In *Behavioural Ecology: an evolutionary approach* (J. R. Krebs and N. B. Davies, eds.). Oxford: Blackwell.

Emlen, S. T. and Oring, L. W. (1977) Ecology, sexual selection and the evolution of mating systems. *Science, N.Y. 197*, 215–23.

Engen, T. and Lipsitt, L. P. (1965) Decrement and recovery of responses to olfactory stimuli in the human neonate. *J. comp. physiol. Psychol. 59*, 312–16.

Enright, J. T. (1980) *The Timing of Sleep and Wakefulness.* Berlin: Springer-Verlag.

Epple, G. (1975) The behaviour of marmoset monkeys (Callithricidae). *Primate Behav. 4*, 195–239.

Epstein, A. N., Fitzsimons, J. T. and Rolls, B. J. (1970) Drinking induced by injection of angiotensin into the brain of the rat. *J. Physiol., Lond. 210*, 457–74.

Erickson, C. J. (1970) Induction of ovarian activity in female ring doves by androgen treatment of castrated males. *J. comp. physiol. Psychol. 71*, 210–15.

Erickson, C. J. and Hutchison, J. B. (1977) Induction of nest material collecting in male Barbary doves by intracerebral androgen. *J. Reprod. Fert. 50*, 9–16.

Erickson, C. J. and Lehrman, D. S. (1964) Effect of castration of male ring doves upon ovarian activity of females. *J. comp. physiol. Psychol. 58*, 164–6.

Erickson, C. J. and Martinez-Vargas, C. (1975) The hormonal basis of cooperative nest-building. In *Neural and Endocrine Aspects of Behaviour in Birds* (P. Wright, P. G. Caryl, and D. M. Vowles, eds.). Amsterdam: Elsevier.

Esser, A. H. Ed., (1971) *Behavior and Environments: the use of Space by Animals and Man.* New York: Plenum.

Esslemont, R. J., Glencross, R. G., Bryant, M. J. and Pope, G. S. (1980) A quantitative study of pre-ovulatory behaviour in cattle (British Friesian heifers). *Appl. Anim. Ethol. 6*, 1–17.

Estes, R. D. (1966) Behaviour and life history of the wildebeest (*Connochaetes taurinus* Burchell). *Nature, Lond. 212*, 999–1000.

Estes, W. K. (1958) Stimulus-response theory of drive. In *Nebraska Symposium on Motivation* (M. R. Jones, ed.). Lincoln: University of Nebraska Press.

Euler, U. S. von (1967) Adrenal medullary secretion and its neural control. In *Neuroendocrinology*, Vol. 2. (L. Martini and W. F. Ganong, eds.). New York: Academic Press.

Evans, H. E. and West Eberhard, M. J. (1970) *The Wasps.* Cambridge Mass.: Harvard University Press.

Evans, L. T. (1953) Tail display in an iguanid lizard, *Leiocephalus carinatus coryi. Copeia 1*, 50–4.

Evans, M. E. (1979) Aspects of the life cycle of the Bewick's swan, based on recognition of individuals at a wintering site. *Bird Study, 26*, 149–62.

Evans, S. M. (1965) Learning in the polychaete *Nereis. Nature, Lond. 207*, 1420.

Evarts, E. V. (1974) Sensorimotor cortex activity associated with movements triggered by visual as compared to somesthetic inputs. In *The Neurosciences: Third Study Program* (F. O. Schmitt and F. G. Worden, eds.). Cambridge Mass.: M.I.T. Press.

Ewbank, R. and Bryant, M. J. (1972) Aggressive behaviour amongst groups of domesticated pigs kept at various stocking rates. *Anim. Behav. 20*, 21–8.

Ewert, J-P. (1967) Aktivierung der Verhaltensfolge beim Beutefang der Erdkröte (*Bufo bufo* L.) durch elektrische Mittelhirn-reizung. *Z. vergl. Physiol. 54,* 455–81.

Ewert, J-P. (1968) Der Einfluss von Zwischenhirndetekten auf die Visuomotorik im Beute-und Fluchtverhalten der Erdkröte (*Bufo bufo* L.). *Z. vergl. Physiol. 61,* 41–70.

Ewert, J-P. (1969). Das Beuteverhalten Zwischenhirn-detekter Erdkröten (*Bufo bufo* L.) gegenuber beweg-ten und ruhenden visuellen Mustern. *Pflügers Arch. ges. Physiol. 306,* 210–18.

Ewert, J-P. (1970) Aufnahme und Verarbeitung visueller Informationen im Beutetang- und Fluchtverhalten der Erdkröte (*Bufo bufo* L.). *Verh. Dtsch. Zool. Ges. Köln 64,* 218–26.

Ewert, J-P. (1971) Single unit response of the toad's (*Bufo americanus*) caudal thalamus to visual objects. *Z. vergl. Physiol. 74,* 81–102.

Ewert, J-P. (1974) The neural basis of visually guided behaviour. *Scient. Am. 230,* (3), 34–42.

Ewert, J-P. (1980) *Neuroethology.* Heidelberg: Springer.

Ewert, J-P. and Borchers, H. W. (1971) Reaktions charakteristik von Neuronen aus dem Tectum opticum und Subtectum der Erdkröte (*Bufo bufo* L.). *Z. vergl. Physiol. 71,* 165–89.

Ewert, J-P. and Borchers, H-W. (1974) Antwort von retinalen Ganglienzellen bei Freibeweglichen Kröten (*Bufo bufo* L.). *J. comp. Physiol. 92,* 117–30.

Ewert, J-P. and Hock, F. (1972) Movement sensitive neurons in the toad's retina. *Expl Brain Res. 16,* 41–59.

Ewert, J-P. and Ingle, D. (1971) Excitatory effects following habituation of prey-catching activity in frogs and toads. *J. comp. physiol. Psychol. 77,* 369–74.

Ewert, J-P. and Rehn, B. (1969) Quantitative analyse der Reiz-reaktions-beziehungen bei visuellen auslösen des Fluchtverhaltens der Wechsel-kröte (*Bufo viridis* Laur). *Behaviour 35,* 212–34.

Ewert, J-P., Speckhardt, I. and Amelang, W. (1970) Visuelle Inhibition and exzitation im Beutefangverhalten der Erdkröte (*Bufo bufo* L.). *Z. vergl. Physiol. 68,* 84–110.

Ewert, J-P. and Traud, R. (1979) Releasing stimuli for antipredator behaviour in the common toad (*Bufo bufo* L.). *Behaviour 68,* 170–80.

Ewert, J-P. and Wietersheim, A. von (1974*a*). Ganglienzellklassen in der retino-tectalen Projektion, der Kröte (*Bufo bufo* L.). *Acta Anat. 88,* 56–66.

Ewert, J-P. & Wietersheim, A. von (1974*b*) Musterauswertung durch Tectum und Thalamus/Praetectum-Neurone im visuellen system der Kröte (*Bufo bufo* L.). *J. comp. Physiol. 92,* 131–48.

Ewert, J-P. & Wietersheim, A. von (1974*c*). Einfluss von Thalamus/Praetectum-Defekten auf die Antwort von Tectum-Neuronen gegenüber visuellen Mustern bei der Kröte (*Bufo bufo* L.). *J. comp. Physiol. 92,* 149–60.

Ewing, L. S. (1972) Hierarchy and its relation to territory in the cockroach *Nauphoeta cinerea. Behaviour 42,* 152–74.

Fabricius, E. (1951) Zur Ethologie junge Anatiden. *Acta. zool. Fenn. 68,* 1–178.

Fagen, R. M. (1976) Exercise, play and physical training in animals. *Perspectives Ethol. 2,* 189–219. New York: Plenum.

Falk, J. L. (1961) The behavioral regulation of water–electrolyte balance. In *Nebraska Symposium on Motivation,* Vol. 9 (M. R. Jones, ed.). Lincoln, Nebraska: University of Nebraska Press.

Fantino, E. and Logan, C. A. (1979) *The Experimental Analysis of Behavior — a Biological Perspective.* San Francisco: Freeman.

Fantz, R. L. (1957) Form preferences in newly hatched chicks. *J. comp. physiol. Psychol. 50,* 422–30.

Fantz, R. L. (1965) Visual perception from birth as shown by pattern selectivity. *Ann. N.Y. Acad. Sci. 118,* 793–814.

F.A.O. (1970) *Production Yearbook No. 24.* Rome: Food and Agriculture Organisation of the United Nations.

Feldberg, W. and Myers, R. D. (1963) A new concept of temperature regulation by amines in the hypothalamus. *Nature, Lond. 200,* 1325.

Fentress, J. C. (1968) Interrupted ongoing behaviour in two species of vole (*Microtus agrestis* and *Clethrionomys britannicus*). *Anim. Behav. 16,* 135–163.

Fentress, J. C. (1972) Development and patterning of movement sequences in inbred mice. In *The Biology of Behavior* J. A. Kiger, ed.). Eugene, Or: Oregon State University Press.

Fentress, J. C. (1973) Development of grooming in mice with amputated fore limbs. *Science, N.Y. 179,* 704–5.

Fentress, J. C. (1976*a*) Dynamic boundaries of patterned behaviour: interaction and self-organisation. In *Growing Points in Ethology* (P. P. G. Bateson and R. A. Hinde, eds.). Cambridge: Cambridge University Press.

Fentress, J. C. Ed., (1976*b*) *Simpler Networks and Behavior.* Sunderland, Mass.: Sinauer.

Fentress, J. C. (1980) How can behaviour be studied from a neuroethological perspective? In *Information Processing in the Nervous System* (H. Pinsker, ed.). New York: Raven Press.

Fentress, J. C. (1981*a*) Order in ontogeny: relational dynamics. In *Behavioral Development* (K. Immelmann, G. W. Barlow, M. Main and L. Petrinovich, eds.). New York: Cambridge University Press.

Fentress, J. C. (1981*b*) Sensory – motor development. In *Sensory and Perceptual Development: Influences of Genetic and Experiential Factors* (R. N. Aslin, J. R. Alberts and M. R. Peterson, eds.). New York: Academic Press.

Fentress, J. C. and Stillwell, F. P. (1973) Grammar of

movement sequence in inbred mice. *Nature, Lond.* *244*, 52–3.

Ferguson, R. S., Albright, J. L., Harrington, R. B., Black, W. C., Donaldson, S. L., Snyder, W. W. and Dietrich, J. P. (1967) Dairy cattle entrance into a milking area. *Am. Zool.* 7, 432.

Fernandez de Molina, A. and Hunsperger, R. W. (1959) Central representation of affective reactions in fore brain and brainstem: electrical stimulation of amygdala, stria terminalis, and adjacent structures. *J. Physiol., Lond.* *145*, 251–65.

Finch, S. (1978) Volatile plant chemicals and their effect on host plant finding by the cabbage root fly *Delia brassicae*. *Entomologia exp. appl.* *24*, 150–9.

Finch, S. and Skinner, G. (1973) Chemosterilization of the cabbage root fly under field conditions. *Ann. appl. Biol.* *73*, 243–58.

Finch, S. and Skinner, G. (1974) Some factors affecting the efficiency of water-traps for capturing cabbage root flies. *Ann. appl. Biol.* *77*, 213–26.

Findlay, J. D. and Beakley, W. R. (1954) Environmental physiology of farm animals. In *Progress in the Physiology of Farm Animals*, Vol. 1 (J. Hammond, ed.). London: Butterworth.

Fisher, A. E. and Coury, J. N. (1962) Cholinergic tracing of a central neural circuit underlying the thirst drive. *Science, N.Y.* *138*, 691–2.

Fisher, J. and Hinde, R. A. (1949) The opening of milk bottles by birds. *Brit. Birds.* *42*, 347–57.

Fisher, R. A. (1930) *The Genetical Theory of Natural Selection.* Oxford: Clarendon Press.

Fisher, R. A. (1958) *The Genetical Theory of Natural Selection*, 2nd edn. New York: Dover.

Fitch, W. M. and Margoliash, E. (1967) Construction of phylogenetic trees. *Science, N.Y.* 155, 279–84.

Fitzsimons, J. T. (1961) Drinking by rats depleted of body fluid without increase in osmotic pressure. *J. Physiol., Lond.* *159*, 297–309.

Fitzsimons, J. T. (1968) La soif extracellulaire. *Ann. nutr. aliment.* *22*, 131–44.

Fitzsimons, J. T. (1969) The role of a renal thirst factor in drinking induced by extracellular stimuli. *J. Physiol., Lond.* *201*, 349–68.

Fitzsimons, J. T. (1972) Thirst. *Physiol., Rev.* *52*, 468–561.

Fitzsimons, J. T. (1979) The Physiology of Thirst and Sodium Appetite. In *Monographs of the Physiological Society*, No. 35. Cambridge: Cambridge University Press.

Fitzsimons, J. T. and Le Magnen, J. (1969) Eating as a regulatory control of drinking in the rat. *J. comp. physiol. Psychol.* *67*, 273–83.

Fitzsimons, J. T. and Simons, B. J. (1968) The effect of angiotension on drinking in the rat. *J. Physiol., Lond.* *198*, 39–41P.

Fleming, A. S. and Rosenblatt, J. S. (1964) Olfactory regulation of maternal behaviour in rats: I. Effects of olfactory bulb removal in experienced and inexperienced lactating and cycling rats. *J. comp. physiol. Psychol.* *86*, 221–32.

Fleshler, M. (1965) Adequate acoustic stimuli for startle reaction in the rat. *J. comp. physiol. Psychol.* *60*, 200–7.

Follett, B. K., Scanes, C. G. and Cunningham, F. J. (1972) A radioimmunoassay for avian luteinising hormone. *J. Endocrinol.* *52*, 359–78.

Forrester, R. C. (1979) Behavioural State and Responsiveness in Domestic Chicks. Ph.D. thesis, University of Reading.

Forrester, R. C. (1980) Sterotypies and the behavioural regulation of motivational state. *Appl. Anim. Ethol.* *6*, 386–7.

Forrester, R. C. and Broom, D. M. (1980) Ongoing behaviour and startle responses of chicks. *Behaviour* *73*, 51–63.

Fox, M. W. (1968) Socialisation, environmental factors and abnormal behavioral development in animals. In *Abnormal Behavior of Animals* (M. W. Fox, ed.). New York: W. B. Saunders.

Frank, F. (1957) The causality of microtine cycles in Germany. *J. Wildl. Mgmt.* *21*, 113–21.

Fraser, A. F. (1968) *Reproductive Behaviour in Ungulates.* London: Academic Press.

Fraser, A. F. (1974) *Farm Animal Behaviour.* London: Baillière Tindall.

Fraser Darling, F. F. (1937) *A Herd of Red Deer.* Oxford: Oxford University Press.

Fraser Darling, F. F. (1938) *Bird Flocks and the Breeding Cycle.* Cambridge: Cambridge University Press.

Frazetta, T. H. (1966) Studies on the morphology and function of the skull in Boidae (Serpentes). Part II. Morphology and function of the jaw apparatus in *Python sebae* and *Python molurus*. *J. Morph.* *118*, 217–96.

Freeland, W. J. and Janzen D. H. (1974) Strategies in herbivory by mammals: the role of plant secondary compounds. *Am. Nat.* *108*, 269–89.

Freeman, B. M. (1971) Stress and the domestic fowl: a physiological appraisal. *World Poult. Sci. J.* *27*, 263–75.

Freeman, B. M. and Vince, M. A. (1974) *Development of the Avian Embryo.* London: Chapman and Hall.

Fretter, V. and Graham, A. (1962) *British Prosobranch Molluscs.* London: Ray Society.

Fretter, V. and Graham, A. (1976) *A Functional Anatomy of Invertebrates.* London: Academic Press.

Friedman, L. and Miller, J. G. (1971) Odor incongruity and chirality. *Science, N.Y.* *172*, 1044–6.

Friedman, M. I. and Stricker, E. M. (1976) The physiological psychology of hunger: a physiological perspective. *Psychol. Rev.* *83*, 409–31.

Friend, T. H. and Polan, C. E. (1978) Competitive order as a measure of social dominance in dairy cattle. *Appl. Anim. Ethol. 4*, 61–70.
Frisch, K. von (1946) Die Tänze der Bienen. *Öst. zool. Z. 1*, 1–48.
Frisch, K. von (1949) Die Polarisation des Himmelslichtes als orientierender Faktor bei den Tänzen der Bienen. *Experientia 5*, 142–8.
Fromme, A. (1941) An experimental study of the factors of maturation and practice in the behavioural development of the embryo of the frog, *Rana pipiens*. *Genet. Psychol. Monogr.* 24, 219–56.
Fry, C. H. (1972) The social organisation of bee-eaters (Meropidae) and cooperative breeding in hot-climate birds. *Ibis 114*, 1–14.
Furuya, Y. (1969) On the fission of troops of Japanese monkeys: 2. general view of troop fission of Japanese moneys. *Primates 10*, 47–69.
Gadbury, J. C. (1975) Some preliminary field observations on the order of entry of cows into herringbone parlours. *Appl. Anim. Ethol. 1*, 275–81.
Gadgil, M. (1971) Dispersal: population consequences and evolution. *Ecology 52*, 253–61.
Gadgil, M. (1972) The function of communal roosts: relevance of mixed roosts. *Ibis 114*, 531–2.
Gadgil, M. and Bossert, W. H. (1970) Life history consequences of natural selection. *Am. Nat. 104*, 1–24.
Galef, B. G. (1976) Social transmission of acquired behaviour: a discussion of tradition and social learning in vertebrates. *Adv. Study Behav. 6*, 77–100.
Gallagher, J. E. (1976) Sexual imprinting: effects of various regimens of social experience on mate preference in Japanese quail *Coturnix coturnix japonica*. *Behaviour 57*, 91–114.
Gallagher, J. E. (1977) Sexual imprinting: a sensitive period in Japanese quail (*Coturnix coturnix japonica*). *J. comp. physiol. Psychol. 91*, 72–8.
Galton, F. (1871) Gregariousness in cattle and men. *MacMillan's Mag., Lond. 23*, 353.
Gandini, G. and Baldwin, P. J. (1978) An encounter between chimpanzees and a leopard in Senegal. *Carnivore 1*, 107–9.
Garcia, J., Ervin, F. R. and Koelling, R. A. (1966) Learning with prolonged delay of reinforcement. *Psychon. Sci. 5*, 121–2.
Garcia, J., Ervin, F. R., York, C. H. and Koelling, R. A. (1967) Conditioning with delayed vitamin injection. *Science, N.Y. 155*, 716–18.
Garrod, D. R. and Feldman, J. D. Eds., (1981) *Development in the Nervous System.* British Society for Developmental Biology Symposium 5. Cambridge: Cambridge University Press.
Gartlan, J. S. (1968) Structure and function in primate society. *Folia Primatol. 8*, 89–120.
Gass, C. L. (1978) Experimental studies of foraging in complex laboratory environments. *Am. Zool. 18*, 729–38.
Gazzaniga, M. S. and Blakemore, C. Eds. (1975) *Handbook of Psychobiology*. New York: Academic Press.
Geist, V. (1971) *Mountain Sheep: a Study in Behavior and Evolution*. Chicago: University of Chicago Press.
Geist, V. (1974) On the relationship of social evolution and ecology in ungulates. *Am. Zool. 14*, 205–20.
Getting, P. A. (1976) Afferent neurons mediating escape swimming of the marine mollusk *Tritonia diomedea*. *J. comp. Physiol. 110*, 271–86.
Gibb, J. A. (1957) Food requirements and other observations on captive tits. *Bird Study 4*, 207–15.
Gibb, J. A. (1960) Populations of tits and goldcrests and their food supply in pine plantations. *Ibis 102*, 163–208.
Gibson, E. J. and Walk, R. D. (1960) The visual cliff. *Scient. Am. 202*, (4), 64–72.
Gilder, P. M. and Slater, P. J. B. (1978) Interest of mice in conspecific male odours is influenced by degree of kinship. *Nature, Lond. 274*, 364–5.
Gill, F. B. and Wolf, L. L. (1975) Economies of feeding territoriality in the golden-winged sunbird. *Ecology 56*, 333–45.
Gillary, H. L. (1966) Stimulation of the blowfly salt receptor. I. NaCl. *J. gen. Physiol. 50*, 337–50.
Glickman, S. E. and Sroges, R. W. (1966) Curiosity of zoo animals. *Behaviour 24*, 151–88.
Globus, A. and Scheibel, A. B. (1966) Loss of dendritic spines as an index of pre-synaptic terminal patterns. *Nature, Lond. 212*, 463–5.
Goethe, F. (1940) Beobachtungen und Versuche über angeborene Schreckreaktionen junger Auerhühner (*Tetrao u. urogallus* L.). *Z. Tierpsychol. 4*, 165–7.
Golani, I., Wolgin, D. and Teitelbaum, P. (1979) A proposed natural geometry of recovery from akinesia in the lateral hypothalamic rat. *Brain Res. 164*, 237–67.
Goldberg, M. E. and Wurtz, R. H. (1972) Activity of superior colliculus in behaving monkey. II. Effect of attention on neuronal responses. *J. Neurophysiol. 35*, 560–74.
Goldsmith, T. H. (1960) The nature of the retinal action potential, and the spectral sensitivity of the ultra violet and green receptor systems of the compound eye of the worker honeybee. *J. gen. Physiol. 43*, 775–99.
Gollub, L. (1977) Conditional reinforcement: schedule effects. In *Handbook of Operant Behaviour* (W. H. Honig and J. E. R. Staddon, eds.). Englewood Cliffs, N.J.: Prentice Hall.
Goodall, J. (1964) Tool-using and aimed throwing in a community of free-living chimpanzees. *Nature, Lond. 201*, 1264–6.
Goodman, E., Jansen, P. and Dewsbury, D. A. (1971)

Midbrain reticular formation lesions: habituation to stimulation and copulatory behaviour in male rats. *Physiol. Behav. 6,* 151–6.

Goss-Custard, J. D. (1970) The responses of redshank (*Tringa totanus* (L) to spatial variations in the density of their prey. *J. Anim. Ecol. 39,* 91–113.

Goss-Custard, J. D. (1977*a*) Optimal foraging and the size selection of worms by redshank, *Tringa totanus,* in the field. *Anim. Behav. 25,* 10–29.

Goss-Custard, J. D. (1977*b*) Predator responses and prey mortality in the redshank *Tringa totanus* (L.) and a preferred prey *Corophium volutator* (Pallas). *J. Anim. Ecol. 46,* 21–36.

Gottlieb, G. (1973). Introduction to behavioral embryology. In *Behavioral Embryology* (G. Gottlieb, ed.). New York: Academic Press.

Gottlieb, G. (1976*a*) Conceptions of prenatal development: behavioral embryology, *Psychol. Rev. 83,* 215–34.

Gottlieb, G. (1976*b*) The roles of experience in the development of behavior and the nervous system. In *Studies on the Development of Behavior and the Nervous System.* Vol. 3. *Neural and Behavioral Specificity* (G. Gottlieb, ed.). New York: Academic Press.

Gottman, J. M. and Notarius, C. (1978) Sequential analysis of observational data using Markov chairs. In *Single Subject Research* (T.R. Kratochwill, ed.). New York: Academic Press.

Gould, J. L. (1975) Honeybee recruitment: the dance-language controversy, *Science, N.Y. 189,* 685–93.

Gould, J. L. (1976) The dance-language controversy. *Rev. Biol.* 51, 211–44.

Grafen, A. and Sibly, R. (1978) A model of mate desertion. *Anim. Behav. 26,* 645–52.

Graham, N. McC. (1962) Energy expenditure of grazing sheep. *Nature, Lond. 196,* 289.

Graham, N. McC. (1965) Some aspects of pasture evaluation. In *Energy Metabolism* (K. L. Blaxter, ed.). London: Academic Press.

Grant, K. A. and Grant, V. (1968) *Hummingbirds and their Flowers.* New York: Columbia University Press.

Gray, J. and Lissmann, H. W. (1946) The coordination of limb movements in the amphibia. *J. exp. Biol. 23,* 133–42.

Green, J., Clemente, C. and Groot, J. de (1957) Rhinencephalic lesions and behavior in cats. *J. comp. Neurol. 108,* 505–45.

Green, M., Green, R. and Carr, W. J. (1966) The hawk-goose phenomenon: a replication and an extension. *Psychon. Sci. 4,* 185–6.

Greenhalgh, J. F. D. (1975) Factors limiting animal production from grazed pasture. *J. Br. Grassld Soc. 30,* 153–60.

Griffin, D. R. and Galambos, R. (1941) The sensory basis of obstacle avoidance by flying bats. *J. exp. Zool. 86,* 481–506.

Grillner, S. (1975) Locomotion in vertebrates: central mechanisms and reflex interaction. *Physiol. Rev. 55,* 247–304.

Grillner, S. (1976) Some aspects of the descending control of the spinal circuits generating locomotor movements. In *Neural Control of Locomotion* (R. M. Herman, S. Grillner, P. S. G. Stein and D. G. Stuart, eds.). New York: Plenum.

Groot P. de (1980) Information transfer in a socially roosting weaver bird (*Quelea quelea;* Ploceinae): an experimental study. *Anim. Behav. 28,* 1249–54.

Grossman, S. P. (1967) *A Textbook of Physiological Psychology.* New York: Wiley.

Grossman, S. P. and Grossman, L. (1963) Food and water intake following lesions or electrical stimulation of the amygdala. *Am. J. Physiol. 205,* 761–5.

Grota, L. J. (1973) Effects of litter size, age of young, and parity on foster mother behavior in *Rattus norvegicus. Anim. Behav. 21,* 78–82.

Groves, P. M. and Thompson, R. F. (1970) Habituation: a dual process theory. *Psychol. Rev. 77,* 419–50.

Grubb, P. and Jewell, P. A. (1966) Social grouping and home range in feral Soay sheep. *Symp. zool. Soc. Lond. 18,* 179–210.

Gruendel, A. D. and Arnold, W. J. (1969) Effects of early social deprivation on reproductive behaviour of male rats. *J. comp. physiol. Psychol.* 67, 123–8.

Guhl, A. M. (1942) Social discrimination in small flocks of the common domestic fowl. *J. comp. Psychol. 34,* 127–48.

Guhl, A. M. (1953) Social behavior of the domestic fowl. *Tech. Bull. Kans. agric. Exp. Stn 73,* 3–48.

Guhl, A. M. (1964) Psychophysiological interrelations in the social behavior of chickens. *Psychol. Bull. 61,* 277–85.

Guhl, A. M. (1968) Social inertia and social stability in chickens. *Anim. Behav. 16,* 219–32.

Guhl, A. M. and Allee, W. C. (1944) Some measurable effects of social organisation in hens. *Physiol, Zool. 17,* 320–47.

Guhl, A. M. and Warren, D. C. (1946) Number of offspring sired by cockerels relating to social dominance in chickens. *Poult. Sci. 25,* 460–72.

Guiton, P. (1966) Early experience and sexual object-choice in the brown leghorn. *Anim. Behav. 14,* 534–8.

Guthrie, E. R. (1952) *The Psychology of Learning.* New York: Harper.

Guyomarc'h, J. C. (1975) Les cycles d'activité d'une couvée naturelle de poussins et leur coordination. *Behaviour* 53, 31–75.

Guyomarc'h, J. C. and Thibout, E. (1969) Rythmes et cycles dans l'emission du chant chez la caille Japonaise (*Coturnix c.j.*) *Rev. Comp. Anim.* 3, 37–49.

Guz, A. (1975) Regulation of respiration in man. *Ann. Rev. Physiol.* 37, 303–24.

Hafez, E. S. E. Ed., (1975) *The Behaviour of Domestic Animals*, 3rd edn. London: Baillière Tindall.

Hafez, E. S. E. and Dyer, I. A. Eds., (1969) *Animal Growth and Nutrition*. Philadelphia: Lea and Febiger.

Hailman, J. P. (1967) The ontogeny of an instinct. The pecking response in chicks of the laughing gull (*Larus atricilla* L.) and related species. *Behaviour Suppl. 15*, 1–159.

Hainsworth, F. R. (1973) On the tongue of a hummingbird: its role in the rate and energetics of feeding. *Comp. Biochem. Physiol. 46A*, 65–78.

Hainsworth, F. R. (1974) Food quality and foraging efficiency: the efficiency of sugar assimilation by hummingbirds. *J. comp. Physiol. 88*, 425–31.

Hainsworth, F. R. (1977) Foraging efficiency and parental care in *Colibri coruscans*. *Condor 79*, 69–75.

Hainsworth, F. R. (1978) Feeding: models of costs and benefits in energy regulation. *Am. Zool. 18*, 701–14.

Hainsworth, F. R. and Wolf, L. L. (1972) Crop volume, nectar concentration and hummingbird energetics. *Comp. Biochem. Physiol. 42A*, 359–66.

Hainsworth, F. R. and Wolf, L. L. (1979) Feeding: an ecological approach. *Adv. Study Behav. 9*, 53–96.

Hall, C. S. (1934) Emotional behaviour in the rat. I. Defaecation and urination as measures of individual differences in emotionality. *J. comp. Psychol. 18*, 385–403.

Hall, E. T. (1966) *The Hidden Dimension*. Garden City, N.Y.: Doubleday.

Hall, K. R. L. (1968) Social organization of old-world monkeys and apes. In *Primates: Study in Adaption and Variability* (Jay, P., ed.). New York: Holt, Rinehart and Winston.

Hall, K. R. L. and DeVore, I. (1965) Baboon social behavior. In *Primate Behavior* (I. DeVore, ed.). New York: Holt, Rinehart and Winston.

Hall, W. C. and Brody, S. (1934) *Res. Bull. Mo. agric. Exp. Stn* No. 208.

Halley, R. J. (1953) The grazing behaviour of South Devon cattle under experimental conditions. *Br. J. Anim. Behav. 1*, 156–7.

Halliday, T. R. (1976) The libidinous newt, an analysis of variations in the sexual behaviour of the male smooth newt, *Triturus vulgaris*. *Anim. Behav. 24*, 398–414.

Halliday, T. R. (1977) The effect of experimental manipulation of breathing behaviour on the sexual behaviour of the smooth newt, *Triturus vulgaris*. *Anim. Behav. 25*, 39–45.

Halliday, T. R. (1978) Sexual selection and mate choice. In *Behavioural Ecology: an Evolutionary Approach* (J. R. Krebs and N. B. Davies, eds.). Oxford: Blackwell.

Halliday, T. R. and Sweatman, H. P. A. (1976) To breathe or not to breathe; the newt's problem. *Anim. Behav. 24*, 551–61.

Hamburg, D. A. and McCown, E. Eds., (1980) *The Great Apes*. Menlo Park, Ca.: Benjamin Cummings.

Hamburger, V. (1968) Origins of integrated behaviour. *Develop. Biol. Suppl. 2*, 251–71.

Hamilton, W. D. (1963) The evolution of altruistic behaviour. *Am. Nat. 97*, 354–6.

Hamilton, W. D. (1964*a*) The genetical evolution of social behaviour. I. *J. theoret. Biol. 7*, 1–16.

Hamilton, W. D. (1964*b*) The genetical evolution of social behaviour. II. *J. theoret. Biol. 7*, 17–32.

Hamilton, W. D. (1971) Geometry for the selfish herd. *J. theoret. Biol. 31*, 295–311.

Hamilton, W. D. (1972) Altruism and related phenomena, mainly in social insects. *Ann. Rev. Ecol. Syst. 3*, 193–232.

Hamilton, W. J. (1963) Success story of the opossum. *Nat. Hist. 72*, (Feb.), 17–25.

Hamilton, W. J. and Gilbert, W. M. (1969) Starling dispersal from a winter roost. *Ecology 50*, 886–98.

Hammel, H. T., Caldwell, F. T. and Abrams, R. M. (1967) Regulation of body temperature in the blue-tongued lizard. *Science, N.Y. 156*, 1260–2.

Hammel, H. T., Hardy, J. D. and Fusco, M. M. (1960) Thermoregulatory responses to hypothalamic cooling in unanaesthetised dogs. *Am. J. Physiol. 198*, 481–6.

Hammel, H. T., Strømme, S. B. and Myhre, K. (1969) Forebrain temperature activates behavioral thermoregulatory response in arctic sculpins. *Science, N.Y. 165*, 83–5.

Hansell, M. H. (1968) The house building behaviour of the caddis-fly *Silo pallipes:* I. the structure of the house and method of house extension. *Anim. Behav. 16*, 558–61.

Harcourt, A. H. (1977) Social Relationships of Wild Mountain Gorilla. Ph.D. thesis, University of Cambridge.

Hardy, A. C. (1956) *The Open Sea*. London: Collins.

Harlow, H. F. (1965) Sexual behavior in the rhesus monkey. In *Sex and Behavior* (F. A. Beach, ed.). New York: Wiley.

Harlow, H. F. (1969) Age-mate or peer affectional system. *Adv. Study Behav. 2*, 333–83.

Harlow, H. F. and Harlow, M. K. (1965) The affectional systems. In *Behavior of Nonhuman Primates*, Vol. 2 (A. M. Schrier, H. F. Harlow and F. Stollnitz, eds.). New York: Academic Press.

Harlow, H. F. and Zimmermann, R. R. (1959) Affectional responses in the infant monkey. *Science, N.Y. 130*, 421–32.

Harré, R. (1974) The conditions for a social psychology of childhood. In *The Integration of a Child into a Social World* (M. P. M. Richards, ed.). Cambridge: Cambridge University Press.

Harris, G. W., Michael, R. P. and Scott, P. P. (1958) Neurological site of action of stilboestrol in eliciting sexual behaviour. *Foundation Symp. on Neurological Basis of Behaviour*. London: Churchill.

Hart, B. (1967) Sexual reflexes and mating behavior in the male dog. *J. comp. physiol. Psychol. 66*, 388–99.

Hart, B. L. (1973) Effects of testosterone propionate and dihydrotestosterone on penile morphology and spinal reflexes of spinal male rats. *Horm. Behav. 4*, 239–46.

Hartline, H. K. and Ratliff, F. (1957) Inhibitory interaction of receptor units in the eye of *Limulus*. *J. gen Physiol. 40*, 357–76.

Hartline, H. K. and Ratliff, F. (1958) Spatial summation of inhibitory influences in the eye of the *Limulus*, and the mutual interaction of receptor units. *J. gen. Physiol. 41*, 1049–66.

Hartline, H. K., Wagner, H. and Ratliff, F. (1956) Inhibition in the eye of *Limulus*. *J. gen. Physiol. 39* 651–73.

Hartmann, E. L. (1973) *The Functions of Sleep*. New Haven, Conn.: Yale University Press.

Harvey, P. H. and Greenwood, P. J. (1978) Anti-predator defence strategies: some evolutionary problems. In *Behavioural Ecology* (J. R. Krebs and N. B. Davies, eds.). Oxford: Blackwell.

Hassell, M. P. (1971) Mutual interference between searching insect parasites. *J. Anim. Ecol. 40*, 473–86.

Hassell, M. P. and May, R. M. (1974) Aggregation of predators and insect parasites and its effect on stability. *J. Anim. Ecol. 43*, 567–94.

Hassell, M. P. and Southwood, T. R. E. (1978) Foraging strategies of insects. *Ann. Rev. Ecol. Syst. 9*, 75–98.

Hausfater, G. (1975) Dominance and reproduction in baboons (*Papio cynocephalus*). *Contrib. Primatol. 7*, 1–150.

Hauske, G. (1967) Stochastische und rhythmische Eigenschaften spontan auftretender Verhaltensweisen von Fischen. *Kybernetik 4*, 26–36.

Hawkes, C. (1971) Behaviour of the adult cabbage rootfly. Field dispersal and behaviour. *Rep. natn. Veg. Res. Stn.* p. 93.

Hay, D. E. and McPhail, J. D. (1975) Mate selection in three-spined sticklebacks. *Can. J. Zool. 53*, 441–50.

Hebb, D. O. (1946) On the nature of fear. *Psychol. Rev. 53*, 250–75.

Hebb, D. O. (1955) Drives and the C.N.S. (conceptual nervous system). *Psychol. Rev. 62*, 243–54.

Hediger, H. (1941) Biologische Gesetzmässigkeiten im Verhalten von Wirbeltieren. *Mitt. Naturf, Ges. Bern.*

Hediger, H. (1950) *Wild Animals in Captivity*. London: Butterworths.

Hediger, H. (1955) *Studies of the Psychology and Behaviour of Captive Animals in Zoos and Circuses*. London: Butterworth.

Hediger, H. (1969) Comparative observations on sleep. *Proc. R. Soc. Med. 62*, 153–6.

Heiligenberg, W. (1974) A stochastic analysis of fish behaviour. In *Motivational Control Systems Analysis* (D. J. McFarland, ed.). London: Academic Press.

Heimer, L. and Larsson, K. (1964) Drastic changes in the mating behavior of male rats following lesions in the junction of diencephalon and mesencephalon. *Experientia 20*, 460–1.

Heimer, L. and Larsson, K. (1967) Impairment of mating behavior in male rats following lesions in the preoptic-anterior hypothalamic continuum. *Brain Res. 3*, 248–63.

Heitman, H., Hahn, L., Kelly, C. F. and Bond, T. E. (1961) Space allotment and performance of growing-finishing swine raised in confinement. *J. Anim. Sci. 20*, 543–6.

Held, R. (1965) Plasticity in sensory-motor systems. *Scient. Am. 213, 5*, 84–94.

Held, R. and Hein, A. V. (1958) Adaptation of disarranged hand eye coordination contingent upon reafferent stimulation. *Percept. Mot. Skills 8*, 87–90.

Held, R. and Hein, A. (1963) Movement-produced stimulation in the development of visually guided behaviour. *J. comp. physiol. Psychol. 56*, 872–6.

Hemsworth, P. H., Beilharz, R. G. and Brown, W. J. (1978) The importance of the courting behaviour of the boar on the success of natural and artificial matings. *Appl. Anim. Ethol. 4*, 341–7.

Hemsworth, P. H., Winfield, C. G. and Mullaney, P. D. (1976) A study of the development of the teat order in piglets. *Appl. Anim. Ethol. 2*, 225–33.

Hendrey, D. P. and Rasche, R. H. (1961) Analysis of a new nonnutritive positive reinforcer based on thirst. *J. comp. physiol. Psychol. 54*, 477–83.

Henry, C. S. (1972) Eggs and rapagula of *Ullulodea* and *Ascaloptynx* (Neuroptera: Ascalophidae): a comparative study. *Psyche, Berl. 79*, 1–22.

Hensel, H. (1974) Thermoreceptors. *Ann. Rev. Physiol. 36*, 233–49.

Henson, O. W. (1965) The activity and function of the middle-ear muscles in echo-locating bats. *J. Physiol., Lond. 180*, 871–87.

Herman, R. M., Grillner, S., Stein, P. S. G. and Stuart, D. G. Eds., (1976) *Neural Control of Locomotion*. New York: Plenum.

Hernández-Peón, R., O'Flaherty, J. J. and Mazzuchelli-O'Flaherty, A. L. (1967) Sleep and other behavioral effects induced by acetylcholine stimulation of basal temporal cortex and striate structres. *Brain, Res. 4*, 243–67.

Hernández-Peón, R., Scherrer, H. and Jouvet, M. (1956) Modification of electric activity in cochlear nucleus during 'attention' in unanaesthetised cats. *Science, N.Y. 123*, 331–2.

Herrnstein, R. J. and Loveland, D. H. (1964) Complex visual concept in the pigeon. *Science, N.Y. 146*, 549–51.

Hess, E. H. (1956) Space perception in the chick. *Scient, Am. 195*, (1), 71–80.

Hickey, W. C. (1961) Growth form of crested wheatgrass as affected by site and grazing. *Ecology 42*, 173–6.

Hidaka, T. and Yamashita, K. (1976) Wing color pattern as the releaser of mating behavior in the swallowtail butterfly *Papilio xuthus* L. (Lepidoptera: Papilionidae). *Appl. Ent. Zool. 10*, 263–7.

Higginbotham, A. C. and Koon, W. E. (1955) Temperature regulation in the Virginia opossum. *Am. J. Physiol. 181*, 69–71.

Hilgard, E. R. (1958) *Theories of Learning*. New York: Appleton-Century-Crofts.

Hillyard, S. A., Hink, R. F., Schwent, V. K. and Picton, T. W. (1973) Electrical signs of selective attention. *Science, N.Y. 182*, 177–9.

Hinde, R. A. (1956) The biological significance of the territories of birds. *Ibis 98*, 340–69.

Hinde, R. A. (1959) Unitary drives. *Anim. Behav. 7*, 130–41.

Hinde, R. A. Ed., (1969) *Bird Vocalisations*. Cambridge: Cambridge University Press.

Hinde, R. A. (1970) *Animal Behaviour: a Synthesis of Ethology and Comparative Psychology*, 2nd edn. New York: McGraw Hill.

Hinde, R. A. Ed., (1972) *Non-Verbal Communication*. Cambridge: Cambridge University Press.

Hinde, R. A. (1973) Constraints on learning – an introduction to the problems. In *Constraints on Learning* (R. A. Hinde and J. Stevenson-Hinde, eds.). London: Academic Press.

Hinde, R. A. (1974) *Biological Bases of Human Social Behaviour*. New York: McGraw Hill.

Hinde, R. A. (1979) *Towards Understanding Relationships*. London: Academic Press.

Hinde, R. A. and Atkinson, S. (1970) Assessing the roles of social partners in maintaining mutual proximity, as exemplified by mother/infant relations in monkeys. *Anim. Behav. 18*, 169–76.

Hinde, R. A., Bell, R. Q. and Steel, E. A. (1963) Changes in sensitivity of the canary brood patch during the natural breeding season. *Anim. Behav. 11*, 553–60.

Hinde, R. A. and Davies, L. (1972*a*) Changes in mother/infant relationship after separation in rhesus monkeys. *Nature, Lond. 239*, 41–2.

Hinde, R. A. and Davies, L. (1972*b*) Removing infant rhesus from mother for 13 days compared with removing mother from infant. *J. Child Psychol. Psychiat. 13*, 227–37.

Hinde, R. A., Rowell, T. E. and Spencer-Booth, Y. (1964) Behaviour of socially living rhesus monkeys in their first six months. *Proc. zool. Soc. Lond. 143*, 609–49.

Hinde, R. A. and Spencer Booth, Y. (1971) Effects of brief separation from mother on rhesus monkeys. *Science, N.Y. 173*, 111–18.

Hinde, R. A. and Steel, E. (1966) Integration of the reproductive behaviour of female canaries. *Symp. Soc. exp. Biol. 20*, 401–26.

Hinde, R. A. and Steel, E. (1978) The influence of daylength and male vocalisations on the estrogen-dependent behaviour of female canaries and budgerigars, with discussion of data from other species. *Adv. Study Behav. 8*, 39–73.

Hinde, R. A. and Stevenson-Hinde, J. Eds., (1973) *Constraints on Learning*. London: Academic Press.

Hinde, R. A. and Stevenson-Hinde, J. (1976) Towards understanding relationships: Dynamic stability. In *Growing Points in Ethology* (P. P. G. Bateson and R. A. Hinde, eds.). Cambridge: Cambridge University Press.

Hinde, R. A., Thorpe, W. H. and Vince, M. A. (1956) The following responses of young coots and moorhens. *Behaviour 9*, 214–42.

Hinde, R. A. and White, L. E. (1974) The dynamics of a relationship – rhesus mother-infant ventro-ventral contact. *J. comp. physiol. Psychol. 86*, 8–23.

Hinsche, G. (1928) Kampfreaktionen bei einheimischen Anuren. *Biol. Zbl. 48*, 577–616.

Hirsch, H. V. B. (1972) Visual perception in cats after environmental surgery. *Expl Brain Res. 15*, 405–23.

Hirsch, H. V. B. and Spinelli, D. N. (1970) Visual experience modifies distribution of horizontally and vertically oriented receptive fields in cats. *Science, N.Y. 168*, 879–71.

Hirsch, H. V. B. and Spinelli, D. N. (1971) Modification of the distribution of receptive field orientation in cats by selective visual exposure during development. *Expl Brain Res. 13*, 509–27.

Hirsch, J. Ed., (1967) *Behavior-genetic analysis*. New York: McGraw Hill.

Hodgson, J., Rodrigues Capriles, J. M. and Fenlon, J. S. (1977) The influence of sward characteristics on the herbage intake of grazing calves. *J. agric. Sci., Camb. 89*, 743–50.

Hodgson, J. and Wilkinson, J. M. (1969) The relationship between live-weight and herbage intake in grazing cattle. *Anim. Prod. 9*, 365–76.

Hogan, J. A. (1981) Homeostasis and Behaviour. In *Analysis of Motivational Processes* (F. M. Toates and T. R. Halliday, eds.). New York: Academic Press.

Hogan, J. A., Kleist, S. and Hutchings, C. S. L. (1970) Display and food as reinforcers in the Siamese fighting fish (*Betta splendens*). *J. comp. physiol. Psychol. 70*, 351–7.

Hogan, J. A. and Roper, T. J. (1978) A comparison of the

properties of different reinforcers. *Adv. Study Behav. 8,* 155–254.

Holling, C. S. (1959) Some characteristics of simple types of predation. *Can. Ent. 91,* 385–98.

Holling, C. S. (1965) The functional response of predators to prey density and its role in mimicry and population regulation. *Mem. ent. Soc. Can. 45,* 1–60.

Holmes, R. T. (1970) Differences in population density, territoriality, and food supply of dunlin on arctic and subarctic tundra. In *Animal Populations in Relation to their Food Resources* (A. Watson, ed.). Oxford: Blackwell.

Holmes, W. (1940) The colour changes and colour patterns of *Sepia officinalis* L. *Proc. zool. Soc. Lond. 110,* 17–35.

Holst, E. von and Mittelstaedt, H. (1950) Das Reafferenzprinzip. *Naturwissenschaften 37,* 464–76.

Holst, E. von and St Paul, U. von (1960) Vom Wirkungsgefüge der Triebe. *Naturwissenschaften 47,* 409–22. Translated: 1963. On the functional organization of drives. *Anim. Behav. 11,* 1–20.

Honig, W. H. and Staddon, J. E. R. (1977) *Handbook of Operant Behaviour.* Englewood Cliffs, N.J.: Prentice Hall.

Hoogland, J. L. and Sherman, P. W. (1976) Advantages and disadvantages of bank swallow (*Riparia riparia*) coloniality. *Ecol. Monog. 46,* 33–58.

Horn, G. (1962) Some neural correlates of perception. In *Viewpoints in Biology,* 1 (J. D. Carthy, ed.). London: Butterworths.

Horn, G. (1963) The response of single units in the striate cortex of unrestrained cats to photic and somaesthetic stimuli. *J. Physiol., Lond. 165,* 80–1.

Horn, G. (1965) Physiological and psychological aspects of selective perception. In *Adv. Study Behav. 1,* 155–215.

Horn, G. (1967) Neuronal mechanisms of habituation. *Nature, Lond. 215,* 707–11.

Horn, H. S. (1968) The adaptive significance of colonial nesting in the Brewer's blackbird. (*Euphagus cynocephalus*). *Ecology 49,* 682–94.

Horridge, G. A. and Burrows, M. (1974) Synapses upon motoneurons of locusts during retrograde degeneration. *Phil. Trans. R. Soc.,* B, *269,* 95–108.

Horvath, T., Kirby, H. W. and Smith, A. A. (1971) Rat's heart-rate and grooming activity in the open field. *J. comp. physiol. Psychol. 76,* 449–53.

Hotta, Y. and Benzer, S. (1970) Genetic dissection of the *Drosophila* nervous system by means of mosaics. *Proc. natn. Acad. Sci. U.S.A. 67,* 1156–63.

Hotta, Y. and Benzer, S. (1972) Mapping of behaviour in *Drosophila* mosaics. *Nature, Lond. 240,* 527–35.

Houston, A. I., Halliday, T. R. and McFarland, D. J. (1977) Towards a model of the courtship of the smooth newt, *Triturus vulgaris,* with special emphasis on problems of observability in the simulation of behaviour. *Med. Biol. Eng. Comput. 15,* 49–61.

Houston, A. I. and McFarland, D. J. (1976) On the measurement of motivational variables. *Anim. Behav. 24,* 459–75.

Howard, H. E. (1920) *Territory in Bird Life.* London: John Murray.

Howell, T. R. and Bartholomew, G. A. (1961) Temperature regulation in Laysan and black-footed albatrosses. *Condor 63,* 185–97.

Hoyle, G. (1975) Identified neurons and the future of neuroethology. *J. exp. Zool. 194,* 51–74.

Hoyle, G. (1976) Approaches to understanding the neurophysiological bases of behavior. In *Simpler Networks and Behavior* (J. C. Fentress, ed.). Sunderland, Mass.: Sinauer.

Hoyle, G. and Willows, A. O. D. (1973) Neuronal basis of behavior in *Tritonia.* II. Relationship of muscular contraction to nerve impulse pattern. *J. Neurobiol. 4,* 239–54.

Hrdy, S. B. (1976) The care and exploitation of non-human primate infants by conspecifics other than the mother. *Adv. Study Behav. 6,* 101–158.

Hsü, F. (1938) Ètude cytologique et comparée sur les sensilla des insectes. *Cellule 47,* 1–60.

Hubel, D. H. (1957) Tungsten microelectrode for recording from single units. *Science, N.Y. 195,* 549–50.

Hubel, D. H. and Wiesel, T. N. (1959) Receptive fields of single neurons in the cats' striate cortex. *J. Physiol., Lond. 148,* 574–91.

Hubel, D. H. and Wiesel, T. N. (1961) Integrative action in the cats' lateral geniculate body. *J. Physiol., Lond. 155,* 385–93.

Hubel, D. H. and Wiesel, T. N. (1962) Receptive fields, binocular interaction and functional architecture in the cats' visual cortex. *J. Physiol., Lond. 160,* 106–54.

Hubel, D. H. and Wiesel, T. N. (1963*a*) Shape and arrangement of columns in the cats' striate cortex. *J. Physiol., Lond. 165,* 559–68.

Hubel, D. H. and Wiesel, T. N. (1963*b*) Receptive fields of cells in striate cortex of very young, visually inexperienced kittens. *J. Neurophysiol. 26,* 994–1002.

Hubel, D. H. and Wiesel, T. N. (1965) Receptive fields and functional architecture in two non-striate visual areas (18 and 19) of the cat. *J. Neurophysiol. 28, 229–89.*

Hubel, D. H. and Wiesel, T. N. (1968) Receptive fields and functional architecture of monkey striate cortex. *J. Physiol., 195,* 215–43.

Huber, F. (1960) Untersuchungen uber die Funktion des Zentralnervensystems und insbesondere des Gehirnes bei der Fortbewegung und der Lauterzeugung bei Grillen. *Z. vergl. Physiol. 44,* 60–132.

Huber, F. (1962) Central nervous control of sound produc-

tion in crickets and some speculations on its evolution. *Evolution 16*, 429–42.
Huber, F. (1974) Neuronal background of species – specific acoustical communication in orthopteran insects (Gryllidae). In *Biology of Brains* (W. B. Broughton, ed.). London: Institute of Biology.
Huber, F. (1978) The insect nervous system and insect behaviour. *Anim. Behav. 26*, 969–81.
Hudson, J. W. (1962) The role of water in the biology of the antelope ground squirrel *Citellus leucurus*. *Univ. Calif. Publ. Zool. 64*, 1–56.
Hudson, S. J. (1977) Multiple fostering of calves onto nurse cows at birth. *Appl. Anim. Ethol. 3*, 57–63.
Hudson, S. J. and Mullard, M. M. (1977) Investigations of maternal bonding in dairy cattle. *Appl. Anim. Ethol. 3*, 271–6.
Hughes, J., Smith, T. W., Kosterlitz, H. W., Fothergill, L. A., Morgan, B. A. and Morris, H. R. (1975) Identification of two related pentapeptides from the brain with potent opiate agonist activity. *Nature, Lond. 258*, 577–9.
Hughes, R. E., Milner, C. and Dale, J. (1964) Selectivity in grazing. In *Grazing in Terrestrial and Marine Environments* (D. J. Crisp, ed.). Oxford: Blackwell.
Hughes, R. N. (1966) Some observations of correcting behaviour in woodlice (*Porcellio scaber*). *Anim. Behav. 14*, 319.
Hughes, R. N. (1967) Turn alternation in woodlice (*Porcellio scaber*). *Anim. Behav. 15*, 282–6.
Hughes, R. N. (1970) Population dynamics of the bivalve *Scrobicularia plana* (da Costa) on an intertidal mudflat in North Wales. *J. Anim. Ecol. 39*, 333–81.
Hulet, E. V., Ercanbrack, S. K., Price, D. A. and Wilson, L. O. (1962) Mating behaviour of the ewe. *J. Anim. Sc. 21*, 870–4.
Hull, C. L. (1943) *Principles of Behavior*. New York: Appleton-Century-Crofts.
Hull, C. L. (1952) *A Behavior System: An Introduction to Behavior Theory Concerning the Individual Organism*. New Haven: Yale University Press.
Humphrey, N. K. (1974) Species and individuals in the perceptual world of monkeys. *Perception 3*, 105–14.
Humphrey, T. (1964) Some correlations between the appearance of human fetal reflexes and the development of the nervous system. In *Progress in Brain Research*, 4 (D. P. Purpura and J. P. Schade, eds.).
Hutchins, M. and Barash, D. (1976) Grooming in primates: implications for its utilitarian function. *Primates 17*, 145–50.
Hutchison, H. G., Woof, R., Mabon, R. M. Saleke, I. and Robb, J. M. (1962) A study of the habits of Zebu cattle in Tanganyika. *J. agric. Sci., Camb. 59*, 301–17.
Hutchison, J. B. (1967) Initiation of courtship by hypothalamic implants of testosterone propionate in castrated doves (*Streptopelia risoria*). *Nature, Lond. 216*, 591––2.
Hutchison, J. B. (1969) Changes in hypothalamic responsiveness to testosterone in male barbary doves (*Streptopelia risoria*). *Nature, Lond. 222*, 176–7.
Hutchison, J. B. (1970) Differential effects of testosterone and oestadiol on male courtship in barbary doves (*Streptopelia risoria*). *Anim. Behav. 18*, 41–51.
Hutchison, J. B. (1974) Post-castration decline in behavioural responsiveness to intra-hypothalamic androgen in doves. *Brain Res. 81*, 169–81.
Hutchison, J. B. (1976) Hypothalamic mechanisms of sexual behaviour with special reference to birds. *Adv. Study Behav. 6*, 159–200.
Hutchison, J. B. Ed., (1978) *Biological Determinants of Sexual Behaviour*. Chichester: Wiley.
Hutchison, J. B. and Poynton, J. C. (1963) A neurological study of the clasp reflex in *Xenopus laevis* (Daudin). *Anim. Behav. 22*, 41–63.
Hutt, C. and Hutt, S. J. (1965) Effects of environmental complexity on stereotyped behaviours of children. *Anim. Behav. 13*, 1–4.
Hutt, S. J., Hutt, C., Lenard, H. G., Bernuth, H. von and Muntjewerff, W. J. (1968) Auditory responsivity in the human neonate. *Nature, Lond. 218*, 888–90.
Iersel, J. J. A. van and Bol, A. C. A. (1958) Preening in two term species. A study on displacement activities. *Behaviour 13*, 1–88.
Ikeda, K., and Kaplan, W. D. (1970*a*) Patterned neural activity of a mutant (*Drosophila melanogaster*). *Proc. natn. Acad. Sci. U.S.A. 66*, 765–72.
Ikeda, K. and Kaplan, W. D. (1970*b*) Unilaterally patterned neural activity of a mutant gynandromorph of *Drosophila melanogaster*. *Am. Zool. 10*, 311.
Ikeda, K. and Wiersma, C. A. G. (1964) Autogenic rhythmicity in the abdominal ganglia of crayfish: the control of swimmeret movements. *Comp. Biochem. Physiol. 12*, 107–15.
Immelmann, K. (1965). Objektfixierung geschlechtlicher Triebhandlungen bei Prachtfinken. *Naturwissenschaften*, *52*, 169.
Immelmann, K. (1972). Sexual and other long-term aspects of imprinting in birds and other species. *Adv. Study Behav. 4*, 147–74.
Immelmann, K. (1977) *Einfuhrung in der Verhaltensforschung*. Berlin: Parey.
Immelmann, K., Barlow, G. W., Main, M. and Petrinovich, L. Eds. (1981) *Behavioural Development: the Bielefeld interdisciplinary project*. New York: Cambridge University Press.
Immelmann, K. and Suomi, S. J. (1981) Sensitive phases in development. In *Behavioral Development: the Bielefeld interdisciplinary project* (K. Immelmann, G. W. Barlow, M. Main and L. Petrinovich, eds.). New York: Cambridge University Press.
Impekoven, M. (1976) Pre-natal parent–young interactions

in birds and their long-term effects. *Adv. Study Behav. 7*, 201–53.

Ingle, D. Ed., (1968) *Central Nervous System and Fish Behavior:* Chicago, University of Chicago Press.

Inkster, I. J. (1957) The mating behaviour of sheep. In *Sheepfarming Annual.* Palmerston North, N. Z.: Massey Agricultural College.

Irving, S. N. and Wyatt, I. J. (1973) Effects of sublethal doses of pesticides on the oviposition behaviour of *Encarsia formosa. Ann. appl. Biol. 75*, 57–62.

Itani, J. (1958) On the acquisition and propagation of a new habit in the natural group of the Japanese monkey at Takasaki-Yama. *Primates 1*, 84–98. (In Japanese with English summary).

Itani, J. (1972) A preliminary essay on the relationship between social organization and incest avoidance in nonhuman primates. In *Primate Socialization* (F. E. Poirier, ed.). New York: Random House.

Ittner, N. R., Bond, T. E. and Kelly, C. F. (1954) Increasing summer gains of livestock with cool water, concentrated roughage, wire corrals and adequate shades. *J. Anim. Sci. 13*, 867–77.

Ivins, J. D. (1952) The relative palatability of herbage plants. *J. Br. Grassld Soc. 7*, 43–54.

James, H. (1959) Flicker: an unconditioned stimulus for imprinting. *Can. J. Psychol. 13*, 59–67.

James, J. W. (1967) The value of social status to cattle and pigs. *Proc. Ecol. Soc. Aust. 2*, 171–81.

Jamieson, W. S. (1975) Studies on the Herbage Intake and Grazing Behaviour of Cattle and Sheep. Ph.D. thesis, University of Reading.

Jarman, P. J. (1974) The social organisation of antelope in relation to their ecology. *Behaviour, 58*, 215–67.

Jay, P. (1965) The common langur in Northern India. In *Primate Behavior* (I. DeVore, ed.). New York: Holt, Rinehart and Winston.

Jay, P. Ed., (1968) *Primates: Studies in Adaptation and Variability.* New York: Holt, Rinehart and Winston.

Jenkins, D., Watson, A. and Miller, G. R. (1963) Population studies of red grouse in north-east Scotland. *J. Anim. Ecol. 32*, 317–76.

Jenkins, P. F. (1978) Cultural transmission of song patterns and dialect development in a free-living bird population. *Anim. Behav. 36*, 50–78.

Jenni, D. A. (1974) Evolution of polyandry in birds. *Am. Zool. 14*, 129–44.

Jennings, T. and Evans, S. M. (1980) Influence of position in the flock and flock size on vigilance in the starling, *Sturnus vulgaris. Anim. Behav. 28*, 634–5.

Jermy, T., Hanson, F. E. and Dethier, V. G. (1968) Induction of specific food preference in lepidopterous larvae. *Entomologia exp. appl. 11*, 211–30.

Jílek, L. and Trojan, S. (1968) *Ontogenesis of the Brain.* Prague: Charles University Press.

Joffe, J. M. (1969) *Prenatal Determinants of Behaviour.* Oxford: Pergamon Press.

Johansson, G. and Lundberg, U. (1978) Psychophysiological aspects of stress and adaptation in technological societies. In *Human Behaviour and Adaptation* (V. Reynolds and N. Blurton Jones, eds.), *Symp. Soc. Study. hum. Biol. 18.* London: Taylor and Francis.

Johnsgard, P. A. (1960) Pair formation mechanisms in *Anas* (Anatidae) and related genera. *Ibis 102*, 616–18.

Johnsgard, P. A. (1965) *Handbook of Waterfowl Behavior.* Ithaca, N.Y.: Cornell University Press.

Johnson, D. F. and Phoenix, C. H. (1976) Hormonal control of female sexual attractiveness, proceptivity and receptivity in rhesus monkeys. *J. comp. physiol. Psychol. 90*, 473–83.

Johnson, E. (1977) Seasonal changes in the skin of mammals. *Symp. zool. Soc. Lond. 39*, 373–404.

Johnson, R. P. (1973) Scent marking in mammals. *Anim. Behav. 21*, 521–35.

Jolly, A. (1966) *Lemur Behavior: A Madagascar Field Study.* Chicago: University of Chicago Press.

Jones, A. R. (1980) Chela injuries in the fiddler crab, *Uca burgersi* Holthuis. *Mar. Behav. Physiol. 7*, 47–56.

Jones, M. R. Ed., (1958) *Nebraska Symposium on Motivation.* Lincoln: University of Nebraska Press.

Jouvet, M. (1967) Mechanism of the states of sleep. A neuropharmacological approach. *Res. Publs. Ass. Res. nerv. ment. Dis. 45*, 86–126.

Joyce, J. P. and Blaxter, K. L. (1965) The effect of wind on heat losses of sheep. In *Energy Metabolism* (K. L. Blaxter, ed.). London: Academic Press.

Kaplan, H. and Hyland, S. (1972) Behavioural development in the Mongolian gerbil. *Anim. Behav. 20*, 147–54.

Katy, S. S. and Elkes, J. Eds., (1961) *Regional Neurochemistry.* Oxford: Pergamon Press.

Katz, B. (1950) Depolarization of sensory terminals and the initiation of impulses in the muscle spindle. *J. Physiol., Lond. 111*, 261–82.

Kaufman, J. H. (1974) The ecology and evolution of social organization in the kangaroo family (Macropodidae). *Am. Zool. 14*, 51–62.

Kavanau, J. L. (1963) Precise monitoring of drinking behavior in small mammals. *J. Mammal. 43*, 345–51.

Kawai, M. (1958) On the system of social ranks in a natural group of Japanese monkeys. *Primates 1*, 111–48.

Kawasaki, M., Ogura, J. H. and Takenouchi, S. (1964) Neurophysiologic observations of normal deglutitions: 1. its relationship to the respiratory cycle; II. its relationship to allied phenomena. *Laryngoscope 74*, 1747–83.

Kay, S. J. Collis, K. A., Anderson, J. C. and Grant, A. J. (1977) The effect of intergroup movement of dairy cows on bulk-milk somatic cell numbers. *J. Dairy Res. 44*, 589–93.

Kear, J. (1964) Colour preference in young Anatidae. *Ibis 106*, 361–9.

Kear, J. (1966) The pecking response of young coots *Fulica atra* and moorhens *Gallinula chloropus*. *Ibis 108*, 118–22.

Keating, M. J. and Gaze, R. M. (1970) The depth distribution of visual units in the contralateral optic tectum following regeneration of the optic nerve in the frog. *Brain Res. 21*, 197–206.

Keenleyside, M. (1955) Aspects of schooling behaviour in fish. *Behaviour 8*, 83–248.

Keiper, R. R. (1970) Studies of stereotypy function in the canary (*Serinus canarius*). *Anim. Behav. 18*, 353–7.

Kendeigh, S. C., Kontogiannis, J. E., Malzac, A. and Roth, R. R. (1969) Environmental regulation of food intake by birds. *Comp. Biochem. Physiol. 31*, 941–57.

Kennedy, D. (1976) Neural elements in relation to network function. In *Simpler Networks and Behavior* (J. C. Fentress, ed.). Sunderland, Mass.: Sinauer.

Kennedy, D., Evoy, W. H. and Hanawalt, J. T. (1966) Release of coordinated behavior in crayfish by single central neurons. *Science, N.Y. 154*, 917–19.

Kenward, R. E. (1978) Hawks and doves: attack success and selection in goshawk flights at woodpigeons. *J. Anim. Ecol. 47*, 449–60.

Keverne, E. B. (1976) Sexual receptivity and attractiveness in the female rhesus monkey. *Adv. Study Behav. 7*, 155–200.

Kiester, A. R. and Slatkin, M. (1974) A strategy of movement and resource utilization. *Theoret. Pop. Biol. 6*, 1–20.

Kiley-Worthington, M. (1977) *Behavioural Problems of Farm Animals*. Stocksfield: Oriel Press.

Kilgour, R. and Scott, T. H. (1959) Leadership in a herd of dairy cows. *Proc. N.Z. Soc. Anim. Prod. 33*, 125–38.

Kinder, E. F. (1927) A study of the nest-building activity of the albino rat. *J. exp. Zool. 47*, 117–61.

King, J. A. (1959) The social behavior of prairie dogs. *Scient. Am. 201*, 128–40.

King, M. G. (1965) Disruptions in the pecking order of cockerels concomitant with degrees of accessibility to feed. *Anim. Behav. 13*, 504–6.

Kitchen, D. W. (1974) Social behaviour and ecology of the pronghorn. *Wildl. Monogr. 38*, 1–96.

Kleiber, M. (1965) Metabolic body size. In *Energy Metabolism* (K. L. Blaxter, ed.). London: Academic Press.

Kleitman, N. (1939) *Sleep and Wakefulness*. Chicago: University of Chicago Press.

Kleitman, N. and Engelmann, T. G. (1953) Sleep characteristics of infants. *J. appl. Physiol. 6*, 269–82.

Klinghammer, E. (1967) Factors influencing choice of mate in birds. In *Early Behavior: Comparative and Developmental Approaches* (H. W. Stevenson, E. H. Hess and H. L. Rheingold). New York: Wiley.

Klopfer, P. H. (1969) *Habitats and Territories*. New York: Basic Books.

Knipling, E. F. (1955) Possibilities of insect control or eradication through the use of sexually sterile males. *J. econ. Ent. 48*, 459–62.

Koford, C. B. (1957) The vicuña and the puma. *Ecol. Monog. 27* 153–219.

Kornhuber, H. H. (1974) Cerebral cortex, cerebellum and basal ganglia: an introduction to their motor functions. In *The Neurosciences: Third Study Program* (F. O. Schmitt and F. G. Worden, eds.). Cambridge, Mass.: M.I.T. Press.

Komisaruk, B. R. (1967) Effects of local brain implants of progesterone on reproductive behavior in ring doves. *J. comp. physiol. Psychol. 64*, 219–24.

Komisaruk, B. R. (1971) Strategies in neuroendocrine neurophysiology. *Am. Zool. 11*, 741–54.

Komisaruk, B. R. Adler, N. T. and Hutchison, J. B., (1972) Genital sensory field: enlargement by estrogen treatment in female rats. *Science, N.Y. 178*, 1295–8.

Konishi, M. (1970) Evolution of design features in the coding of species specificity. *Am. Zool. 10*, 67–72.

Kop, P. P. A. M. and Heuts, B. A. (1973) An experiment on sibling imprinting in the jewel fish (*Hemichromis bimaculatus* (Gill 1862, Cichlidae). *Rev. Comp. Anim. 7*, 6376.

Kortlandt, A. (1940*a*) Eine Ubersicht der angeborenen Verhaltensweisen des Mitteleuropaischen Komorans (*Phalacrocrocorax carbo sinensis* Shaw & Nodd), ihre Funktion, ontogenetische Entwicklung, und phylogenetische Herkunft. *Arch. néerl. Zool. 4*, 401–42.

Kortlandt, A. (1940*b*) Wechselwirkung zwischen Instinkten. *Archs neérl. Zool. 4*, 442–520.

Kow, L-M. and Pfaff, D. W. (1973) Estrogen effect on pudendal nerve receptive field size in the female rat. *Anat. Rec. 175*, 362–3.

Kozak, W., Rodieck, R. W. and Bishop, P. O. (1965) Response of single units in lateral geniculate nucleus of cat to moving visual patterns. *J. Neurophysiol. 28*, 19–47.

Krebs, D. L. (1970) Altruism – an examination of the concept and a review of the literature. *Psychol. Bull. 73*, 258–302.

Krebs, J. R. (1971) Territory and breeding density in the great tit, *Parus major* L. *Ecology 52*, 2–22.

Krebs, J. R. (1973) Behavioral aspects of predation. In *Perspectives in Ethology*, Vol. 1 (P. P. G. Bateson and P. H. Klopfer, eds.). New York: Plenum Press.

Krebs, J. R. (1974) Colonial nesting and social feeding as strategies for exploiting food resources in the great blue heron (*Ardea herodias*). *Behaviour 51*, 99–134.

Krebs, J. R. (1978) Optimal foraging: decision rules for predators. In *Behavioural Ecology* (J. R. Krebs and N. B. Davies, eds.). Oxford: Blackwell.

Krebs, J. R., Ashcroft, R. and Webber, M. (1978) Song repertoires and territory defence in the great tit (*Parus major*). *Nature, Lond. 271*, 539–42.

Krebs, J. R. and Davies, N. B. Eds., (1978) *Behavioural Ecology: An Evolutionary Approach*. Oxford: Blackwell.

Krebs, J. R., Ericksen, J. T., Webber, M. I. and Charnov, E. L. (1977) Optimal prey selection in the great tit (*Parus major*). *Anim. Behav. 25*, 30–8.

Krebs, J. R., MacRoberts, M. H. and Cullen, J. M. (1972) Flocking and feeding in the great tit *Parus major* – an experimental study. *Ibis 114*, 507–30.

Krebs, J. R., Ryan, J. and Charnov, E. L. (1974) Hunting by expectation or optimal foraging? A study of patch use by chickadees. *Anim. Behav. 22*, 953–64.

Kreithen, M. L. and Keeton, W. T. (1973) Detection of polarised light by the homing pigeon, *Columba livia*. *J. comp. Physiol. 89*, 83–92.

Kruijt, J. P. (1964) Ontogeny of social behaviour in Burmese red jungle fowl (*Gallus gallus spadiceus Bonaterre*). *Behaviour Suppl. 12*, 1–201.

Kruijt, J. P. (1971) Early experience and the development of social behaviour in jungle fowl. *Psychiat. Neurol. Neurochir. 74*, 7–20.

Kruijt, J. P. and Hogan J. A. (1967) Social behaviour on the lek in the black grouse, *Lyrurus tetrix tetrix* (L.). *Ardea 55*, 203–40.

Kruuk, H. (1964) Predators and anti-predator behaviour of the black headed gull *Larus ridibundus*. *Behaviour Suppl. 11*, 1–129.

Kruuk, H. (1972) *The Spotted Hyena*. Chicago: University of Chicago Press.

Kuffler, S. W. (1953) Discharge patterns and functional organization of mammalian retina. *J. Neurophysiol. 16*, 37–68.

Kühme, W. (1965) Communal food distribution and division of labour in African hunting dogs. *Nature, Lond. 205*, 443–4.

Kullmann, E. J. (1972) Evolution of social behaviour in spiders (Araneae: Eresidae and Theridiidae). *Am. Zool. 12*, 419–26.

Kummer, H. (1957) Soziales Verhalten einer Mantelpavian-Gruppe. *Beih. schweiz. Z. Psychol. ihre Anwend. 33*, 1–91.

Kummer, H. (1968) *Social Organization of Hamadryas Baboons*. Chicago: University of Chicago Press.

Kuo, Z. Y. (1932) Ontogeny of embryonic behavior in Aves. V. The reflex concept in the light of embryonic behaviour in birds. *Psychol. Rev. 39*, 499–515.

Kupfermann, I. and Weiss, K. R. (1978) The command neuron concept. *Behav. Brain Sci. 1*, 3–39.

Kutsch, W. (1969) Neuromuskulare Aktivität bei verschiedenen Verhaltensweisen von drei Grillenarten. Z. *vergl. Physiol. 63*, 335–78.

Kydd, D. D. (1966) The effect of intensive sheep stocking over a five-year period on the development and production of the sward. I. Sward structure and botanical composition. *J. Br. Grassld Soc. 21*, 284–8.

Labov, J. B. (1979) Factors influencing infanticidal behavior in mated male house mice (*Mus musculus*). *Behav. Ecol. Sociobiol.* 6, 297–303.

Lacey, J. I. (1967) Somatic patterning and stress: some revisions of activation theory. In *Psychological Stress: Issues in Research* (M. H. Appley and R. Trumbull, eds.). New York: Appleton Century Crofts.

Lacey, J. I., Kagan, J., Lacey, B. C. and Moss, H. A. (1963) The visceral level: situational determinants and behavioural correlates of autonomic response patterns. In *Expression of the Emotions in Man* (P. H. Knapp, ed.). New York: International University Press.

Lack, D. (1954) *The Natural Regulation of Animal Numbers*. Oxford: Oxford University Press.

Lack, D. (1968) *Ecological Adaptions for Breeding in Birds*. London: Methuen.

Lannoy, J. de (1967) Zur Prägung von Instinkthandlungen (Untersuchungen an Stockenten *Anas platyrhynchos* L. und kolbenenten *Netta ruffina* Pallas). *Z. Tierpsychol. 24*, 162–200.

Larkin, S. and McFarland, D. (1978) The cost of changing from one activity to another. *Anim. Behav. 26*, 1237–46.

Larsen, H. J. (1963) The feeding habits of grazing and green feeding cows. *J. Anim. Sci. 22*, 1134–5.

Las, A. (1980) Male courtship persistence in the greenhouse whitefly, *Trialeurodes vaporariorum* Westwood (Homoptera: Aleyrodidae). *Behaviour 72*, 107–26.

Lashley, K. S. (1938) Experimental analysis of instinctive behaviour. *Psychol. Rev. 45*, 445–71.

Lasiewski, R. C. (1963) Oxygen consumption of torpid, resting, active and flying hummingbirds. *Physiol. Zool. 36*, 122–40.

Lazarus, J. (1972) Natural selection and the functions of flocking in birds: a reply to Murton. *Ibis 114*, 556–8.

Lazarus, J. (1978; Vigilance, flock size and domain of danger size in the white-fronted goose. *Wildfowl 29*, 135–45.

Lazarus, J. (1979) The early warning function of flocking in birds: an experimental study with captive *Quelea*. *Anim. Behav. 27*, 855–65.

Leaver, J. D. (1975) Utilization of grassland by dairy cows. *Proc. Easter Sch. agric. Sci. Univ. Nott. 23*, 307–27.

Le Boeuf, B. J. (1974) Male–male competition and reproductive success in elephant seals. *Am. Zool. 14*, 163–76.

Lees, A. D. (1948) The sensory physiology of the sheep-tick *Ixodes ricinus* L. *J. exp. Biol. 25*, 145–207.

Lees, A. D. and Milne, A. (1951) The seasonal and diurnal activities of individual sheep-ticks (*Ixodes ricinus* L.). *Parasitology 41*, 189–208.

Lehman, U. (1976*a*) Stochastic principles in the temporal

control of activity behaviour. *Int. J. Chronobiol. 4,* 223–66.

Lehmann, U. (1976*b*) Short-term and circadian rhythms in the behaviour of the vole, *Microtus agrestis* (L.). *Oecologia 23,* 185–99.

Lehrman, D. S. (1965) Interaction between internal and external environments in the regulation of the reproductive cycle of the ring dove. In *Sex and Behavior* (F. A. Beach, ed.). New York: Wiley.

Lehrman, D. S., Brody, P. and Wortis, R. P. (1961) The presence of the mate and of nesting material as stimuli for the development of incubation behavior and for gonadotropin secretion in the ring dove (*Streptopelia risoria*). *Endocrinology 68,* 507–16.

Leiman, A. L., Seil, F. J. and Kelly, J. M. (1975) Maturation in electrical activity of cerebral neocortex in tissue culture. *Expl Neurol. 48,* 275–91.

Le Magnen, J. (1971) Advances in studies on the physiological control and regulation of food intake. In *Progress in Physiological Psychology,* Vol. 4 (E. Stellar and J. M. Sprague, eds.). New York: Academic Press.

Le Magnen, J. and Devos, M. (1970) Metabolic correlates of the meal onset in the free food intake of rats. *Physiol. Behav. 5,* 805–14.

Le Magnen, J. and Tallon, S. (1966) La periodicité spontanée de la prise d'aliments ad libitum du rat blanc. *J. Physiol., Paris 58,* 323–49.

Lemon, R. E. and Chatfield, C. (1971) Organisation of song in cardinals. *Anim. Behav. 19,* 1–17.

Le Neindre, P. and Garel, J. P. (1976) Existence d'une période sensible pour l'établissement du comportement maternel de la vache après la mise-bas. *Biol. Behav. 1,* 217–21.

Lenhardt, M. L. (1977) Vocal contour clues in maternal recognition of goat kids. *Appl. Anim. Ethol. 3,* 211–19.

Lettvin, J. Y., Maturana, H. R., McCulloch, W. S. and Pitts, W. H. (1959) What the frog's eye tells the frog's brain. *Proc. Inst. Radio Engr. 47,* 1940–51.

Lettvin, J. Y., Maturana, H. R., Pitts, W. H. and McCulloch, W. S. (1961) Two remarks on the visual system of the frog. In *Sensory Communication* (W. A. Rosenblith, ed.). Cambridge: M.I.T. Press and New York: Wiley.

Leventhal, A. G. and Hirsch, H. V. B. (1975) Cortical effect of early selective exposure to diagonal lines. *Science, N.Y. 190,* 902–4.

Levi, L. Ed., (1965) *Emotions: Their Parameters and Measurements.* New York: Raven Press.

Levick, W. R. (1967) Receptive fields and trigger features of ganglion cells in the visual streak of the rabbits' retina. *J. Physiol., Lond. 188,* 285–307.

Levin, D. A. (1976) The chemical defenses of plants to pathogens and herbivores. *Ann. Rev. Ecol. Syst. 7,* 121–59.

Levine, S., Glick, D. and Nakane, P. K. (1967) Adrenal and plasma corticosterone and vitamin A in rat adrenal glands during postnatal development. *Endocrinology 80,* 910–14.

Levine, S., Jones, L. E. (1965) Adrenocorticotrophic hormone (ACTH) and passive avoidance learning. *J. comp. physiol. Psychol. 59,* 357–60.

Levins, R. and MacArthur, R. (1969) An hypothesis to explain the incidence of monophagy. *Ecology 50,* 910–11.

Levy, D. M. (1944) On the problem of movement restraint. *Am. J. Orthopsychiat. 14,* 644–71.

Lewis, D. B. and Gower, D. M. (1980) *Biology of Communication.* Glasgow: Blackie.

Lill, A. and Wood-Gush, D. G. M. (1975) Potential ethological isolating mechanisms and assortative mating in the domestic fowl. *Behaviour 25,* 16–44.

Lillywhite, P. G. (1977) Single photon signals and transduction in an insect eye. *J. comp. Physiol. 122,* 189–200.

Limbaugh, C. (1961) Cleaning symbioses. *Scient. Am. 205, 2,* 42–9.

Lincoln, G. A., Youngson, R. W. and Short, R. V. (1970) The social and sexual behaviour of the red deer. *J. Reprod. Fert., Suppl. 11,* 71–103.

Lindauer, M. (1954) Temperaturregulierung und Wasserhaushalt in Bienenstaat. *Z. vergl. Physiol. 36,* 391–432.

Lindauer, M. (1961) *Communication Among Social Bees.* Cambridge Mass.: Harvard University Press.

Lindsay, D. R. (1965) The importance of olfactory stimuli in the mating behaviour of the ram. *Anim. Behav. 13,* 75–8.

Lindsay, D. R., Dunsmore, D. G., Williams, J. D. and Syme, G. J. (1976) Audience effects on the mating behaviour of rams. *Anim. Behav. 24,* 818–21.

Lindsley, D. B. (1951) Emotion. In *Handbook of Experimental Psychology* (S. S. Stevens, ed.). New York: Wiley.

Lipsitt, L. P. (1967) Learning in the human infant. In *Early Behavior: Comparative and Developmental Approaches* (H. W. Stevenson, E. H. Hess and H. L. Rheingold, eds.). New York: Wiley.

Lisk, R. D. (1962) Diencephalic placement of estradiol and sexual receptivity in the female rat. *Am. J. Physiol. 203,* 493–6.

Lisk, R. D. (1967) Neural localization for androgen activation of copulatory behavior in the male rat. *Endocrinology 80,* 754–61.

Lisk, R. D., Russel, J. A., Kohler, S. G. and Hanks, J. B. (1973) Regulation of hormonally mediated maternal nest structure in the mouse (*Mus musculus*) as a function of neonatal hormone manipulation. *Anim. Behav. 21,* 296–301.

Lissmann, H. W. (1951) Continuous electrical signals from

the tail of a fish, *Gymnarchus niloticus*. *Nature, Lond.* *167*, 201–2.

Lissmann, H. W. (1964*a*) The neurological basis of the locomotory rhythm in the spinal dogfish (*Scyllium canicula, Acanthias vulgaris*): I. Reflex behaviour. *J. exp. Biol.* *23*, 143–61.

Lissmann, H. W. (1964*b*) The neurological basis of the locomotory rhythm in the spinal dogfish (*Scyllium canicula, Acanthias vulgaris*): II. The effect of deafferentation. *J. exp. Biol.* *23*, 162–76.

Lloyd, J. A. and Christian, J. J. (1967) Relationship of activity and aggression to density in two confined populations of house mice *Mus musculus*). J. Mammal. 48, 262–9.

Loewenstein, W. R., and Rathkamp, R. (1958) The sites for mechano-electric conversion in a Pacinian corpuscle. *J. gen. Physiol.* *41*, 1245–65.

Logan, F. A. (1965) Decision making by rats. *J. comp. physiol. Psychol.* *59*, 1–12.

Lorenz, K. (1935) Der Kumpan in der Umwelt des Vogels. *J. Orn., Lpz.* *83*, 137–213 and 289–394. (Translated (1937): The companion in the birds' world. *Auk* *54*, 245–73).

Lorenz, K. (1937) Uber die Bildung des Instinktbegriffes. *Naturwissenschaften* *25*, 289–300, 307–18, 324–31.

Lorenz, K. (1965) *Evolution and Modification of Behavior*. Chicago: University of Chicago Press.

Lorenz, K. Z. (1939) Vergleichende Verhaltensforschung. *Zool. Anz. Suppl.* *12*, 69–102.

Lorenz, K. Z. and Tinbergen, N. (1938) Taxis und Instinktbewegung in der Eirollbewegung der Graugans. *Z. Tierpsychol.* *2*, 1–29.

Lovari, S. and Hutchison, J. B. (1975) Behavioural transitions in the reproductive cycle of Barbary doves (*Streptopelia risoria* L.). *Behaviour* *53*, 126–50.

Loy, J. (1971) Estrus behavior of free-ranging rhesus monkeys. *Primates* *12*, 1–31.

Lubbock, J. (1882) *Ants, Bees, and Wasps*. London: Kegan Paul, Trench & Co.

Lüscher, M. (1961) Air-conditioned termite nests. *Scient. Am.* *205*, (1), 138–45.

MacArthur, R. H. and Pianka, E. R. (1966) On the optimal use of a patching environment. *Am. Nat.* *102*, 381–3.

MacArthur, R. H. and Wilson, E. O. (1967) *The Theory of Island Biogeography*. Princeton: Princeton University Press.

McBride, G. (1963) The test order and communication in young pigs. *Anim. Behav.* *11*, 53–6.

McBride, G., Arnold, G. W., Alexander, G. and Lynch, J. J. (1967) Ecological aspects of behaviour of domestic animals. *Proc. Ecol. Soc. Aust.* *2*, 133–65.

McBride, G., Foenander, F. and Slee, C. (1970) The development of social and sexual behaviour in the domestic fowl. *Rev. comport. anim.* *4*, 51–7.

McBride, G., Parer, I. P. and Foenander, F. (1969) The social organization and behaviour of the feral domestic fowl. *Anim. Behav. Monogr.* *2*, 125–81.

McCance, R. A. (1936) Experimental sodium chloride deficiency in man. *Proc. R. soc.*, B, *119*, 245–68.

McCleery, R. H. (1978) Optimal behaviour sequences and decision making. In *Behavioural Ecology an Evolutionary Approach* (J. R. Krebs and N. B. Davies, eds.). Oxford: Blackwell.

McCosker, J. E. (1977) Fright posture of the plesiopid fish *Calloplesiops altivelis:* an example of Batesian mimicry. *Science, N.Y.* *197*, 400–1.

MacDonald, D. W. (1977) The behavioural ecology of the red fox *Vulpes vulpes:* a study of social organisation and resource exploitation. D. Phil. thesis, University of Oxford.

MacDonald, D. W. (1980) Patterns of scent marking with urine and faeces amongst carnivore communities. *Symp. zool. Soc. Lond.* *45*, 107–39.

MacDonald, R. W. (1979) Some observations and field experiments on the urine marking behaviour of the red fox, *Vulpes vulpes* L. *Z. Tierpsychol.* *51*, 1–22.

McDougall, W. (1923) *An Outline of Psychology*. London: Methuen.

McFarland, D. J. (1965*a*) The effect of hunger on thirst motivated behaviour in the Barbary dove. *Anim. Behav.* *13*, 286–300.

McFarland, D. J. (1965*b*) Hunger, thirst and displacement pecking in the Barbary dove. *Anim. Behav.* *13*, 293–300.

McFarland, D. J. (1965*c*) Control theory applied to the control of drinking in the Barbary dove. *Anim. Behav.* *13*, 478–92.

McFarland, D. J. (1966) On the causal and functional significance of displacement activities. *Z. Tierpsychol*, *23*, 217–35.

McFarland, D. J. (1969) Mechanisms of behavioural disinhibition. *Anim. Behav.* *17*, 238–42.

McFarland, D. J. (1971) *Feedback Mechanisms in Animal Behaviour*. London: Academic Press.

McFarland, D. J. (1973) Stimulus relevance and homeostasis. In *Constraints on Learning* (R. A. Hinde and J. Stevenson-Hinde, eds.). London: Academic Press.

McFarland, D. J. Ed., (1974*a*) *Motivational Control Systems Analysis*. London: Academic Press.

McFarland, D. J. (1974*b*) Time-sharing as a behavioral phenomenon. *Adv. Study Behav.* *5*, 201–25.

McFarland, D. J. (1977) Decision making in animals. *Nature, Lond.* *269*, 15–21.

McFarland, D. J. and Budgell P. (1970) The thermoregulatory role of feather movements in the barbary dove. (*Streptopelia risoria*). *Physiol. Behav.* *5*, 763–71.

McFarland, D. J. and Lloyd, I. H. (1973) Time-shared feeding and drinking. *Q. Jl. exp. Psychol.* *25*, 48–61.

McFarland, D. J. and Sibly, R. M. (1972) 'Unitary drives' revisited. *Anim. Behav.* *20*, 548–63.

McFarland, D. J. and Sibly, R. M. (1975) The behavioural final common path. *Phil. Trans. R. Soc.*, B, *270*, 265–93.

McFarland, D. J. and Wright, P. (1969) Water conservation by inhibition of food intake. *Physiol. Behav. 4*, 95–9.

Machlis, L. (1977) An analysis of the temporal patterning of pecking in chicks. *Behaviour 63*, 1–70.

McKinney, F. (1965) The comfort movements of Anatidae. *Anim. Behav.* 25, 121–217.

McKinney, F. (1975) The evolution of duck displays. In *Function and Evolution in Behaviour* (G. P. Baerends, C. G. Beer and A. Manning, eds.). Oxford: Oxford University Press.

Mackintosh, N. J. (1973) Stimulus selection: learning to ignore stimuli that predict no change in reinforcement. In *Constraints on Learning* (R. A. Hinde and J. Stevenson-Hinde, eds.). London: Academic Press.

McLoughlin, V. L. (1980) The effects of the social and physical environment on the behaviour of gerbil families. Ph.D. thesis, University of Reading.

McManus, J. (1971) Early postnatal growth and the development of temperature regulation in the Mongolian gerbil (*Meriones unguiculatus*). *J. Mammal. 52*, 782–92.

Macnair, M. R. and Parker, G. A. (1978) Models of parent–offspring conflict. II. Promiscuity. *Anim. Behav. 26*, 111–22.

Macnair, M. R. and Parker, G. A. (1979) Models of parent–offspring conflict. III. Intra-brood conflict. *Anim. Behav.* 27, 1202–9.

Magoun, H. W., Harrison, F., Brobeck, J. R. and Ranson, S. W. (1938) Activation of heat loss mechanisms by local heating of the brain. *J. Neurophysiol. 1*, 101–14.

Mainardi, D., Marsan, M. and Pasquali, A. (1965) Causation of sexual preferences of the house mice. The behaviour of mice reared by parents whose odour was artificially altered. *Atti. Soc. ital. Sci. nat. 104*, 325–38.

Maldonado, H. (1970) The deimatic reaction in the praying mantis *Stagmatoptera biocellata*. *Z. vergl. Physiol.* 68, 60–71.

Malmo, R. B. (1959) Activation: a neurophysiological dimension. *Psychol. Rev. 66*, 367–86.

Manning, A. (1976) The place of genetics in the analysis of behaviour. In *Growing Points in Ethology* (P. P. G. Bateson and R. A. Hinde, eds.). Cambridge: Cambridge University Press.

Manning, A. (1979) *An Introduction to Animal Behaviour*, 3rd edn. London: Arnold.

Marler, P. R. (1955) Characteristics of some animal calls. *Nature, Lond. 176*, 6–8.

Marler, P. R. (1957) Specific distinctiveness in the communication signals of birds. *Behaviour 11*, 13–39.

Marler, P. R. (1976*a*) Social organization, communication and graded signals: the chimpanzee and the gorilla. In *Growing Points in Ethology* (P. P. G. Bateson and R. A. Hinde, eds.). Cambridge: Cambridge University Press.

Marler, P. R. (1976*b*) Sensory templates in species-specific behaviour. In *Simpler Networks and Behaviour* (J. C. Fentress, ed.). Sunderland, Mass.: Sinauer.

Marler, P. R. (1978) The vocal ethology of primates: implications for psychophysics and psychophysiology. *Rec. Adv. Primatol. 1*, 795–801.

Marler, P. R. and Hamilton, W. J. (1966) *Mechanisms of Animal Behavior*. New York: Wiley.

Marler, P. R. and Tamura, M. (1964) Culturally transmitted patterns of vocal behaviour in sparrows. *Science, N.Y. 146*, 1483–6.

Marsh, R. and Campling, R. C. (1970) Fouling of pastures by dung. *Herb. Abstr. 40*, 123–40.

Marten, G. C. (1969) *Measurement and Significance of Forage Palatability. Proc. Nat. Conf. Forage Quality Evaluation and Utilisation.* Lincoln, Neb.: Nebraska Center for Continuing Education.

Martini, L. and Ganong W. F. Eds., (1967) *Neuroendocrinology*, Vol. 2. New York: Academic Press.

Mason, J. W. (1968) A review of psychoendocrine research on the pituitary adrenal cortical system. *Psychosom. Med. 30*, 576–607.

Mason, J. W. (1975) Emotion as reflected in patterns of endocrine integration. In *Emotions: Their Parameters and Measurement* (L. Levi, ed.). New York: Raven Press.

Mason, J. W., Tolson, W. W., Brady, J. V., Tolliver, G. A. and Gilmore, G. I. (1968). *Psychosom. Med., 30,* 775.

Mason, W. A. (1960) The effects of social restriction on the behavior of rhesus monkeys: I. Free social behavior. *J. comp. physiol. Psychol. 53*, 582–9.

Mason, W. A. (1961) The effects of social restriction on the behavior of rhesus monkeys: III. Dominance tests. *J. comp. physiol. Psychol. 54*, 694–9.

Mason, W. A. and Berkson, G. (1975) Effects of maternal mobility on the development of rocking and other behaviors in rhesus monkeys: a study with artificial mothers. *Devl. Psychobiol. 8*, 197–211.

Matthews, G. V. T. (1968) *Bird Navigation*, 2nd edn. Cambridge: Cambridge University Press.

Maturana, H. R. (1964) Functional organization of the pigeon retina. In *Information Processing in the Nervous System* (R. W. Gerard, ed.). Amsterdam: Excerpta Medica Foundation.

Maturana, H. R. and Frenk, S. (1963) Directional movement and horizontal edge detectors in the pigeon retina. *Science, N.Y. 142*, 977–9.

Maturana, H. R., Lettvin, J. Y., McCulloch, W. S. and Pitts, W. H. (1960) Anatomy and physiology of vision in the frog (*Rana pipiens*) *J. gen. Physiol. 43*, 129–75.

Mayer, J. (1955) The regulation of energy intake and the bodyweight: the glucostatic theory and the lipostatic hypothesis. *Ann. N.Y. Acad. Sci. 63*, 15–43.

Mayer, J. and Marshall, N. B. (1956). Specificity of gold thioglucose for ventromedial hypothalamic lesions and obesity. *Nature, Lond. 178*, 1399–1400.

Maynard Smith, J. (1965) The evolution of alarm cells. *Am. Nat. 99*, 59–63.

Maynard Smith, J. (1971) What use is sex? *J. theoret. Biol. 30*, 319–35.

Maynard Smith, J. (1976*a*) Group selection. *Q. Rev. Biol. 51*, 277–83.

Maynard Smith, J. (1976*b*) Evolution and the theory of games. *Am. Scient. 64*, 41–5.

Maynard Smith, J. (1977) Parental investment: a prospective analysis. *Anim. Behav. 25*, 1–9.

Maynard Smith, J. (1978*a*) The ecology of sex. In *Behavioural Ecology* (J. R. Krebs and N. B. Davies, eds.). Oxford: Blackwell.

Maynard Smith, J. (1978*b*) *The Evolution of Sex*. Cambridge: Cambridge University Press.

Maynard Smith, J. and Price, G. R. (1973) The logic of animal conflict. *Nature, Lond. 246*, 15–18.

Maynard Smith, J. and Ridpath, M. G. (1972) Wife sharing in the Tasmanian native hen, *Tribonyx mortierii*: a case of kin selection? *Am. Nat. 106*, 447–52.

Mayr, E. (1963) *Animal Species and Evolution*. Cambridge Mass.: Harvard University Press.

Mech, L. D. (1970) *The Wolf: the Ecology and Behavior of an Endangered Species*. Garden City, N.Y.: Doubleday.

Meddis, R. (1975) On the function of sleep. *Anim. Behav. 23*, 676–91.

Meese, G. B. and Baldwin, B. A. (1975) Effects of olfactory bulb on maternal behaviour in sows. *Appl. Anim. Ethol. 1*, 379–86.

Meese, G. B. and Ewbank, R. (1973*a*) Exploratory behaviour and leadership in the domestic pig. *Br. vet. J. 129*, 251–9.

Meese, G. B. and Ewbank, R. (1973*b*) The establishment and nature of the dominance hierarchy in the domesticated pig. *Anim. Behav. 21*, 326–34.

Meifert, D. W., LaBreque, G. C., Smith, C. N. and Morgan, P. B. (1967) Control of house flies on some West Indian Islands with metepa, apholate and trichlorphon baits. *J. econ. Ent. 60*, 480–5.

Menzel, E. W. (1974) A group of young chimpanzees in a one-acre field. In *Behaviour of Nonhuman Primates*, Vol. 5 (A. M. Schrier and F. Stollnitz, eds.).

Menzel, E. W., Davenport, R. K. and Rogers, C. M. (1963) The effects of environmental restriction upon the chimpanzee's responsiveness to objects. *J. comp. physiol. Psychol. 56*, 78–85.

Menzel, E. W. and Halperin, S. (1975) Purposive behaviour as a basis for objective communication between chimpanzees. *Science, N.Y. 189, 652–4*.

Metz, H. (1974) Stochastic models for the temporal fine structure of behaviour sequences. In *Motivational Control Systems Analysis* (D. J. McFarland, ed.). London: Academic Press.

Metz, J. H. M. (1975) Time patterns of feeding and rumination in domestic cattle. *Meded. LandbHoogesch. Wageningen* 75–12, 1–66.

Meyer, J. S., Novak, M. A., Bowman, R. E. and Harlow, H. F. (1975) Behavioral and hormonal effects of attachment objects in surrogate-peer-reared and mother-reared infant rhesus monkeys. *Devl. Psychobiol. 8*, 425–35.

Meyer-Holzapfel, M. (1968) Abnormal behavior of zoo animals. In *Abnormal Behavior of Animals* (M. W. Fox, ed.). New York: W. B. Saunders.

Michael, C. R. (1966*a*) Receptive fields of directionally selective units in the optic nerve of the ground squirrel. *Science, N.Y. 152*, 1092–5.

Michael, C. R. (1966*b*) Receptive fields of opponent color units in the optic nerve of the ground squirrel. *Science, N.Y. 152*, 1095–7.

Michael, R. P. (1961) An investigation of the sensitivity of circumscribed neuronal areas to hormonal stimulation. In *Regional Neurochemistry* (S. S. Katy and J. Elkes, eds.). Oxford: Pergamon Press.

Michael, R. P. and Crook, J. H. Eds., (1973) *Comparative Ecology and Behaviour of Primates*. London: Academic Press.

Michael, R. P. and Keverne, E. B. (1968) Pheromones: their role in the behaviour of sexual states in primates. *Nature, Lond. 218*, 746–9.

Miles, F. A. (1970) Centrifugal effects in the avian retina. *Science, N.Y. 170*, 992–5.

Miles, F. A. (1972) Centrifugal control of the avian retina. III. Effects of electrical stimulation of the isthmo optic tract on the receptive field properties of retinal ganglian cells. *Brain Res. 48*, 115–29.

Milinski, M. (1977) Do all members of a swarm suffer the same predation? *Z. Tierpsychol. 45*, 373–88.

Miller, L. A. and Olesen, J. (1979) Avoidance behaviour in green lacewings. 1. Behaviour of free flying green lacewings to hunting bats and ultrasound. *J. comp. Physiol. 131*, 113–20.

Miller, N. E. (1959) Liberalization of basic S-R concepts: extensions to conflict behavior, motivation and social learning. In *Psychology: a Study of a Science*, Vol. II (S. Koch, ed.). New York: McGraw Hill.

Miller, R. E. and Ogawa, N. (1962) The affect of adrenocorticotrophic hormone (ACTH) on avoidance conditioning in adrenalectomized rats. *J. comp. physiol. Psychol. 55*, 211–13.

Milner, P. M. (1970) *Physiological Psychology*. New York: Holt, Rinehart and Winston.

Mitchell, D. E., Giffin, F., Muir, D., Blakemore, C. and Sluyters, R. C. van (1976) Behavioural compensation of cats after early rotation of one eye. *Expl Brain Res. 25*, 109–13.

Mittelstaedt, H. (1957) Prey capture in mantids. In *Recent Advances in Invertebrate Physiology* (B. T. Scheer, ed.). Eugene, Oregon: University of Oregon Publications.

Mittelstaedt, H. (1962) Control systems of orientation in insects. *Ann. Rev. Ent. 7*, 177–98.

Moffat, C. B. (1931) A pied wagtail roost in Dublin. *Br. Birds 24*, 364–6.

Mogenson, G. J. and Calaresu, F. R. (1978) In *Hunger Models: Computable Theory of Feeding Control* (D. A. Booth, ed.). London: Academic Press.

Moltz, H. (1975) Maternal behaviour: some hormonal, neural and chemical determinants. In *The Behaviour of Domestic Animals*, 3rd edn (E. S. E. Hafez, ed.). London: Baillière Tindall.

Moltz, H., Levin, R. and Leon, M. (1969) Differential effects of progesterone on the maternal behaviour primiparous and multiparous rats. *J. comp. physiol. Psychol. 67*, 36–40.

Moltz, H., Lubin, M., Leon, M. and Numan, M. (1970) Hormonal induction of maternal behaviour in the ovariectomised nulliparous rat. *Physiol. Behav. 5*, 1373–7.

Moltz, H. and Wiener, E. (1966) Effects of ovariectomy on maternal behavior of primiparous rats. *J. comp. physiol. Psychol. 62*, 382–7.

Moody, M. F. and Parriss, J. R. (1961) The discrimination of polarised light by *Octopus:* a behavioural and morphological study. *Z. vergl. Physiol. 44*, 268–91.

Moore, F. R. (1977) Flocking behaviour and territorial competitors. *Anim. Behav. 25*, 1063–5.

Moreng, R. E. and Shaffner, C. S. (1951) Lethal internal temperatures for the chicken, from fertile egg to mature bird. *Poult. Sci. 30*, 255–66.

Morgan, M. J., Fitch, M. D., Holman, J. G. and Lea, S. E. G. (1976) Pigeons learn the concept of an 'A'. *Perception 5*, 57–66.

Morgan, P. D., Boundy, C. A. P., Arnold, G. W. and Lindsay, D. R. (1975) The roles played by the senses of the ewe in the location and recognition of lambs. *Appl. Anim. Ethol. 1*, 139–50.

Moruzzi, G. and Magoun, H. W. (1949) Brain stem reticular formation and activation of the E.E.G. *Electroenceph. clin. Neurophysiol. 1*, 445–73.

Mountcastle, V. B. (1957) Modality and topographic properties of single neurons of cats' somatic sensory cortex. *J. Neurophysiol. 20*, 508–34.

Moyer, K. E. (1958) The effect of adrenalectomy on anxiety motivated behaviour. *J. genet. Psychol. 92*, 11–16.

Mueller, H. C. and Parker, P. G. (1980) Naive ducklings show different cardiac responses to hawk than to goose models. *Behaviour*

Muller, H. J. (1932) Some genetic aspects of sex. *Am. Nat. 66*, 118–38.

Munn, N. L. (1950) *Handbook of Physiological Research on the Rat*. Boston: Houghton Mifflin Co.

Murphy, J. V. and Miller, R. E. (1955) The affect of adrenocorticotrophic hormone (ACTH) on avoidance conditioning in the rat. *J. comp. physiol. Psychol. 48*, 47–9.

Murray, R. W. (1962) The response of the ampullae of Lorenzini of elasmobranchs to electrical stimulation. *J. exp. Biol. 39*, 119–28.

Murton, R. K. (1971*a*) The significance of a specific search image in the feeding behaviour of the wood-pigeon. *Behaviour 40*, 10–42.

Murton, R. K. (1971*b*) Why do some bird species feed in flocks? *Ibis 113*, 534–6.

Murton, R. K., Isaacson, A. J. and Westwood, N. J. (1963) The feeding ecology of the woodpigeon. *Br. Birds*, *56*, 345–75.

Murton, R. K., Isaacson, A. J. and Westwood, N. J. (1966) The relationships between woodpigeons and their clover food supply and the mechanism of population control. *J. appl. Ecol. 3*, 55–96.

Myers, R. D. and Sharpe, L. G. (1968) Temperature in the monkey: transmitter factors released from the brain during thermoregulation. *Science, N.Y. 161*, 572–3.

Mylrea, P. J. and Beilharz, R. G. (1964) The manifestation and detection of oestrus in heifers. *Anim. Behav. 12*, 25–30.

Nelson, J. B. (1965) The behaviour of the gannet. *Brit. Birds 58*, 233–88.

Nelson, J. B. (1978) *The Sulidae: Gannet and Boobies*. Oxford: Oxford University Press.

Nelson, K. (1964) The temporal patterning of courtship behaviour in the glandulocaudine fishes. *Behaviour 24*, 75–113.

Nelson, K. (1965) After effects of courtship in the male three-spined stickleback. *Z. vergl. Physiol. 50*, 569–97.

Nice, M. M. (1941) The role of territory in bird life. *Am. Midl. Nat. 26*, 441–87.

Nisbet, I. C. T. (1973) Courtship-feeding, egg-size and breeding success in common terns. *Nature, Lond. 241*, 141–2.

Nishida, T. and Kawanaka, K. (1972) Inter unit-group relationships among wild chimpanzees in the Mahali Mountains. *Kyoto Univ. Afr. Stud. 7*, 131–69.

Noirot, E. (1972) Ultrasounds and maternal behaviour in small rodents. *Devl. Psychobiol. 5*, 371–87.

Noirot, E. (1974) Nest building by the virgin female mouse

exposed to ultrasound from inaccessible pups. *Anim. Behav. 22*, 410–20.

Noirot, E. and Pye, J. D. (1969) Sound analysis of ultrasonic distress calls of mouse pups as a function of their age. *Anim. Behav. 17*, 340–9.

Norris, K. S. Ed., (1966) *Whales, Dolphins and Porpoises*. Berkeley, Ca.: University of California Press.

Norris, M. L. and Adams, C. E. (1972) The growth of the Mongolian gerbil from birth to maturity. *J. Zool., Lond. 166*, 277–82.

Nottebohm, F. (1967) The role of sensory feedback in the development of avian vocalizations. *Proc. 14th Ornith. Cong., Oxford*. Oxford: Blackwell.

Nottebohm, F. (1968) Auditory experience and song development in the chaffinch (*Fringilla coelebs*): ontogeny of a complex motor pattern. *Ibis 110*, 549–68.

Nottebohm, F., Stokes, T. M. and Leonard, C. M. (1976) Central control of song in the canary. *J. comp. Neurol. 165*, 457–86.

Novin, D. (1962) The relation between electrical conductivity of brain tissue and thirst in the rat. *J. comp. physiol. Psychol. 55*, 145–54.

Noyes, J. S. (1974) The biology of the leek moth. *Acrolepia assectella* (Zeller) Ph.D. thesis, University of London.

Numan, M. (1974) Medial preoptic area and maternal behaviour in the female rat. *J. comp. physiol. Psychol. 87*, 746–59

Oates, J. F. (1977) The guereza and its food. In *Primate Ecology* (T. H. Clutton-Brock, ed.). London: Academic Press.

Obara, Y. (1970) Studies on the mating behaviour of the white cabbage butterfly, *Pieris rapae crucivora* Boisduval. III. Near-ultra-violet reflection as the signal of intra specific communication. *Z. vergl. Physiol. 69*, 99–116.

Obara, Y. and Hidaka, T. (1968) Recognition of the female by the male on the basis of ultra-violet reflection, in the white cabbage butterfly, *Pieris rapae crucivora* Boisduval. *Proc. Japan Acad. 44*, 829–32.

O'Connor, R. J. (1978) Brood reduction in birds: selection for infanticide and suicide? *Anim. Behav. 26*, 79–96.

Oldroyd, H. (1964) *The Natural History of Flies*. London: Weidenfeld and Nicholson.

Olds, J. S. and Milner, P. (1954) Positive reinforcement produced by electrical stimulation of septal area and other regions of rat brain. *J. comp. physiol. Psychol. 47*, 419–27.

Olesen, J. and Miller, L. A. (1979) Avoidance behaviour in green lacewings. II. Flight muscle activity. *J. comp. Physiol. 131*, 121–8.

Oomura, Y., Kimura, K., Ooyama, H., Maeno, T., Iki, M. and Kumioshi, M. (1964) Reciprocal activities of the ventromedial and lateral hypothalamic areas in cats. *Science, N.Y. 143*, 484–5.

Orians, G. H. (1969) On the evolution of mating systems in birds and mammals. *Am. Nat. 103*, 589–603.

Orians, G. H. and Willson, M. F. (1964) Interspecific territories of birds. *Ecology 45*, 736–45.

Otto, D. (1971) Untersuchungen zur zentralnervosen Kontrolle der Lauterzeugung von Grillen. *Z. vergl. Physiol. 74*, 227–71.

Packard, A. and Sanders, G. D. (1971) Body patterns of *Octopus vulgaris* and maturation of the response to disturbance. *Anim. Behav. 19*, 780–90.

Packer, C. (1975) Male transfer in olive baboons. *Nature, Lond. 255*, 219–20.

Packer, C. (1977) Reciprocal altruism in *Papio anubis*. *Nature, Lond. 265*, 441–3.

Padilla, S. C. (1935) Further studies on the delayed pecking of chicks. *J. comp. Psychol. 20*, 413–43.

Pain, B. F. and Broom, D. M. (1978) The effects of injected and surface-spread slurry on the intake and behaviour of dairy cows. *Anim. Prod. 26*, 75–83.

Pain, B. F., Leaver, J. D. and Broom, D. M. (1974) Effects of cow slurry on herbage production, intake by cattle and grazing behaviour. *J. Br. Grassld Soc. 29*, 85–91.

Palka, Y. S. and Sawyer, C. H. (1966) The effects of hypothalamic implants of ovarian steroids on oestrus behaviour in rabbits. *J. Physiol., Lond. 185*, 251–69.

Panksepp, J. (1978) Analysis of feeding patterns: data reduction and theoretical implications. In *Hunger Models: Computable Theory of Feeding Control* (D. A. Booth, ed.). London: Academic Press.

Parker, G. A. (1970) The reproductive behaviour and the nature of sexual selection in *Scatophaga stercoraria* L. II. The fertilization rate and the spatial and temporal relationship of each sex around the site of mating and oviposition. *J. Anim. Ecol. 39*, 205–28.

Parker, G. A. (1974) The reproductive behaviour and the nature of sexual selection in *Scatophaga stercoraria* L. IX. Spatial distribution of fertilization rates and evolution of male search strategy within the reproductive area. *Evolution 28*, 93–108.

Parker, G. A. (1978) Searching for mates. In *Behavioural Ecology: an Evolutionary Approach* (J. R. Krebs and N. B. Davies, eds.). Oxford: Blackwell.

Parker, G. A. and Macnair, M. R. (1978) Models of parent-offspring conflict. I. Monogamy. *Anim. Behav. 26*, 97–110.

Parker, G. A. and Macnair, M. R. (1979) Models of parent–offspring conflict. IV. Suppression: evolutionary retaliation by the parent. *Anim. Behav. 27*, 1210–35.

Partridge, B. L. (1980) The effect of school size on the structure and dynamics of minnow schools. *Anim. Behav. 28*, 68–77.

Partridge, B. L. and Pitcher, T. J. (1979) Evidence against

a hydrodynamic function for fish schools. *Nature, Lond.* *279*, 418–19.
Partridge, B. L. and Pitcher, T. J. (1980) The sensory basis of fish schools: relative roles of lateral line and vision. *J. comp. Physiol.* *135*, 315–25.
Partridge, L. (1974) Habitat selection in titmice. *Nature, Lond.* *247*, 573–4.
Partridge, L. (1978) Habitat selection. In *Behavioural Ecology: an Evolutionary Approach* (K. R. Krebs and N. B. Davies, eds.). Oxford: Blackwell.
Patterson, I. J. (1965) Timing and spacing of broods in the black-headed gull *Larus ridibundus*. *Ibis* *107*, 433–59.
Patterson, I. J., Dunnet, G. M. and Fordham, R. A. (1971) Ecological studies of the rook (*Corvus frugilegus* L.) in North-east Scotland: dispersion. *J. appl. Ecol.* *8*, 815–33.
Patterson, R. L. S. (1968) *Investigations on Boar Taint*. Dublin: European Association for Animal Production, Commission on Pig Production.
Pattie, F. A. (1936) The gregarious behaviour of normal chicks and chicks hatched in isolation. *J. comp. Psychol.* *21*, 161–78.
Pavlov, I. P. (1924) Lectures on the work of the cerebral hemispheres, No. 2. In *I. P. Pavlov, Selected Works* (Kh. S. Koshtoyants, ed.). Moscow: Foreign Languages Publishing House.
Payne, W. J. A., Laing, W. I. and Raivoka, E. N. (1951) Grazing behaviour of dairy cattle in the tropics. *Nature, Lond.* *167*, 610–11.
Pearson, K. G. and Iles, J. F. (1970) Discharge patterns of coxal levator and depressor motoneurones of the cockroach, *Periplaneta americana*. *J. exp. Biol.* *52*, 139–65.
Peeke, H. V. S. and Herz, M. J. (1973) *Habituation* (2 Vols.). New York: Academic Press.
Pelwijk, J. J. ter and Tinbergen, N. (1937) Eine reizbiologische Analyse einiger Verhaltensweisen von *Gasterosteus aculeatus* L. *Z. Tierpsychol.* *1*, 193–204.
Penfield, W. and Jasper, H. H. (1954) *Epilepsy and the functional anatomy of the human brain*. Boston: Little Brown.
Pettersson, M. (1956) Diffusion of a new habit among greenfinches. *Nature, Lond.* *177*, 709–10.
Pfaff, D., Lewis, C., Diakow, C. and Keiner, M. (1973) Neurophysiological analysis of mating behavior responses as hormone-sensitive reflexes. *Prog. physiol. Psychol.* *5*, 253&97.
Phillips, R. E. and Peck, F. W. (1975) Brain organisation and neuro-muscular control of vocalisation in birds. In *Neural and Endocrine Aspects of Behaviour in Birds* (P. Wright, P. G. Caryl and D. M. Vowles, eds.). New York: Elsevier.
Phillips, R. E. and Youngren, O. M. (1971) Brain stimulation and species-typical behaviour: activities evoked by electrical stimulation of the brains of chickens (*Gallus gallus*). *Anim. Behav.* *19*, 757–79.
Phillips, R. E. and Youngren, O. M. (1973) Electrical stimulation of the brain as a tool for study of animal communication. Behavior evoked in mallard ducks (*Anas platyrhynchos*). *Brain. Behav. Evol.* *8*, 253–86.
Phillips, R. E. and Youngren, O. M. (1976) Pattern generator for repetitive avian vocalization: preliminary localization and functional characterization. *Brain. Behav. Evol.* *13*, 165–78.
Phoenix, C. H., Copenhaver, K. H. and Brenner, R. M. (1976) Scanning electron microscopy of penile papillae in intact and castrated male rats. *Horm. Behav.* *7*, 217–22.
Pianka, E. R. (1970) On *r*- and K-selection. *Am. Nat.* *104*, 592–7.
Pinel, J. P. J. and Treit, D. (1978) Burying as a defensive response in rats. *J. comp. physiol. Psychol.* *92*, 708–12.
Pinniger, D. B. (1974) A laboratory simulation of residual populations of stored product pests and an assessment of their susceptibility to a contact insecticide. *J. Stored Prod. Res.* *10*, 217–23.
Pinniger, D. B. (1975) The behaviour of insects in the presence of insecticides: the effect of fenitrothion and malathion on resistant and susceptible strains of *Tribolium castaneum* Herbot. *Proc. 1st Int. Wkg. Conf. Stored-Prod. Ent., Savannah Ga. 1974*, 301–8.
Pinsker, H. Ed., (1980) *Information Processing in the Nervous System*. New York: Raven Press.
Pirenne, M. H. (1956) Physiological mechanisms of vision and the quantum nature of light. *Biol. Rev.* *31*, 194–241.
Pitcher, T. J. (1973) The 3-dimensional structure of schools in the minnow (*Phoxinus phoxinus* L). *Anim. Behav.* *21*, 673–86.
Pitcher, T. J. (1979) Sensory information and the organization of behaviour in a shoaling cyprinid fish. *Anim. Behav.* *27*, 126–49.
Pitcher, T. J., Partridge, B. L. and Wardle, C. S. (1976) A blind fish can school. *Science, N.Y.* *194*, 963–5.
Pittendrigh, C. S., Caldarola, P. C. and Cosbey, E. S. (1973) A differential effect of heavy water on temperature-dependent and temperature-independent aspects of circadian system of *Drosophila pseudoobscura*. *Proc. natn. Acad. Sci. U.S.A.* *70*, 2037–41.
Poindron, P. and Carrick, M. J. (1976) Hearing recognition of the lamb by its mother. *Anim. Behav.* *24*, 600–2.
Poirier, F. E., Ed., (1972) *Primate Socialization*. New York: Random House.
Pomeranz, B. and Chung, S. H. (1970) Dendritic-tree anatomy codes form-vision physiology in tadpole retina. *Science, N.Y.* *170*, 983–4.

Porter, R. W., Cavanaugh, E. B., Critchlow, B. V. and Sawyer, C. H. (1957) Localised changes in electrical activity of the hypothalamus in estrus cats following vaginal stimulation. *Am. J. Physiol. 189*, 145–51.

Portmann, A. (1953) *Das Tier Als Sociales Wesen*. Zurich: Rhein Verlag.

Posner, M. I. (1975) Psychobiology of attention. In *Handbook of Psychobiology* (M. S. Gazzaniga and C. Blakemore, ed.). New York: Academic Press.

Potts, G. W. (1973) The ethology of *Labroides dimidiatus* (Cur. & Vol.) (Labridae, Pisces) on Aldabra. *Anim. Behav. 21*, 250–91.

Powell, G. C. and Nickerson, R. C. (1965) Aggregation amongst juvenile king crabs (*Paralithodes camtschatica*, Tilesius) Kodiak, Alaska. *Anim. Behav. 13*. 374–80.

Powell, G. V. N. (1974) Experimental analysis of the social value of flocking by starlings (*Sturnus vulgaris*) in relation to predation and foraging. *Anim. Behav. 22*, 501–5.

Powers, B. and Valenstein, E. S. (1972) Sexual receptivity: facilitation by medial preoptic lesions in female rats. *Science, N.Y. 175*, 1003–5.

Prechtl, H. F. R. (1965) Problems of behavioural studies in the newborn infant. *Adv. Study Behav. 1*, 75–99.

Pretorius, P. S. (1970) Effect of aggressive behaviour on production and reproduction in the angora goat. *Agroanim. 2*, 161–4.

Pulliam, H. R. (1974) On the theory of optimal diets. *Am. Nat. 108*, 59–75.

Pusey, A. E. (1980) Inter-community transfer of chimpanzees in Gombe National Park. In *The Great Apes* (D. A. Hamburg and E. McCown, eds.). Menlo Park, Ca.: Benjamin Cummings.

Putkonen, P. T. S. (1967) Electrical stimulation of the avian brain. Behavioral and autonomic reactions from the archistriatum, vertomedial forebrain and the diencephalon in the chicken. *Annls Acad. scient. Fenn. Ser. A., V Medica. 130*, 1–95.

Pyke, G. H. (1978*a*) Optimal foraging: movement patterns of bumblebees between inflorescences. *Theoret. Pop. Biol. 13*, 72–98.

Pyke, G. H. (1978*b*) Optimal foraging in hummingbirds: testing the marginal value theorem. *Am. Zool. 18*, 739–52.

Pyke, G. H., Pulliam, H. R. and Charnov, E. L. (1977) Optimal foraging: a selective review of theories and tests. *Q. Rev. Biol.* 52, 137–54.

Radakov, D. V. (1947) Methodological principles for study of fish schooling behaviour. *J. Ichthyol. 9*, 159–65.

Ralls, K. (1971) Mammalian scent marking, *Science, N.Y. 171*, 443–9.

Ramon y Cajal, S. (1909) *Histologie der Système nerveux de l'homme et des vertébrés*. Paris: Maloine.

Ramsay, O. (1951) Familial recognition in domestic birds. *Auk 68*, 1–16.

Randall, W. C. (1943) Factors influencing the temperature regulation of birds. *Am. J. Physiol. 139*, 56–63.

Rasch, E., Swift, H., Riesen, A. H. and Chow, K. L. (1961) Altered structure and composition of retinal cells in dark-reared mammals. *Expl Cell Res. 25*, 348–63.

Ratliff, F. and Hartline, H. K. (1959) The responses of *Limulus* optic nerve fibers to patterns of illumination on the receptor mosaic. *J. gen. Physiol. 42*, 1241–55.

Read, J. S. and Haines, C. P. (1976) The functions of the female sex phenomones of *Ephestia cautella* (Walker) (Leipidoptera, Phyticidae). *J. stored Prod. Res. 12*, 49–53.

Redican, W. K. (1975) Facial expressions in non-human primates. *Primate Behav. 4*, 104–94.

Reese, E. S. (1975) A comparative field study of the social behaviour and related ecology of reef fishes of the family Chaetodontidae. Z. *Tierpsychol. 37*, 37–61.

Renner, M. (1957) Neue Versuche über den Zeitsinn der Honigbiene. *Z. vergl. Physiol. 40*, 85–118.

Reynolds, V. (1978) *Introduction*. In *Human Behaviour and Adaptation*, Part II (V. Reynolds and N. Blurton-Jones, eds.). *Symp. Soc. Study hum. Biol. 18*, London: Taylor and Francis.

Rheingold, H. L. (1963) *Maternal Behavior in Mammals*. New York: Wiley.

Rhijn, J. G. van (1977*a*) Processes in feathers caused by bathing in water. *Ardea 65*, 126–47.

Rhijn, J. G. van (1977*b*) The patterning of preening and other comfort behaviour in a herring gull. *Behaviour 63*, 71–109.

Ribbands, C. R. (1953) *The Behaviour and Social Life of Honeybees*. London: Bee Research Association.

Richard, A. (1970) A comparative study of the activity patterns and behaviour of *Alouatta villosa* and *Ateles geoffroyi*. *Folia Primatol. 12*, 241–63.

Richards, M. P. M. (1966) Infantile handling in rodents: a reassessment in the light of recent studies of maternal behaviour. *Anim. Behav. 14*, 582.

Richards, M. P. M. Ed., (1974*a*) *The Integration of a Child into a Social World*. Cambridge: Cambridge University Press.

Richards, S. M. (1974*b*) The concept of dominance and methods of assessment. *Anim. Behav. 22*, 914–30.

Richmond, G. and Sachs, B. D. (1981) Grooming in Norway rats: the development and adult expression of a complex motor pattern. *Behaviour 75*, 82–96.

Richter, C. P. (1936) Increased salt appetite in adrenalectomised rats. *Am. J. Physiol. 115*, 155–61.

Richter, C. P. (1942–3) Total self regulatory functions in animals and human beings. *Harvey Lect. Ser. 38*, 63–103.

Riechert, S. (1981) Spider interaction strategies: communication vs. coercion. In *Spider Communication: Mechanisms and Ecological Significance* (P. N. Witt and J. Rovner, eds.). Princeton, N.J.: Princeton University Press.

Roberts, B. L. (1969) Spontaneous rhythms in the motoneurons of spinal dogfish (*Scyliorhinus canicula*). *J. mar. Biol. Ass. U.K. 49*, 33–49.

Roberts, B. L. and Russell, I. J. (1972) The activity of lateral line efferent neurons in stationary and swimming dogfish. *J. exp. Biol. 57*, 435–48.

Roberts, W. W. (1958) Rapid escape learning without avoidance learning motivated by hypothalamic stimulation in cats. *J. comp. physiol. Psychol. 51*, 391–9.

Roberts, W. W. (1962) Fear-like behavior elicited from dorsomedial thalamus of cat. *J. comp. physiol. Psychol. 55*, 191–7.

Roberts, W. W., Steinberg, M. L. and Mears, L. W. (1967) Hypothalamic mechanisms for sexual, aggressive, and other motivational behaviors in the opossum (*Didelphis virginiana*). *J. comp. physiol. Psychol. 64*, 1–15.

Robson, E. A. (1962) The swimming response and its pacemaker in the anemone, *Stomphia coccinea*. *J. exp. Biol. 38*, 685–94.

Rodieck, R. W. and Stone, J. (1965) Analysis of receptive fields of cat retinal ganglion cells. *J. Neurophysiol. 28*, 833–49.

Roeder, K. D. (1962) The behaviour of free flying moths in the presence of artificial ultrasonic pulses. *Anim. Behav. 10*, 300–4.

Roeder, K. D. (1963) *Nerve Cells and Insect Behavior*. Cambridge, Mass.: Harvard University Press.

Roeder, K. D. and Treat, A. E. (1961) The detection and evasion of bats by moths. *Am. Scient. 49*, 135–48.

Rosen, J. and Hart, F. M. (1963) Effects of early isolation upon adult timidity and dominance in *Peromyscus*. *Psychol. Rep. 13*, 47–50.

Rosenblatt, J. S. (1965) Effects of experience on sexual behavior in male cats. In *Sex and Behavior* (F. A. Beach, ed.). New York: Wiley.

Rosenblatt, J. S. (1967) Nonhormonal basis of maternal behavior in the rat. *Science, N.Y. 156*, 1512–14.

Rosenblatt, J. S. (1976) Stages in the early behavioural development of altricial young of selected species of non-primate mammals. In *Growing Points in Ethology* (P. P. G. Bateson and R. A. Hinde, eds.). Cambridge: Cambridge University Press.

Rosenblatt, J. S., Siegel, H. I. and Mayer, A. D. (1979) Progress in the study of maternal behavior in the rat: hormonal sensory and developmental aspects. *Adv. Study Behav. 10*, 226–311.

Rosenblith, W. A. Ed., (1961) *Sensory Communication*. Cambridge, Mass.: M.I.T. Press and New York: Wiley.

Rossi, P. J. (1968) Adaption and negative after effect to lateral optical displacement in newly hatched chicks. *Science, N.Y. 160*, 430–2.

Rossi, P. J. (1972) Population density and food dispersion on the development of prism-induced after effects in newly-hatched chicks. *Devl. Psychobiol. 5*, 239–48.

Roth, L. M. and Eisner, T. (1962) Chemical defences of anthropods. *Ann. Rev. Ent. 7*, 107–36.

Rowell, C. H. F. (1961) Displacement grooming in the chaffinch. *Anim. Behav. 9*, 38–63.

Rowell, T. E. (1966*a*) Hierarchy in the organization of a captive baboon group. *Anim. Behav. 14*, 430–43.

Rowell, T. E. (1966*b*) Forest living baboons in Uganda. *J. Zool., Lord. 147*, 344–64.

Rowell, T. E. (1974) The question of social dominance. *Behav. Biol. 11*, 131–54.

Rowell, T. E., Hinde, R. A. and Spencer-Booth, Y. (1964) 'Aunt' – infant interaction in captive rhesus monkeys. *Anim. Behav. 12*, 219–26.

Roy, J. H. B., Shillam, K. W. G. and Palmer, J. (1955) The outdoor rearing of calves on grass with special reference to growth rate and grazing behaviour. *J. Dairy Res. 22*, 252–69.

Royama, T. (1970) Factors governing the hunting behaviour and selection of food by the great tit (*Parus major* L.). *J. Anim. Ecol. 39*, 619–68.

Rozin, P. (1967) Specific aversions as a component of specific hungers. *J. comp. physiol. Psychol. 64*, 237–42.

Rozin, P. (1968) Specific aversions and neophobia as a consequence of vitamin deficiency and/or poisoning in half-wild and domestic rats. *J. comp. physiol. Psychol. 66*, 82–8.

Rozin, P. (1976) The selection of foods by rats, humans and other animals. *Adv. Study Behav. 6*, 21–76.

Rozin, P. and Kalat, J. W. (1971) Specific hungers and poison avoidance as adaptive specializations of learning. *Psychol. Rev. 78*, 459–89.

Ruckebusch, Y. (1975) The hypnogram as an index of adaptations of farm animals to changes in their environment. *Appl. Anim. Ethol. 2*, 3–18.

Ruelle, J. E. (1964) L'Architecture du nid de *Macrotermes natalensis* et son sens fonctionnel. In *Etudes sur les Termites africains* (A. Bouillon, ed.). Leopoldville: University of Leopoldville.

Ruiter, L. de (1956) The measurement of the "prey value" of preys. *Archs néerl. Zool. 11*, 524–6.

Rumbaugh, D. M. Ed., (1974) *Gibbon and Siamang*, Vol. 3. Basel: Karger.

Rusak, B. and Zucker, I. (1975) Biological rhythms and animal behavior. *Ann. Rev. Psychol. 26*, 137–71.

Rusak, B. and Zucker, I. (1979) Neural regulation of circadian rhythms. *Physiol. Rev. 59*, 449–526.

Russek, M. (1971) Hepatic receptors and the neurophysiological mechanisms controlling feeding behavior. In *Neurosciences Research*, Vol. 4 (S. Ehrenpreis and O. C. Solnitzky, eds.). New York: Academic Press.

Russell, E. M. and Pearce, G. A. (1971) Exploration of novel objects by marsupials. *Behaviour 40*, 312–22.

Russell, G. F. and Hills, J. I. (1971) Odor differences between enantiomeric isomers. *Science, N.Y. 172*, 1043–4.

Russell, W. M. S., Mead, A. P. and Hayes, J. S. (1954) A basis for the quantitative study of the structure of behaviour. *Behaviour 6*, 153–205.

Rzóska, J. (1953) Bait shyness, a study in rat behaviour. *Br. J. Anim. Behav. 1*, 128–35.

Saladin, K. D. (1979) Behavioural parasitology and perspectives on miracidial host-finding. *Z. Parasitenk. 60*, 197–210.

Salk, L. (1973) The role of the heart beat in the relations between mother and infant. *Scient. Am. 228*, (5), 24–9.

Salzen, E. A. (1962) Imprinting and fear. *Symp. zool. Soc. Lond. 8*, 197–217.

Sambraus, H. H. (1971) Das Sexualverhalten des Hausrindes speziell des Stieres. *Z. Tierpsychol., Beiheft 6* 1–54.

Sambraus, H. H. and Osterkorn, K. (1974) The social stability of a herd of cattle. *Z. Tierpsychol. 35*, 418–24.

Sanchez Riviello, M. and Shaw, J. G. (1966) Use of field bait-stations in chemosterilant control of the Mexican fruit-fly. *J. econ. Ent. 59*, 753–4.

Sand, A. (1938) The function of the ampullae of Lorenzini, with some observations on the effect of temperature on sensory rhythms. *Proc. R. Soc., B, 125*, 524–53.

Sandow, J. D. and Bailey, W. J. (1978) An experimental study of defensive stridulation in *Mygalopsis ferruginea* Redtenbacher (Orthoptera: Tettigoniidae). *Anim. Behav. 26*, 1004–11.

Sargeant, T. D. (1969) Behavioural adaptations of cryptic moths. III. Resting attitudes of two bark-like species *Melanolophia canadaria*, and *Catocala ultronia*. *Anim. Behav. 17*, 670–2.

Sargent, A. B. (1972) Red fox spatial characteristics in relation to waterfowl predation. *J. Wildl. Mgmt. 36*, 225–36.

Sassenrath, E. N. (1970) Increased adrenal responsiveness related to social stress in rhesus monkeys. *Horm. Behav. 1*, 283–98.

Saunders, D. S. (1976) *An Introduction to Biological Rhythms*. Glasgow: Blackie.

Schadé, J. P., Corner, M. A. and Peters, J. J. (1965) Some aspects of the electro-ontogenesis of sleep patterns. *Prog. Brain Res. 18*, 70–8.

Schaffer, H. R. Ed., (1971) *The Origins of Human Social Relations*. London: Academic Press.

Schake, L. M. and Riggs, J. K. (1966) Diurnal and nocturnal activities of lactating beef cows in confinement. *J. Anim. Sci. 25*, 254.

Schaller, G. B. (1963) *The Mountain Gorilla: Ecology and Behavior*. Chicago: University of Chicago Press.

Schaller, G. B. (1965) The behavior of the mountain gorilla. In *Primate Behavior* (I. DeVore, ed.). New York: Holt, Rinehart and Winston.

Schaller, G. B. (1967) *The Deer and the Tiger: a Study of Wildlife in India*. Chicago: University of Chicago Press.

Schaller, G. B. (1972) *The Serengeti Lion: a Study of Predator-Prey Relations*. Chicago: University of Chicago Press.

Schaller, G. B. and Emlen, J. T. (1962) The ontogeny of avoidance behaviour in some precocial birds. *Anim. Behav. 10*, 370–81.

Scheer, B. T. Ed., (1957) *Recent Advances in Invertebrate Physiology*. Eugene, Oregon: University of Oregon Publications.

Schein, M. W. (1963) On the irreversibility of imprinting. *Z. Tierpsychol. 20*, 462–7.

Schein, M. W. and Fohrman, M. H. (1955) Social dominance relationships in a herd of dairy cattle. *Br. J. Anim. Behav. 3*, 45–55.

Schein, M. W., Hyde, C. E. and Fohrman, M. H. (1955) The effect of psychological disturbances on milk production in dairy cattle. *Proc. Ass. sth. agric. Wkrs* 52nd Conv., Louisville, Ky, p. 79.

Schein, M. W., McDowell, R. E., Lee, D. H. K. and Hyde, C. E. (1957) Heat tolerances of Jersey and Sindhi-Jersey crossbreeds in Louisiana and Maryland. *J. Dairy Sci. 40*, 1405–15.

Schilstra, A. J. (1978) Simulation of feeding behaviour: comparison of deterministic and stochastic models incorporating a minimum of presuppositions. In *Hunger Models: Computable Theory of Feeding Control* (D. A. Booth, ed.). London: Academic Press.

Schjelderup-Ebbe, T. (1922) Beiträge zur Socialpsychologie des Haushuhns. *Z. Psychol. 88*, 225–52.

Schleidt, W. M. (1955) Untersuchungen über die Auslösung des Kollern beim Truthahn (*Meleagris galapavo*). *Z. Tierpsychol. 11*, 417–35.

Schleidt, W. M. (1961) Reaktionen von Truthünern auf fliegende Raubvögel und Versuche zur Analyse ihrer AAM's. *Z. Tierpsychol. 18*, 534–60.

Schleidt, W. M. (1964*a*) Über die Spontaneitat von Erbkoordinationen. *Z. Tierpsychol. 21*, 235–57.

Schleidt, W. M. (1964*b*) Über das Wirkungsgefuge von Balzbewegungen des Truthahnes. *Naturwissenschaften. 51*, 445–6.

Schleidt, W. M. (1965) Gaussian interval distributions in spontaneously occurring innate behaviour. *Nature, Lond. 206*, 1061–2.

Schleidt, W. M. (1970) Precocial sexual strutting behaviour in turkeys (*Meleagris gallopavo* L.). *Anim. Behav. 18*, 760–1.

Schleidt, W. M. (1974) How 'fixed' is the fixed action pattern? *Z. Tierpsychol. 36*, 184–211.

Schleidt, W. M. (1981) Stereotyped feature variables are essential constituents of behaviour patterns.
Schleidt, W. M. and Shalter, M. D. (1973) Stereotype of a fixed action pattern during ontogeny in *Coturnix coturnix coturnix*. *Z. Tierpsychol. 33*, 35–7.
Schloeth, R. (1961) Das Sozialleben des Camargue-rindes. *Z. Tierpsychol. 18*, 574–627.
Schmidt-Koenig, K. (1975) *Migration and Homing in Animals*. Berlin: Springer.
Schmidt-Nielsen, K. (1979) *Animal Physiology: Adaptation and Environment*, 2nd edn. Cambridge: Cambridge University Press.
Schmidt-Nielsen, K., Bretz, W. L. and Taylor, C. R. (1970) Panting in dogs: unidirectional air flow over evaporative surfaces. *Science, N.Y. 169*, 1102–4.
Schmitt, F. O. and Worden, F. G. Eds., (1974) *The Neurosciences: Third Study Program*. Cambridge, Mass.: M.I.T. Press.
Schneider, D. (1954) Beitrag zur einer Analyse des Beute-und Fluch einheimischer Anuren. *Biol. Zbl. 73*, 225–82.
Schneider, D. (1957) Elektrophysiologische Untersuchungen von Chemo-und Mechanorezeptoren der Antene des Seidenspinners *Bombyx mori* L. *Z. vergl. Physiol. 8*, 15–30.
Schneider, D. (1969) Insect olfaction: deciphering system for chemical messages. *Science, N.Y. 163*, 1031–6.
Schneirla, T. C., Rosenblatt, J. S. and Tobach, E. (1963) Maternal behavior in the cat. In *Maternal Behavior in Mammals* (H. L. Rheingold, ed.). New York: Wiley.
Schoener, T. W. (1968) Sizes of feeding territories among birds. *Ecology 49*, 123–41.
Schoener, T. W. (1971) Theory of feeding strategies. *Ann. Rev. Ecol. Syst. 2*, 369–404.
Scholander, P. F., Flagg, W., Hock, R. J. and Irving, L. (1953) Studies on the physiology of frozen plants and animals in the arctic. *J. cell. comp. Physiol. 42*, 1–56.
Schoonhoven, L. M. and Meerman, J. (1978) Metabolic cost of changes in diet and neutralization of allelochemics. *Entomologia exp. appl. 24*, 489–93.
Schrameck, J. E. (1970) Crayfish swimming: alternating motor output and giant fibre activity. *Science, N.Y. 169*, 698–700.
Schrier, A. M., Harlow, H. F. and Stollnitz, F. Eds., (1965) *Behavior of Nonhuman Primates*, Vol. 2. New York: Academic Press.
Schürch, A. and Wenk, C. Eds., (1970) *Energy Metabolism of Farm Animals*. Zurich: Juris Verlag.
Schutz, F. (1965) Sexuelle Prägung bei Anatiden. *Z. Tierpsychol. 22*, 50–103.
Scott, J. P. (1945) Social behavior, organization and leadership in a small flock of domestic sheep. *Comp. psychol. Monogr. 18*, 1–29.
Seabrook, M. F. (1972) To determine the influence of the herdsman's personality on milk yield. *J. agric. labour Sci. 1*, 45–59.
Seabrook, M. F. (1977) Cowmanship. *Fmrs' Wkly Extra*, Dec. 23, 26 pp.
Seath, D. M. and Miller, G. D. (1946) Effect of warm weather on grazing performance of milking cows. *J. Dairy Sci, 29*, 199–206.
Seath, D. M. and Miller, G. D. (1947) Effect of hay feeding in summer on milk production and grazing performance of dairy cows. *J. Dairy Sci. 39*, 921–5.
Seay, B. (1966) Maternal behaviour in primiparous and multiparous rhesus monkeys. *Folia Primatol. 4*, 146–68.
Sebeok, T. A. (1977) *How Animals Communicate*. Bloomington: Indiana University Press.
Seghers, B. H. (1974) Schooling behaviour in the guppy (*Poecilia reticulata*): an evolutionary response to predation. *Evolution 28*, 486–9.
Selman, I. E., McEwan, A. D. and Fisher, E. W. (1970*a*) Studies on natural suckling in cattle during the first eight hours post partum. II. Behavioural studies (calves). *Anim. Behav. 18*, 284–9.
Selman, I. E., McEwan, A. D., and Fisher, E. W. (1970*b*) Serum immune globulin concentrations of calves left with their dams for the first two days of life. *J. comp. Path. 80*, 419–27.
Selman, I. E., McEwan, A. D. and Fisher, E. W. (1971) Studies on dairy calves allowed to suckle their dams at fixed times post partum. *Res. vet. Sci. 12*, 1–6.
Selye, H. (1950) *The Physiology and Pathology of Exposure to Stress*. Montreal: Acta.
Sevenster, P. (1973) Incompatibility of response and reward. In *Constraints on Learning* (R. A. Hinde and J. Stevenson-Hinde, eds.). London: Academic Press.
Sewell, G. D. (1968) Ultrasound in rodents. *Nature, Lond. 217*, 682–3.
Sharpe, R. M. (1975) The influence of the sex of littermates on subsequent maternal behaviour in *Rattus norvegicus*. *Anim. Behav. 23*, 551–9.
Shaw, E. (1962) The schooling of fishes. *Scient. Am. 206*, 128–38.
Shea, J. D. C. (1976) Variations in early experience and social behaviour in the domestic fowl. *Biol. Behav. 2*, 135–47.
Shepher, J. (1971) Mate selection among second generation kibbutz adolescents and adults: incest avoidance and negative imprinting. *Arch. sex. Behav. 1*, 293–307.
Sherman, P. W. (1977) Nepotism and the evolution of alarm calls. *Science, N.Y. 197*, 1246–53.
Sherrington, C. S. (1898) Decerebrate rigidity and reflex coordination of movements. *J. Physiol., Lond. 22*, 319.
Sherry, D. F. (1981) Parental care and the development of

thermoregulation in red junglefowl. *Behaviour,* *76,* 250–279.
Shettleworth, S. J. (1972) Constraints on learning. *Adv. Study Behav. 4,* 1–68.
Shettleworth, S. J. (1973) Food reinforcement and the organization of behaviour in golden hamsters. In *Constraints on Learning* (R. A. Hinde and J. Stevenson-Hinde, eds.). London: Academic Press.
Shillito, E. and Alexander, G. (1975) Mutual recognition amongst ewes and lambs of four breeds of sheep (*Ovis aries*). *Appl. Anim. Ethol. 1,* 151–65.
Shorten, M. (1954) The reaction of the brown rat towards changes in its environment. In *Control of Rats and Mice,* Vol. 2. *Rats* (D. Chitty, ed.). Oxford: Oxford University Press.
Sibly, R. (1975) How incentive and deficit determines feeding tendency. *Anim. Behav. 23,* 437–46.
Sibly, R. M. and McCleery, R. H. (1976) The dominance boundary method of determining motivational state. *Anim. Behav. 24,* 108–24.
Sibly, R. and McFarland, D. (1974) A state–space approach to motivation. In *Motivational Control Systems Analysis* (D. J. McFarland, ed.). London: Academic Press.
Sibly, R. and McFarland, D. (1976) On the fitness of behaviour sequences. *Am. Nat. 110,* 601–17.
Sidman, R. L. (1968) Development of interneuronal connections in brains of mutant mice. In *Physiological and Biochemical Aspects of Nervous Integration* (F. D. Carlson, ed.). Englewood Cliffs, N.J.: Prentice Hall.
Sidman, R. L. (1972) Cell proliferation, migration and interaction in the developing mammalian nervous system. In *The Neurosciences: Second Study Program* (F. O. Schmitt, ed.). New York: Rockefeller University Press.
Sidman, R. L. (1974) Cell-cell recognition in the developing central nervous system. In *The Neurosciences: Third Study Program* (F. O. Schmitt and F. G. Worden, eds.). Cambridge, Mass.: M.I.T. Press.
Sidman, R. L. and Green, M. C. (1970) 'Nervous, a new mutant mouse with cerebellar disease. *Symp. Centre Nat. Rech. Sci.* Orleans-la Source, 49–61.
Siegfried, W. R. and Underhill, L. G. (1975) Flocking as an anti-predator strategy in doves. *Anim. Behav. 23,* 504–8.
Signoret, J. P. (1965) Untersuchungen der geschlechtlichen Verhaltensweise der Sau. *Int. Koll. K. Marx Univ., Leipsig,* 15–21.
Signoret, J. P. (1975) Influence of the sexual receptivity of a teaser ewe on the mating preference in the ram. *Appl. Anim. Ethol. 1,* 229–32.
Signoret, J. P., Baldwin, B. A., Fraser, D. and Hafez, E. S. E. (1975) In *The Behaviour of Domestic Animals* (E. S. E. Hafez, ed.). London: Baillière Tindall.
Signoret, J. P. and Mesnil du Buisson, F. du (1961) Etude du comportement de la truie en oestrus. *Int. Congr. Anim. Reprod.* 4. The Hague, pp. 171–5.
Silberglied, R. E. and Taylor, O. R. (1978) Ultraviolet reflection and its behavioural role in the courtship of the sulfur butterflies *Colias eurytheme* and *C. philodice* (Lepidoptera, Pieridae). *Behav. Ecol. Sociobiol. 3,* 203–43.
Simmons, K. E. L. (1965) Pattern of dispersion of the white wagtail and other birds outside the breeding season. *Bull. B.O.C. 85,* 161–8.
Simonds, P. E. (1965) The bonnet macaque in South India. In *Primate Behavior* (I. DeVore, ed.). New York: Holt, Rinehart and Winston.
Simpson, M. J. A. (1968) The display of the siamese fighting fish *Betta spendens. Anim. Behav. Monogr. 1,* 1–73.
Simpson, M. J. A. (1973) The social grooming of male chimpanzees. In *Comparative Ecology and Behaviour of Primates* (J. H. Crook and R. P. Michael, eds.). London: Academic Press.
Simpson, M. J. A. (1979) Daytime rest and activity in socially living rhesus monkey infants. *Anim. Behav. 27,* 602–12.
Skinner, B. F. (1953) *Science and Human Behavior.* New York: Macmillan.
Skutch, A. F. (1935) Helpers at the nest. *Auk 52,* 257–73.
Skutch, A. F. (1961) Helpers among birds. *Condor 63,* 198–226.
Slater, P. J. B. (1974*a*) Bouts and gaps in the behaviour of zebra finches, with special reference to preening. *Rev. comport. anim. 8,* 47–61.
Slater, P. J. B. (1974*b*) The temporal pattern of feeding in the zebra finch. *Anim. Behav. 22,* 506–15.
Slater, P. J. B. (1975) Temporal patterning and the causation of bird behaviour. In *Neural and Endocrine Aspects of Behaviour in Birds* (P. Wright, P. C. Caryl and D. M. Vowles, eds.). Amsterdam: Elsevier.
Slater, P. J. B. (1978) *Sex Hormones and Behaviour.* London: Arnold.
Slater, P. J. B. and Ollason, J. C. (1971) The temporal pattern of behaviour in isolated male zebra finches: transition analysis. *Behaviour 42,* 248–69.
Slotnick, B. M. (1967) Disturbances of maternal behaviour in the rat following lesions of the cingulate cortex. *Behaviour 29,* 204–36.
Sluckin, W. Ed., (1979) *Fear in Animals and Man.* New York: Van Nostrand Reinhold.
Smith, C. C. (1977) Feeding behaviour and social organization in howling monkeys. In *Primate Ecology* (T. H. Clutton-Brock, ed.). London: Academic Press.
Smith, C. C. and Fretwell, S. D. (1974) The optimal balance between size and number of offspring. *Am. Nat. 108,* 499–506.
Smith, F. V. (1962) Perceptual aspects of imprinting. *Symp. zool. Soc. Lond. 8,* 171–91.

Smith, J. C. and Roll, D. L. (1967) Trace conditioning with X-rays as an aversive stimulus. *Psychon. Sci. 9,* 11–12.

Smith, J. N. M. (1974*a*) The food searching behaviour of two European thrushes. I. Description and analysis of search paths. *Behaviour 48,* 276–302.

Smith, J. N. M. (1974*b*) The food searching behaviour of two European thrushes. II. The adaptiveness of the search patterns. *Behaviour 49,* 1–61.

Smith, J. N. M. and Dawkins, R. (1971) The hunting behaviour of individual great tits in relation to spatial variations in their food density. *Anim. Behav. 19,* 695–706.

Smith, M. (1951) *The British Amphibians and Reptiles.* London: Collins.

Smith, R. H. (1979) On selection for inbreeding in polygynous animals. *Heredity 43,* 205–11.

Smith, W. and Hale, E. B. (1959) Modification of social rank in the domestic chicken. *J. comp. physiol. Psychol. 52,* 373–5.

Smythe, N. (1970) On the existence of 'pursuit invitation' signals in mammals. *Am. Nat. 104,* 491–4.

Snow, B. K. and Snow, D. W. (1972) Feeding niches of hummingbirds in a Trinidad valley. *J. Anim. Ecol. 41,* 471–85.

Snow, D. W. (1968) The singing assemblies of little hermits. *The Living Bird 7,* 47–55.

Snyder, F. (1966) Toward an evolutionary theory of dreaming. *Am. J. Psychiat. 123,* 121–36.

Soffié, M., Thinès, G., Marneffe, G. de (1976) Relation between milking order and dominance value in a group of dairy cows. *Appl. Anim. Ethol. 2,* 271–6.

Sokolov, E. N. (1960) Neuronal models and the orienting reflex. In *The Central Nervous System and Behavior* (M. A. Brazier, ed.). New York: Macy Foundation.

Sollberger, A. (1965) *Biological Rhythm Research.* Amsterdam: Elsevier.

Soltysik, S., Jaworska, K., Kowalska, M. and Radom, S. (1961) Cardiac responses to simple acoustic stimuli in dogs. *Acta Biol. exp., Vars. 21,* 235–52.

Sonnemann, P. and Sjölander, S. (1977) Effects of cross-fostering on the sexual imprinting of the female zebra finch *Taeniopygia guttata. Z. Tierpsychol. 45,* 337–48.

Spence, K. W. (1951) Theoretical interpretations of learning. In *Comparative Psychology* (C. P. Stone, ed.). New York: Prentice Hall.

Spencer-Booth, Y. (1970) The relationships between mammalian young and conspecifics other than mothers and peers: a review. *Adv. Study Behav. 3,* 120–94.

Spencer-Booth, Y. and Hinde, R. A. (1971) Effects of 6 days separation from mother on 18- to 32-week-old rhesus monkeys. *Anim. Behav. 19,* 174–91.

Spencer-Booth, Y., Hinde, R. A. and Bruce, M. (1965) Social companions and the mother-infant relationship in rhesus monkeys. *Nature, Lond. 208,* 301.

Spieth, H. T. (1968) Evolutionary implications of sexual behaviour in *Drosophila. Evol. Biol. 2,* 157–93.

Spurway, H. and Haldane, J. B. S. (1953) The comparative ethology of vertebrate breathing. I. Breathing in newts, with a general survey. *Anim. Behav. 6,* 8–34.

Squires, V. R. (1974) Grazing distribution and activity patterns of Merino sheep on a saltbush community in south-east Australia. *Appl. Anim. Ethol. 1,* 17–30.

Squires, V. R. and Daws, G. T. (1975) Leadership and dominance relationships in Merino and Border Leicester sheep. *Appl. Anim. Ethol. 1,* 263–74.

Staddon, J. E. R. (1972) A note on the analysis of behavioural sequences in *Columba livia. Anim. Behav. 20,* 284–92.

Stallcup, J. A. and Woolfenden, G. E. (1978) Family status and contributions to breeding by Florida scrub jays. *Anim. Behav. 26,* 1144–56.

Stamps, J. A. and Barlow, G. W. (1973) Variation and stereotypy in the displays of *Anolis aeneus* (Sauria: Iguanidae). *Behaviour 47,* 67–94.

Stamps, J. A. and Metcalf, R. A. (1980) Parent–offspring conflict. In *Sociobiology: beyond nature/nurture* (G. W. Barlow, ed.). A.A.A.S. Selected Symposium 35.

Steel, E. A. and Hinde, R. A. (1964) Effect of exogenous oestrogen on brood patch development of intact and ovariectomised canaries. *Nature, Lond, 202,* 718–9.

Stellar, E. and Hill, J. H. (1952) The rats' rate of drinking as a function of water deprivation. *J. comp. physiol. Psychol. 45,* 96–102.

Stellar, E. and Sprague, J. M. (1971) *Progress in Physiological Psychology,* Vol. 4. New York: Academic Press.

Stephan, F. K. and Zucker, I. (1972) Circadian rhythms in drinking behavior and locomotor activity of rats are eliminated by hypothalamic lesions. *Proc. natn. Acad. Sci. U.S.A. 69,* 1583–6.

Stephens, D. B. (1974) Studies on the effect of social environment on the behaviour and growth rates of artificially-reared British Friesian male calves. *Anim. Prod. 18,* 23–34.

Stephens, D. B. and Toner, J. N. (1975) Husbandry influences on some physiological parameters of emotional responses in calves. *Appl. Anim. Ethol. 1,* 233–43.

Stern, J. J. and Hoffman, B. M. (1970) Effects of social isolation until adulthood on maternal behaviour in guinea pigs. *Psychon. Sci. 21,* 15–16.

Stevens, S. S. Ed., (1951) *Handbook of Experimental Psychology.* New York: Wiley.

Stevenson-Hinde, J. (1972) Effects of early experience and testosterone on song as a reinforcer. *Anim. Behav. 23,* 430–5.

Stevenson-Hinde, J. (1973) Constraints on reinforcement. In *'Constraints on Learning'* (R. A. Hinde and J. Stevenson-Hinde, eds.). London: Academic Press.

Stiles, F. G. (1978) Ecological and evolutionary implications of bird pollination. *Am. Zool. 18,* 715–27.
Stobbs, T. H. (1973*a*) The effect of plant structure on the intake of tropical pastures. I. Variation in the bite size of grazing cattle. *Aust. J. agric. Res. 24,* 809–19.
Stobbs, T. H. (1973*b*) The effect of plant structure on the intake of tropical pastures. II. Differences in sward structure, nutritive value, and bite size of animals grazing *Setaria anceps* and *Chloris gayana* at various stages of growth. *Aust. J. agric. Res. 24,* 821–9.
Stobbs, T. H. (1974*a*) Components of grazing behaviour of dairy cows on some tropical and temperate pastures. *Proc. Aust. Soc. Anim. Prod. 10,* 299–302.
Stobbs, T. H. (1974*b*) Rate of biting by Jersey cows as influenced by the yield and maturity of pasture swards. *Trop. Grassld 8,* 81–6.
Stobbs, T. H. (1975) The effect of plant structure on the intake of tropical pasture. III. Influence of fertilizer nitrogen on the size of bite harvested by Jersey cows grazing *Sertaria anceps* CV. Kazungula swards. *Aust. J. agric. Res. 26,* 997–1007.
Stobbs, T. H. and Cowper, L. J. (1972) Automatic measurement of the jaw movements of dairy cows during grazing and rumination. *Trop. Grassld 6,* 107–12.
Stoddart, D. M. (1980) *The Ecology of Vertebrate Olfaction.* London: Chapman and Hall.
Stokols, D. (1972) On the distinction between density and crowding: some implications for future research. *Psychol. Rev. 79,* 275–7.
Stonehouse, B. and Perrins, C. (1977) *Evolutionary Ecology.* London: MacMillan.
Struhsaker, T. T. (1967*a*) Social structure among vervet monkeys (*Cercopithecus aethiops*). *Behaviour 29,* 83–121.
Struhsaker, T. T. (1967*b*) Auditory communications among vervet monkeys (*Cercopithecus aethiops*). In *Social Communication Among Primates* (S. A. Altmann, ed.). Chicago: University of Chicago Press.
Struhsaker, T. T. and Leland, L. (1979) Ecology of five sympatric monkey species in the Kibale forest, Uganda. *Adv. Study Behav. 9,* 159–228.
Stryker, M. P. and Sherk, H. (1975) Modification of cortical orientation selectivity in the cat by restricted visual experience: a re-examination. *Science, N.Y. 190,* 904–6.
Sugiyama, Y. (1976) Life history of male Japanese monkeys. *Adv. Study Behav. 7,* 255–84.
Surtees, G. (1966) Locomotory behaviour of pyrethrum-resistant and susceptible strains of grain weevil, *Sitophilus granarius* (L.) Coleoptera, Curculionidae. *Anim. Behav. 14,* 201–3.
Sylva, K., Bruner, J. S. and Genova, P. (1974) The role of play in the problem-solving of children 3–5 years old. In *Play: its Role in Development and Evolution* (J. S. Bruner, A. Jolly and K. Sylva, eds.). Harmondsworth: Penguin.
Syme, G. J. (1974) Competitive orders as measures of social dominance. *Anim. Behav. 22,* 931–40.
Syme, G. J. and Syme, L. A. (1979) *Social Structure in Farm Animals.* Amsterdam: Elsevier.
Syme, L. A., Syme, G. J., Waite, T. G. and Pearson, A. J. (1975) Spatial distribution and social status in a small herd of dairy cows. *Anim. Behav. 23,* 609–14.
Taggart, P., Hedworth-Whitty, R., Carruthers, M. and Gordon, P. D. (1976) Observations on electrocardiogram and plasma catecholamines during dental procedures: the forgotten vagus. *Br. med. J. 2,* 787–9.
Taghert, P. H. and Willows, A. O. D. (1978) Control of a fixed action pattern by single, central neurons in the marine mollusk, *Tritonia diomedea. J. comp. Physiol. 123,* 253–9.
Tallarico, R. B. (1961) Studies of visual depth perception: choice behavior of newly hatched chicks on a visual cliff. *Percept. Mot. Skills 12,* 259–62.
Tallarico, R. B. and Farrell, W. M. (1964) Studies of visual depth perception: an effect of early experience on chicks on the visual cliff. *J. comp. physiol. Psychol. 57,* 94–6.
Taylor, C. R. and Rountree, V. J. (1973) Temperature regulation and heat balance in running cheetahs: a strategy for sprinters. *Am. J. Physiol. 224,* 848–51.
Taylor, R. H. (1962) The Adèlie penguin *Pygoscelis adeliae* at Cape Royds. *Ibis 104,* 176–204.
Teitelbaum, H. and Milner, P. M. (1963) Activity changes following partial hippocampal lesions in rats. *J. comp. physiol. Psychol. 56,* 284–9.
Teitelbaum, P. and Epstein, A. N. (1962) The lateral hypothalamic syndrome: recovery of feeding and drinking after lateral hypothalamic lesions. *Psychol. Rev. 69,* 74–90.
Terkel, J. and Rosenblatt, J. S. (1968) Maternal behavior induced by maternal blood plasma injected into virgin rats. *J. comp. physiol. Psychol. 65,* 479–82.
Terkel, J. and Rosenblatt, J. S. (1971) Aspects of nonhormonal maternal behavior in the rat. *Horm. Behav. 2,* 161–71.
Terkel, J. and Rosenblatt, J. S. (1972) Humoral factors underlying maternal behavior at parturition: cross transfusion between freely moving rats. *J. comp. physiol. Psychol. 80,* 365–71.
Teyler, T. J. (1971) Effects of restraint on heart-rate conditioning in rats as a function of UCS location. *J. comp. physiol. Psychol. 77,* 31–7.
Thomas, G. (1974) The influences of encountering a food object on subsequent searching behaviour in *Gasterosteus aculeatus* L. *Anim. Behav. 22,* 941–52.
Thompson, R. D., Grant, C. V., Pearson, E. W. and Cor-

ner, G. W. (1968) Differential heart-rate response of starlings to sound stimuli of biological origin. *J. Wildl. Mgmt. 32*, 888–93.
Thompson, R. F. (1975) *Introduction to Physiological Psychology*. New York: Harper and Row.
Thompson, W. A., Vertinsky, I. and Krebs, J. R. (1974) The survival value of flocking in birds: a simulation model. *J. Anim. Ecol. 43*, 785–820.
Thornhill, R. (1976) Sexual selection and paternal investment in insects. *Am. Nat. 110*, 153–63.
Thorpe, W. H. (1958*a*) The learning of song patterns by birds, with especial reference to the song of the chaffinch, *Fringilla coelebs*. *Ibis 100*, 535–70.
Thorpe, W. H. (1958*b*) Further studies on the process of song learning in the chaffinch *Fringilla coelebs gengleri*. *Nature, Lond. 182*, 554–7.
Thorpe, W. H. (1961) *Bird Song*. Cambridge: Cambridge University Press.
Thorpe, W. H. (1963) *Learning and Instinct in Animals*, 2nd edn. London: Methuen.
Tinbergen, J. M. and Drent, R. H. (1980) The starling as a successful forager. In *Bird Problems in Agriculture* (E. N. Wright, I. R. Inglis and C. J. Feare, eds.). Croydon: British Crop Protection Council.
Tinbergen, L. (1960) The natural control of insects in pinewoods. I. Factors influencing the intensity of predation by songbirds. *Archs néerl. Zool. 13*, 265–343.
Tinbergen, N. (1935) Ueber die Orientierung des Bienenwolfes (*Philanthus triangulum* Fabr.) II. Die Bienenjagd. *Z. vergl. Physiol. 21*, 699–716.
Tinbergen, N. (1940) Die Übersprungsbewegung. *Z. Tierpsychol. 4*, 1–10.
Tinbergen, N. (1951) *The Study of Instinct*. Oxford: Oxford University Press.
Tinbergen, N., Impekoven, M. and Franck, D. (1967) An experiment on spacing out as a defence against predation. *Behaviour 28*, 307–21.
Tinbergen, N. and Kuenen, D. J. (1939) Uber die auslosenden und die richtunggebenden Reizsituationen der Sperrbewegung von jungen Drosseln (*Turdus m. merula* L. & *T. e. ericetorum* Turton). *Z. Tierpsychol. 3*, 37–60.
Tinbergen, N., Meeuse, B. J. D., Boerema, L. K., and Varossieau, W. (1942) Die Balz des Samtfalters, *Eumenis semele* (L). *Z. Tierpsychol. 5*, 182–226.
Tinbergen, N. and Perdeck, A. C. (1950) On the stimulus situation releasing the begging response in the newly hatched herring gull chick (*Larus argentatus argentatus* Pont.). *Behaviour 3*, 1–39.
Toates, F. M. and Archer, J. (1978) A comparative review of motivational systems using classical control-theory. *Anim. Behav. 26*, 368–80.
Toates, F. M. and Booth, D. A. (1974) Control of food intake by energy supply. *Nature, Lond. 251*, 710–11.
Toates, F. M. and Halliday, T. R. Eds., (1981) *Analysis of Motivational Processes*. London: Academic Press.
Tolhurst, B. E. and Vince, M. A. (1976) Sensitivity to odours in the embryo of the domestic fowl. *Anim. Behav. 24*, 772–9.
Tolman, E. C. (1932) *Purposive Behavior in Animals and Men*. New York: Appleton-Century.
Tolman, E. C. (1951) A psychological model. In *Toward a General Theory of Action* (T. Parsons and E. A. Shils, eds.). Cambridge, Mass.: Harvard University Press.
Tomkins, T. and Bryant, M. J. (1972) Mating behaviour in a small flock of lowland sheep. *Anim. Prod. 15*, 203–210.
Tonndorf, J. and Khanna, S. M. (1966) Some properties of sound transmission in the middle and outer ears of cats. *J. acoust. Soc. Amer. 41*, 513–21.
Toutain, P. L. and Ruckebusch, Y. (1973) Sommeil paradoxal et environnement. *C. r. Séanc. Soc. Biol. 167*, 550–4.
Towbin, E. J. (1949) Gastric distension as a factor in the satiation of thirst in esophagostomized dogs. *Am. J. Physiol. 159*, 533–41.
Trammel, K., Roelofs, W. L. and Glass, E. H. (1974) Sex-pheromone trapping of males for control of red-banded leafroller in apple orchards. *E. econ. Ent. 67*, 159–64.
Treisman, A. (1964) Monitoring and storage of irrevelant messages in selective attention. *J. verb. Learn. verb. Behav. 3*, 449–59.
Treisman, M. (1975*a*) Predation and the evolution of gregariousness. I. Models for concealment and evasion. *Anim. Behav. 23*, 779–800.
Treisman, M. (1975*b*) Predation and the evolution of gregariousness. II. An economic model for predator–prey interaction. *Anim. Behav. 23*, 801–25.
Trivers, R. L. (1971) The evolution of reciprocal altruism. *Q. Rev. Biol. 46*, 35–57.
Trivers, R. L. (1972) Parental investment and sexual selection. In *Sexual Selection and the Descent of Man, 1871–1971* (B. G. Campbell, ed.). Chicago: Aldine Publishing Company.
Trivers, R. L. (1974) Parent–offspring conflict. *Am. Zool. 14*, 249–64.
Trivers, R. L. and Hare, H. (1976) Haplodiploidy and the evolution of the social insects. *Science, N.Y. 191*, 249–63.
Tugendhat, B. (1960) The normal feeding behaviour of the three-spined stickleback (*Gasterosteus aculeatus* L.) *Behaviour 15*, 284–318.
Tumlinson, J. H., Gueldner, R. C., Hardee, D. D., Thompson, A. C., Hedin, P. A. and Minyard, J. P. (1970) The boll weevil sex attractant. In *Chemicals Controlling Insect Behavior* (M. Beroza, ed.). New York: Academic Press.

Tyler, P. S. and Binns, T. J. (1973) Laboratory evaluation of insecticides against susceptible and malathion resistant strains of *O. surinamensis* (L.) (Coleoptera, Sylvanidae). *Minist. Agric. Fish. Fd, Pest Control Lab., Res. Rep.* No. 9.

Tyler, S. J. (1972) The behaviour and social organization of the New Forest ponies. *Anim. Behav. Monogr. 5,* 85–196.

Uexküll, J. von (1940) *Bedeutungslehre,* Vol.10. Leipzig: Bios.

Underwood, E. J. (1956) *Trace Elements in Human and Animal Nutrition.* New York: Academic Press.

Uttal, W. R. (1973) *The Psychobiology of Sensory Coding.* New York: Harper and Row.

Valenstein, E. S., Riss, W. and Young, W. C. (1955) Experimental and genetic factors in the organization of sexual behavior in male guinea pigs. *J. comp. physiol. Psychol. 48,* 397–403.

Van Lawick, H. and Van Lawick-Goodall, J. (1971) *Innocent Killers.* Boston: Houghton Mifflin.

Van Lawick-Goodall, J. (1968) Behaviour of free-living chimpanzees in the Gombe Streat Reserve. *Anim. Behav. Monogr. 1,* 161–311.

Vehrencamp, S. L. (1977) Relative fecundity and parental effort in communally nesting anis, *Crotophaga sulcirostris. Science, N.Y. 197,* 403–5.

Verner, J. (1964) Evolution of polygamy in the long-billed marsh wren. *Evolution 18,* 252–61.

Verner, J. and Willson, M. F. (1966) The influence of habitats on mating systems of North American passerine birds. *Ecology 47,* 143–7.

Verplanck, W. S. and Hayes, J. R. (1953) Eating and drinking as a function of maintenance schedule. *J. comp. physiol. Psychol. 46,* 327–33.

Vidal, J-M. (1976) L'Empreinte chez les animaux. *Recherche 63,* 24–35.

Vince, M. A. (1961) 'String pulling' in birds: III The successful response in greenfinches and canaries. *Behaviour 17,* 103–29.

Wachtel, M. A., Bekoff, M. and Fuenzalida, C. E. (1978) Sparring by male deer during rutting: class participation, seasonal changes and the nature of asymmetric contests. *Biol. Behav. 3,* 319–30.

Wald, G. (1959) The photoreceptor process in vision. In *Handbook of Physiology. Neurophysiology,* Vol. 1. (J. Field, ed.), pp. 671–92. Washington, D.C.: American Physiological Association.

Wald, G. (1968) Molecular basis of visual excitation *Science, N.Y. 162,* 230–9.

Walk, R. D. and Gibson, E. J. (1961) A comparative and analytical study of visual depth perception. *Psychol. Monogr. 75,* 1–44.

Wallis, S. J. (1979) The sociology of *Cercocebus albigigena johnsoni Lydekker,* an arboreal rain-forest primate. Ph.D. thesis, University of London.

Walser, E. S. (1977) Maternal behaviour in mammals. *Symp. zool. Soc. Lond. 41,* 313–31.

Walther, F. R. (1977) Sex and activity dependency of distances between Thomson's gazelles (*Gazella thomsoni* Günther 1884). *Anim. Behav. 25,* 713–19.

Ward, P. (1965) Feeding ecology of the black-faced dioch *Quelea quelea* in Nigeria. *Ibis 107,* 173–214.

Ward, P. (1971) The migration patterns of *Quelea quelea* in Africa. *Ibis 113,* 275–97.

Ward, P. (1972). The functional significance of mass drinking flights by sandgrouse: Pteroclididae. *Ibis 114,* 533–6.

Ward, P. and Zahavi, A. (1973) The importance of certain assemblages of birds as 'information centres' for food-finding. *Ibis 115,* 517–34.

Warner, G. F. (1970) Behaviour of two species of grapsid crab during intraspecific encounters. *Behaviour 36,* 9–19.

Warner, G. F. (1971) On the ecology of a dense bed of the brittle-star *Ophiothrix fragilis. J. mar. Biol. Ass. U.K. 51,* 267–82.

Warner, G. F. (1977) *The Biology of Crabs.* London: Elek.

Waterhouse, A. (1978) The effects of pen conditions on the development of calf behaviour. *Appl. Anim. Ethol. 4,* 285–6.

Waterman, T. H. (1961) Light sensitivity and vision. In *The Physiology of the Crustacea.* Vol. 2 (T. H. Waterman, ed.), pp. 1–64. New York: Academic Press.

Watson, A. (1967) Population control by territorial behaviour in red grouse. *Nature, Lond. 215,* 1274–5.

Watson, A. Ed., (1970) *Animal Populations in Relation to their Food Resources.* Oxford: Blackwell.

Watson, A. and Moss, R. (1979) Population cycles in the Tetraonidae. *Ornis fenn. 56,* 87–109.

Watson, A. and Moss, R. (1980) Advances in our understanding of the population dynamics of red grouse from a recent fluctuation in numbers. *Ardea* in press.

Watts, C. R. and Stokes, A. W. (1971) The social order of turkeys. *Scient. Am. 224, 6,* 112–18.

Webb, W. B. (1974) Sleep as an adaptive response. *Percept. Mot. Skills. 38,* 1023–7.

Wecker, S. C. (1963) The role of early experience in habitat selection by the prairie deermouse *Peromyscus maniculatus bairdi. Ecol. Monogr. 33,* 307–25.

Weiskrantz, L. and Cowey, A. (1963) The aetiology of food reward. *Anim. Behav. 11,* 225–34.

Weiss, B. and Laties, V. G. (1961) Behavioural thermoregulation. *Science, N.Y. 133,* 1338–44.

Welch, R. A. S. and Kilgour, R. (1971) Mismothering in Romney sheep. *Proc. N.Z. Soc. Anim. Prod. 31,* 41.

Welty, J. C. (1934) Experiments in group behaviour of fishes. *Physiol. Zool. 7,* 85–128.

Welty, J. C. (1962) *The Life of Birds*. Philadelphia: Saunders.
Wendell-Smith, C. P. (1964) Effect of light deprivation on the post-natal development of the optic nerve. *Nature, Lond. 204*, 707.
Werner, E. E. and Hall, D. J. (1974) Optimal foraging and the size selection of prey by the bluegill sunfish (*Lepomis macrochirus*). *Ecology. 55*, 1216–32.
Westby, G. W. M. (1975) Further analysis of the individual discharge characteristics predicting social dominance in the electric fish, *Gymnotus carapo*. *Anim. Behav. 23*, 249–60.
Westoby, M. (1974) An analysis of diet selection by large generalist herbivores. *Am. Nat. 108*, 290–304.
White, L. E. and Hinde, R. A. (1975) Some factors affecting mother–infant relations in rhesus monkeys. *Anim. Behav. 23*, 527–42.
Whitman, C. O. (1919) The behavior of pigeons. In *Posthumous Works of Charles Otis Whitman, III* (H. A. Carr, ed.). Washington: Carnegie Institute.
Whitten, W. K. (1956) Modification of the oestrous cycle of the mouse by external stimuli associated with the male. *J. Endocrinol. 13*, 399–404.
Wickler, W. (1968) *Mimicry in Plants and Animals*. London: Weidenfeld and Nicholson.
Wiepkema, P. R. (1968) Behaviour changes in CBA mice as a result of one gold thioglucose injection. *Behaviour 32*, 179–210.
Wiepkema, P. R. (1971) Positive feedbacks as work during feeding. *Behaviour 39*, 266–73.
Wierzbowski, S. (1978) The sexual behaviour of experimentally underfed bulls. *Appl. Anim. Ethol. 4*, 55–60.
Wiesel, T. N. and Hubel, D. H. (1963) Single cell responses in striate cortex of kittens deprived of vision in one eye. *J. Neurophysiol. 26*, 1003–17.
Wiesner, B. P. and Sheard, N. M. (1933) *Maternal Behaviour in the Rat*. Edinburgh: Oliver and Boyd.
Wildey, K. B. (1977) The effectiveness of three contact insecticides against a susceptible and a malathion-resistant strain of the saw-toothed grain beetle (*Oryzaephilus surinamensis*). *Proc. Br. Crop Prot. Conf. 1977*, 169–77.
Wiley, R. H. (1973) Territoriality and non-random mating in sage grouse *Centrocercus urophasianus*. *Anim. Behav. Monogr. 6*, 85–169.
Wiley, R. H. (1974) Evolution of social organization and life history patterns among grouse. *Q. Rev. Biol. 49*, 201–27.
Williams, C. G., Siegel, P. B. and Cross, W. B. (1977) Social strife in cockerel flocks during the formation of peck rights. *Appl. Anim. Ethol. 3*, 35–45.
Williams, G. C. (1964) Measurement of consociation among fishes and comments on the evolution of schooling. *Misc. Publs Mus. Biol. Univ. Mich. 2*, 349–84.
Williams, G. C. (1966) *Adaptation and Natural Selection*. Princeton, N.J.: Princeton University Press.
Williams, G. C. (1979) The question of adaptive sex ratio in outcrossed vertebrates. *Proc. R. Soc.*, B, *205*, 567–80.
Willows, A. O. D. (1967) Behavioral acts elicited by stimulation of single, identifiable brain cells. *Science, N.Y. 157*, 570–4.
Willows, A. O. D., Dorsett, D. A. and Hoyle, G. (1973*a*) The neuronal basis of behavior in *Tritonia*. 1. Functional organization of the central nervous system. *J. Neurobiol. 4*, 207–37.
Willows, A. O. D., Dorsett, D. A. and Hoyle, G. (1973*b*) The neuronal basis of behavior in *Tritonia*. III. Neuronal mechanism of a fixed action pattern. *J. Neurobiol. 4*, 255–85.
Willows, A. O. D. and Hoyle, G. (1969) Neuronal network triggering a fixed action pattern. *Science, N.Y. 166*, 1549–51.
Wilson, D. M. (1966) Insect walking. *Ann. Rev. Ent. 11*, 103–22.
Wilson, E. O. (1971) *The Insect Societies*. Cambridge, Mass.: Harvard University Press.
Wilson, E. O. (1975) *Sociobiology. The New Synthesis*. Cambridge, Mass.: Harvard University Press.
Wilson, J. A. (1979) *Principles of Animal Physiology*. New York: Macmillan.
Wilson, R. S. (1964) Autonomic changes produced by noxious and innocuous stimulation. *J. comp. physiol. Psychol. 58*, 290–5.
Winchester, C. F. and Morris, M. J. (1956) Water intakes of cattle. *J. Anim. Sci. 15*, 722–40.
Windle, W. F. (1944) Genesis of somatic motor function in mammalian embryos: a synthesizing article. *Physiol. Zool. 17*, 247–60.
Wisniewski, E. W. and Albright, J. L. (1978) Parlor entrance behavior of dairy cattle trained to enter a herringbone parlor with conditioning methods. *Proc. Int. Symp. Machine Milking, Louisville Kentucky U.S.A.* 460–6.
Wolf, L. L. (1978) Aggressive social organization in nectarivorous birds. *Am. Zool. 18*, 765–78.
Wolf, L. L. and Hainsworth, F. R. (1971) Time and energy budgets of territorial hummingbirds. *Ecology. 52*, 980–8.
Wolf, L. L. and Hainsworth, F. R. (1977) Temporal patterning of feeding by hummingbirds. *Anim. Behav. 25*, 976–89.
Wolf, L. L. and Hainsworth, F. R. (1978) In *Chemical Zoology*, Vol. 10 (A. Brush, ed.). New York: Academic Press.
Wolf, L. L., Hainsworth, F. R. and Gill, F. B. (1975) Foraging efficiencies and time budgets in nectar feeding birds. *Ecology. 56*, 117–28.
Wolf, L. L., Hainsworth, F. R. and Stiles, F. G. (1972)

Energetics of foraging: rate and efficiency of nectar extraction by hummingbirds. *Science, N.Y. 176,* 1351–2.

Wolford, J. H. and Ringer, R. K. (1962) Adrenal weight, adrenal ascorbic acid, adrenal cholesterol and differential leucocyte counts as physiological indicators of 'stressor' agents in laying hens. *Poult. Sci. 41,* 1521–9.

Wood, M. T. (1977) Social grooming patterns in two herds of monozygotic twin dairy cows. *Anim. Behav. 25,* 635–42.

Wood-Gush, D. G. M., Duncan, I. J. H. and Fraser, D. (1975) In *The Behaviour of Domestic Animals,* 3rd edn (E. S. E. Hafez, ed.). London: Baillière Tindall.

Wood-Gush, D. G. M., Duncan, I. J. H. and Savory, C. J. (1978) Observations on the social behaviour of domestic fowl in the wild. *Biol. Behav. 3,* 193–205.

Woodhead, S. and Bernays, E. A. (1978) The chemical basis of resistance of *Sorghum bicolor* to attack by *Locusta migratoria. Entomologia exp. appl. 24,* 123–44.

Woodland, D. J., Jaafar, Z. and Knight, M. L. (1980) The 'pursuit deterrent' function of alarm signals. *Am. Nat. 115,* 748–53.

Woods, A. (1974) *Pest Control: A Survey.* London: McGraw Hill.

Woodworth, R. S. (1918) *Dynamic Psychology.* New York: Columbia University Press.

Wrangham, R. W. (1975) The behavioural ecology of chimpanzees in Gombe National Park, Tanzania. Ph.D. thesis, University of Cambridge.

Wrangham, R. W. (1980) Sex differences in chimpanzee dispersion. In *The Great Apes* (D. A. Hamburg and E. McCown, eds.). Menlo Park, Ca.: Benjamin Cummings.

Wright, E. N., Inglis, I. R. and Feare, N. C. J. Eds., (1980) *Bird Problems in Agriculture.* Croydon: British Crop Protection Council.

Wright, J. W. (1976) Effect of hunger on the drinking behaviour of rodents adapted for mesic and xeric environments. *Anim. Behav. 24,* 300–4.

Wright, P., Caryl, P. G. and Vowles, D. M. Eds., (1975) *Neural and Endocrine Aspects of Behaviour in Birds.* Amsterdam: Elsevier.

Wright, R. H. (1957) A theory of olfaction and of the action of mosquito repellants. *Can. Ent. 89,* 518–28.

Wynne-Edwards, V. C. (1962) *Animal Dispersion in Relation to Social Behaviour.* Edinburgh: Oliver and Boyd.

Young, B. A. (1970) Application of carbon dioxide entry rate technique to measurement of energy expenditure by grazing cattle. In *Energy Metabolism of Farm Animals* (A. Schürch and C. Wenk, eds.). Zürich: Juris Verlag.

Young, B. A. and Corbett, J. L. (1972*a*) Maintenance energy requirement of grazing sheep in relation to herbage availability. I. Calorimetric estimates. *Aust. J. agric. Res. 23,* 57–76.

Young, B. A. and Corbett, J. L. (1972*b*) Maintenance energy requirement of grazing sheep in relation to herbage availability. II. Observations on grazing intake. *Aust. J. agric. Res. 23,* 77–85.

Zach, R. and Falls, J. B. (1976*a*) Ovenbird (Aves: Parulidae) hunting behavior in a patchy environment: an experimental study. *Can. J. Zool. 54,* 1863–79.

Zach, R. and Falls, J. B. (1976*b*) Foraging behavior, learning and exploration by captive ovenbirds (Aves: Parulidae). *Can. J. Zool. 54,* 1880–93.

Zach, R. and Falls, J. B. (1976*c*) Do ovenbirds (Aves: Parulidae) hunt by expectation? *Can. J. Zool. 54,* 1894–903.

Zack, S. (1978*a*) The effects of foreleg amputation on head grooming behaviour in the praying mantis, *Sphodromantis lineola. J. comp. Physiol. 125,* 253–8.

Zack, S. (1978*b*) Head grooming behaviour in the praying mantis. *Anim. Behav. 26,* 1107–19.

Zahavi, A. (1971*a*) The function of pre-roost gatherings and communal roosts. *Ibis 113,* 106–9.

Zahavi, A. (1971*b*) The social behaviour of the white wagtail *Motacilla alba alba* wintering in Israel. *Ibis 113,* 203–11.

Zahavi, A. (1974) Communal nesting by the Arabian babbler. A case of individual selection. *Ibis 116,* 84–7.

Zahavi, A. (1977) Reliability in communication systems and the evolution of altruism. In *Evolutionary Ecology* (B. Stonehouse and C. Perrins, eds.). London: MacMillan.

Zarrow, M. X., Denenberg, V. H., Haltmeyer, G. C. and Brumaghin, J. T. (1967) Plasma and adrenal corticosterone levels following exposures of the 2-day-old rat to various stressors. *Proc. Soc. exp. Biol. Med. 125,* 113–16.

Zarrow, M. X., Gandelman, R. and Denenberg, V. H. (1971) Prolactin: is it an essential hormone for maternal behaviour in the mammal? *Horm. Behav. 2,* 343–54.

Zeigler, H. P. (1976) Feeding behaviour of the pigeon. *Adv. Study Behav. 7,* 285–389.

Zeki, S. M. (1978) Functional specialisation in the visual cortex of the rhesus monkey. *Nature, Lond. 274,* 423–8.

Zippelius, H. M. and Schleidt, W. M. (1956) Ultraschalloute bei jungen Mäusen. *Naturwissenschaften 21,* 502.

Zolman, J. F. and Martin, R. C. (1967) Instrumental aversive conditioning in newly hatched domestic chicks. *Psychon. Sci. 8,* 183–4.

Zucker, I. (1966) Effects of an antiandrogen on the mating behaviour of male guinea pigs and rats. *J. Endocrinol. 35,* 209–10.

Author Index

Subject index